PRACTICE MA[obscured] LEVELS A, B, [obscured]

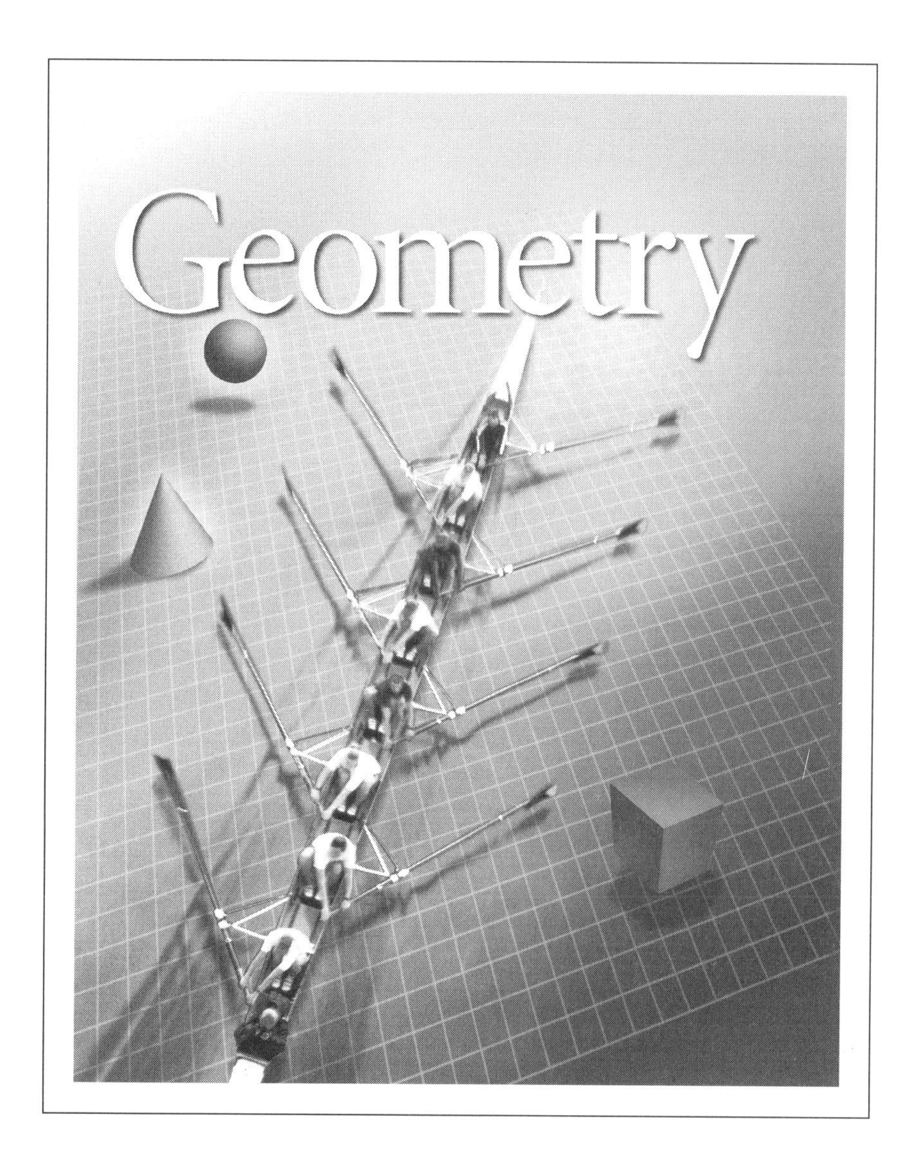

HOLT, RINEHART AND WINSTON

A Harcourt Classroom Education Company

Austin • New York • Orlando • Atlanta • San Francisco • Boston • Dallas • Toronto • London

To the Student

Practice Masters Levels A, B, and C consist of three levels of exercises graded by level of difficulty. There is a full page for each of the three levels of exercises for each lesson in the *Pupil's Edition*. Level A exercises are the least difficult, designed for students to practice the lesson objectives through the use of examples. Level B exercises are middle-range exercises that students can handle with the use of current examples together with some prior knowledge. Level C exercises are the most challenging exercises, relevant to the lesson and appropriate for the student who has mastered the lesson.

Printed in the United States of America

ISBN 0-03-064811-4

1 2 3 4 5 6 7 066 05 04 03 02 01 00

Table of Contents

NAME ______________________ CLASS ____________ DATE ____________

Practice Masters Level A

1.1 The Building Blocks of Geometry

For Exercises 1–3, refer to the figure at the right.

1. Name all the segments in the rectangle.

____

2. Name the rays that form $\angle D$.

____

3. Name each of the angles in the rectangle using three different methods.

____

For Exercises 4–6, use the space provided to neatly draw and label the figure described. Use a straightedge.

4. a ray with endpoint Q that goes through point M ______________

5. $\angle PAN$ ______________

6. $\overline{RS}$ ______________

State whether each object could best be modeled by a point, a line, or a plane.

7. a laser ______________

8. the top of your desk ______________

9. a town on a map ______________

10. a sidewalk intersection ______________

For Exercises 11–12, refer to the figure at the right.

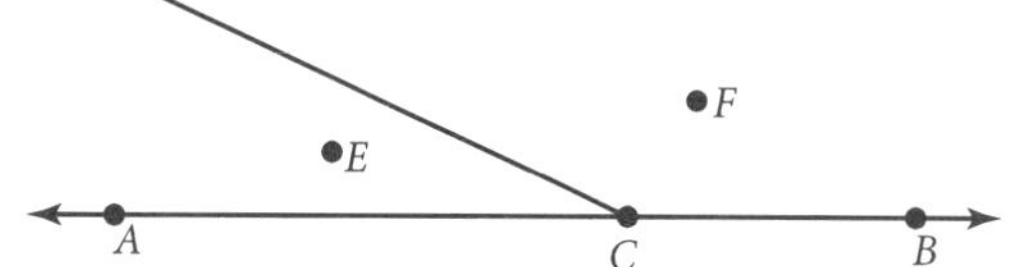

11. Name three collinear points. ______________

12. Name a point in the interior of $\angle ACD$. ______

Classify each statement as true or false, and explain your reasoning in each false case.

13. In geometry, a postulate is a statement which can be proven. ______________

14. Two lines can intersect at more than one point. ______________

Practice Masters Level B

1.1 The Building Blocks of Geometry

For Exercises 1–4, refer to the figure at the right.

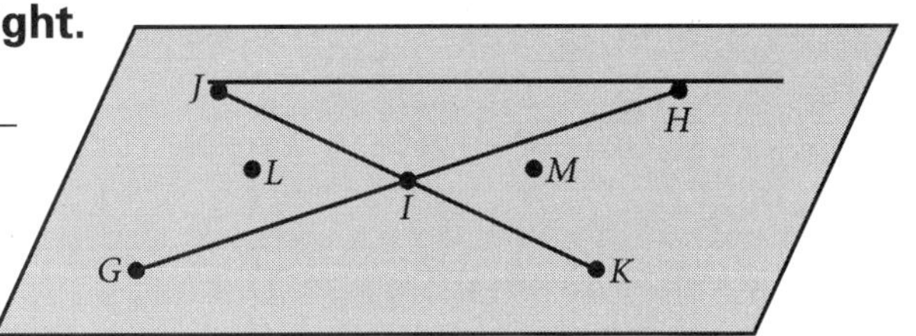

1. Name a point in the interior of $\angle JIG$. ______

2. Name four angles in the figure: ______________________

3. Name three collinear points in the figure: ______________________

4. How many different planes contain points H, I, and M? ______________________

Classify each statement as true or false, and explain your reasoning in each false case.

5. One line contains exactly two points. ______________________

6. The intersection of two planes is a line. ______________________

7. Three lines can intersect at more than one point. ______________________

Name a familiar object that can be modeled by each of the following:

8. a point ______________________ 9. a line ______________________

Use the space provided to neatly draw and label the figure described. Use a straightedge.

10. $\angle POR$ ______________________

11. $\overline{KL}$ ______________________

12. 2 lines that intersect at P ______________________

For Exercises 13–15, refer to the figure at the right.

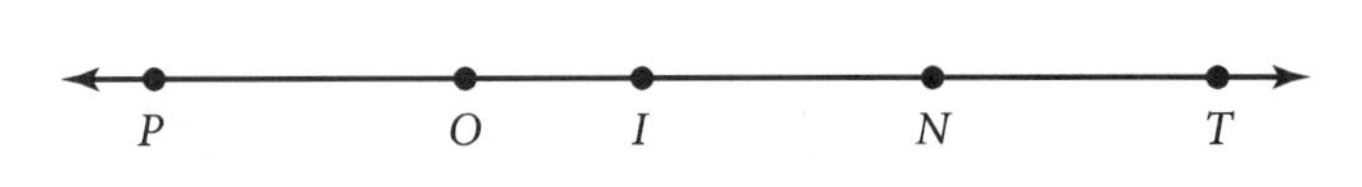

13. Name three segments in the figure. ______________________

14. *True or False:* $\overline{ON}$ is the same as $\overrightarrow{NO}$. ______________________

15. How many different rays can be named in the figure? ______________________

Practice Masters Level C
1.1 The Building Blocks of Geometry

For Exercises 1–3, refer to the figure at the right.

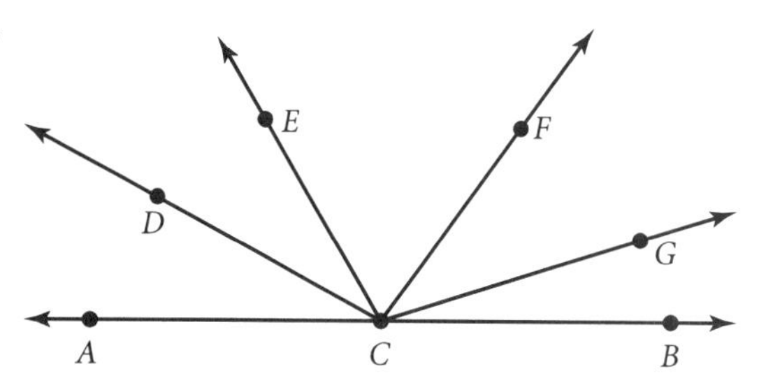

1. How many angles appear in the figure? ______

2. Name three collinear points. ______________

3. C is called the ______ of the angles.

For Exercises 4–7, refer to the figure at the right.

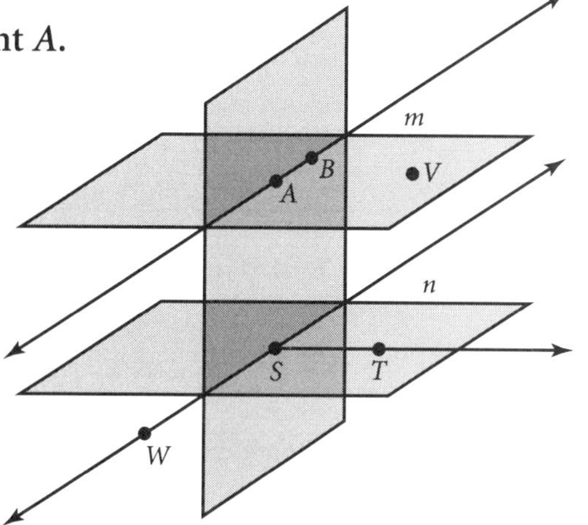

4. Name two pairs of points that are coplanar with point A.

____

5. How many planes in the figure contain line m?

____

6. *True or False:* $\overline{ST}$ and line n are coplanar. ______

7. Name the intersection of n and $\overline{ST}$. __________

For Exercises 8–9, use the space provided to neatly draw and label the figure described. Use a straightedge.

8. two planes that intersect at line l

9. a plane that contains $\angle QRS$

______________________________ ______________________________

For Exercises 10–14, classify each statement as true or false, and explain your reasoning.

10. A line can be defined as a perfectly straight figure that extends forever. ______________

11. If points A, B, and C are collinear, then $\overline{AB}$ is the same as $\overline{AC}$. ______________

12. Three planes must intersect each other at exactly two lines. ______________

13. Two points can name a plane. ______________

14. If X and Y are in plane Q, then $\overline{XY}$ is in plane Q. ______________

NAME ______________________ CLASS __________ DATE __________

Practice Masters Level A

1.2 Measuring Length

For Exercises 1–3, find the lengths determined by the points on the number line.

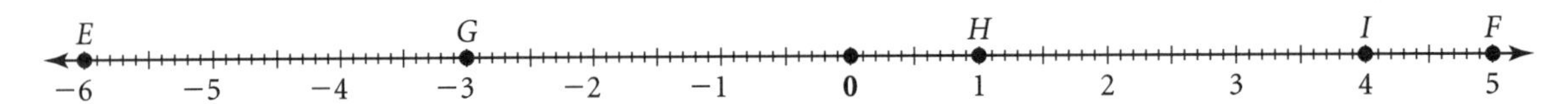

1. HI __________ 2. GI __________ 3. EG __________

4. On the number line below, plot points A and B so that $AB = 4$.

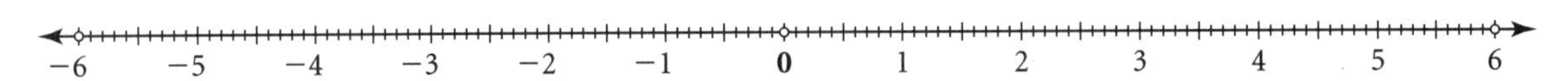

In Exercises 5–7, point *A* is between points *C* and *T* on $\overline{CT}$. Sketch each figure and find the missing lengths.

5. $CA = 12, AT = 5, CT =$ ______

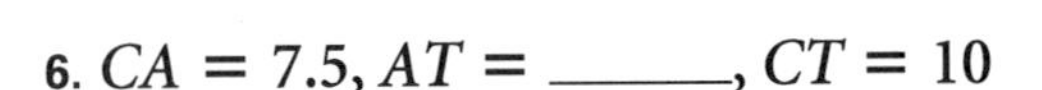

6. $CA = 7.5, AT =$ ______, $CT = 10$

7. $CA =$ ______, $AT = 8.7, CT = 9.4$

8. Name all congruent segments in the figure at the right.

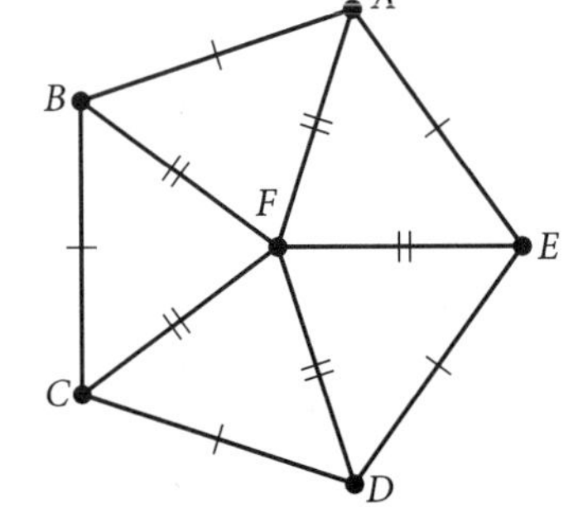

9. Your family is driving to Texas for vacation. As you drive along I-30, a straight highway, you notice the mileage sign shown at the right. Use the sign to determine the distance between Hope, AR, and New Boston, TX.

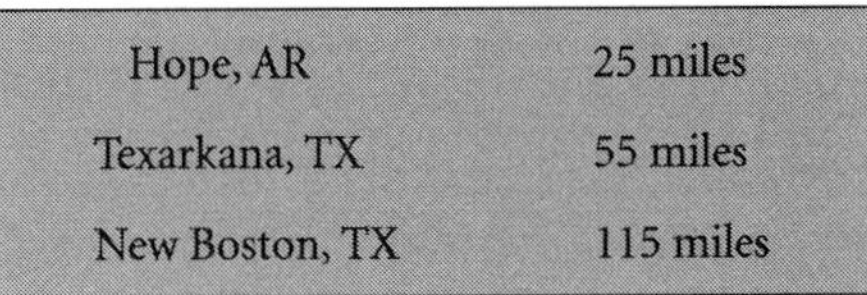

Hope, AR	25 miles
Texarkana, TX	55 miles
New Boston, TX	115 miles

____

In the number line below, $AC = 8.1$. Find the indicated values.

10. x __________ 11. BC __________

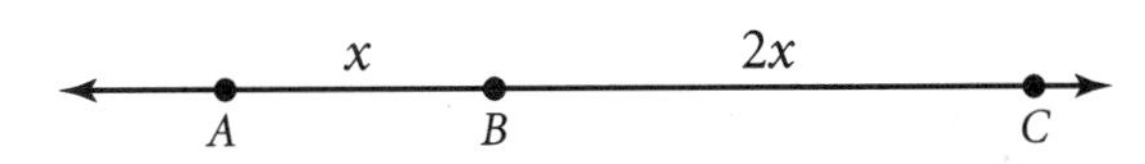

NAME ______________________ CLASS ______________ DATE ______________

Practice Masters Level B

1.2 Measuring Length

For Exercises 1–4, find the lengths determined by the points on the number line.

1. BC ______________

2. AB ______________

3. Point R is not shown. If $BR = 4$, locate and plot the possible coordinates of R. ______________

For Exercises 4–5, use the figure at the right.

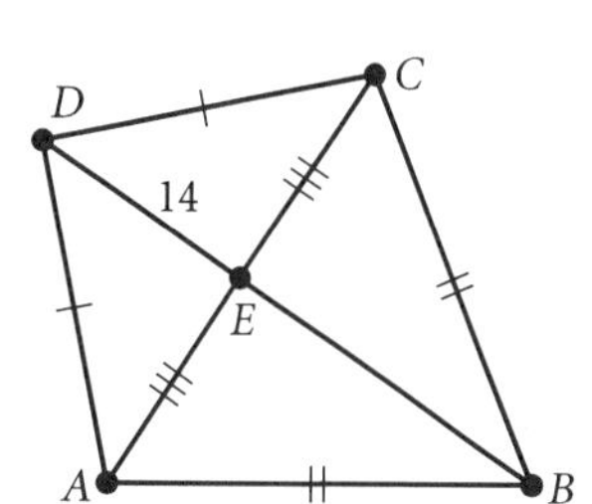

4. Name all congruent segments in the figure.

____

5. If EB is 1.5 times greater than ED, find EB.

____

Point *R* is between points *A* and *T* on $\overline{AT}$. Sketch a figure for each set of values, and find the missing lengths.

6. $AR =$ ______, $RT = 42.5$, $AT = 51.3$

7. $AR = 0.39$, $RT =$ ______, $AT = 0.72$

8. When you left your home in Memphis, TN, this morning, your odometer read "11,279". You are traveling along a straight highway to Nashville, TN. You are now in Jackson, TN, and you see a sign that says "Nashville-128 miles". Your odometer now reads "11,365". Use this information to find the distance between Memphis and Nashville. ______________

If $DE = 91$, find the indicated values.

9. x ______________ 10. DC ______________ 11. CE ______________

If $CJ = 60$, find the indicated values.

12. x ______________ 13. CI ______________ 14. IJ ______________

NAME ______________ CLASS ______________ DATE ______________

Practice Masters Level C

1.2 Measuring Length

For Exercises 1–4, find the lengths determined by the points on the number line.

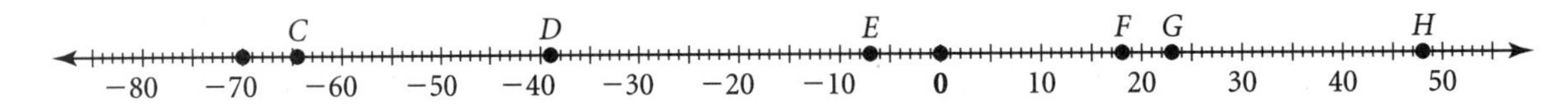

1. CE ______________ 2. EH ______________

3. If $DR = 43$, locate and plot the possible coordinates of R. ______________

4. Name the congruent segments on the number line. ______________

For Exercises 5–6, use the figure at the right.

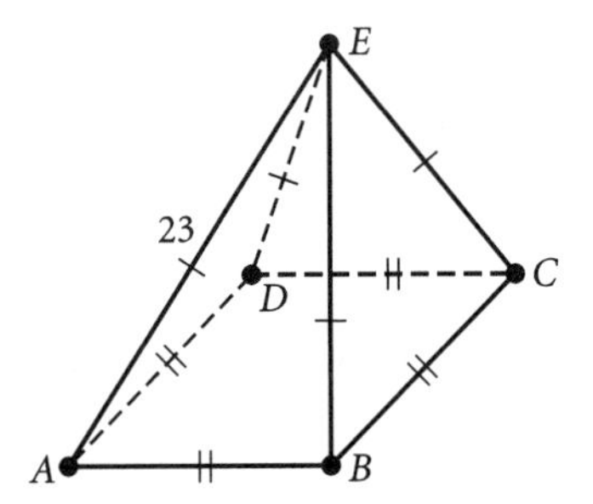

5. Name all congruent segments in the figure.

____

6. If AE is 1.5 times greater than AD, find AD.

____

7. When you left your home in Memphis, TN, this morning, your odometer read "28,974.6". You are traveling along a straight highway to Nashville, TN. You are now in Jackson, TN, and you see a sign that says "Nashville - 128 miles". Your odometer now reads "29,060". Use this information to find the distance between Memphis and Nashville.

____

If $CD = 5x - 7$, find the indicated values.

8. x ______________ 9. CE ______________ 10. CD ______________

If the ratio of $\frac{RS}{ST} = \frac{5}{7}$, find the indicated values.

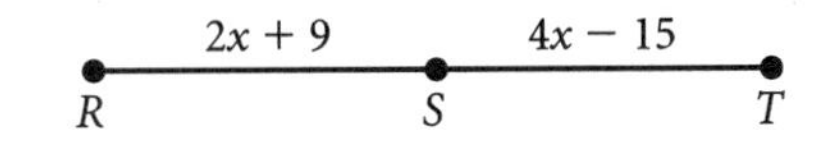

11. x ______________ 12. RS ______________

13. ST ______________ 14. RT ______________

NAME ________________ CLASS ________ DATE ________

Practice Masters Level A

1.3 Measuring Angles

For Exercises 1–3, use a protractor to find the measures of the indicated angles. You may extend the rays if necessary.

1. m$\angle K$ = ________
2. m$\angle Q$ = ________
3. m$\angle Z$ = ________

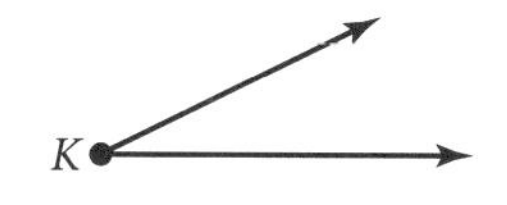

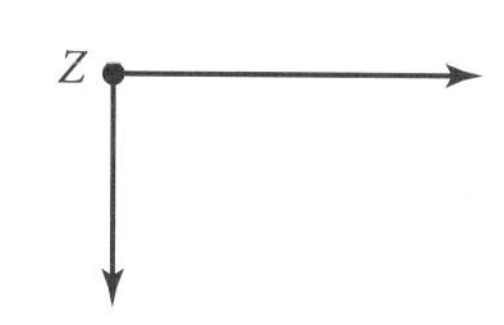

For Exercises 5–6, use a protractor to sketch an angle of the indicated size. Be sure to label your angle.

4. m$\angle PAN = 52°$
5. m$\angle ROX = 160°$

For Exercises 6–8, classify each statement as true or false, and explain your reasoning in each false case.

6. If two angles are complementary, then they form a linear pair. ________
7. Supplementary angles are always congruent. ________
8. Two acute angles can be complementary. ________

For Exercises 9–11, refer to the figure at the right.

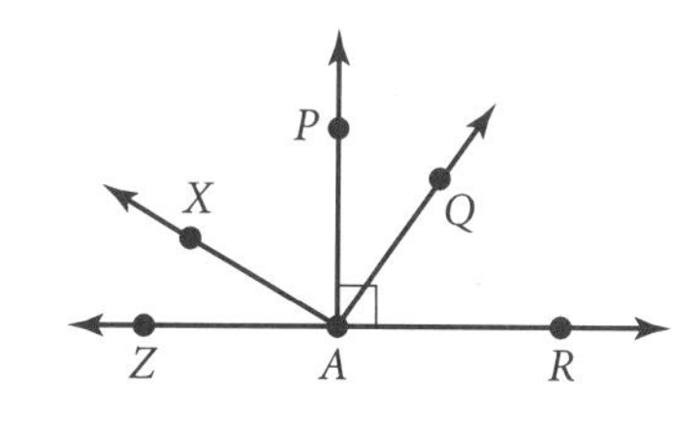

9. If m$\angle PAQ = 28°$, find m$\angle QAR$. ________
10. If m$\angle XAP = 56°$, find m$\angle XAR$. ________
11. Name three pairs of supplementary angles in the figure.

____

In the figure at the right, m$\angle EAI = x + 15$, and m$\angle IAO = x - 11$.

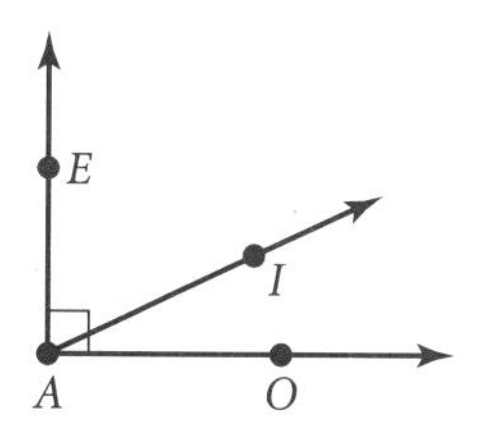

12. Find x. ________
13. Find m$\angle EAI$. ________
14. Find m$\angle IAO$. ________
15. $\angle EAI$ and $\angle IAO$ are called ________ angles.

NAME ______________________ CLASS ____________ DATE ____________

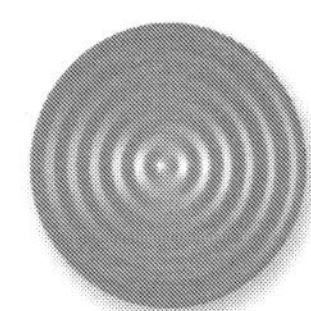

Practice Masters Level B

1.3 Measuring Angles

For Exercises 1–4, use a protractor to find the measures of the indicated angles. You may extend the rays if necessary.

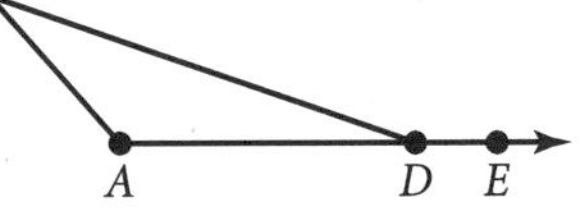

1. $m\angle CAD$ ____________
2. $m\angle CDA$ ____________
3. $m\angle ACD$ ____________
4. $m\angle CDE$ ____________
5. In the figure above, what is the relationship between $\angle CDE$ and $\angle CDA$? ____________
6. Draw a figure where $m\angle ABC = 115°$ and $m\angle DBC = 55°$.

____

If $\angle BAT$ and $\angle TAZ$ form a linear pair and $m\angle BAT = 5x - 6$, and $m\angle TAZ = 7x + 18$, find the measure of each angle.

7. $m\angle BAT =$ ____________
8. $m\angle TAZ =$ ____________

If $\angle LFO \cong \angle EFR$, find the measures of the indicated angles.

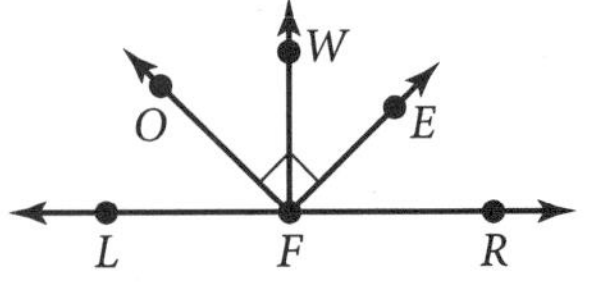

9. $m\angle LFO =$ ____________
10. $m\angle EFR =$ ____________
11. What is the angle between the minute and hour hands on a clock at 2:30? ____________
12. An angle has measure 3 times that of its complement. What is the measure of the angle? ____________

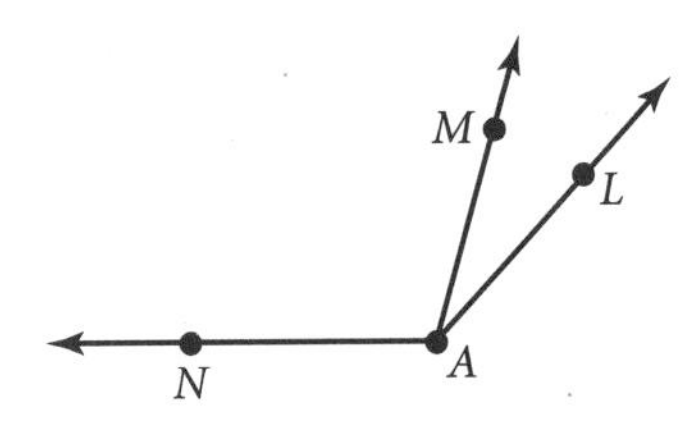

In the figure at the left, $m\angle MAN = 17x + 3$, $m\angle MAL = 9(x - 3)$, and $m\angle NAL = 3(7x + 2)$.

13. Find x. ____________
14. Find $m\angle MAN$. ____________
15. Find $m\angle MAL$. ____________
16. Find $m\angle NAL$. ____________

Use the figure at the right to find the indicated measures.

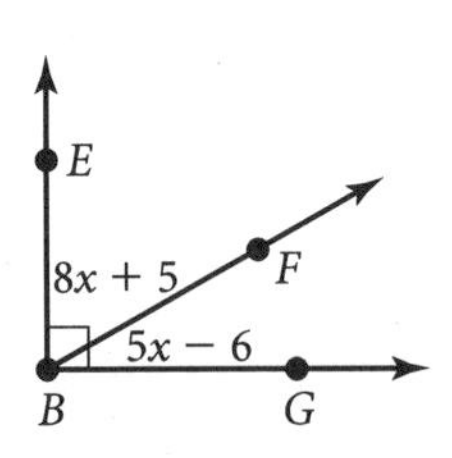

17. x ____________
18. $m\angle EBG$ ____________
19. $m\angle EBF$ ____________
20. $m\angle GBF$ ____________

NAME ______________________ CLASS __________ DATE __________

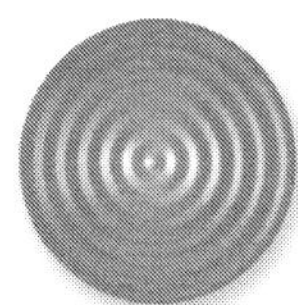

Practice Masters Level C

1.3 Measuring Angles

1. In the space at the right, draw a figure containing the following angles.

   $m\angle ABC = 48°$ $m\angle CDA = 72°$

   $m\angle BCD = 110°$ $m\angle DAB = 130°$ __________

For Exercises 2–7, use a protractor to find the measures of the indicated angles. You may extend the rays if necessary.

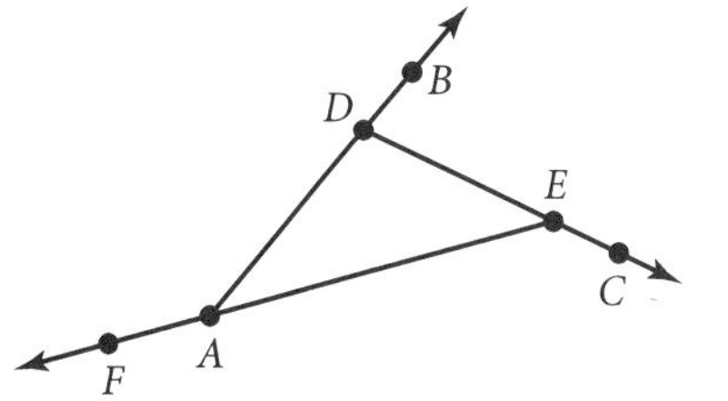

2. $m\angle ADE$ __________ 3. $m\angle DEA$ __________

4. $m\angle EAD$ __________ 5. $m\angle FAD$ __________

6. $m\angle BDE$ __________ 7. $m\angle CEA$ __________

8. Find the sum of the measures in Exercises 2–4, then find the sum of the angles for Exercises 5–7. Describe any pattern you discover. __________

____

For Exercises 9–13, use the figure at the right to find the indicated measures.

D

6x − 17 11x + 10

A C B

9. x __________ 10. $m\angle ACD$ __________

11. $m\angle DCB$ __________ 12. $m\angle ACB$ __________

13. Describe the relationship between $\angle ACD$ and $\angle BCD$. __________

14. The ratio of an angle with its complement is $\frac{5}{7}$. Find the measure of the angle. __________

15. The supplement of an angle is 9 times greater than the measure of the complement of the angle. Find the measure of the angle. __________

In the figure at the right, $m\angle DAF = 18x - 3$. Find the indicated measures.

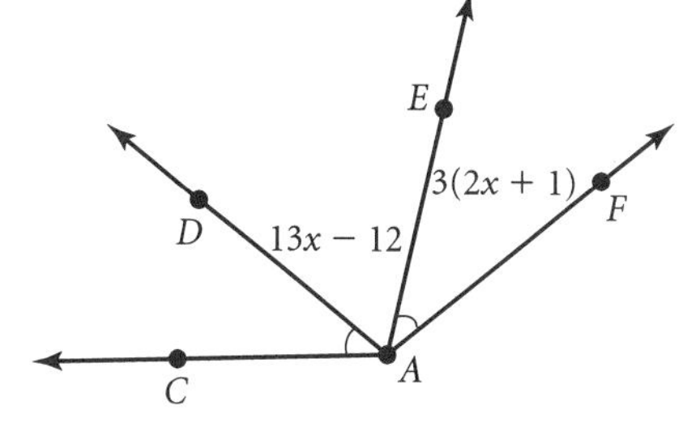

16. x __________ 17. $m\angle FAE$ __________

18. $m\angle DAE$ __________ 19. $m\angle DAF$ __________

20. $m\angle CAF$ __________

NAME ______________________ CLASS ____________ DATE ____________

Practice Masters Level A

1.4 Exploring Geometry by Using Paper Folding

For Exercises 1–3, use the figure at the right to complete the following statements.

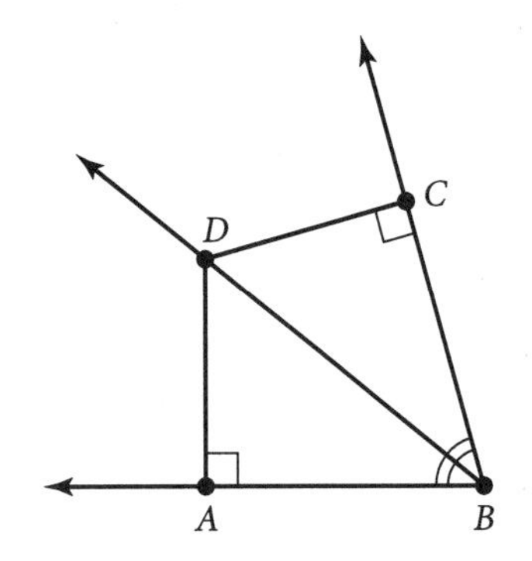

1. $\overline{BD}$ is called the ______________________ of $\angle ABC$.

2. If $m\angle CBD = 37°$, then $m\angle ABC =$ ____________

3. $\overline{CD}$ and $\overline{AD}$ are ______________________

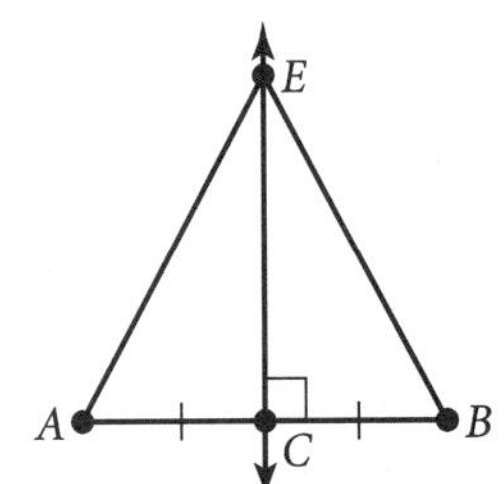

If $AB = 9x - 1$, and $AC = 3x + 7$, find the following:

4. x ____________ 5. AC ____________

6. CB ____________ 7. AB ____________

Construct all the geometric figures below by folding a sheet of paper.

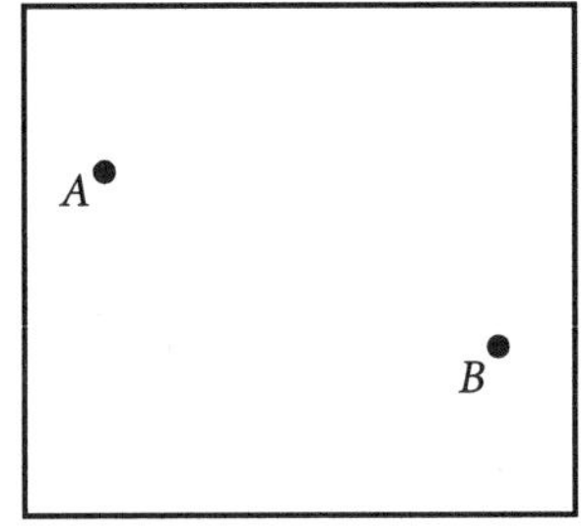

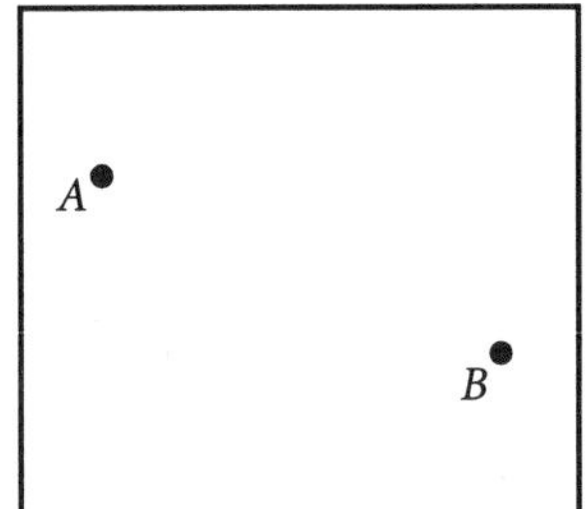

8. Describe how to construct line l through A and B.

____

9. Describe how to construct two lines perpendicular to line l: line m through point A and line n through point B.

____

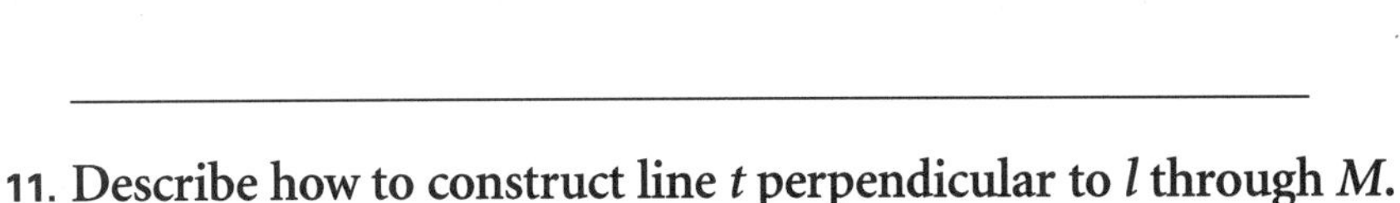

10. Describe how to construct point M, the midpoint of $\overline{AB}$.

____

11. Describe how to construct line t perpendicular to l through M.

____

12. Write a conjecture about the relationship between lines m, t, and n.

____

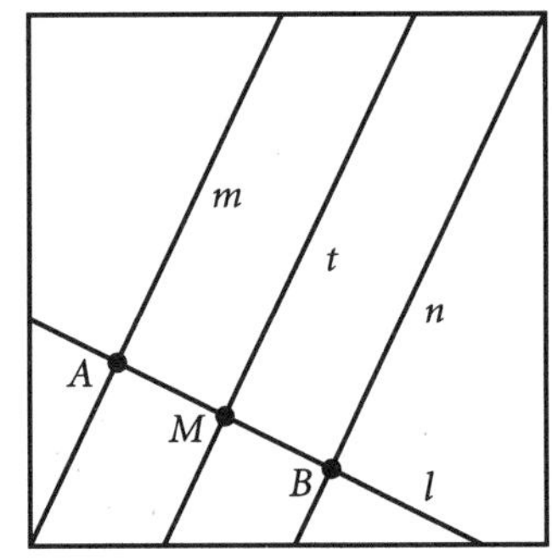

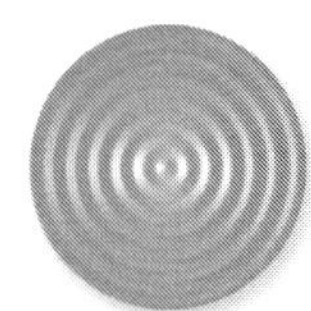

Practice Masters Level B

1.4 Exploring Geometry by Using Paper Folding

Use the figure at the right for Exercises 1–11.

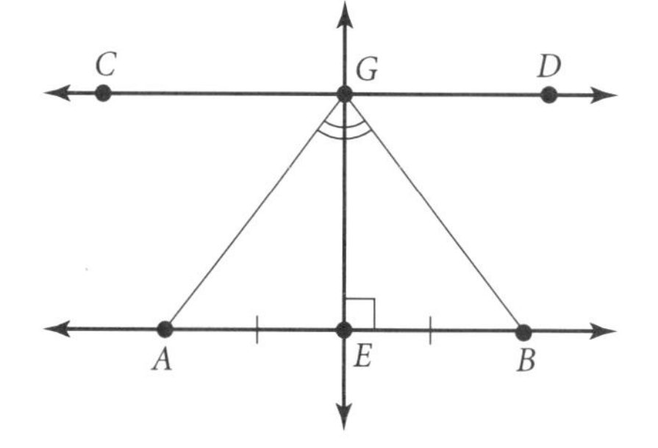

1. The distance from G to $\overline{AB}$ is ______________________

2. *True or False:* $CG = DG$ ______________________

3. Describe the relationship between $\overline{GE}$ and $\overline{AB}$.

   ____

4. Describe the relationship between $\overline{GE}$ and $\angle AGB$.

   ____

5. Explain why $GA = GB$. ____

If m$\angle AGE = 5x - 1$, m$\angle AGB = 9x + 6$, and $AE = 3x + 9$, find the following measurements:

6. x ______________ 7. m$\angle BGE$ ______________ 8. m$\angle AGB$ ______________

9. AE ______________ 10. EB ______________ 11. AB ______________

Construct all of the geometric figures below by folding a sheet of paper.

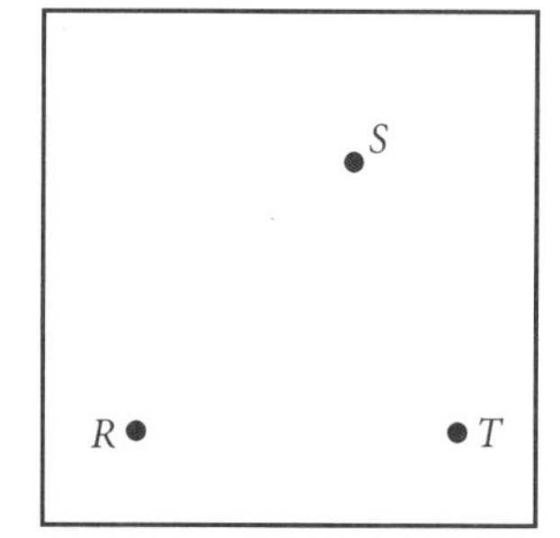

12. Fold three lines, the first containing R and S, the second containing S and T, and the third containing R and T.

13. Construct the perpendicular bisectors of $\overline{RS}$, $\overline{ST}$, and $\overline{RT}$. What do you notice about the lines you just constructed?

    ____

Label the point where the bisectors intersect as X. Carefully measure the following distances:

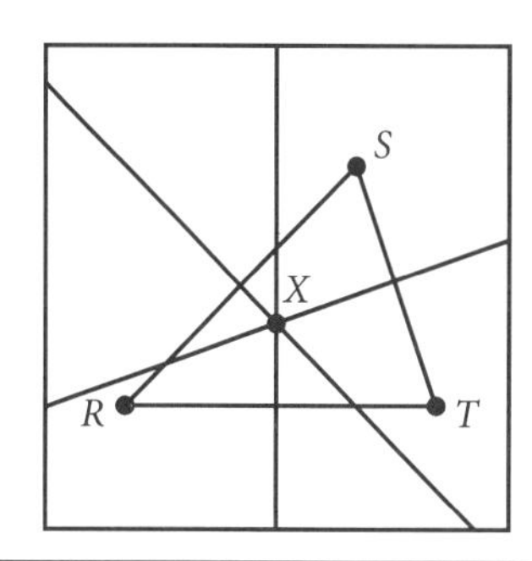

14. XS ______________ 15. XR ______________ 16. XT ______________

17. What do you notice about the measurements you made in Exercises 14–16? ______________________________

NAME ______________________ CLASS __________ DATE __________

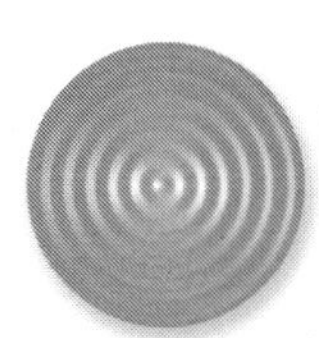

Practice Masters Level C

1.4 Exploring Geometry by Using Paper Folding

Use paper folding and the figure at the right for Exercises 1–3. Put marks on the figure to indicate $\overline{DB}$ is the perpendicular bisector of $\overline{AC}$.

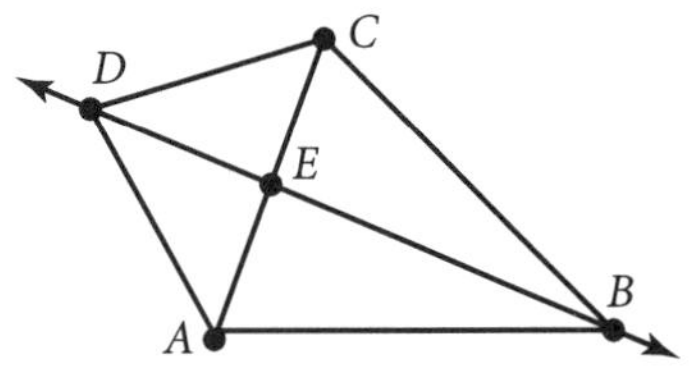

1. Name all pairs of congruent segments. ______________________

2. Describe why each pair of segments from Exercise 1 is congruent.

____

3. Which angles do you think are congruent? Why? ______________________

____

In Exercise 4–5, construct the geometric figures below by folding a sheet of paper.

4. Describe how to construct two segments, $\overline{RS}$ and $\overline{TU}$, so that each segment is the perpendicular bisector of the other, and $\overline{TU}$ is longer than $\overline{RS}$.

____

5. Connect the endpoints of $\overline{RS}$ to the endpoints of $\overline{TU}$, then measure the segments formed. What do you notice?

____

6. Find the error in the figure at the right assuming $AB = 14x$. Explain your answer.

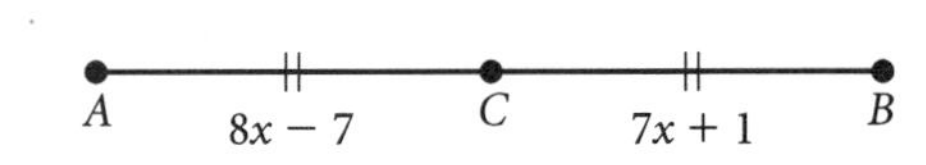

____

In the figure at the right, $\overline{AT}$ is the angle bisector of $\angle MAN$. Find the following:

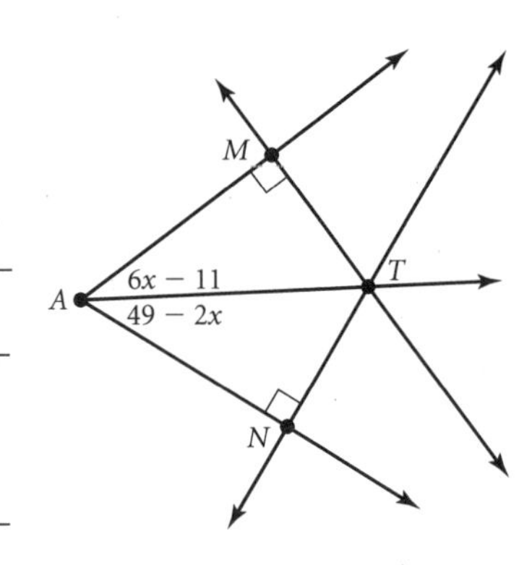

7. x __________ 8. $m\angle MAT$ __________ 9. $m\angle MAN$ __________

10. Explain why $MT = NT$. ______________________

11. Sketch $\overline{MN}$ on the figure. How are $\overline{MN}$ and $\overline{AT}$ related?

____

NAME ______________________ CLASS ____________ DATE ____________

Practice Masters Level A

1.5 Special Points in Triangles

In Exercises 1–3, construct the circumscribed circle of each triangle.

1. 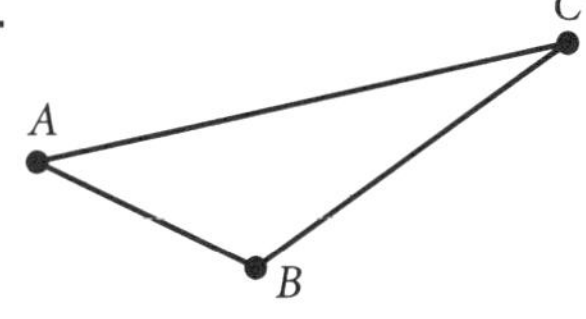

2. 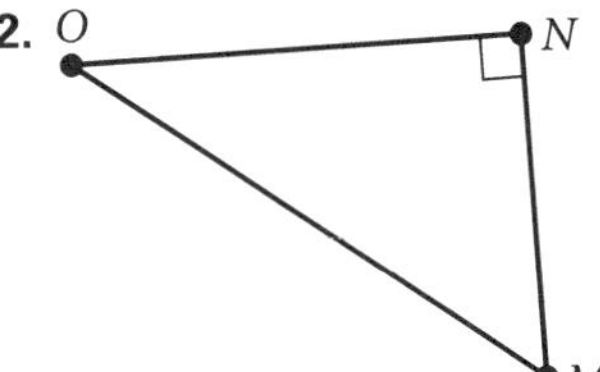

3. 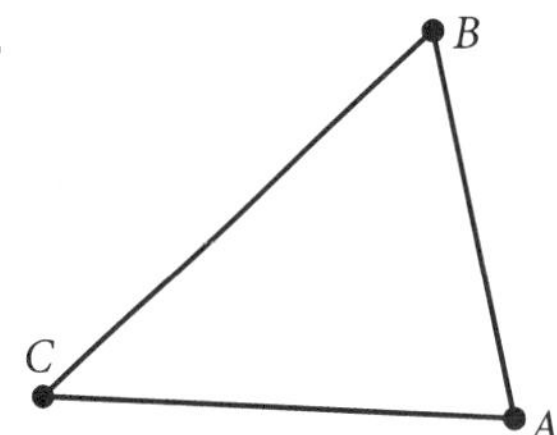

4. What do you notice about the locations of the circumcenters you constructed in Exercises 1–3? ______________________________

____

5. Construct the incircle of $\triangle PQR$.

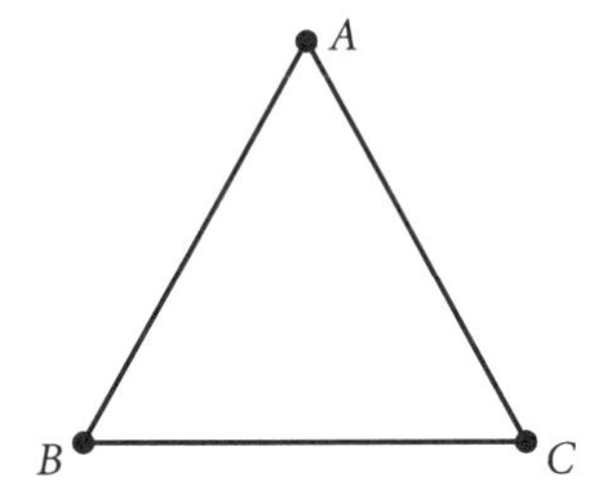

6. Construct the circle that passes through the three points below.

For Exercises 7–11, refer to the figure at the right. Classify each statement as true or false, and explain your reasoning in each false case.

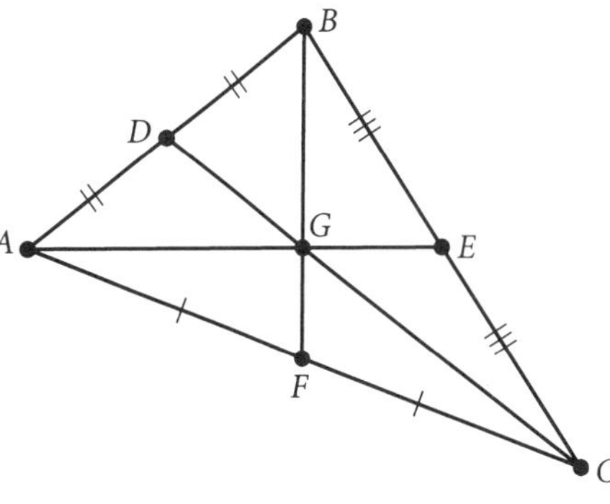

7. $\angle BAE \cong \angle CAE$ ______________________

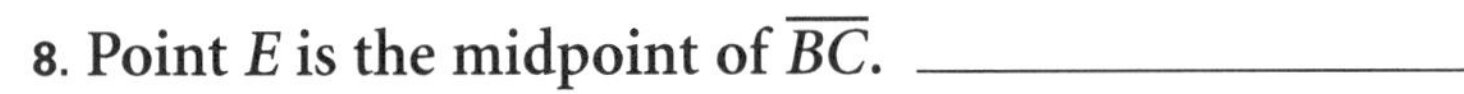

8. Point E is the midpoint of $\overline{BC}$. ______________

9. Point G is the circumcenter of $\triangle ABC$. ______________

10. $\overline{FB}$, $\overline{AE}$, and $\overline{CD}$ are altitudes of $\triangle ABC$. ______________

11. $\overline{FB}$, $\overline{AE}$, and $\overline{CD}$ are concurrent. ______________

NAME ______________________ CLASS __________ DATE __________

Practice Masters Level B

1.5 Special Points in Triangles

Each of the following statements is true sometimes. In the space provided, sketch an example of when the statement is true, and an example of when it is false. Be sure to label your drawings.

Statement: In $\triangle ABC$, the bisector of $\angle A$ is perpendicular to $\overline{BC}$.

1. True example

2. False example

Statement: The circumcenter of $\triangle RAT$ is outside of the triangle.

3. True example

4. False example

5. A portion of a circle is shown at the right. Choose three points on the circle and draw a triangle to connect them. Then construct the circumscribed circle around the triangle to complete the figure.

Trace the given figures on folding paper. Then construct the indicated geometric figures.

6. Construct the circumcircle of both $\triangle BCD$ and $\triangle EFG$.

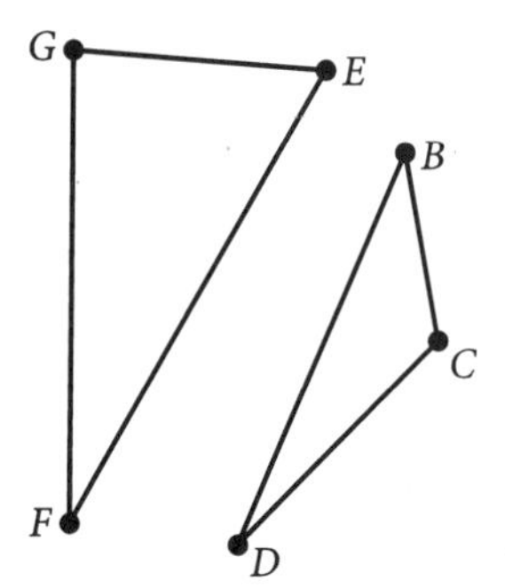

7. Construct the medians and incircle of $\triangle ABC$.

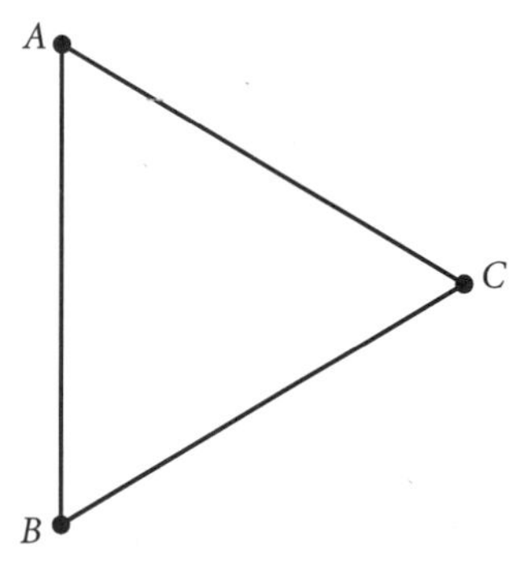

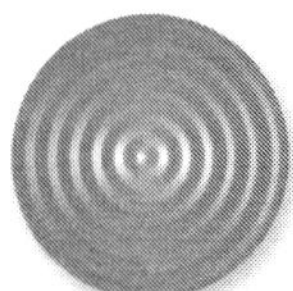

Practice Masters Level C

1.5 Special Points in Triangles

Construct the indicated geometric figures. In each case, label the intersection point on $\overline{AB}$ as point *M*, the intersection point on $\overline{BC}$ as *N*, and the point on $\overline{AC}$ as *P*.

1. angle bisectors of $\triangle ABC$

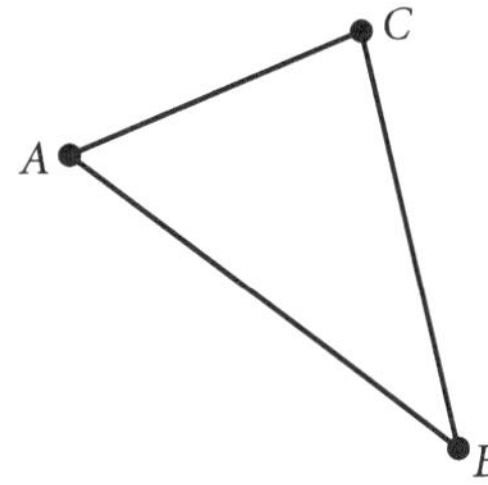

Figure 1

2. medians of $\triangle ABC$

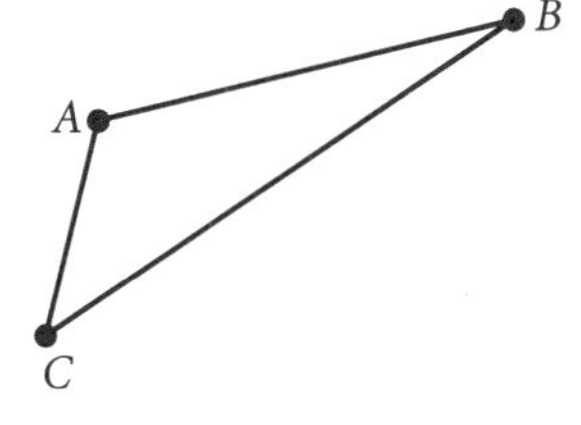

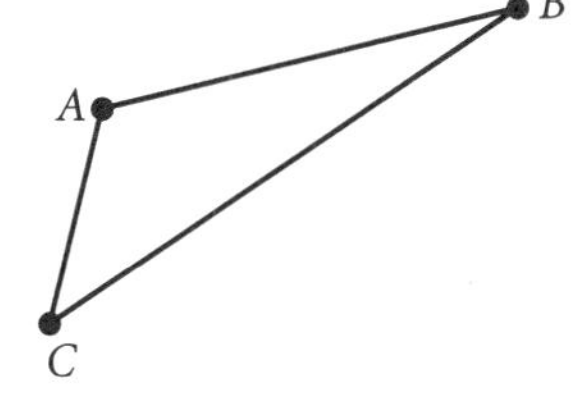

Figure 2

3. circumcircle of $\triangle ABC$

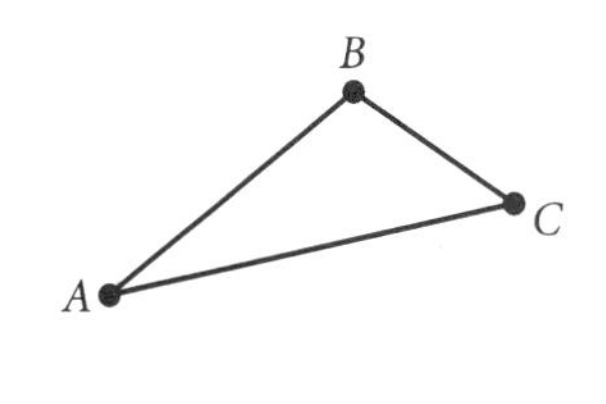

Figure 3

4. For each of the triangles in Exercises 1–3, carefully measure the given segments and complete the following table.

	AM	*MB*	*BN*	*NC*	*CP*	*PA*
Figure 1						
Figure 2						
Figure 3						

Use the measurements from Exercise 4 to calculate the following.

		$\frac{AM}{MB}$	$\frac{BN}{NC}$	$\frac{CP}{PA}$	$\frac{AM}{MB} \cdot \frac{BN}{NC} \cdot \frac{CP}{PA}$
5.	Figure 1				
6.	Figure 2				
7.	Figure 3				

8. What interesting result did you observe in Exercises 5–7? ______

______

Complete each statement with *always, sometimes,* or *never.*

9. The circumcenter of a right triangle is the midpoint of the longest side of the triangle. ______

10. The centroid is inside the triangle. ______

NAME ______________________ CLASS ____________ DATE ____________

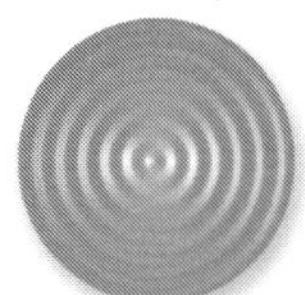

Practice Masters Level A

1.6 Motion in Geometry

In Exercises 1–3, determine whether each description represents a reflection, a rotation, or a translation.

1. the motion of the blades of a ceiling fan ____________

2. riding your skateboard down a straight sidewalk ____________

3. the image you see in a clean window ____________

Identify each motion as a reflection, rotation, or translation.

4. ____________

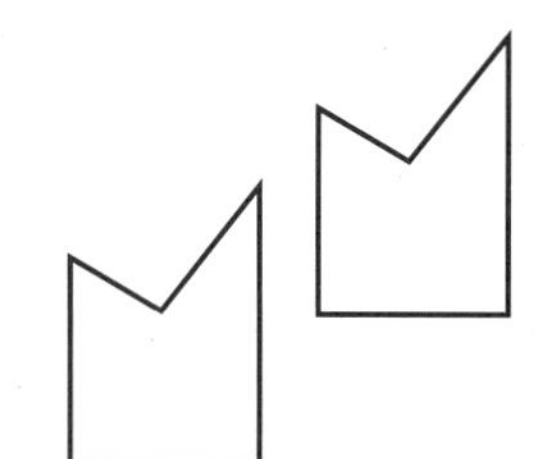

5. ____________

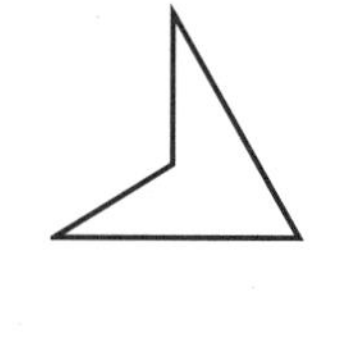

6. ____________

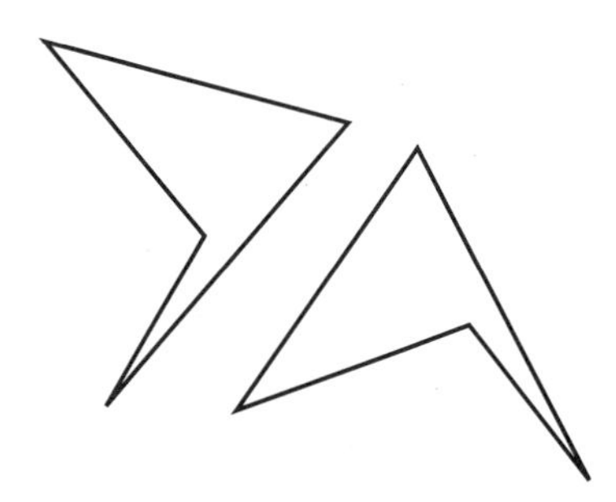

Trace the figures on folding paper.

7. reflect $\triangle AMP$ across line m

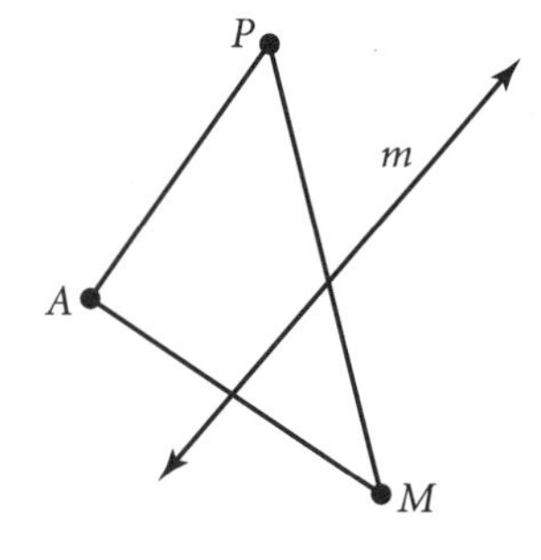

8. rotate $\triangle KIT$ about point P

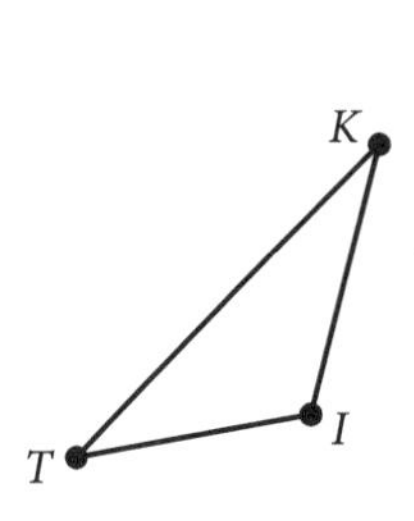

9. translate $\triangle ABC$ along line l

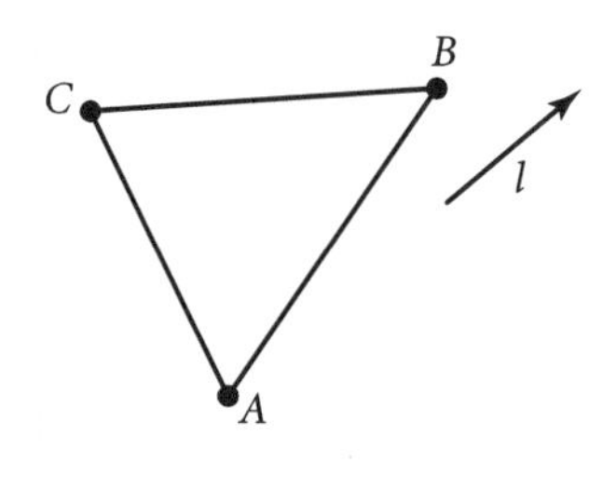

Classify each statement as true or false. Explain your reasoning in each false case.

10. A figure reflected across a line is congruent to its preimage. ____________

11. If a point is translated along a line, then the line is the perpendicular bisector of the segment that connects the point with its image. ____________

12. A figure rotated about a fixed point is congruent to its preimage. ____________

Practice Masters Level B

1.6 Motion in Geometry

In Exercises 1–4, give an example from everyday life, other than the descriptions in your textbook, that represents the given transformation.

1. a reflection ______________________

2. a rotation ______________________

3. a translation ______________________

4. a glide reflection ______________________

5. Draw the translation line in the figure below.

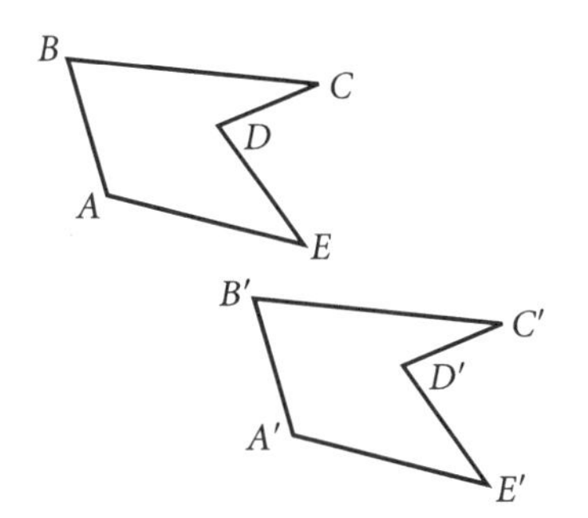

6. Draw the reflection line in the figure below.

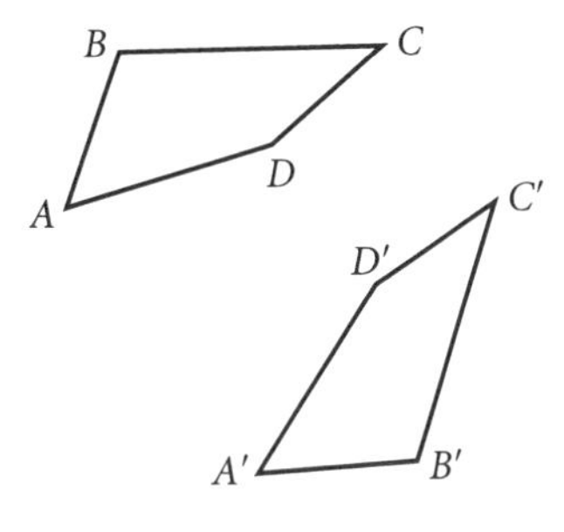

7. In Exercise 5, what is the relationship between $\overline{AD}$ and $\overline{A'D'}$? ______________________

8. In Exercise 6, what is the relationship between the line of reflection and $\overline{BB'}$? ____________

For Exercises 9 and 10, perform the indicated transformation. Use tracing paper if necessary.

9. Translate $\triangle ABC$ along line l, then reflect your drawing across line m.

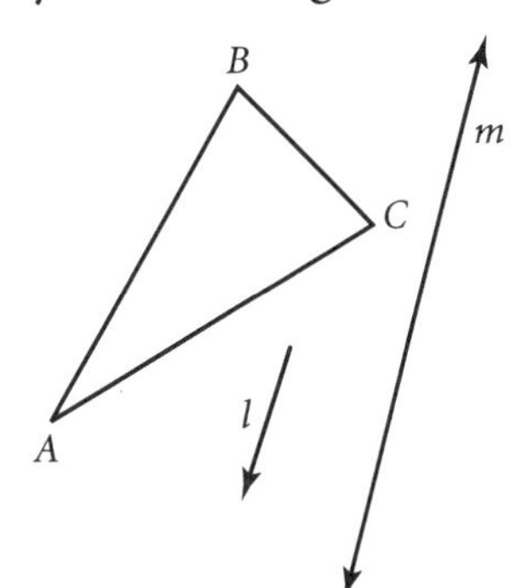

10. Rotate $\triangle SPY$ about point O.

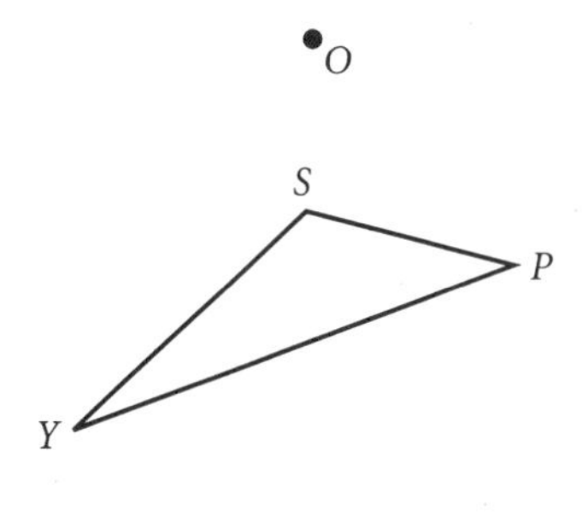

11. Name the type of transformation you performed in Exercise 9. ______________________

NAME ______________________ CLASS ____________ DATE ____________

Practice Masters Level C

1.6 Motion in Geometry

Using the figure at the right, reflect *FLAG* as directed. Trace the figure onto tracing paper if necessary.

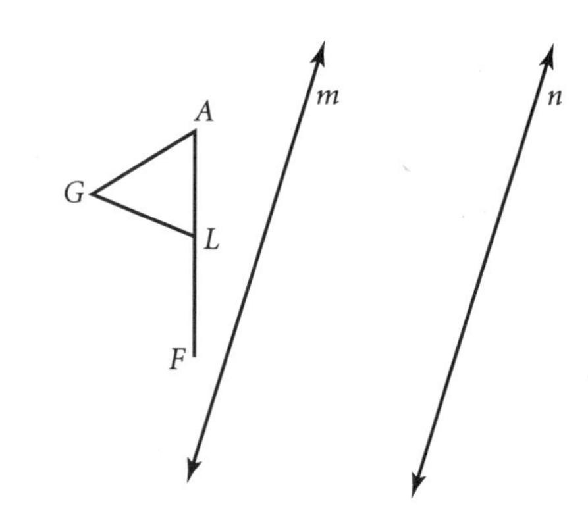

1. Reflect *FLAG* over line *m*. Label the image $F'L'A'G'$.
2. Reflect your figure from Exercise 1 over line *n*. Label the image $F''L''A''G''$.
3. Identify the transformation relating *FLAG* to $F''L''A''G''$.

   ____

4. Measure the distance between lines *n* and *m*. ____________
5. Measure the distance between *F* and F''. ____________
6. How are the distances in Exercise 4 and Exercise 5 related? ____________

Using the figure at the right, reflect *PORT* as directed. Trace the figure onto tracing paper if necessary.

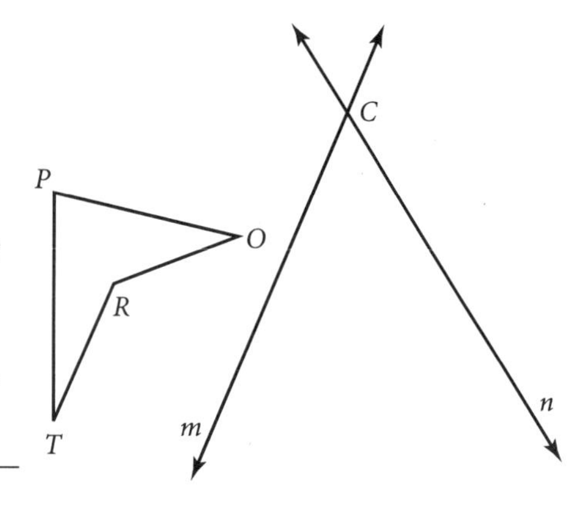

7. Reflect *PORT* over line *m*. Label the image $P'O'R'T'$.
8. Reflect your figure from Exercise 7 over line *n*. Label the image $P''O''R''T''$.
9. Identify the transformation relating *PORT* to $P''O''R''T''$.

   ____

10. Measure $\angle OCO''$. ____________
11. Measure the acute angle formed by lines *n* and *m*. ____________
12. How are the angle measures in Exercise 10 and Exercise 11 related? ____________

For Exercises 13–16, use the figure at the right.

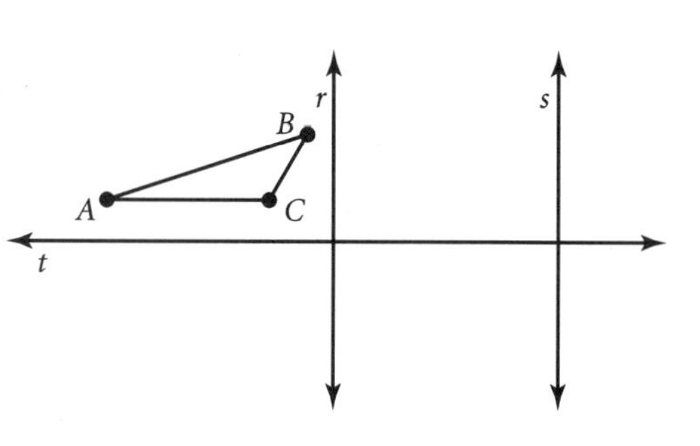

13. Reflect $\triangle ABC$ over line *t*.
14. Reflect your image from Exercise 13 over line *r*.
15. Reflect your image from Exercise 14 over line *s*. Label this final image as *DEF*.

16. Identify the transformation relating $\triangle ABC$ to $\triangle DEF$. ____________

NAME ______________________ CLASS ______________ DATE ______________

Practice Masters Level A

1.7 Motion in the Coordinate Plane

Explain how you would plot the following points in a coordinate plane.

1. $(0, -3)$ ______________________

2. $(-6, -2)$ ______________________

3. $(4, -12)$ ______________________

4. $(-7, 0)$ ______________________

Point *A* is located at $(-2, 3)$ and point *X* is located at $(3, -2)$.

5. What is the *x*-coordinate of point *A*? ______

6. What is the *y*-coordinate of point *X*? ______

Use the given rule to translate each triangle on the grid provided.

7. $T(x, y) = (x - 3, y + 2)$

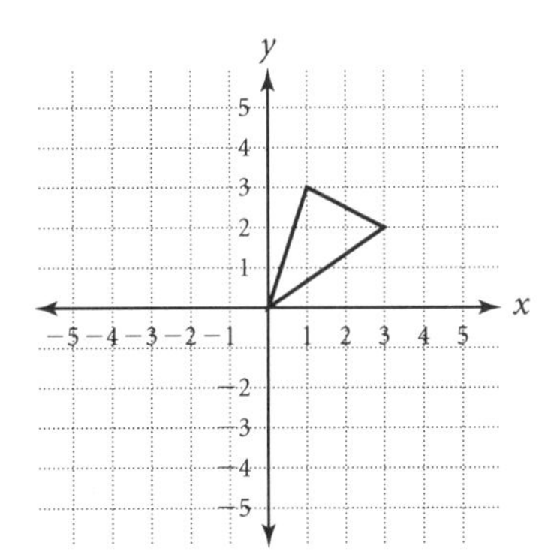

8. $G(x, y) = (x - 4, y - 4)$

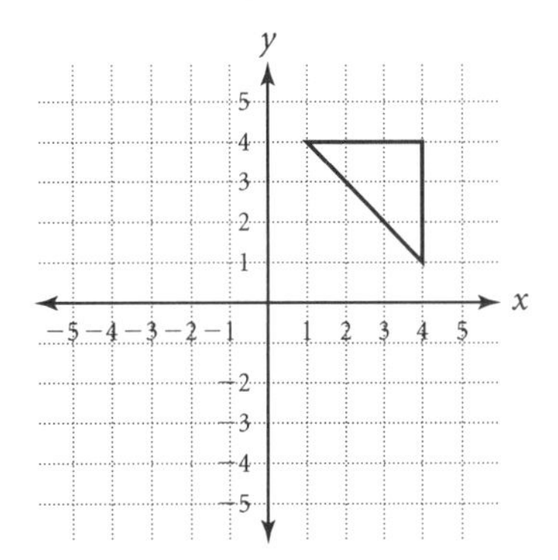

Write the rule in the form $T(x, y) = (?, ?)$ that describes the transformation pictured. In each picture, $\triangle ABC$ is the preimage, and $\triangle A'B'C'$ is the image.

9. $T(x, y) =$ ______________________

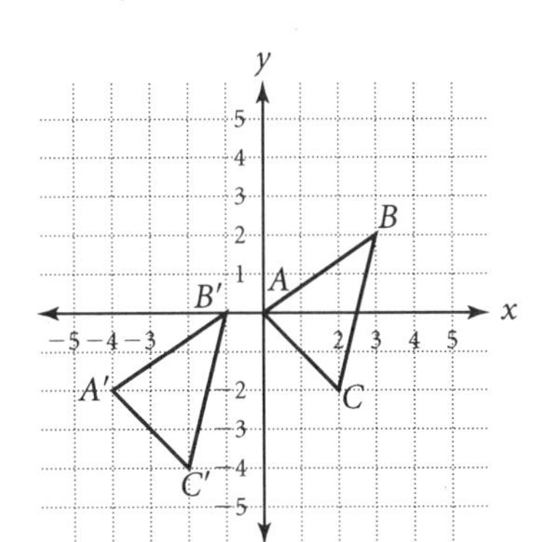

10. $T(x, y) =$ ______________________

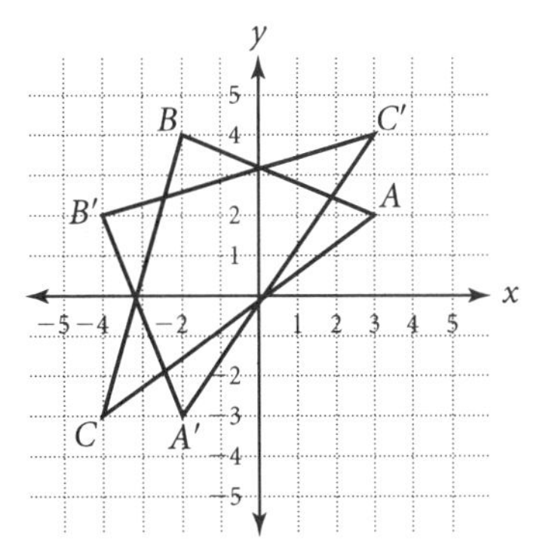

11. Identify the type of transformation pictured in Exercise 10. ______________________

NAME ______________________ CLASS ____________ DATE ____________

Practice Masters Level B

1.7 Motion in the Coordinate Plane

The coordinates of $\triangle ABC$ are $A(-2, -1)$, $B(4, 1)$, and $C(-1, 3)$. Plot the points on the grid provided, and connect them to form a triangle. Use the given rule to transform the figure.

1. $T(x, y) = (-x, -y)$

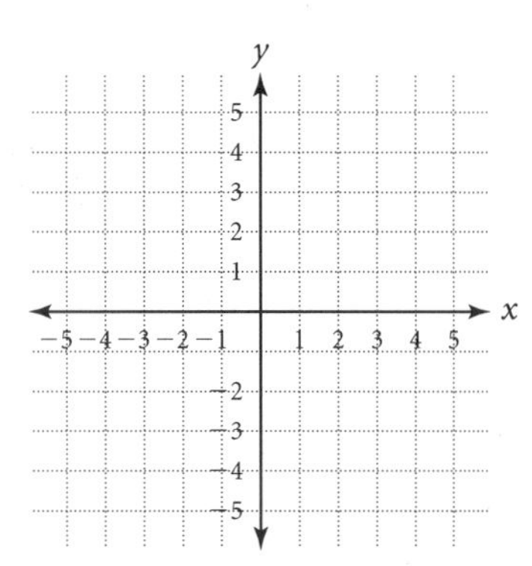

2. $P(x, y) = (x + 3, y + 1)$

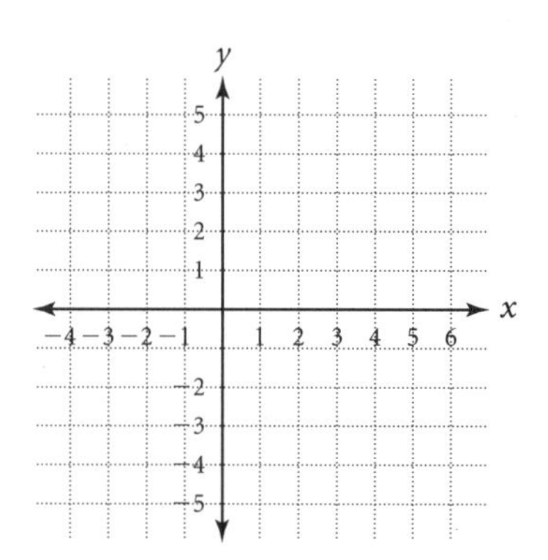

3. In Exercise 1, what is the measure of $\angle AOA'$, if O is the origin? ____________

4. In Exercise 2, what is the slope of $\overline{CC'}$? ____________

For Exercises 5–6, write a rule in the form $P(x, y) = (?, ?)$ that describes the given transformation.

5. a preimage figure reflected across the y-axis ____________

6. a preimage figure translated 4 units to the left and 3 units up ____________

7. Your aunt must take the bus to work every day. To reach the bus stop, she leaves her home, travels north for 4 blocks, east for 3 blocks, then turns left and travels one more block. If north-south is on the y-axis, and east-west is on the x-axis, write a rule in the form $T(x, y) = (?, ?)$ that describes her travel. ____________

For Exercises 8–11, use the figure at the right.

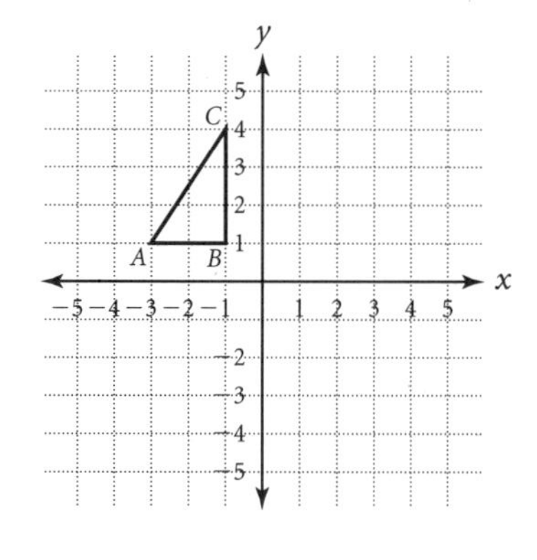

8. What are the coordinates of triangle ABC? ____________

9. How do the coordinates change if you reflect $\triangle ABC$ over the x-axis? ____________

10. Translate the reflected image described in Exercise 9 six units to the right. Label this new image $A'B'C'$.

11. Write a rule that describes the relationship between $\triangle ABC$ and $\triangle A'B'C'$. ____________

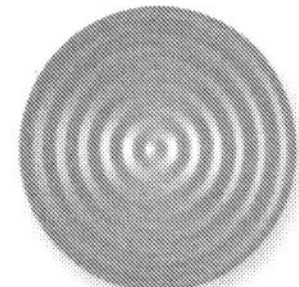

Practice Masters Level C

1.7 Motion in the Coordinate Plane

For Exercises 1–7, use the $\triangle TED$ in the grid provided at the right. The line $y = x$ has been drawn for you.

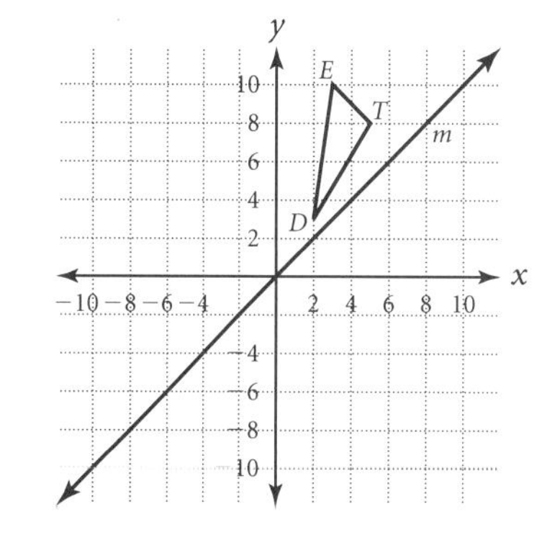

1. Reflect $\triangle TED$ across line m. Label your figure $T'E'D'$.

2. Write a rule in the form $R(x, y) = (?, ?)$ for the transformation.

   ____

3. Reflect $\triangle T'E'D'$ over the y-axis. Label this figure $T''E''D''$.
4. Find $m\angle DOD''$, $m\angle EOE''$ and $m\angle TOT''$.

   ____

5. What angle does the line m form with the y-axis? ______________________
6. Describe the transformation that relates $\triangle TED$ to $\triangle T''E''D''$. ______________________
7. Write a rule for the transformation that relates $\triangle TED$ to $\triangle T''E''D''$. ______________________

For Exercises 8–11, use the figure at the right.

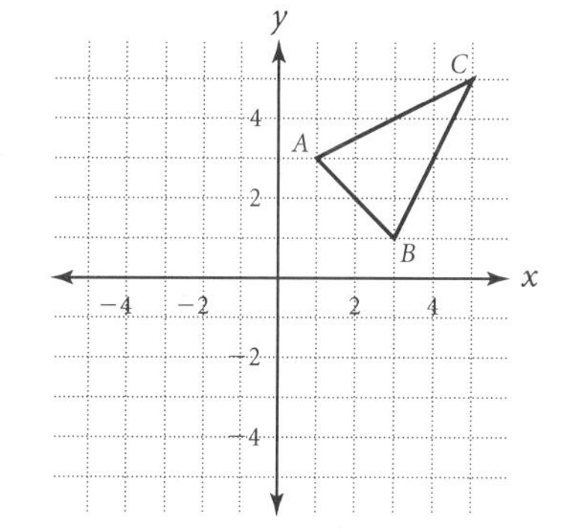

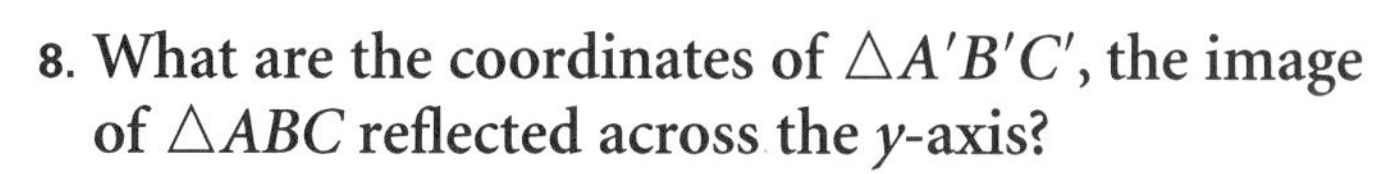

8. What are the coordinates of $\triangle A'B'C'$, the image of $\triangle ABC$ reflected across the y-axis?

   ____

9. Translate $\triangle A'B'C'$ 4 units down. Label this new image $A''B''C''$.
10. Write a rule that describes the relationship between $\triangle ABC$ and $\triangle A''B''C''$.

    ____

11. Identify the transformation that relates $\triangle ABC$ to $\triangle A''B''C''$. ______________________

In Exercises 12–16, describe the result of applying each rule to a figure in a coordinate plane.

12. $H(x, y) = (x + 2, -y)$ ______________________
13. $B(x, y) = (-y, -x)$ ______________________
14. $K(x, y) = (-x, -y)$ ______________________
15. $P(x, y) = (y, x - 8)$ ______________________
16. $I(x, y) = (x, y)$ ______________________

NAME ______________________ CLASS ______________ DATE ______________

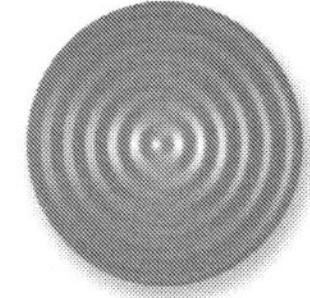

Practice Masters Level A

2.1 An Introduction to Proofs

A *diagonal* of a polygon is a line segment that connects non-adjacent vertices of the polygon. A polygon can be separated into triangles by drawing all possible diagonals from one vertex. Draw the diagonals that will separate the following polygons into triangles, then record your results in the table at the right. Exercise 2 demonstrates how to draw the diagonals.

1. 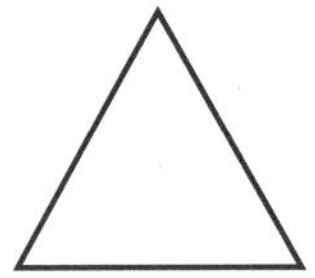

2. 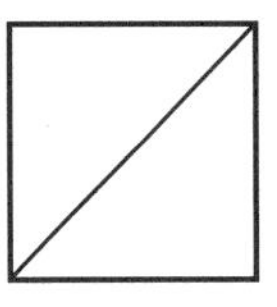

3. 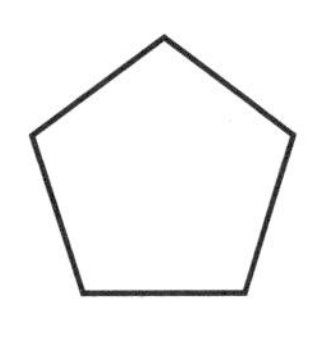

4. 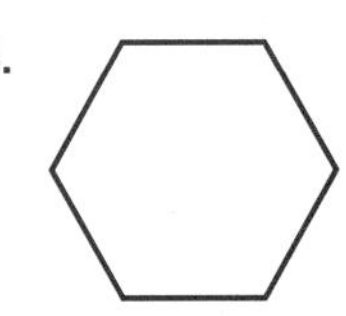

5. 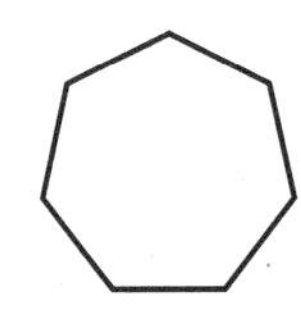

6. 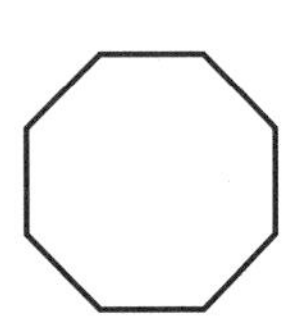

Number of sides	Number of triangles
3	
4	2
5	
6	
7	
8	

7. What is the pattern? ______________________

8. If this pattern continues, into how many triangles can a polygon with 10 sides be separated? ______________

9. Write an expression for the number of triangles that can be drawn in a polygon with n sides. ______________

The figure at the right was constructed by reflecting point *C* over line *l*, then drawing the segments between points *A*, *C*, and *C′*. A student wrote a conjecture stating that $\triangle ACC'$ is an isosceles triangle. Use this figure for Exercises 10–12.

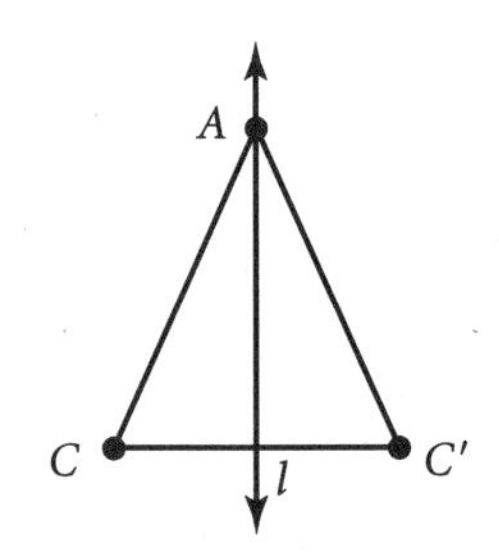

10. Test the conjecture by measuring AC and AC'.

    $AC =$ ______________ $AC' =$ ______________

11. Do you think the student's conjecture is correct? Why or why not? ______________

    ____

Logical arguments that ensure true conclusions are called *proofs.*

12. Write another conjecture about the figure. Explain how you could prove your conjecture is true.

    ____

    ____

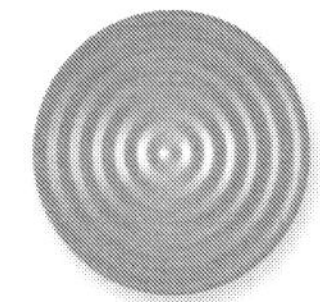

Practice Masters Level B

2.1 An Introduction to Proofs

A *diagonal* of a polygon is a line segment that connects non-adjacent vertices of the polygon. How many diagonals can be drawn in a polygon that has 3, 4, 5, or 6 vertices? Draw them. Record your data in the table.

1. 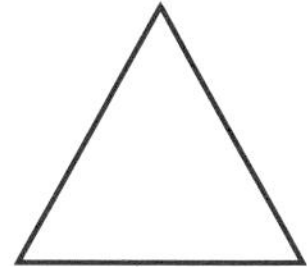

2. 

3. 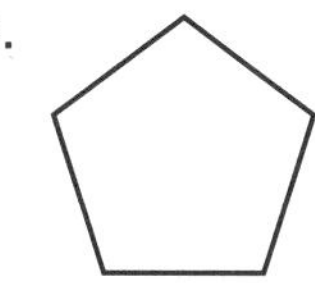

4. 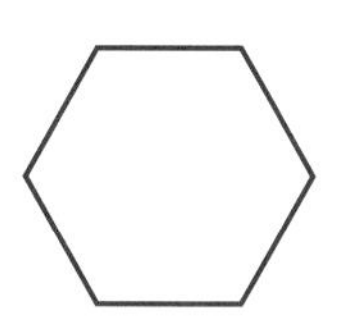

Number of sides	Number of diagonals
3	0
4	
5	
6	

5. What is the pattern? ________________

6. If this pattern continues, how many diagonals can be drawn in a polygon with 10 sides? ________________

7. Write an expression for the number of diagonals that can be drawn in polygon with n sides. ________________

A student wrote a conjecture about the figure at the right stating that the diagonals of a rectangle are perpendicular bisectors of each other.

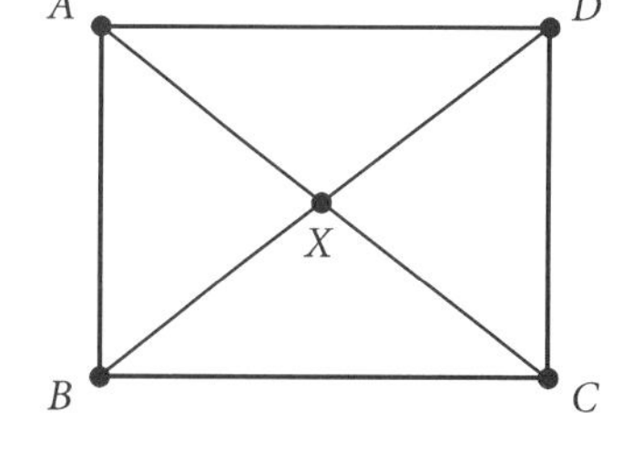

8. What measurements could you make to test this conjecture? Label and record your measurements in the space provided:

____

____

9. Do you think the student's conjecture is correct? Why or why not? ________________

____

10. Write another conjecture about the figure. ________________

Explain how you could prove your conjecture is true.

____

____

____

NAME ______________________ CLASS __________ DATE __________

Practice Masters Level C

2.1 An Introduction to Proofs

Use the table to answer Exercises 1–5. The numbers in the table are the powers of 2: $2^1 = 2, 2^2 = 4, 2^3 = 8, \ldots$

A	B	C	D
2	4	8	16
32	64	128	256

1. Fill in the missing entries in the table.
2. Look at the columns in the table. Describe the pattern.

____

____

3. Without calculating the value, determine the column in which 2^{23} will occur. __________
4. Explain in your own words what it means to prove a statement. __________

____

5. Describe how you could prove the conjecture that the diagonals of a parallelogram bisect each other.

____

____

____

A child is building a large cube out of small cubes. Each new layer of the cube is a different color and completely covers the previous layer. The child first uses red, then green, yellow, and blue cubes, in that order.

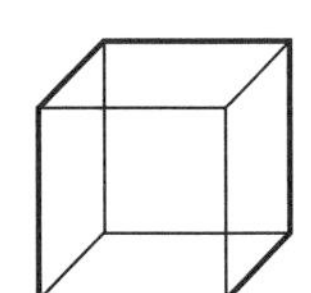

6. How many green cubes will the child need to build the second layer? __________
7. How many small cubes will this new cube contain? Explain your reasoning.

____

____

Layer #	Number of additional cubes	Total number of small cubes
1 (red)		1
2 (green)		
3 (yellow)		

8. Complete the table at right to find the number of cubes needed for the yellow layer.
9. If the child continues to add on to the cube, after how many layers will there be 1331 cubes? __________
10. If the last layer is blue, how many total cubes will the large cube contain? __________

NAME ______________________ CLASS __________ DATE __________

Practice Masters Level A

2.2 An Introduction to Logic

For Exercises 1–4, refer to the following statement:

All dogs are mammals.

1. Rewrite the statement as a conditional. ______________________

2. Identify the hypothesis and conclusion of the conditional.

   Hypothesis: ______________________

   Conclusion: ______________________

3. Draw an Euler diagram that illustrates this conditional.

   ____

4. Write the converse of the conditional you wrote in Exercise 1. If the converse is false, give a counterexample to show that it is false.

   ____

   ____

For Exercises 5–6, refer to the given hypothesis and conclusion and the figure at the right.

Hypothesis: ***m*** **is the perpendicular bisector of** $\overline{AB}$**.**
Conclusion: $PA = PB$

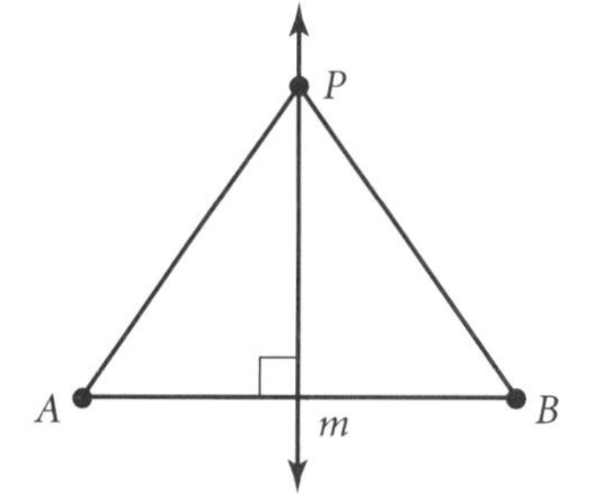

5. Write a conditional with the given hypothesis and conclusion.

   ____

6. Write the converse of the conditional you wrote in Exercise 5. If the converse is false, give a counterexample to show that it is false.

   ____

7. Arrange the three statements below into a logical chain. Then write the conditional statement that follows from the logic.
   If I go shopping, I will buy a new umbrella.
   If it rains on Saturday, then I am going shopping.
   If I buy a new umbrella, then I won't get wet.

   ____

8. "If-then" statements are called: ______________________

9. Write the logical notation for a conditional statement. ______________________

NAME ______________________ CLASS __________ DATE __________

Practice Masters Level B

2.2 An Introduction to Logic

For Exercises 1–4, refer to the following statement:

All rectangles are parallelograms.

1. Rewrite the statement as a conditional. ______________________

2. Identify the hypothesis and conclusion of the conditional.

   Hypothesis: ______________________

   Conclusion: ______________________

3. Draw an Euler diagram that illustrates this conditional.

   ____

4. Write the converse of the conditional you wrote in Exercise 1.
   If the converse is false, give a counterexample to show that it is false.

   ____

   ____

For Exercises 5–6, refer to the given hypothesis and conclusion.

Hypothesis: a **and** b **are even integers.** **Conclusion:** $a + b$ **is an even integer.**

5. Write a conditional statement using the given hypothesis and conclusion.
   Is your conditional *true* or *false*? Explain. Give a counterexample if it is false.

   ____

   ____

6. Write the converse of the conditional you wrote in Exercise 5. Is the converse *true* or *false*? Explain. Give a counterexample if it is false.

   ____

   ____

7. Draw a logical conclusion from the following statements:
   If you finish your chores on time, you may use the car tonight.
   You finish your chores on time. ______________________

8. Give three examples of a conditional statement, and explain why they are conditional statements. ______________________

   ____

NAME ______________________ CLASS __________ DATE __________

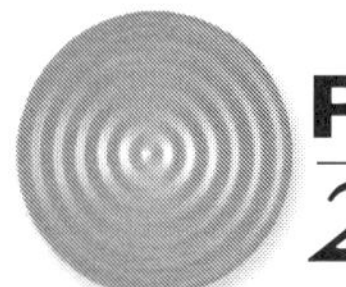

Practice Masters Level C

2.2 An Introduction to Logic

For Exercises 1–2, refer to the given hypothesis and conclusion.

Hypothesis: $x^2 = 9$ Conclusion: $x = 3$

1. Write a conditional statement using the given hypothesis and conclusion. Why is it a conditional? Is your conditional *true* or *false*? Explain. Give a counterexample if it is false.

____

____

2. Write the converse of the conditional you wrote in Exercise 1. Is the converse *true* or *false*? Explain. Give a counterexample if it is false.

____

____

For Exercises 3–4, refer to the given hypothesis and conclusion:

Hypothesis: $MA = MB$ Conclusion: M, A, and B are collinear.

3. Write a conditional statement using the given hypothesis and conclusion. Is your conditional *true* or *false*? Explain. Give a counterexample if it is false.

____

____

4. Write the converse of the conditional you wrote in Exercise 3. Is the converse *true* or *false*? Explain. Give a counterexample if it is false.

____

____

____

5. Use the given statements to draw a conclusion. Draw an Euler diagram if necessary.

   If you are eighteen, you can vote.
   You vote in today's election.

____

____

____

NAME ______________________ CLASS __________ DATE __________

Practice Masters Level A

2.3 Definitions

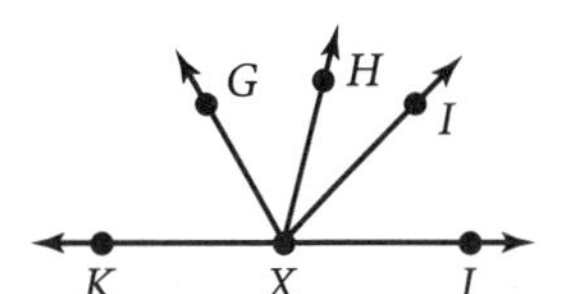

1. Name all pairs of adjacent angles in the figure at the right.

____

____

The following are kites:

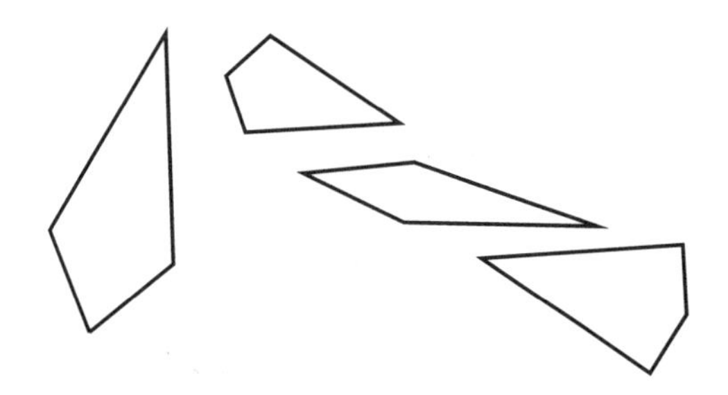

The following are not kites:

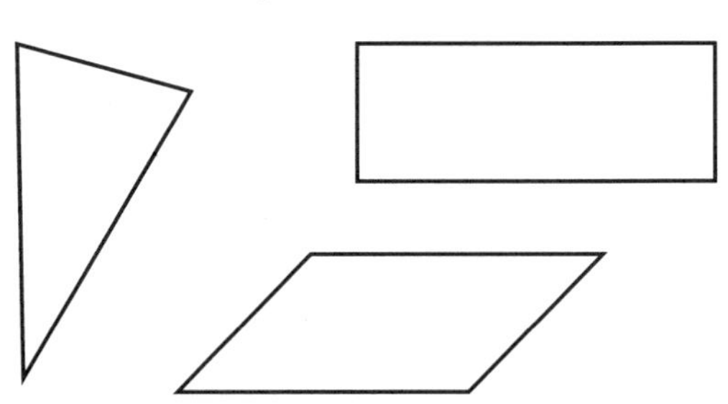

2. Draw a Euler diagram to represent the definition of a kite. ______________________________

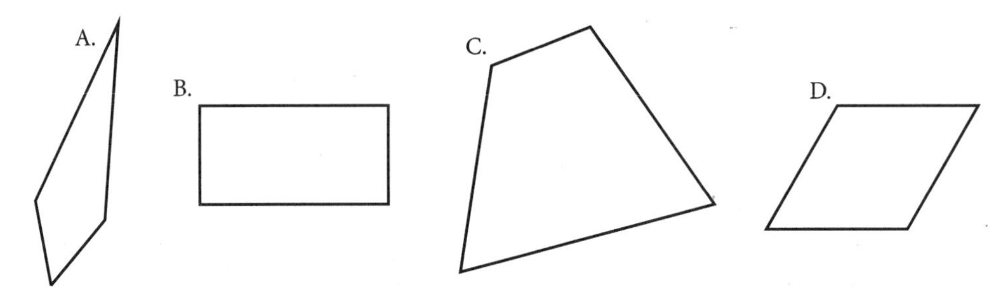

3. Which of the figures above is a kite? Write a definition for a kite. ______________

____

Use the steps in Exercises 4–7 to determine whether the given sentence is a definition.

A triangle is formed by three segments.

4. Write the sentence as a conditional statement. ______________________

5. Write the converse of the conditional. ______________________

6. Write a biconditional statement. ______________________

7. Decide whether the statement is a definition, and explain your reasoning. ____________

____

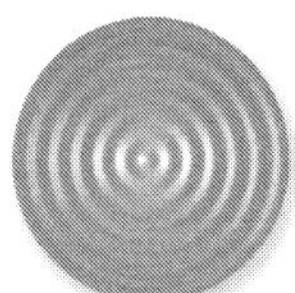

Practice Masters Level B

2.3 Definitions

1. Name all pairs of adjacent angles in the figure at the right.

____

____

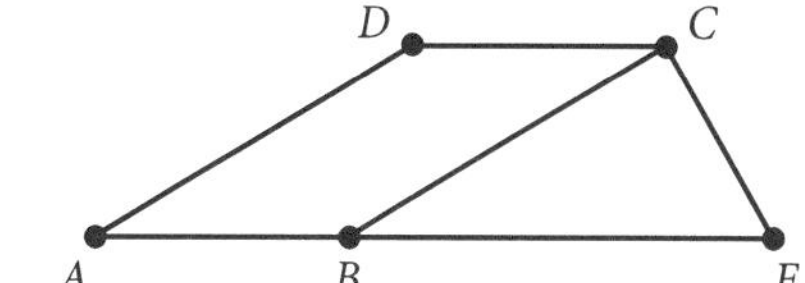

The following are equilateral polygons:

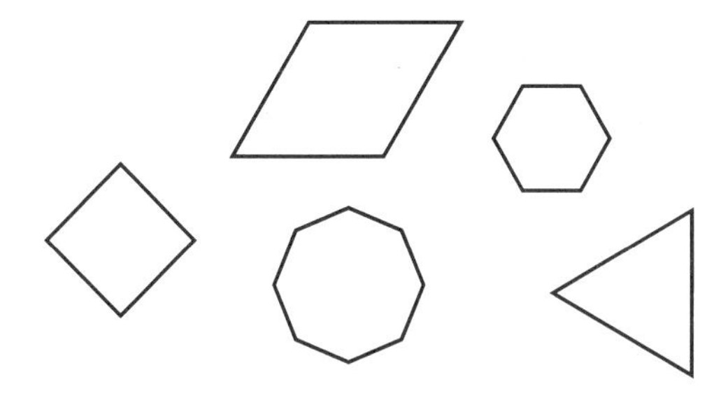

The following are not equilateral polygons:

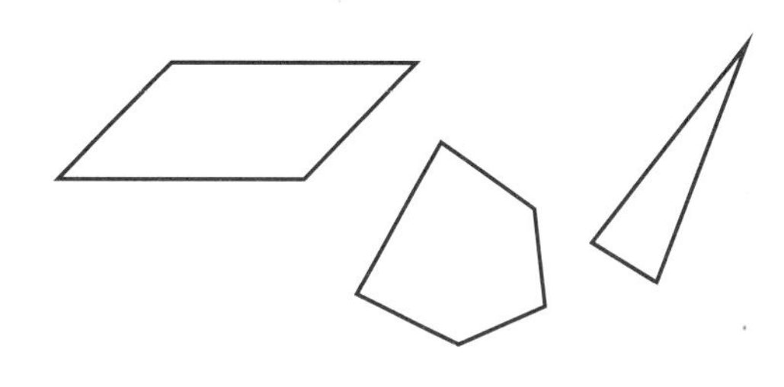

2. Which of the figures below are equilateral? ______________________

A.

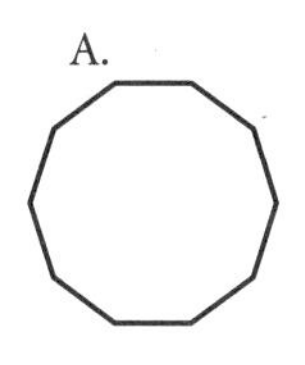

B.

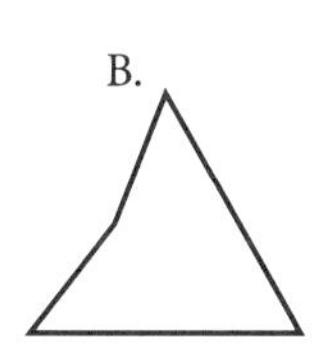

C.

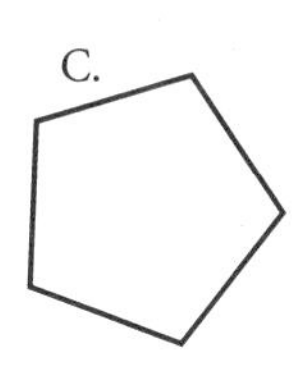

D.

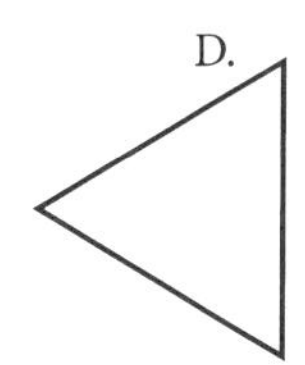

3. Draw a Euler diagram, and write a definition of equilateral polygons.

____

____

Use the steps in Exercises 4–7 to determine whether the given sentence is a definition.

A square is a figure with four congruent sides.

4. Write the sentence as a conditional statement: ______________________

5. Write the converse of the conditional: ______________________

6. Write a biconditional statement: ______________________

7. Decide whether the statement is a definition, and explain your reasoning.

____

NAME ______________________ CLASS ____________ DATE ____________

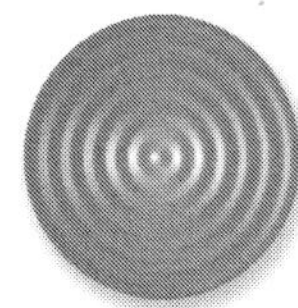

Practice Masters Level C

2.3 Definitions

1. Name all pairs of adjacent angles in the figure at the right.

____

____

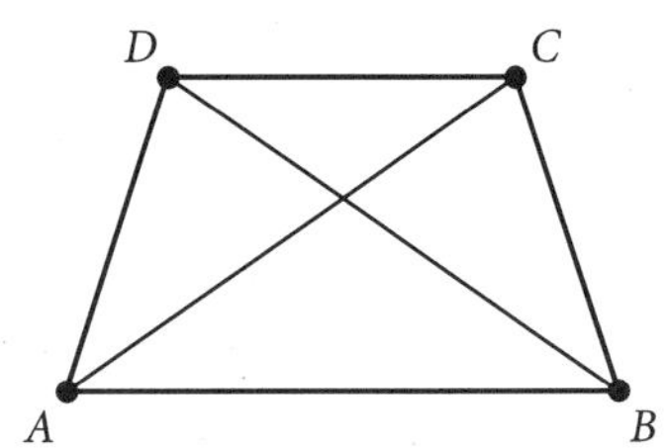

The following are trapezoids: The following are not trapezoids:

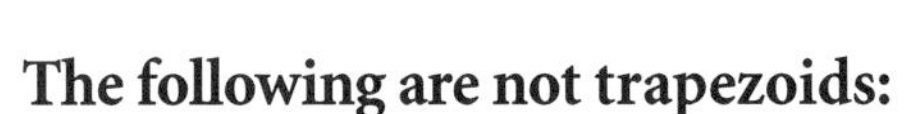

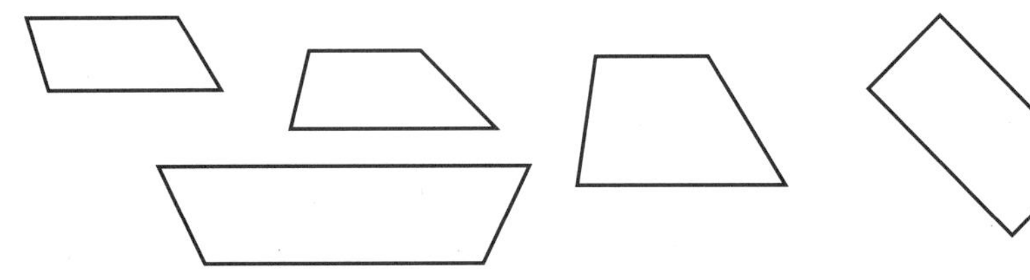

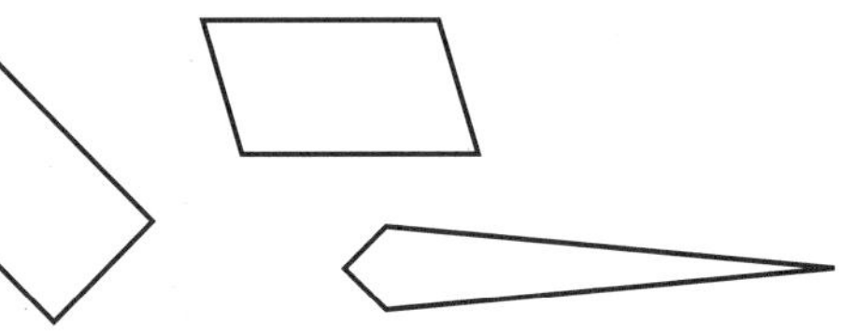

2. Which of the figures below are trapezoids? ______________________

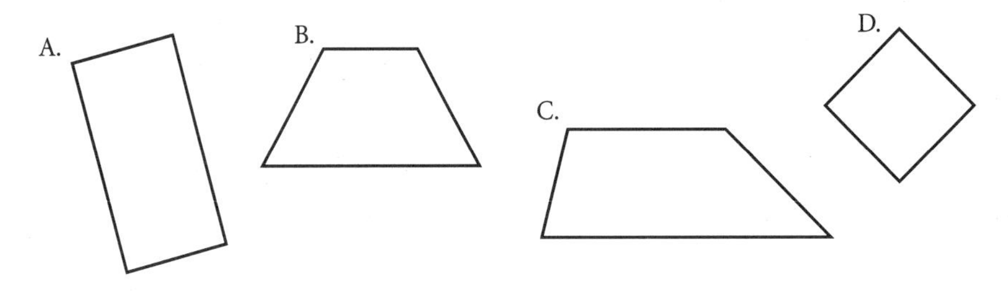

3. Draw a Euler diagram, and write a definition of a trapezoid.

____

____

Use the steps in Exercises 4–7 to determine whether the given sentence is a definition.

Linear pairs are supplementary, adjacent angles.

4. Write the sentence as a conditional statement. ______________________

5. Write the converse of the conditional. ______________________

6. Write a biconditional statement. ______________________

7. Decide whether the statement is a definition, and explain your reasoning. ____________

____

NAME ______________________ CLASS ____________ DATE ____________

Practice Masters Level A

2.4 Building a System of Geometry Knowledge

Match each property with its definition.

_______ 1. Addition Property

_______ 2. Symmetric Property

_______ 3. Substitution Property

_______ 4. Multiplication Property

_______ 5. Division Property

_______ 6. Reflexive Property

_______ 7. Subtraction Property

_______ 8. Transitive Property

a. If $a = b$, then $ac = bc$.

b. If $a = b$, then $a - c = b - c$.

c. For all real numbers a, $a = a$.

d. If $a = b$, you may replace a with b in any true equation containing a and the resulting equation will still be true.

e. If $a = b$ and $c \neq 0$, then $\frac{a}{c} = \frac{b}{c}$.

f. If $a = b$, then $a + c = b + c$.

g. For all real numbers a and b, if $a = b$, then $b = a$.

h. For all real numbers a and b, if $a = b$ and $b = c$, then $a = c$.

Refer to the diagram at right, in which m$\angle$*NAG* = m$\angle$*EAL*. Use the Overlapping Angles Theorem to complete the following:

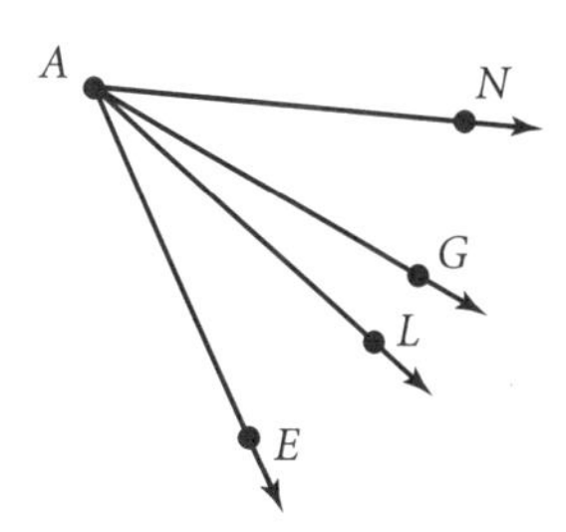

9. $m\angle NAG + m\angle GAL = m\angle GAL +$ ____________

If m$\angle$*NAG* = 24°, and m$\angle$*NAL* = 36°, find the following:

10. $m\angle GAL$ ____________ 11. $m\angle AEN$ ____________

Complete the proof below:

Given: $m\angle 1 = m\angle 2$
$m\angle T + m\angle 3 + m\angle 2 = 180°$
$m\angle T + m\angle 1 + m\angle 4 = 180°$

Prove: $m\angle 3 = m\angle 4$

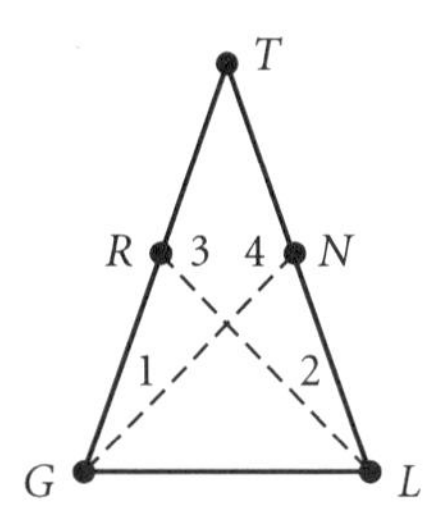

Statements	Reasons
$m\angle T = m\angle T$	12.
$m\angle 1 = m\angle 2$ $m\angle T + m\angle 3 + m\angle 2 = 180°$ $m\angle T + m\angle 1 + m\angle 4 = 180°$	Given
$m\angle T + m\angle 3 + m\angle 2 = m\angle T + m\angle 1 + m\angle 4$	13.
$m\angle 3 + m\angle 2 = m\angle 1 + m\angle 4$	14.
$m\angle 3 + m\angle 1 = m\angle 1 + m\angle 4$	15.
$m\angle 3 = m\angle 4$	16.

NAME ______________________ CLASS ______________ DATE ______________

Practice Masters Level B

2.4 Building a System of Geometry Knowledge

Identify the Properties of Equality that justify the indicated steps.

Statements	Reasons
$3x + 12 = 5x$	Given
$12 = 2x$	**1.**
$6 = x$	**2.**

For Exercises 5–8, use the figure at the right. If $CE = FD$ and $CD = 11x–21$, find the following:

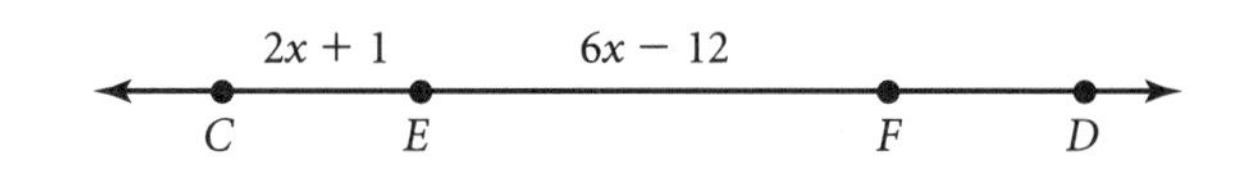

3. x ______________ **4.** CE ______________

5. EF ______________ **6.** CD ______________

For Exercises 9–12, use the figure at the right. $\angle NXG \cong \angle LXE$, $\angle AXN \cong \angle GXL$.

7. $m\angle NXG + m\angle GXL =$ ______________

If $m\angle AXN = 2(3x + 4)$, and $m\angle GXL = 8x–9$, find the following:

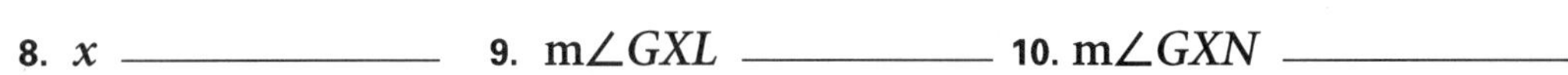

8. x ______________ **9.** $m\angle GXL$ ______________ **10.** $m\angle GXN$ ______________

Fill in the blanks in the following proof:

Given: ΔRDA and ΔCTB are equilateral triangles.
$RD = TC$

Prove: $AC = DB$

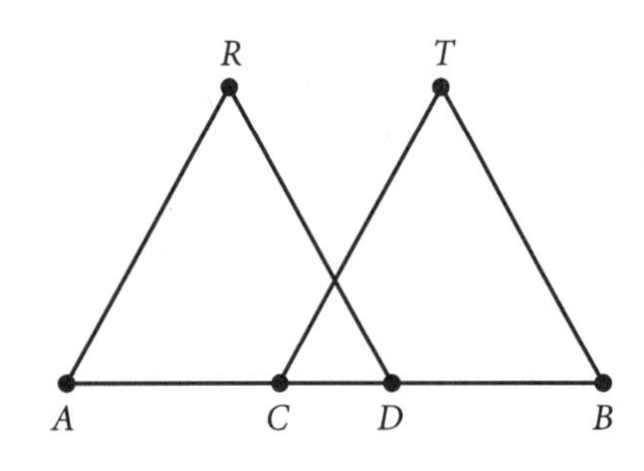

Statements	Reasons
11. $RD =$ _____ $=$ _____	Definition of equilateral triangle
12. $TC =$ _____ $=$ _____	Definition of equilateral triangle
$RD = TC$	**Given**
$AD = CB$	**13.**
$AC + CD = AD$ $CD + DB = CB$	Segment Addition Postulate
$AC + CD = CD + DB$	**14.**
$CD = CD$	**15.**
$AC = DB$	**16.**

NAME ______ CLASS ______ DATE ______

Practice Masters Level C

2.4 Building a System of Geometry Knowledge

In Exercises 1–4, use the Properties of Equality to fill in the missing reasons in the proof:

Statements	Reasons
$a = 6$	Given
$a^2 = 6a$	**1.**
$a^2 - 6a = 0$	**2.**
$a(a - 6) = 0$	Distributive Property
$\frac{a(a-6)}{(a-6)} = \frac{0}{(a-6)}$	**3.**
$a = 0$	Simplify
$6 = 0$	**4.**

5. What startling fact did you "prove" in Exercises 1–4? ______

6. What property of equality is violated in this proof? ______

For Exercises 7–12, refer to the figure at right. The figure was formed by reflecting $\triangle ABC$ over $\overline{AD}$.

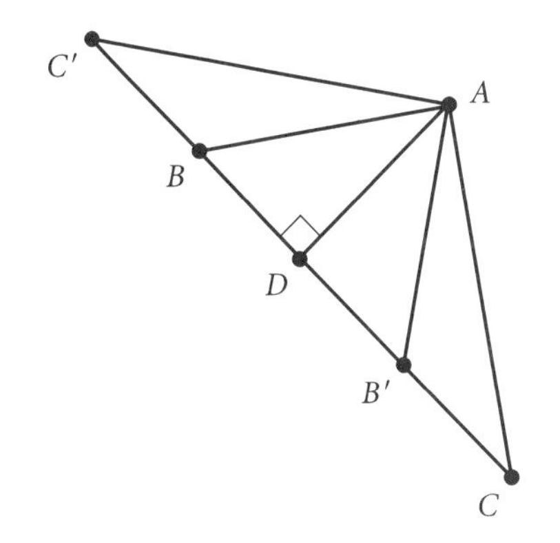

Given: $m\angle BAC = m\angle B'AC' = 90°$
$\overline{AD}$ bisects $\angle BAB'$
$m\angle BAD \approx 37°$
$m\angle C' + m\angle ABC' + m\angle BAC' = 180°$
$m\angle C + m\angle AB'C + m\angle CAB' = 180°$
$m\angle C + m\angle CAD = m\angle C' + m\angle C'AD = 90°$

Find the following:

7. $m\angle BAB'$ ______
8. $m\angle BAC'$ ______
9. $m\angle AB'D$ ______
10. $m\angle B'AD$ ______
11. $m\angle AB'C$ ______
12. $m\angle C'$ ______

13. Write a paragraph proof to show that the solution to $\frac{1}{4}x - 7 = 2$ is $x = 36$.

______

______

______

______

______

NAME ______________________ CLASS ______________ DATE ______________

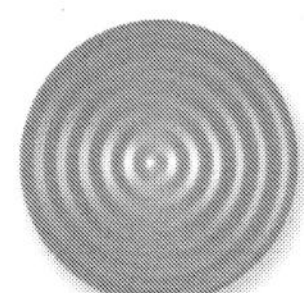

Practice Masters Level A

2.5 Conjectures That Lead to Theorems

Complete the two-column proof.

Given: $\angle 3$ and $\angle 4$ are vertical angles.
Prove: $\angle 3 \cong \angle 4$

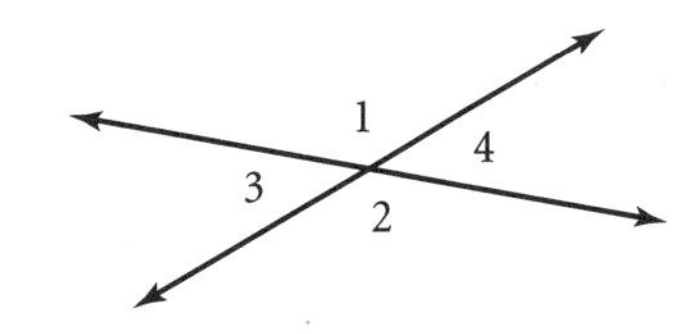

Statements	Reasons
1.	1. Given
2. $m\angle 1 + m\angle 4 = 180°$ $m\angle 2 + m\angle 3 = 180°$	2.
3.	3. Definition of Supplementary Angles
4.	4.

For Exercises 5–9, use the figure at right in which $\angle C \cong \angle G$ and $m\angle G + m\angle WAG = 90°$.

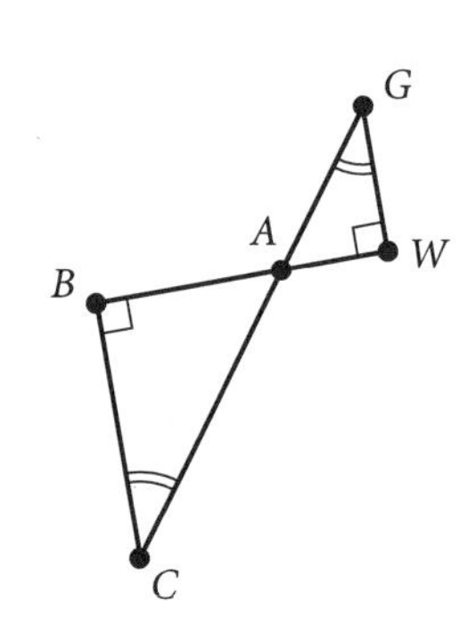

5. What is the relationship between $\angle BAC$ and $\angle WAG$?

____

If $m\angle WAG = (7x + 4)°$, and $m\angle CAB = (9x - 10)°$, find the following:

6. x ______________ 7. $m\angle GAW$ ______________

8. $m\angle CAB$ ______________ 9. $m\angle G$ ______________

For Exercises 10–13, use the figure at the right.
If $\angle CDE$ is 7 times larger than $\angle CDB$, find the following:

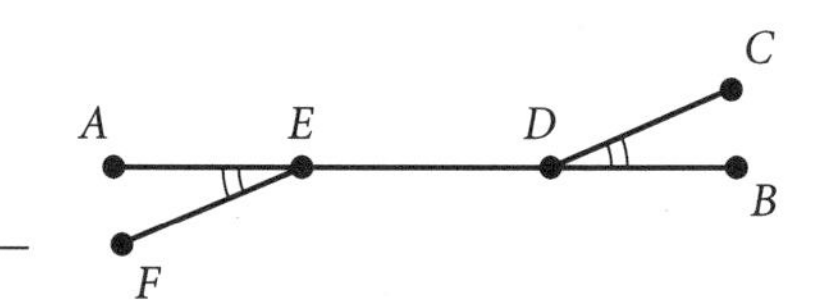

10. $m\angle CDB$ ________ 11. $m\angle CDE$ ________ 12. $m\angle FED$ ________

13. Explain how you found your answer to Exercise 11.

____

Tell whether each argument is an example of induction or deduction. Explain your reasoning.

14. James looked outside and decided to wear his coat to school. Therefore, it was snowing.

____

15. If Sarah pays for her insurance, then she can get her driver's license. Sarah shows you her driver's license. Therefore, Sarah paid for her insurance.

____

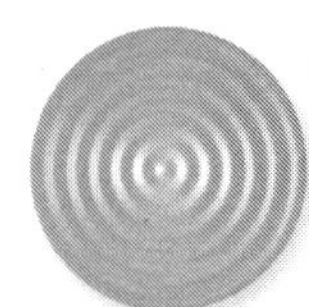

Practice Masters Level B

2.5 Conjectures That Lead to Theorems

Complete the two-column proof.

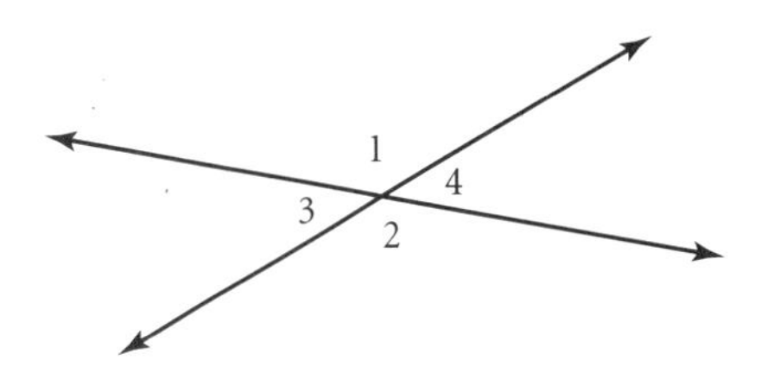

Given: $\angle 3$ and $\angle 4$ are vertical angles.
Prove: $\angle 3 \cong \angle 4$

Statements	Reasons
1.	1.
2.	2.
3.	3.
4.	4.

For Exercises 5–12, use the figure at the right. $\angle 1 \cong \angle 8$ and $m \parallel n$.

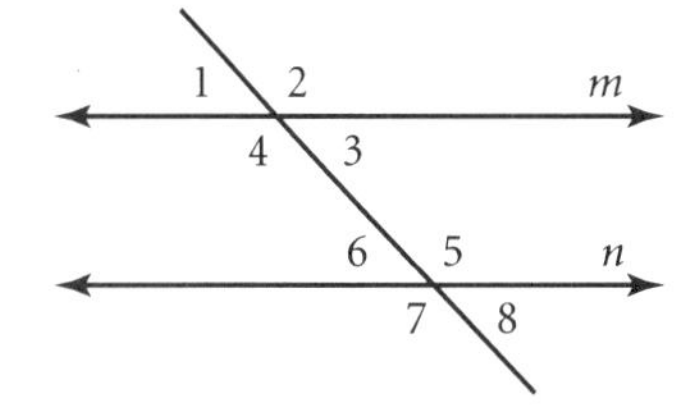

5. Name all the pairs of vertical angles.

____

6. Explain why $m\angle 2 \cong m\angle 7$. ______________________

____

7. What angles are congruent to $\angle 1$? ______________________

8. What angles are congruent to $\angle 2$? ______________________

If $m\angle 1 = 7x + 8$, and $m\angle 3 = 4(3x - 5)$, find the following:

9. x ______________ 10. $m\angle 1$ ______________

11. $m\angle 5$ ______________ 12. $m\angle 6$ ______________

You carefully measure all the angles of a figure that looks like a rectangle. You discover that all of the angles are 90° and conclude that all rectangles have four 90° angles.

13. Did you use inductive or deductive reasoning? ______________________

14. Is your conclusion a "proof"? Explain. ______________________

____

____

____

NAME ______________________ CLASS __________ DATE __________

Practice Masters Level C

2.5 Conjectures That Lead to Theorems

1. Write a paragraph proof.

   Given: $\angle 3$ and $\angle 4$ are vertical angles.
   Prove: $\angle 3 \cong \angle 4$

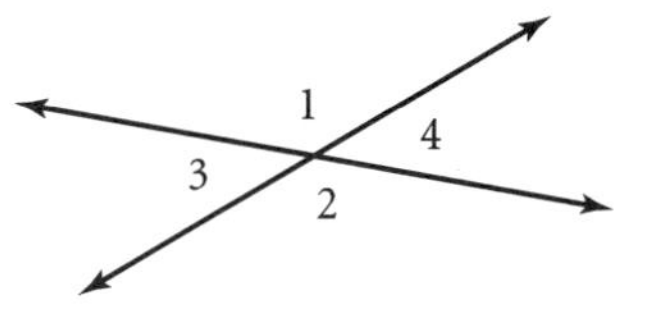

____

____

____

____

____

Use the figure at right for Exercises 2–9.

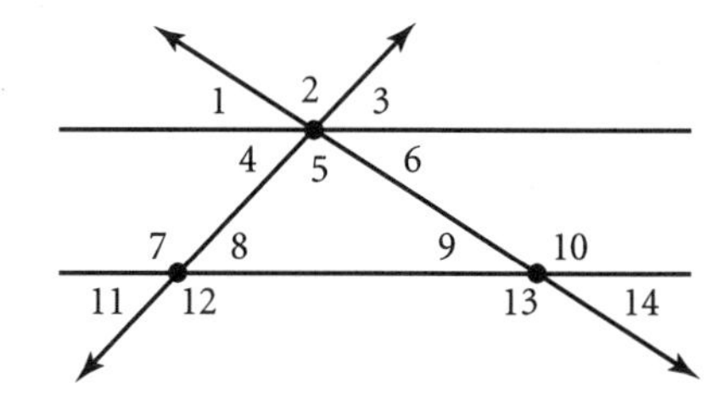

Given: $\angle 8 \cong \angle 4$;
$m\angle 5 + m\angle 8 + m\angle 9 = 180°$;
$m\angle 2 = 102.16$;
$m\angle 8 = 7x + 19$;
$m\angle 11 = 32x - 83$

Use the Vertical Angles Theorem and the Congruent Supplements Theorem to find the following:

2. x __________ 3. $m\angle 4$ __________ 4. $m\angle 5$ __________ 5. $m\angle 9$ __________

6. $m\angle 10$ __________ 7. $m\angle 12$ __________ 8. $m\angle 1$ __________ 9. $m\angle 3$ __________

10. Lewis Carroll, best known as the author of *Alice's Adventures in Wonderland,* loved mathematics and logic puzzles. The following famous puzzle was found in his diary:

    "The Dodo says that the Hatter tells lies.
    The Hatter says that the March Hare tells lies.
    The March Hare says that both the Dodo and the Hatter tell lies."

    Who is telling the truth? Did you use inductive or deductive reasoning? Explain your answer.

____

____

____

____

NAME ______________________ CLASS ____________ DATE ____________

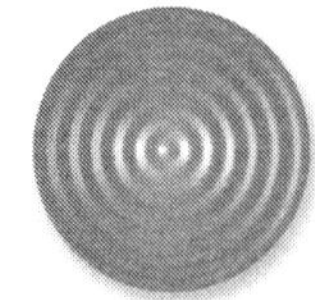

Practice Masters Level A

3.1 Symmetry in Polygons

Draw all of the axes of symmetry for each figure.

1. 

2. 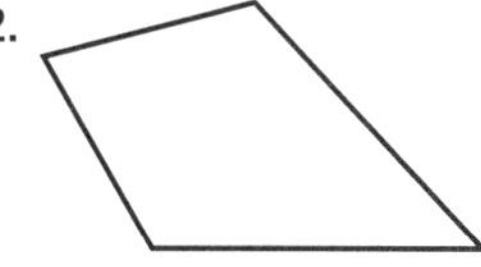

3. 

Each figure below shows part of a shape with reflectional symmetry. Complete each figure.

4. 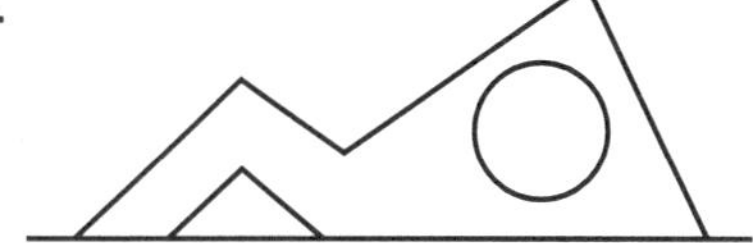

5. 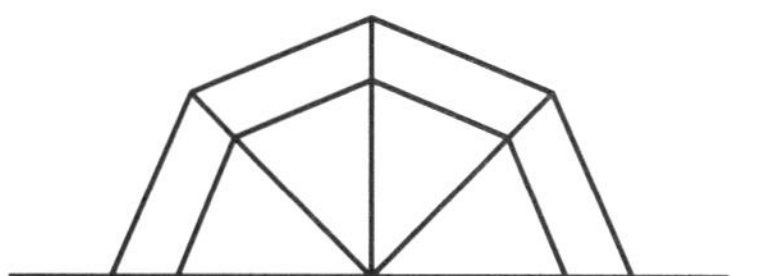

6. 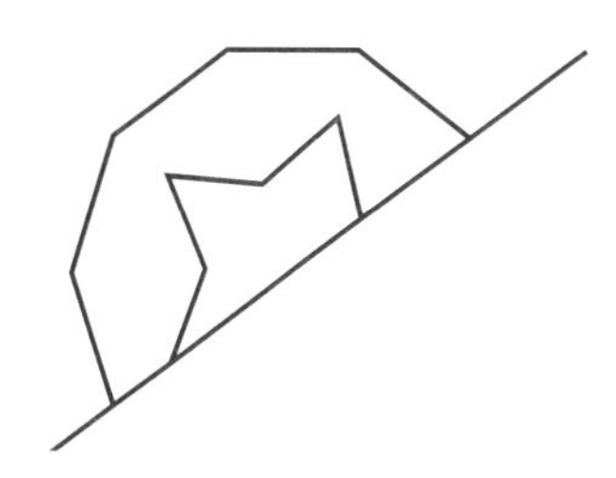

7. Which of the completed shapes from Exercises 4–6 also have rotational symmetry? ____________

Match each term with its definition.

_____ 8. polygon

_____ 9. reflectional symmetry

_____ 10. rotational symmetry

_____ 11. regular polygon

_____ 12. central angle of a regular polygon

_____ 13. axis of symmetry

_____ 14. center of a regular polygon

a. a polygon that is both equiangular and equilateral

b. the point that is equidistant from all vertices of the polygon

c. an angle whose vertex is the center of the circle and whose sides pass through two consecutive vertices

d. the line over which an image is reflected

e. a plane figure formed from three or more segments such that each segment intersects exactly two other segments

f. the reflected image across a line coincides exactly with the preimage

g. an image that is the same as the preimage after a rotation of any degree measure other than 0° or multiple of 360°

NAME ______________________ CLASS ____________ DATE ____________

Practice Masters Level B

3.1 Symmetry in Polygons

Examine each figure below. Determine whether it has reflectional symmetry, rotational symmetry, or both. If it has reflectional symmetry, draw all of the axes of symmetry. If it has rotational symmetry, mark the center of rotation and find the measure of the central angle.

1. ______________________

____

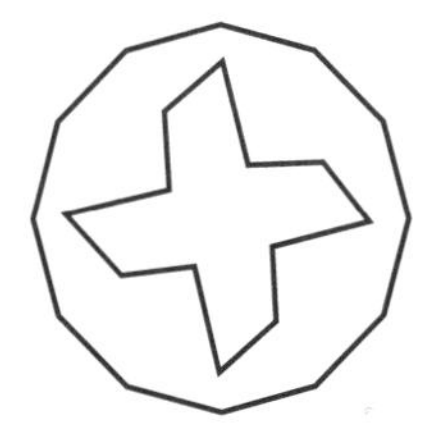

2. ______________________

____

3. ______________________

____

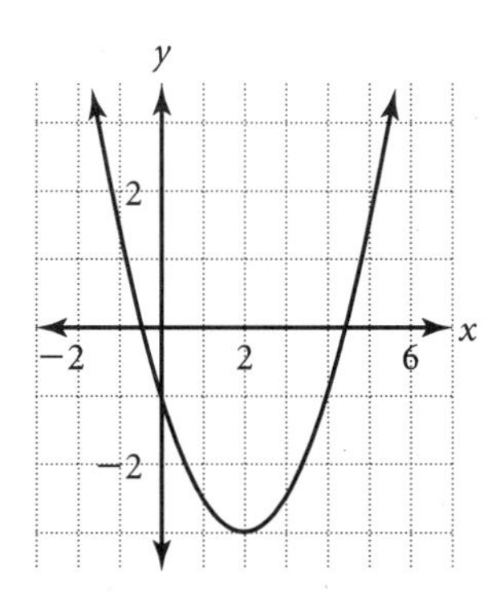

4. The following figure shows part of a shape with reflectional symmetry. Complete the figure.

5. The following figure is part of a shape with 6-fold rotational symmetry. Complete the figure.

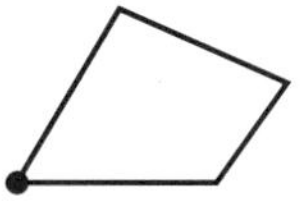

Classify each statement as true or false. Explain your reasoning in each false case.

6. A parallelogram has both reflectional and rotational symmetry. ______________
7. The axis of symmetry of a segment is its perpendicular bisector. ______________
8. All equilateral polygons are regular. ______________
9. A regular n-gon has $\frac{360}{n}$-fold rotation symmetry. ______________
10. An equilateral triangle has three axes of symmetry. ______________
11. A polygon can be formed from two segments. ______________
12. The center of a polygon is equidistant from each vertex. ______________

Practice Masters Level C

3.1 Symmetry in Polygons

Examine each figure below. Determine whether it has reflectional symmetry, rotational symmetry, or both. If it has reflectional symmetry, draw all of the axes of symmetry. If it has rotational symmetry, mark the center of rotation.

1. ______________________
____

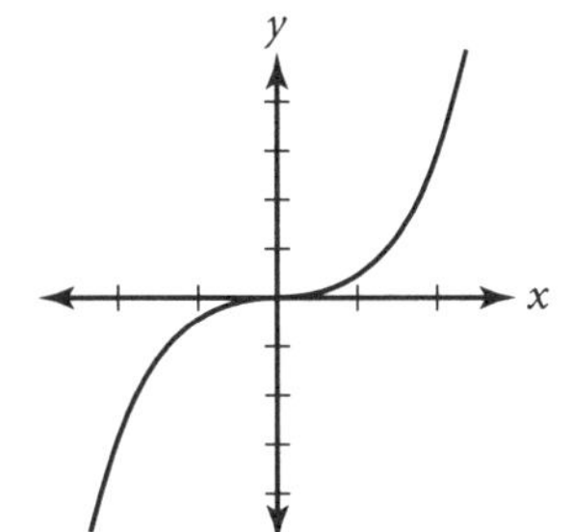

2. ______________________
____

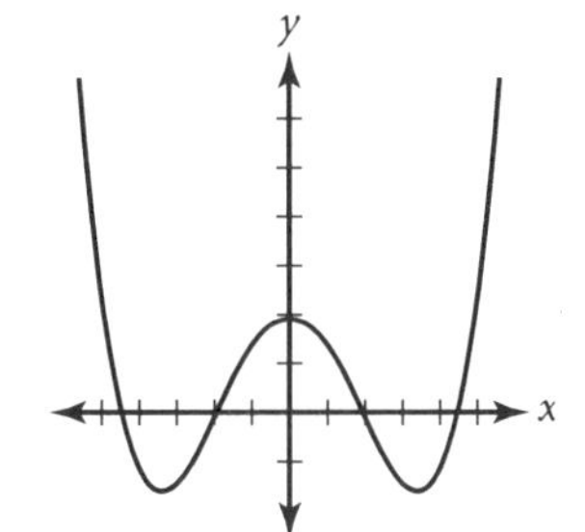

3. ______________________
____

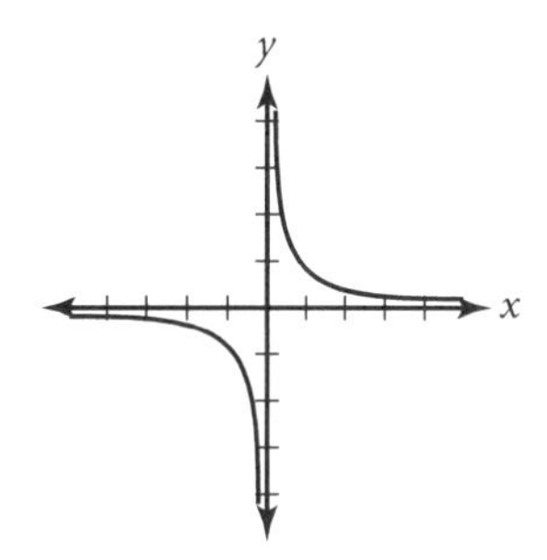

In each figure below, reflect over the given axis of symmetry, or rotate as directed to complete the figure.

4. The figure has 2-fold rotation symmetry about the origin.

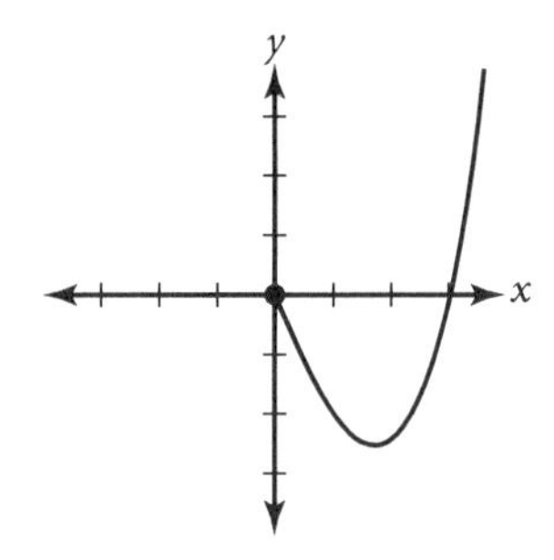

5. The axis of symmetry is the y-axis.

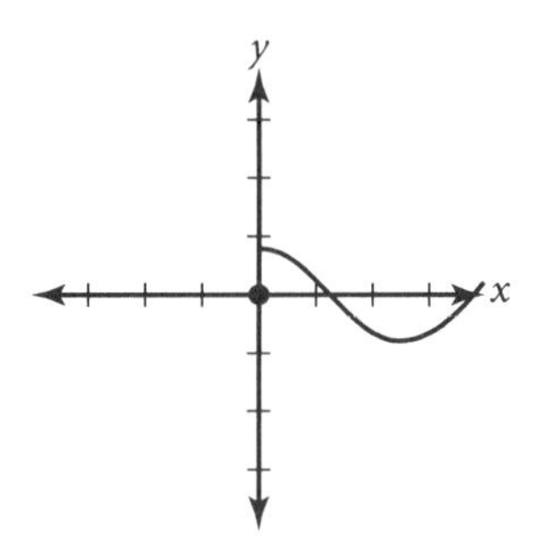

6. The figure has reflectional symmetry.

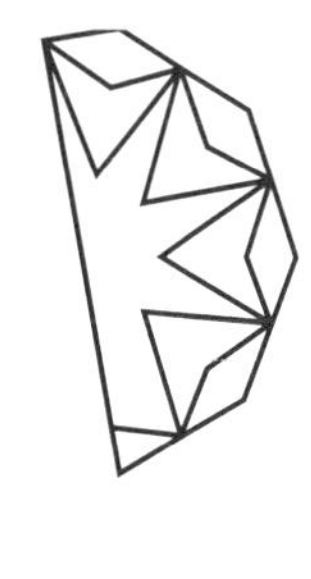

For Exercises 7–13, use the regular polygons pictured at the right to complete the table and answer the question.

	Number of sides	Number of axes of symmetry	Measure of central angle
7.	3		
8.	4		
9.	5		
10.	7		
11.	8		
12.	9		

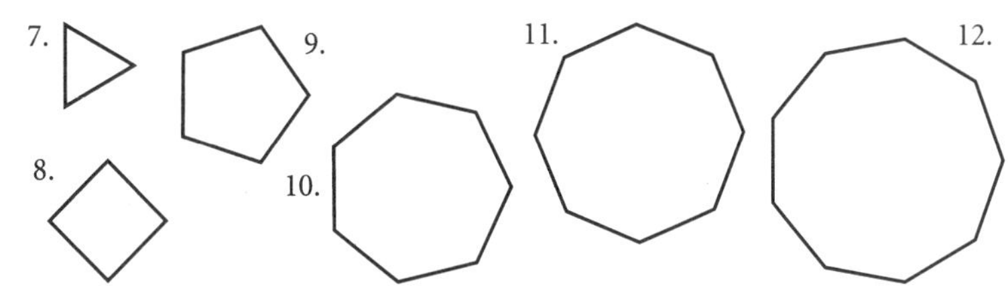

13. Write a conjecture about the location of the axes of symmetry. ______________________
____
____

NAME ______________________ CLASS ______________ DATE ______________

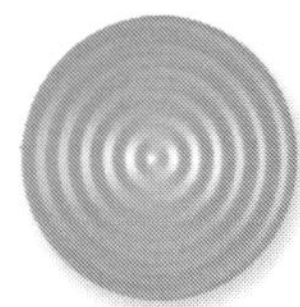

Practice Masters Level A

3.2 Properties of Quadrilaterals

Match each term with its definition.

______ 1. quadrilateral

______ 2. parallelogram

______ 3. rhombus

______ 4. rectangle

______ 5. square

______ 6. trapezoid

a. a quadrilateral with four congruent sides and four right angles

b. a quadrilateral with four right angles

c. a quadrilateral with two pairs of parallel sides

d. a quadrilateral with four congruent sides

e. a quadrilateral with only one pair of parallel sides

f. any four sided polygon

In parallelogram *ABCD*, $BC = 12$, $BE = 11.7$, $m\angle ACB = 71°$, $m\angle DAB = 120°$. Find the indicated measures.

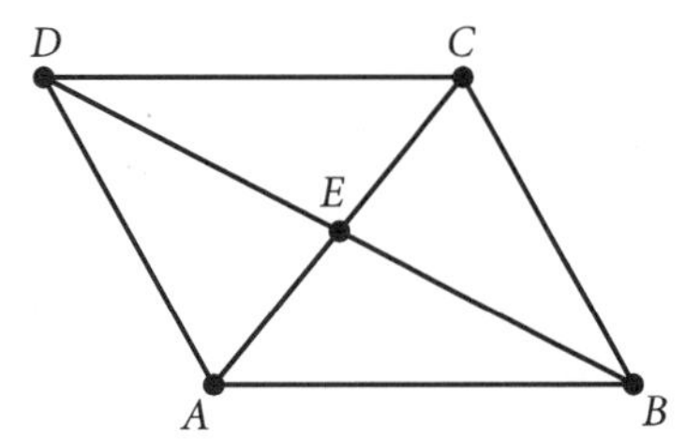

7. $m\angle DCA$ ______________ 8. BD ______________

9. $m\angle DCB$ ______________ 10. AD ______________

11. $m\angle DAC$ ______________ 12. ED ______________

In rectangle *RECT*, $RE = 72$, $AC = 80.5$, $m\angle CAT = 53.13°$, $m\angle AEC = 26.57°$. Find the indicated measures.

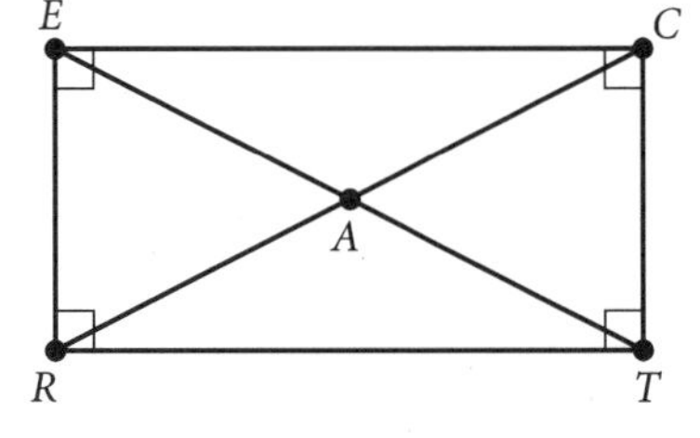

13. $m\angle EAR$ ______________ 14. $m\angle REA$ ______________

15. $m\angle RAT$ ______________ 16. AR ______________

17. ET ______________ 18. CT ______________

In rhombus *RHMB*, $RH = 9$, $m\angle BRM = 35.3°$, $m\angle BRH = 70.6°$. Find the indicated measures.

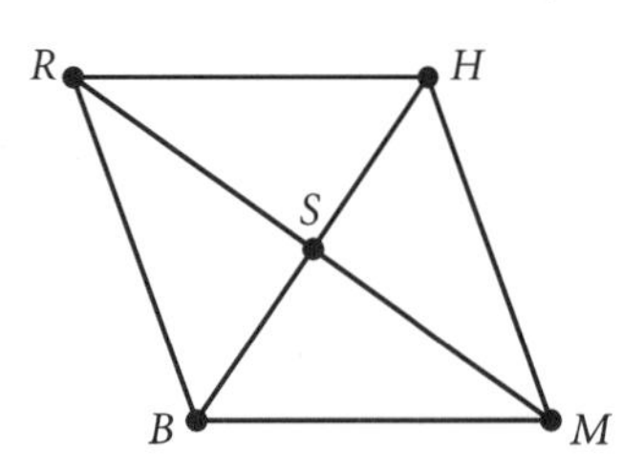

19. $m\angle HMB$ ______________ 20. HM ______________

21. $m\angle MSB$ ______________ 22. $m\angle HRM$ ______________

NAME ______________________ CLASS ____________ DATE ____________

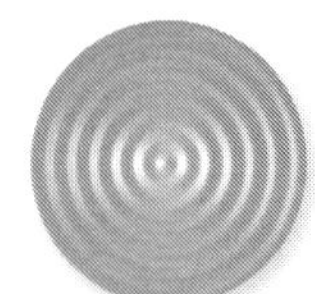

Practice Masters Level B

3.2 Properties of Quadrilaterals

Fill in the blank so that the sentence is true.

1. A ______________ is a quadrilateral with only one pair of parallel sides.
2. A square is a quadrilateral with ______ congruent sides and ______ right angles.
3. A rhombus is a ______________ with four ______________ sides.
4. A ______________ is a quadrilateral with two pairs of parallel sides.
5. Any four-sided polygon is a ______________________.
6. A rectangle is a quadrilateral with ______________________.

In parallelogram *ABCD*, m∠*DAB* = 11*x* + 1, m∠*ABC* = 2(7*x* + 2), m∠*CDB* = 6*x* + 1, m∠*DCA* = 5*x* − 1. Find the following measures.

7. x ______________
8. m∠DAB ______________
9. m∠DCB ______________
10. m∠ADC ______________
11. m∠ACB ______________
12. m∠ADB ______________

In rectangle *RECT*, *RA* = 7*x*, *RC* = 16.5*x* − 6. Find the following measures.

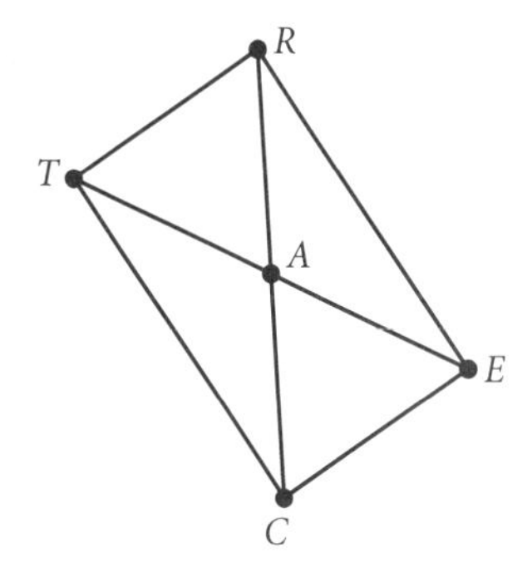

13. x ______________
14. AC ______________
15. RC ______________
16. AT ______________
17. AE ______________
18. TE ______________

Use the definitions of quadrilaterals and your conjectures from Activities 1–4 in your text to decide whether each statement is true or false. If the statement is false, give a counterexample.

19. If a quadrilateral is equilateral, then it is a square. ______________
20. If a quadrilateral is a rectangle, then it is equiangular. ______________
21. Every square is a rhombus. ______________
22. If a quadrilateral has perpendicular diagonals, then it is a rhombus. ______________

Practice Masters Level C

3.2 Properties of Quadrilaterals

For Exercises 1–3, use the markings and your conjectures about quadrilaterals from the textbook to identify the following figures.

1. ______________ 2. ______________ 3. ______________

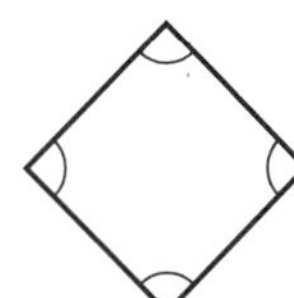

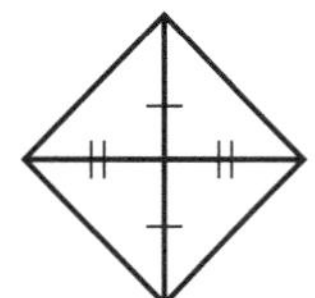

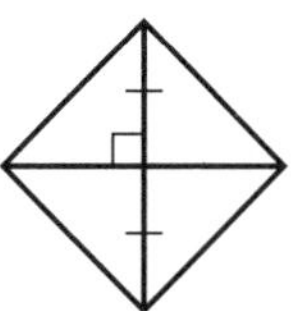

Use the figure at right for Exercises 4–5. Quadrilateral *ACBD* was formed by reflecting scalene triangle *ABC* across $\overline{AB}$.

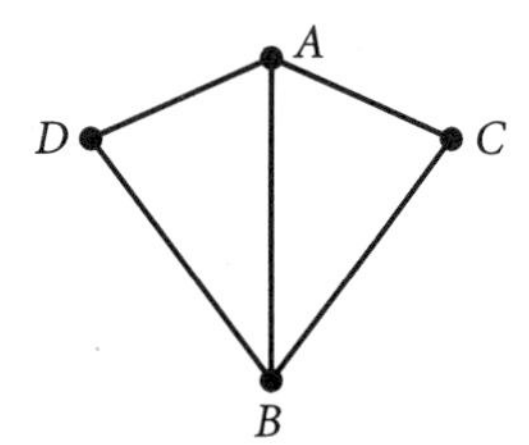

4. What type of special quadrilateral is *ACBD*? Explain.

____

5. How is $\overline{AB}$ related to $\overline{DC}$ (not shown)?

____

In the figure at right, $\triangle ABC$ is isosceles, with $AB = BC$.

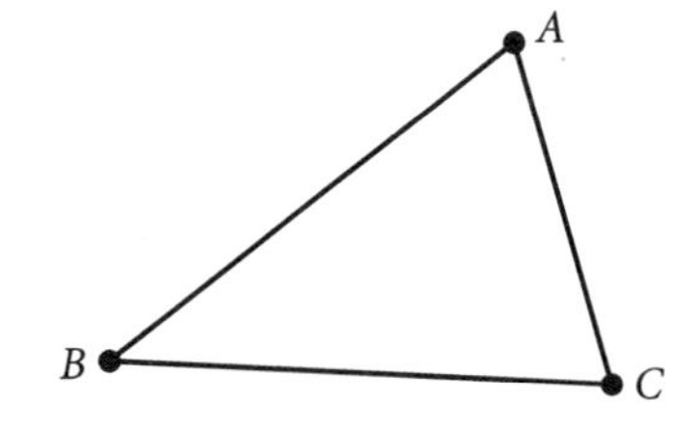

6. What type of quadrilateral results if $\triangle ABC$ is reflected across $\overline{AC}$? Explain.

____

7. What type of quadrilateral results if $\triangle ABC$ is reflected across $\overline{AB}$? Explain.

____

Each figure below shows part of a shape with the given rotational symmetry. Complete each shape and identify the resulting quadrilateral. Explain your answer.

8. 4-fold ______________________

____

9. 2-fold ______________________

____

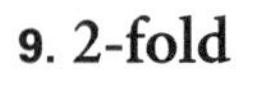

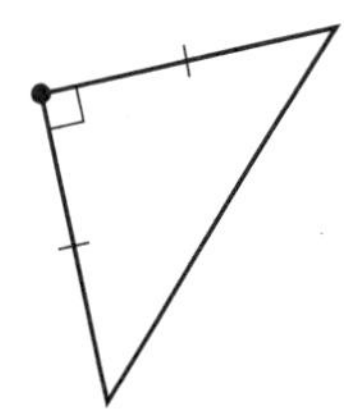

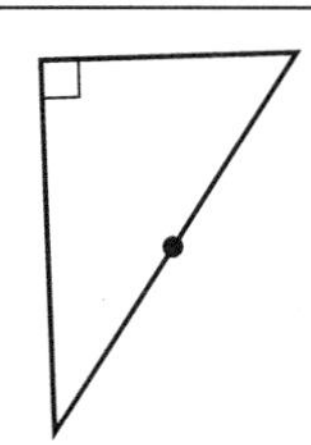

Practice Masters Level A

3.3 Parallel Lines and Transversals

Match each term with its definition.

_____ 1. transversal

_____ 2. alternate interior angles

_____ 3. alternate exterior angles

_____ 4. same-side interior angles

_____ 5. corresponding angles

a. two nonadjacent interior angles that lie on opposite sides of a transversal

b. two nonadjacent exterior angles that lie on opposite sides of a transversal

c. two nonadjacent angles, one interior and one exterior, that lie on the same side of a transversal

d. interior angles that lie on the same side of a transversal

e. a line, ray, or segment that intersects two or more coplanar lines, rays, or segments, each at a different point

In the figure at the right, $r \parallel s$, $m\angle 2 = 40°$, and $m\angle 4 = 60°$. Find the indicated measures.

6. $m\angle 1$ __________
7. $m\angle 3$ __________
8. $m\angle 5$ __________
9. $m\angle 6$ __________
10. $m\angle 7$ __________
11. $m\angle 8$ __________
12. $m\angle 9$ __________
13. $m\angle 10$ __________
14. $m\angle 11$ __________
15. $m\angle 12$ __________
16. $m\angle 13$ __________
17. $m\angle 14$ __________

Complete the proof.

Given: $l \parallel m$
Line p is a transversal.

Prove: $\angle 1 \cong \angle 2$

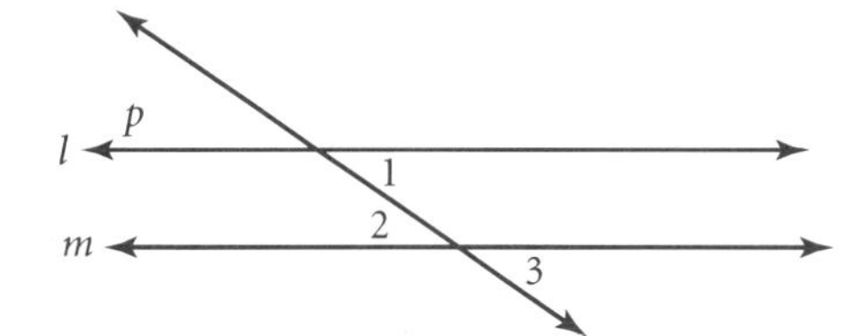

Statements	Reasons
Line p is parallel to line m. Line p is a transversal.	18.
$\angle 1 \cong \angle 3$	19.
$\angle 3 \cong \angle 2$	20.
$\angle 1 \cong \angle 2$	21.

NAME ______________________ CLASS __________ DATE __________

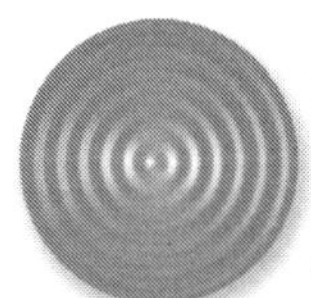

Practice Masters Level B

3.3 Parallel Lines and Transversals

Draw a figure for each vocabulary word. Label all lines and angles.

1. alternate interior angles
2. transversal
3. same-side interior angles
4. alternate exterior angles
5. corresponding angles

In the figure at the right, $\angle B \cong \angle C$, $m\angle BAC = 40°$, $m\angle B = 70°$, $m\angle BAD = 18°$, and $\overline{FD} \parallel \overline{CA}$. Find the indicated measures.

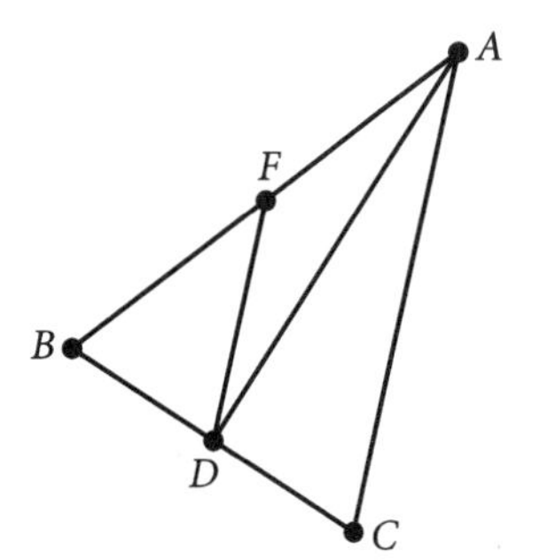

6. $m\angle DAC$ ______ 7. $m\angle C$ ______

8. $m\angle FDA$ ______ 9. $m\angle DFB$ ______

10. $m\angle BDF$ ______ 11. $m\angle ADC$ ______

Use the figure at the right, in which $r \parallel s$, $m \parallel n$, for Exercises 12–21. In Exercises 12–17, give the theorem or postulate that justifies each statement.

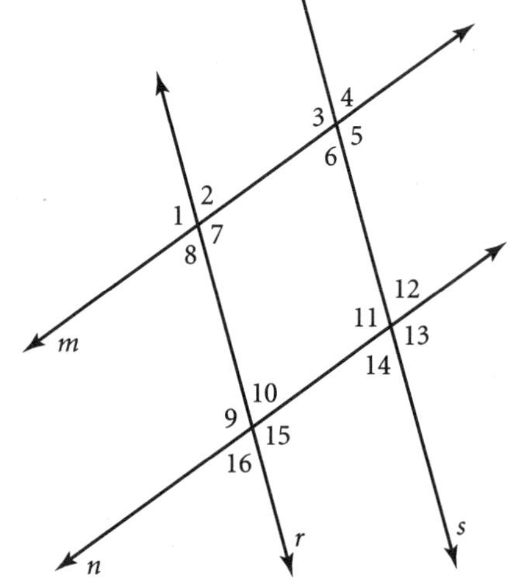

12. $\angle 8 \cong \angle 10$ ______

13. $\angle 14 \cong \angle 12$ ______

14. $m\angle 10 + m\angle 15 = 180°$ ______

15. $\angle 1 \cong \angle 9$ ______

16. $m\angle 2 + m\angle 3 = 180°$ ______

17. $\angle 3 \cong \angle 13$ ______

In Exercises 18–21, complete the two-column proof:
Given: $r \parallel s$, $m \parallel n$ **Prove:** $\angle 4 \cong \angle 16$

Statements	Reasons
$r \parallel s$, $m \parallel n$	18.
$\angle 4 \cong \angle 14$	19.
$\angle 14 \cong \angle 16$	20.
$\angle 4 \cong \angle 16$	21.

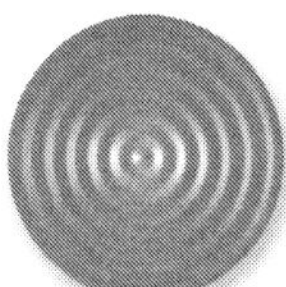

Practice Masters Level C

3.3 Parallel Lines and Transversals

Explain in your own words the definition of each vocabulary word.

1. alternate interior angles ______________________

2. transversal ______________________

3. same-side interior angles ______________________

4. alternate exterior angles ______________________

5. corresponding angles ______________________

In trapezoid *TRAP* at the right, $m\angle APR = 2x^2$, $m\angle PRT = 6x + 8$, $m\angle T = 3x^2 + 10$. Find the indicated measures.

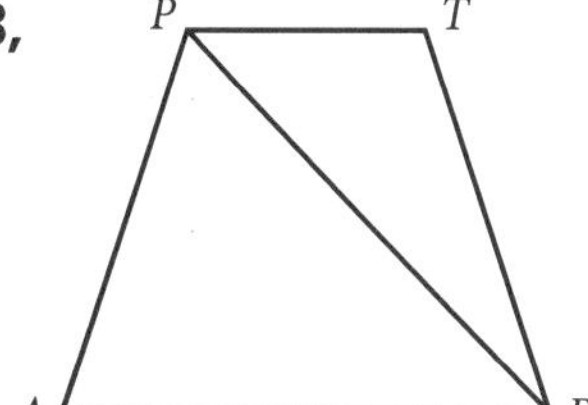

6. x __________

7. $m\angle PRT$ __________

8. $m\angle T$ __________

9. Write a two-column or paragraph proof to prove that same-side exterior angles are supplementary.

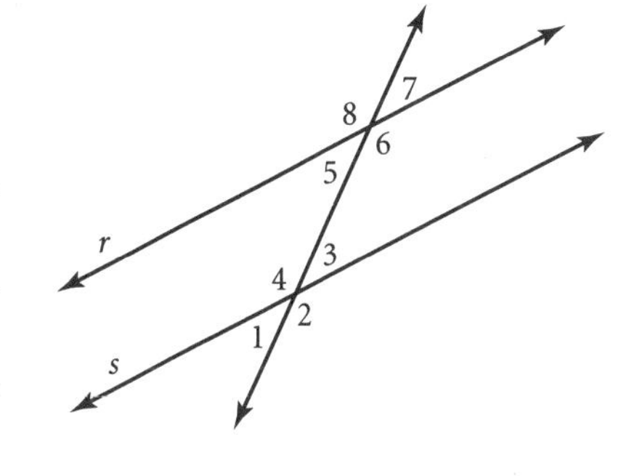

____

____

____

____

Write a two-column proof:
Given: Parallelogram $ABCD$; $\angle E \cong \angle A$
Prove: $\angle E \cong \angle DCB$

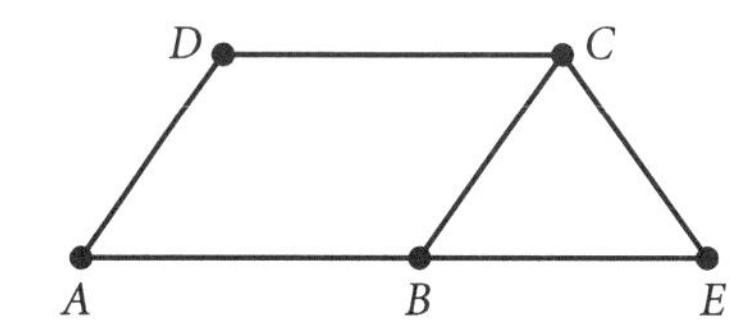

Statements	Reasons
10.	
11.	
12.	

NAME ______________________ CLASS __________ DATE __________

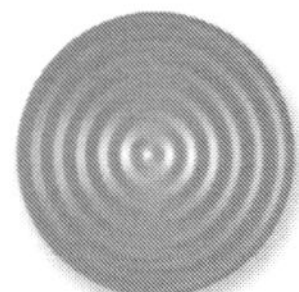

Practice Masters Level A

3.4 Proving That Lines are Parallel

Use the figure at right to complete the two-column proof:

Given: $\angle 4 \cong \angle 14$; $m\angle 11 + m\angle 8 = 180°$
Prove: $r \parallel s$

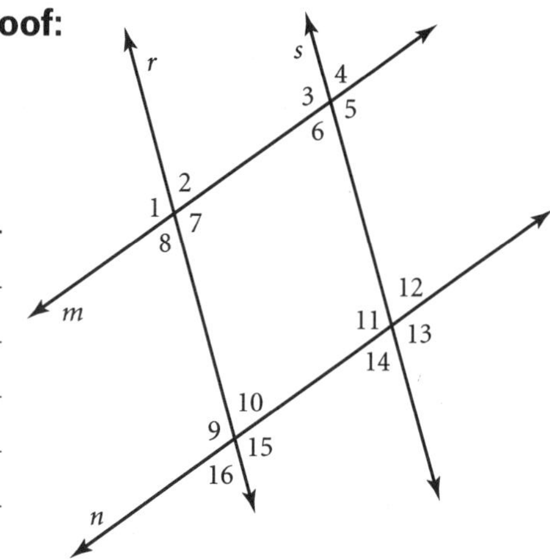

Statements	Reasons
$m\angle 4 \cong m\angle 14$	1.
$m \parallel n$	2.
$m\angle 11 + m\angle 8 = 180°$	3.
$m\angle 8 + m\angle 9 = 180°$	4.
$m\angle 9 \cong m\angle 11$	5.
$r \parallel s$	6.

For Exercises 7–10, refer to the diagram at right, and fill in the name of the appropriate theorem or postulate.

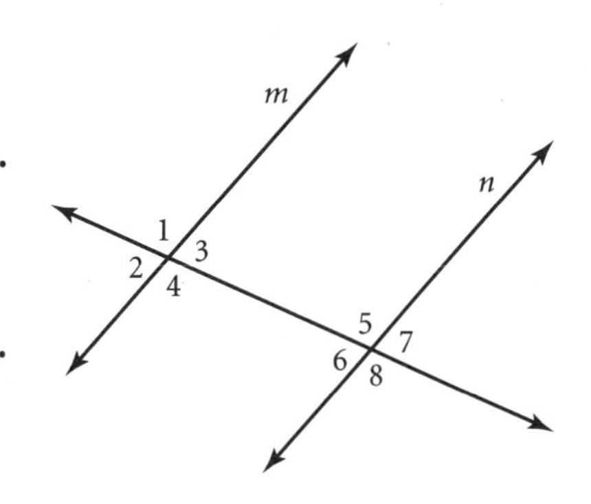

7. If $m\angle 3 = m\angle 6$, then $m \parallel n$ by the Converse of the

______________________________________.

8. If $m\angle 2 = m\angle 6$, then $m \parallel n$ by the Converse of the

______________________________________.

9. If $m\angle 2 = m\angle 7$, then $m \parallel n$ by the Converse of the

______________________________________.

10. If $\angle 3$ and $\angle 5$ are supplementary, then $m \parallel n$ by the Converse of the ______________________.

For Exercises 11–12, use the figure at right.

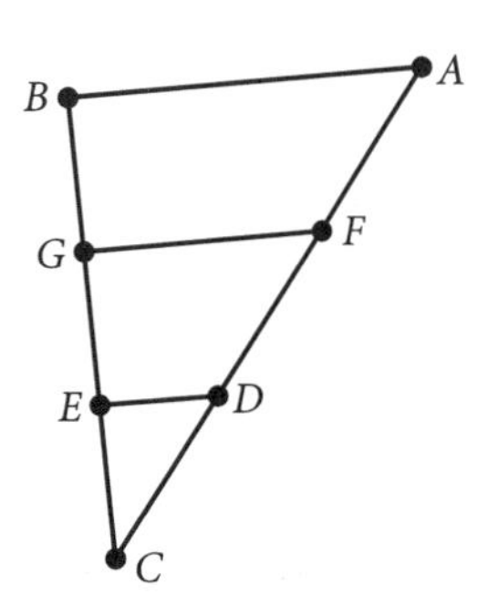

11. If $\overline{BA} \perp \overline{BC}$ and $\overline{ED} \perp \overline{EC}$, what is the relationship between $\overline{BA}$ and $\overline{ED}$? Explain.

____

12. If $\overline{DE} \parallel \overline{BA}$ and $\overline{GF} \parallel \overline{DE}$, what is the relationship between $\overline{BA}$ and $\overline{GF}$? Explain.

____

Practice Masters Level B

3.4 Proving That Lines are Parallel

Use the figure at right to complete the two-column proof:

Given: $\angle 4 \cong \angle 16$; $m\angle 4 + m\angle 1 = 180°$
Prove: $m \parallel n$

Statements	Reasons
$m\angle 4 + m\angle 3 = 180°$	**1.**
$m\angle 4 + m\angle 1 = 180°$	**2.**
$m\angle 1 = m\angle 3$	**3.**
$r \parallel s$	**4.**
$m\angle 2 = m\angle 4$	**5.**
$m\angle 2 = m\angle 8$	**6.**
$m\angle 4 = m\angle 8$	**7.**
$m\angle 4 = m\angle 16$	**8.**
$m\angle 8 = m\angle 16$	**9.**
$m \parallel n$	**10.**

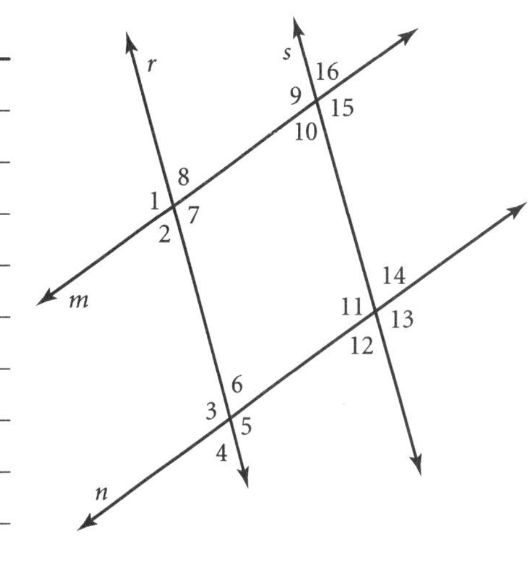

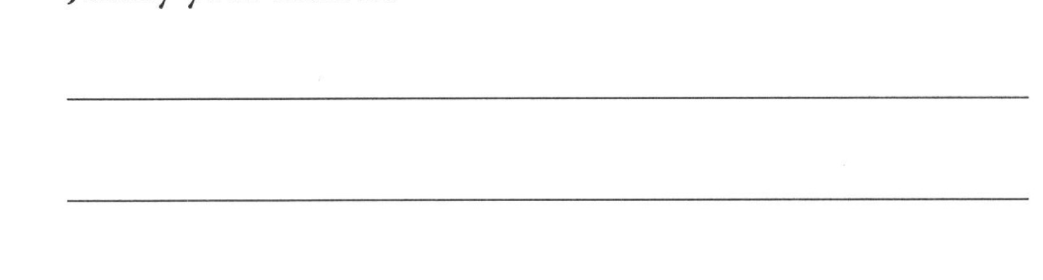

11. In the figure at right, $m\angle 1 = 3x + 14$, $m\angle 2 = 9x + 14$, and $m\angle 3 = 30x + 14$. Determine whether or not $r \parallel s$. Justify your answer.

____

____

____

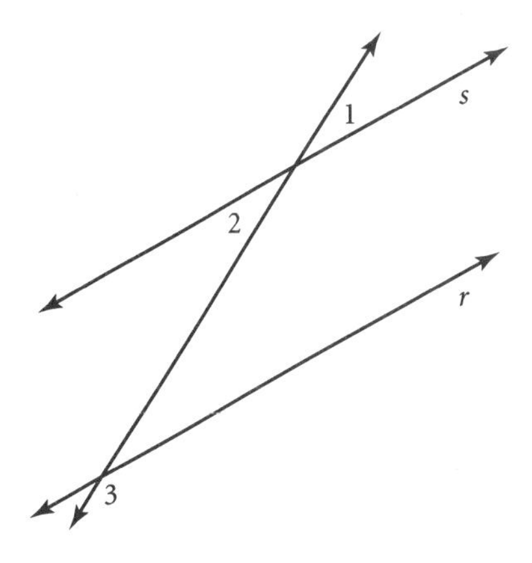

Use the figure at right for the statements in Exercises 12–15. What conclusion can you draw from each statement? Justify your answer.

12. $m\angle 1 = m\angle 4$ ______________________________

13. $m \perp t$ and $m \perp q$ ______________________________

14. $s \parallel q$ and $t \parallel q$ ______________________________

15. $m\angle 3 = m\angle 1$ ______________________________

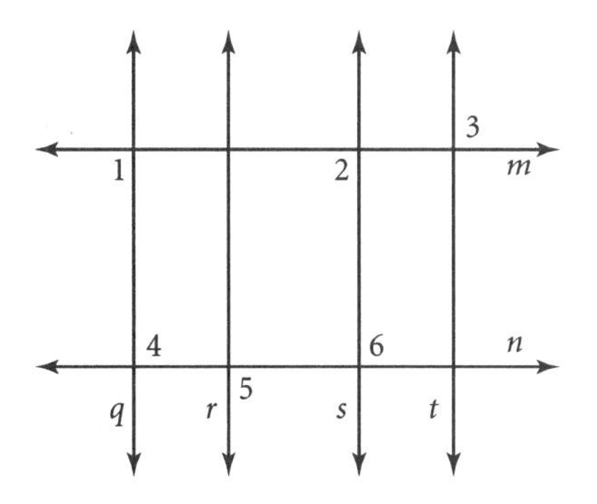

NAME ______________________ CLASS ______________ DATE ______________

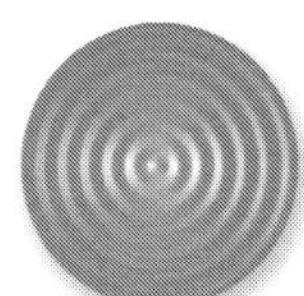

Practice Masters Level C

3.4 Proving That Lines are Parallel

For Exercises 1–3, use the figure at the right.

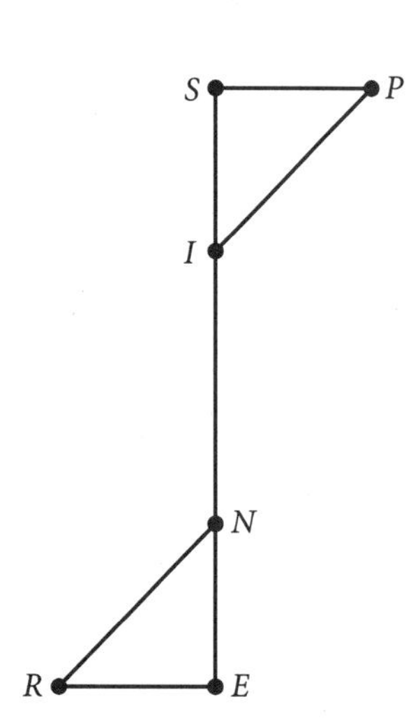

1. If $\angle PSI$ and $\angle REN$ are right angles, what can you conclude? Justify your answer.

____

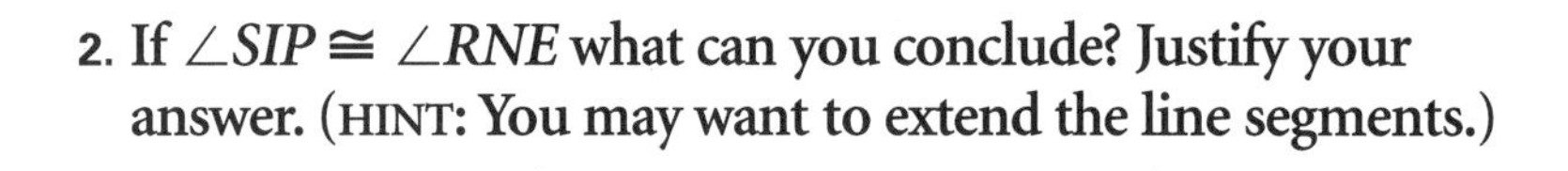

2. If $\angle SIP \cong \angle RNE$ what can you conclude? Justify your answer. (HINT: You may want to extend the line segments.)

____

3. Write a conjecture about the relationship between $\angle P$ and $\angle R$. Why do you think your conjecture is true?

____

In the figure at right, $m \parallel n$, $m\angle 1 = 16x + 10$, $m\angle 2 = 24x - 21$, and $m\angle 3 = 102 - 8x$.

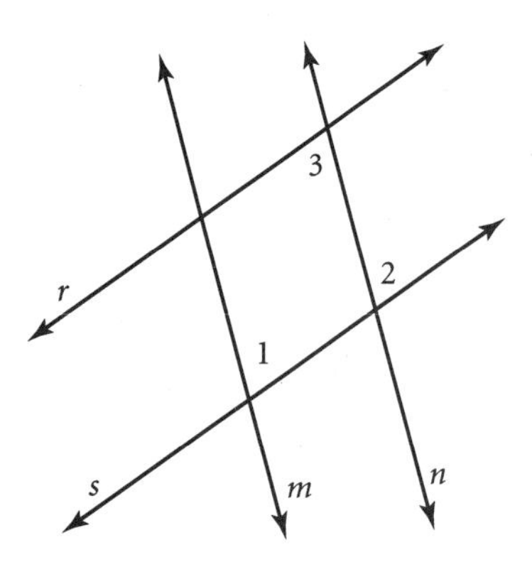

4. Determine whether or not $r \parallel s$. Justify your answer.

____

____

In each figure below, decide what lines or segments are parallel. Justify your conclusion. If not enough information is given, write *cannot be determined.*

5. ____

____

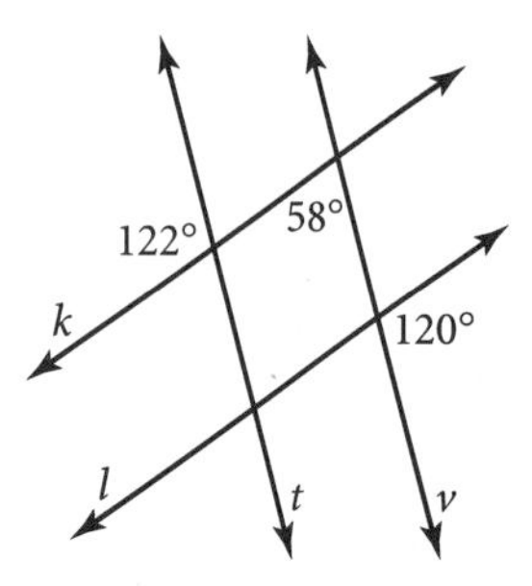

6. ____

____

$\triangle ABC$ is equilateral.

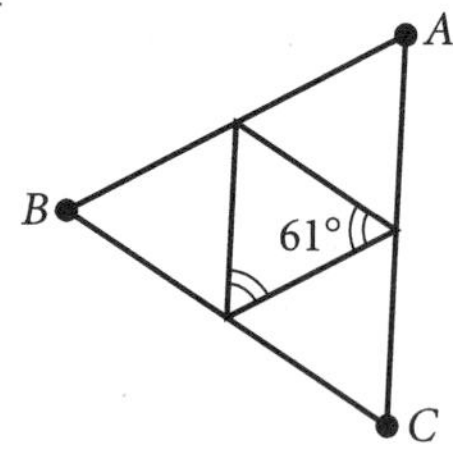

NAME ______________________ CLASS ____________ DATE ____________

Practice Masters Level A

3.5 The Triangle Sum Theorem

For Exercises 1–3, use the figure at the right.

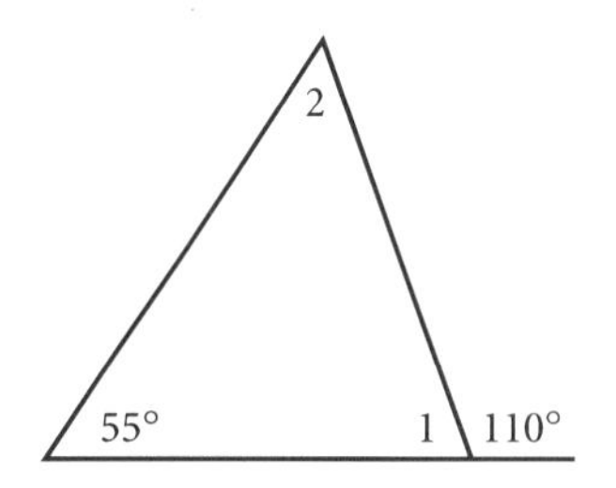

1. m∠1 = ______________________

2. m∠2 = ______________________

3. The angle which measures 110° is called an ____________.

Use the rectangle at the right for Exercises 4–6. m∠4=40°

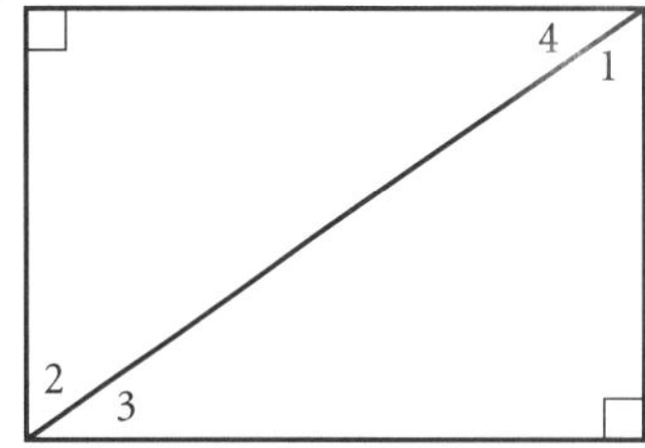

4. m∠1 ______________________

5. m∠2 ______________________

6. m∠3 ______________________

In △*PQR*, m∠*P* = (3*x* − 5)°, m∠*Q* = (7*x* − 2)°, and m∠*R* = (5*x* + 7)°. Find the indicated measures.

7. x ____________ 8. m∠P ____________

9. m∠Q ____________ 10. m∠R ____________

In the figure at the right, ∠C ≅ ∠BAC, and m∠*BAD* = 113°. Find the indicated measures.

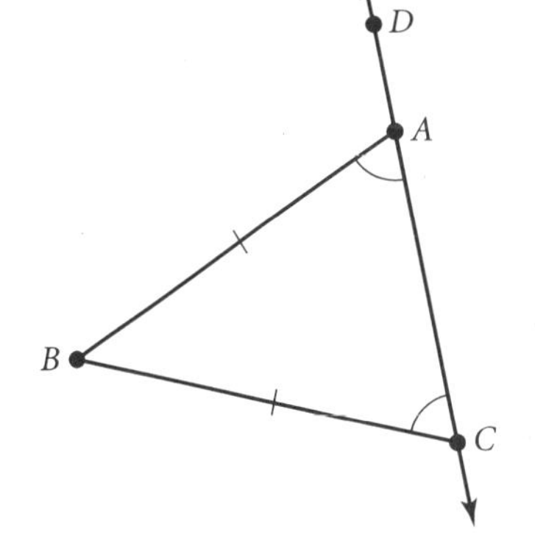

11. m∠BAC ______________________

12. m∠ACB ______________________

13. m∠ABC ______________________

14. How many lines can be drawn parallel to $\overline{AC}$ through B? Why?

____

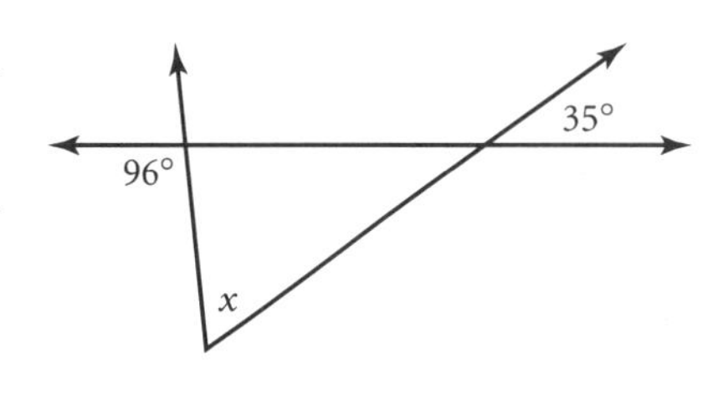

15. In the figure at the right, find x. ____________

NAME ______________________ CLASS ____________ DATE ____________

Practice Masters Level B

3.5 The Triangle Sum Theorem

In the figure at the right, $m\angle BAD = 25°$, $m\angle BCD = 35°$, $m\angle CDA = 135°$, and $m\angle BCA = m\angle BAC$. Find the indicated measures.

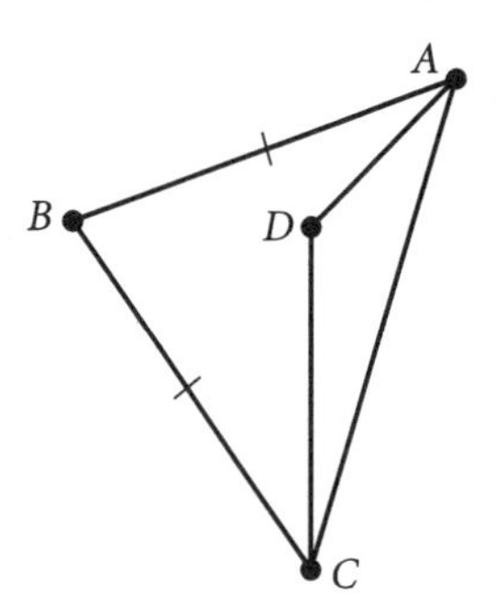

1. $m\angle ABC$ ____________________

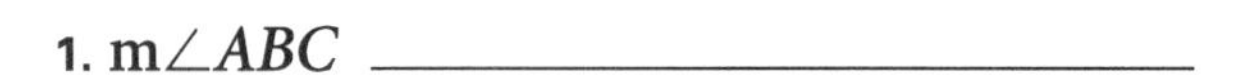

2. $m\angle DCA$ ____________________

3. $m\angle DAC$ ____________________

Figure *ABCDE* at the right is a regular pentagon. Find the indicated measures.

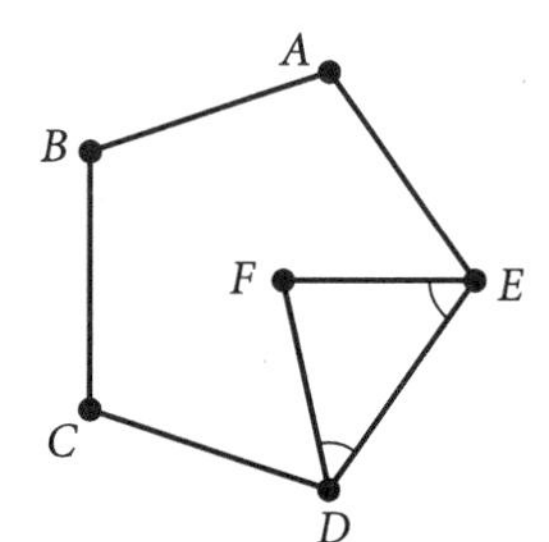

4. $m\angle EFD$ ____________________

5. $m\angle FED$ ____________________

6. $m\angle FDE$ ____________________

In the figure at right, find the following:

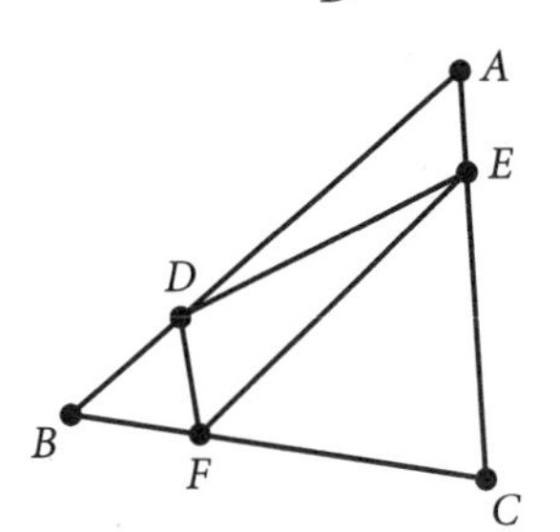

7. $m\angle ADE + m\angle AED + m\angle CEF + m\angle EFC + m\angle FDB + m\angle DFB$ [HINT: Add the angles of each triangle and simplify.]

____

Quadrilateral *ABCD* at the right is a rectangle. Find the indicated measures.

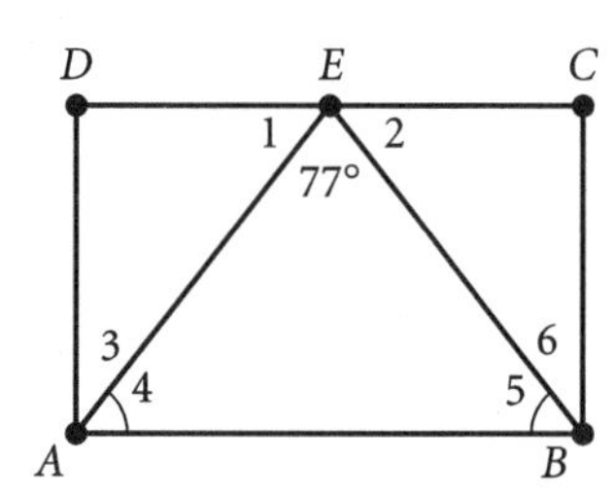

8. $m\angle 1$ ________ 9. $m\angle 2$ ________ 10. $m\angle 3$ ________

11. $m\angle 4$ ________ 12. $m\angle 5$ ________ 13. $m\angle 6$ ________

In $\triangle ABC$ at the right, $m\angle BAC = 4x + 6$, $m\angle ABC = 6x + 24$, and $m\angle BCA = 4x - 25$. Find the indicated measures.

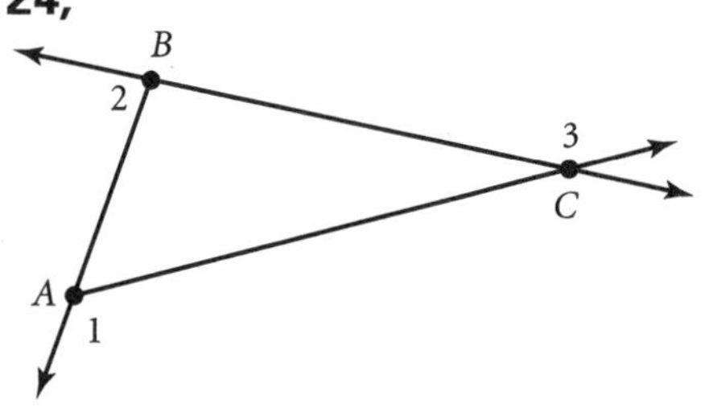

14. $m\angle 1$ ____________________

15. $m\angle 2$ ____________________

16. $m\angle 3$ ____________________

17. How many lines can be drawn parallel to $\overline{CB}$ through A? Why?

____

NAME ______________________ CLASS ____________ DATE ____________

Practice Masters Level C

3.5 *The Triangle Sum Theorem*

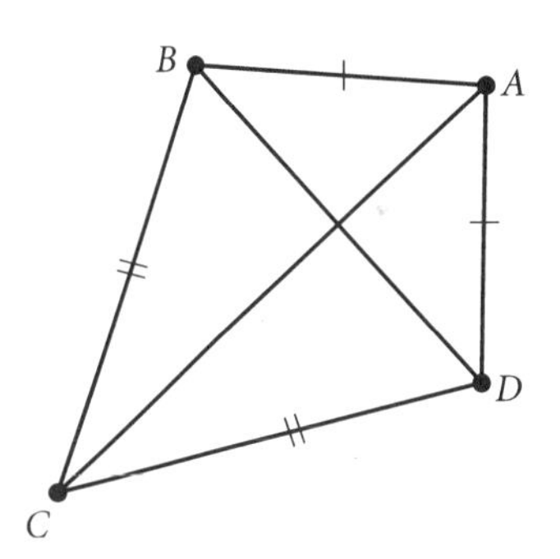

For Exercises 1–6, use the quadrilateral at the right, in which m∠*BAC* = 3*x* − 2, m∠*ABC* = 8*x* − 23, and m∠*BCA* = 61 − 2*x*. Find the indicated measures.

1. m∠ACD ____________ 2. m∠ABD ____________

3. m∠CDA ____________ 4. m∠BCD ____________

5. m∠BAD ____________ 6. m∠BDC ____________

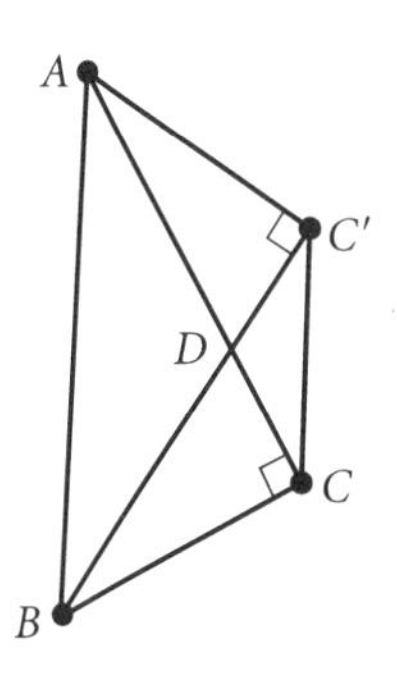

In trapezoid *AC′CB* at the right, m∠*DC′C* = m∠*DCC′*, and m∠*DAB* = m∠*DBA*. If m∠*DBA* = 31°, find the indicated measures.

7. m∠ABC ______________________

8. m∠ADB ______________________

9. m∠DCC' ______________________

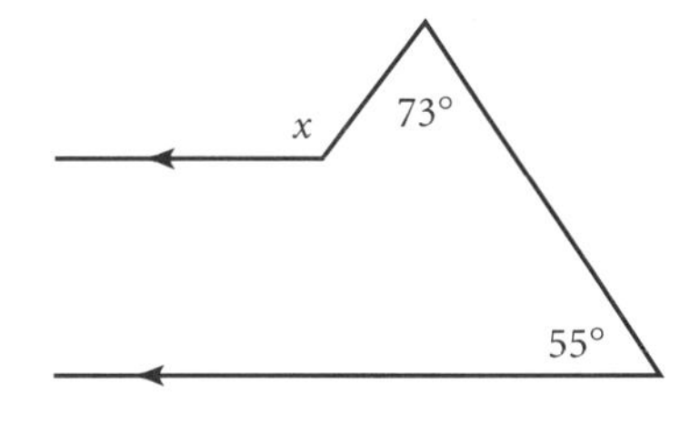

10. In the figure at the right, find x. ____________

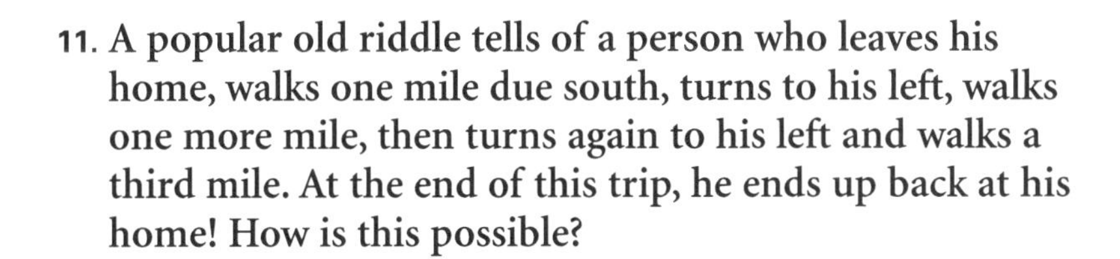

11. A popular old riddle tells of a person who leaves his home, walks one mile due south, turns to his left, walks one more mile, then turns again to his left and walks a third mile. At the end of this trip, he ends up back at his home! How is this possible?

____

____

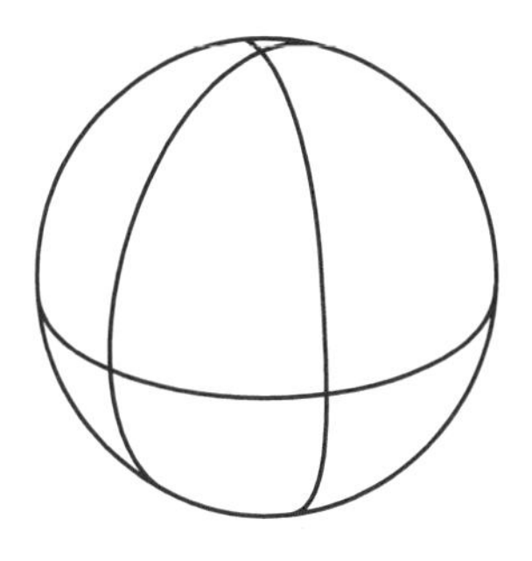

12. In the sphere at the right, three great circles are drawn so that they form a "triangle". Make a conjecture about the sum of the angles of this triangle. Explain your reasoning.

____

NAME ______________________ CLASS ______________ DATE ______________

Practice Masters Level A

3.6 Angles in Polygons

In Exercises 1–3, find the indicated angle measures, *x*.

1. ______________ 2. ______________ 3. ______________

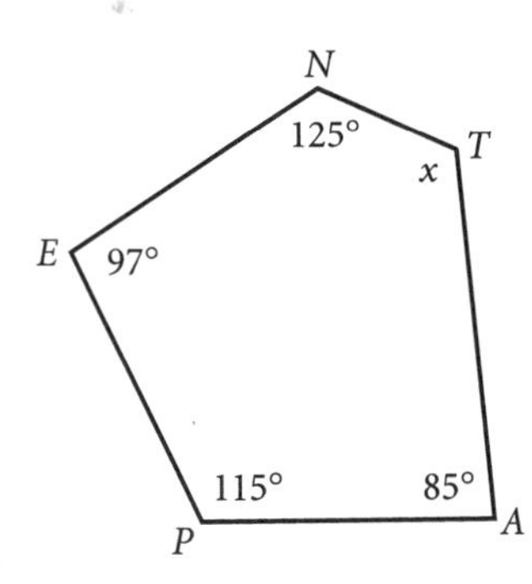

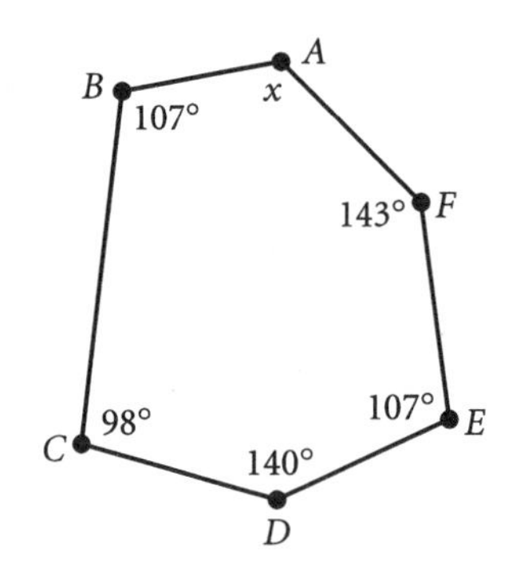

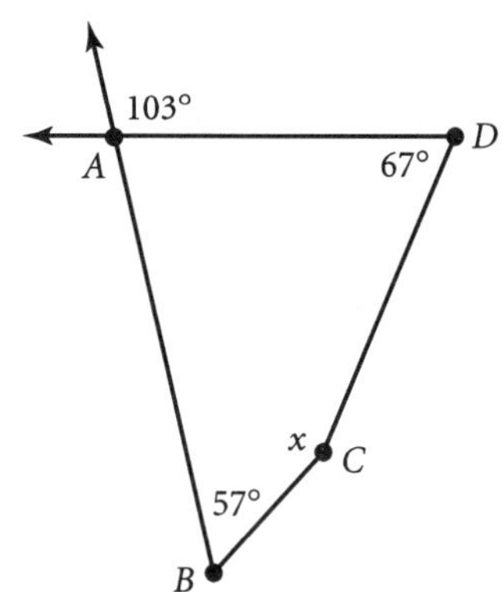

For each polygon, determine the measure of an interior angle and the measure of an exterior angle.

4. a regular octagon ______________ 5. a regular decagon ______________

For Exercises 6–7, an interior angle measure of a regular polygon is given. Find *n*, the number of sides of the polygon.

6. 140° ______________ 7. 156° ______________

For Exercises 8–9, an exterior angle measure of a regular polygon is given. Find *n*, the number of sides of the polygon.

8. 30° ______________ 9. 20° ______________

For Exercises 10–12, use the figure at the right to find the indicated measures.

10. $m\angle D$ ______________

11. $m\angle C$ ______________

12. $m\angle B$ ______________

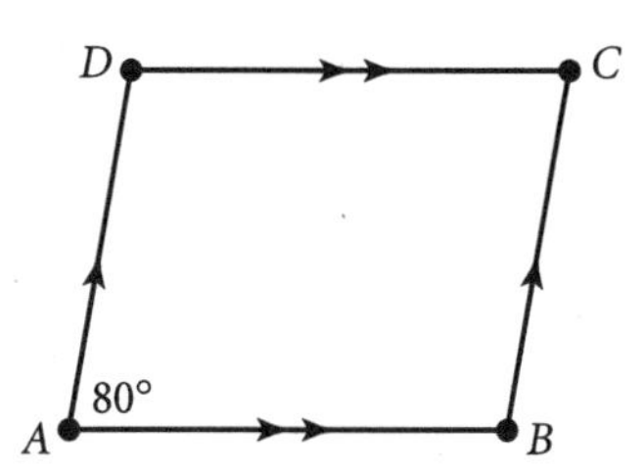

A regular polygon has an exterior angle measure of $(x + 3)°$ and an interior angle measure of $(13x - 33)°$.

13. Find the measure of each angle. ______________

14. How many sides does this polygon have? ______________

NAME ______________________ CLASS ____________ DATE ____________

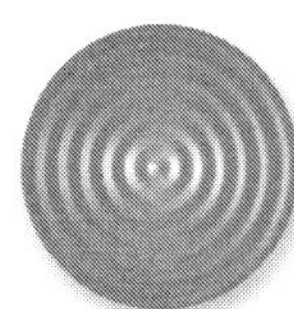

Practice Masters Level B

3.6 Angles in Polygons

For Exercises 1–3, find the indicated angle measure, *x*.

1. ________________ **2.** ________________ **3.** ________________

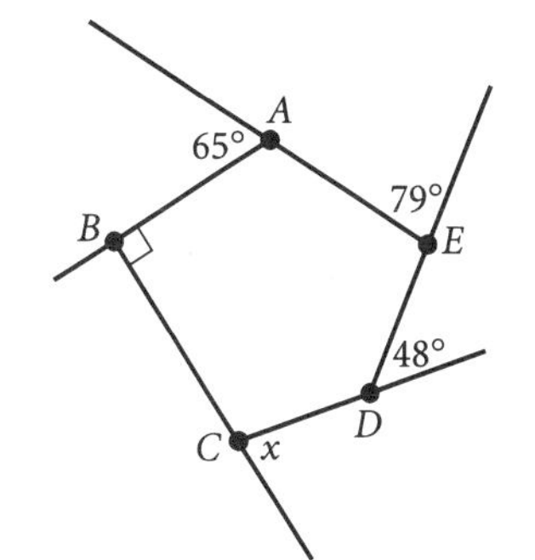

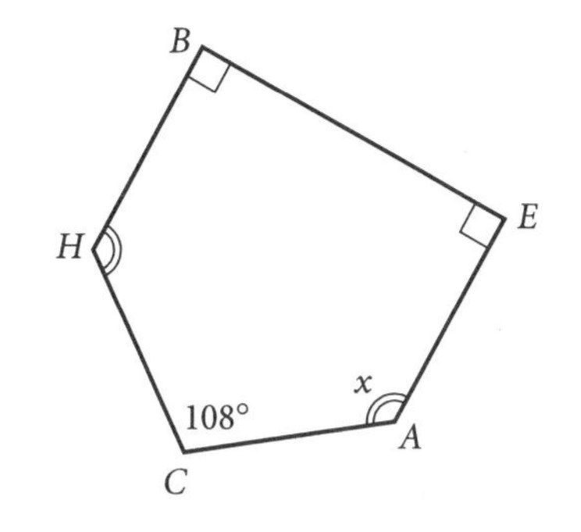

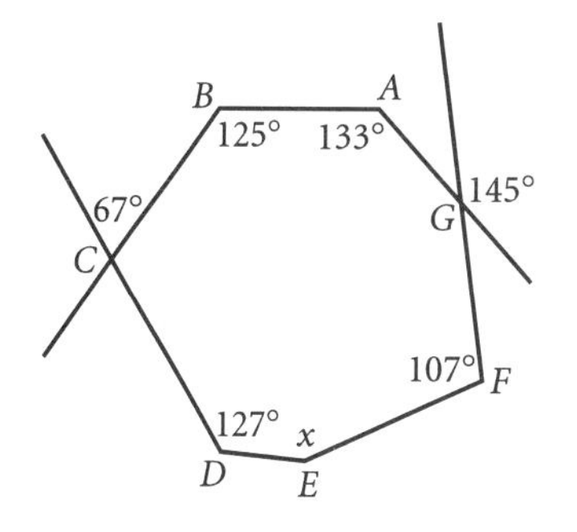

In the figure at the right, m∠*A* = 4*x* + 7, m∠*B* = 4*x* − 18, m∠*C* = 5(*x* − 1), m∠*D* = 2*x* + 1, and m∠*E* = 7*x* − 39. Find the indicated measures.

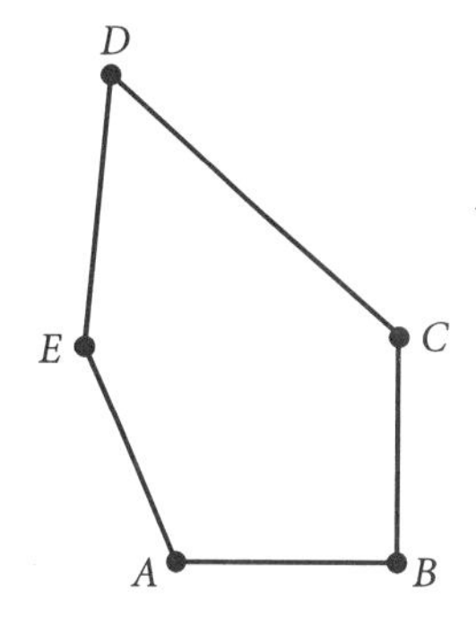

4. x ________________ **5.** m∠A ________________

6. m∠B ________________ **7.** m∠C ________________

8. m∠D ________________ **9.** m∠E ________________

In the figure at the right, m∠1 = 5*x* + 11, m∠4 = 3*x* + 1, m∠6 = 8*x* − 19, and m∠7 = 3*x* − 13. Find the indicated measures.

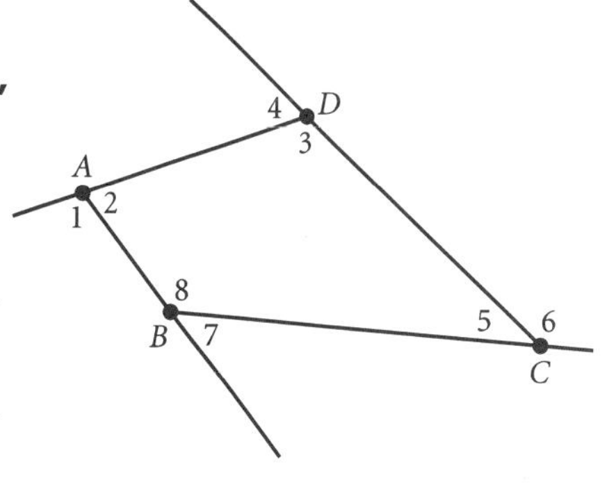

10. m∠2 ________________ **11.** m∠3 ________________

12. m∠8 ________________ **13.** m∠5 ________________

14. What are the interior and exterior angle measures of a regular nonagon? ________________

15. How many sides does a regular polygon with interior angle measure of 168° have? ________________

16. How many sides does a regular polygon with exterior angle measure of 20° have?

NAME ______________________ CLASS ____________ DATE ____________

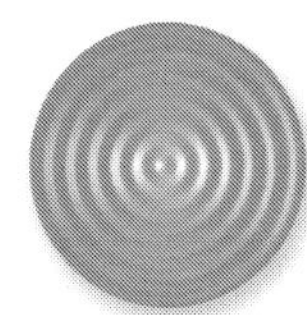

Practice Masters Level C

3.6 *Angles in Polygons*

For Exercises 1–4, determine the number of sides of the regular polygon described.

1. The measure of one interior angle is twice the measure of the exterior angle. ________
2. The measure of one interior angle is half the measure of the exterior angle. ________
3. The measure of one interior angle is three times the measure of the exterior angle. ________
4. The ratio of the exterior angle measure to the interior angle measure is 2:3. ________

In the figure at the right, $m\angle A = 7x + 6y$, $m\angle B = 38y$, $m\angle C = 13x + 3y$, $m\angle D = 19x - 9y$, and $m\angle E = 15x$. If $m\angle A = 73°$, $m\angle C = 103°$, find the indicated measures.

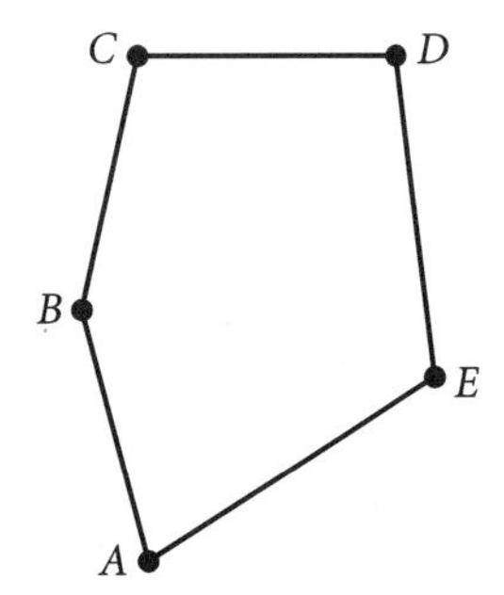

5. x ________
6. y ________
7. $m\angle B$ ________
8. $m\angle D$ ________
9. $m\angle E$ ________

The sphere at the right shows 2 lines of longitude and 4 lines of latitude. Use the sphere for Exercises 10 and 11.

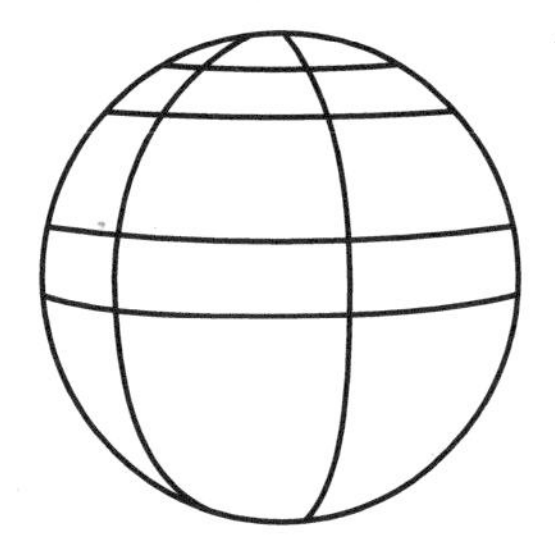

10. Describe the shape formed by the intersections of the latitude and longitude lines. Write a conjecture about the angle-sum of the shapes.

____

____

11. Do you think "corresponding" angles are congruent on the sphere?

____

In the figure at the right, $m\angle A = 45°$, $m\angle JFG = 100°$, $m\angle FJI = 112°$, $m\angle GHI = 91°$, and $m\angle C = 44°$. Find the indicated measures.

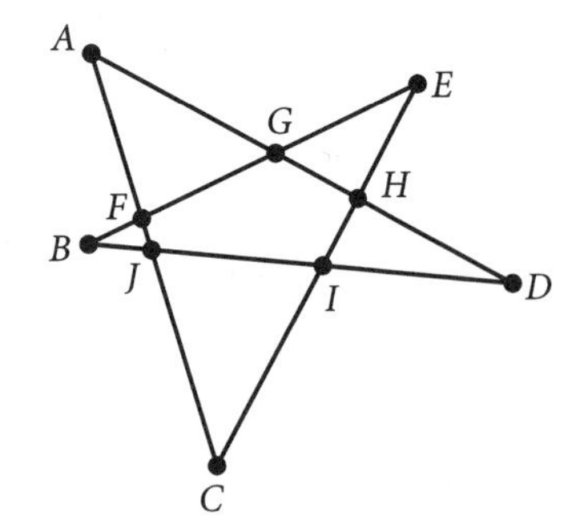

12. $m\angle B$ ________
13. $m\angle FGH$ ________
14. $m\angle DHI$ ________
15. $m\angle HIJ$ ________
16. $m\angle D$ ________
17. $m\angle E$ ________

NAME ______________________ CLASS ____________ DATE ____________

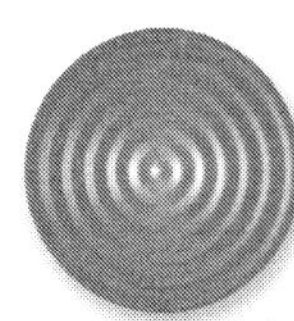

Practice Masters Level A

3.7 Midsegments of Triangles and Trapezoids

Find the indicated measures.

1. AC ____________

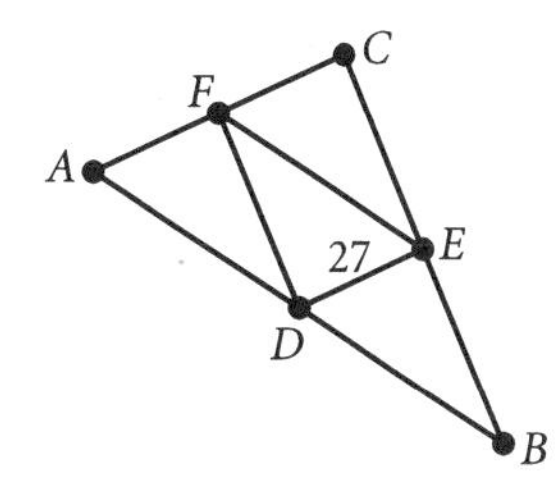

2. AB ____________

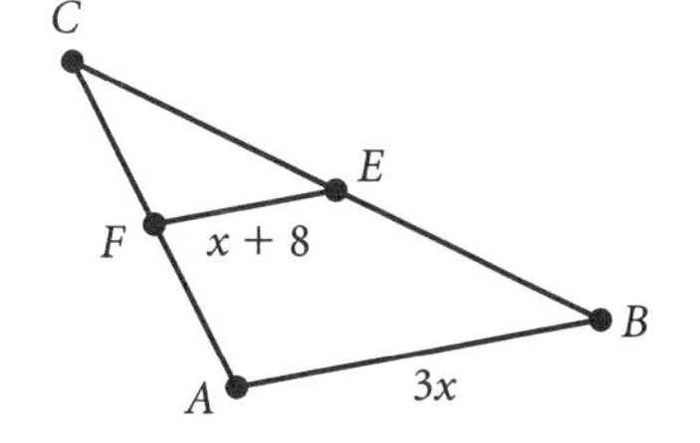

3. DC ____________

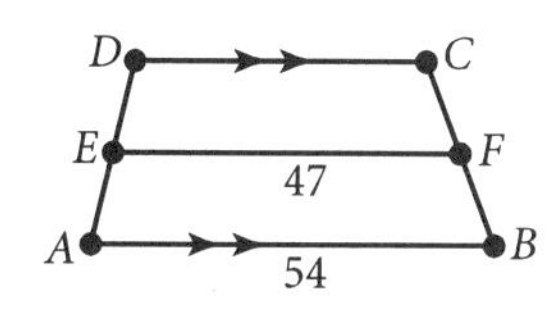

4. HI ____________

DE ____________

GF ____________

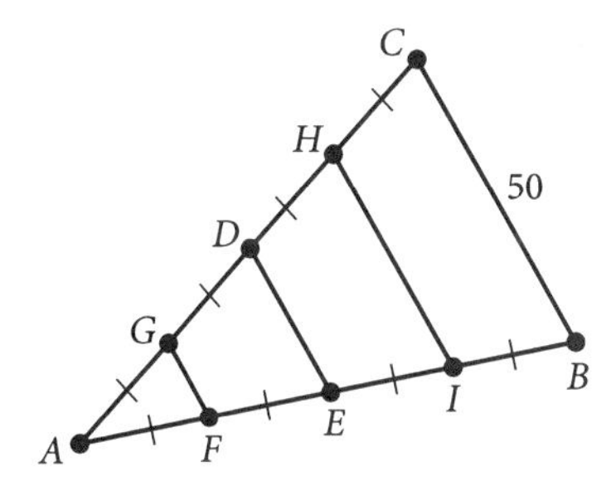

5. FG ____________

AB ____________

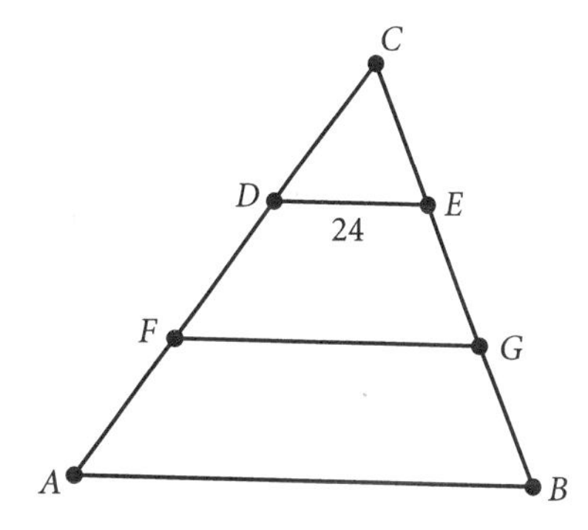

Use the figure at the right for Exercises 6–9. △*FED* was formed by joining the midpoints of △*ABC*.

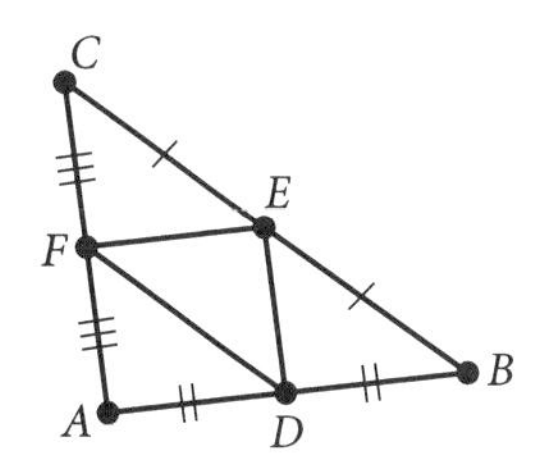

6. $\overline{FE}$, $\overline{ED}$, and $\overline{FD}$ are called ____________.

What type of quadrilateral is each of the following? Explain your answer.

7. $FEBD$ ____________

8. $EBAF$ ____________

9. Name all sets of congruent segments. ____________

NAME ______________________ CLASS ____________ DATE ____________

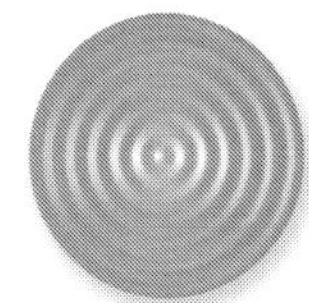

Practice Masters Level B

3.7 Midsegments of Triangles and Trapezoids

1. Neatly copy the figure at right on a piece of paper. Use paper-folding to find the midpoints of each side, then make folds to connect the midpoints. Cut out the new triangles you have formed. Make a conjecture about the small triangles. Explain your answer.

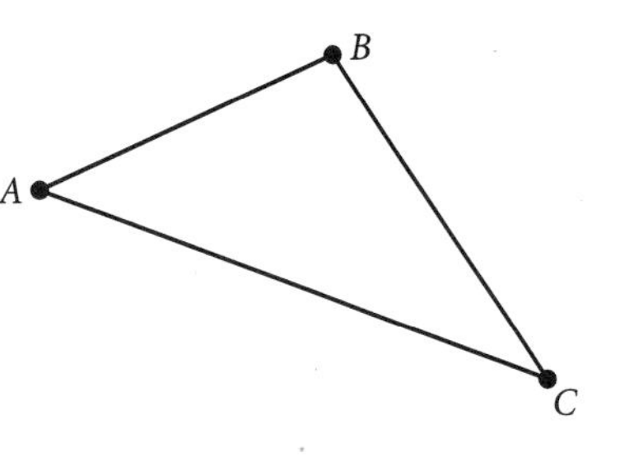

____

____

____

For Exercises 2–9, use the figure at the right, in which *D, E,* and *F* are midpoints. Find the indicated measures.

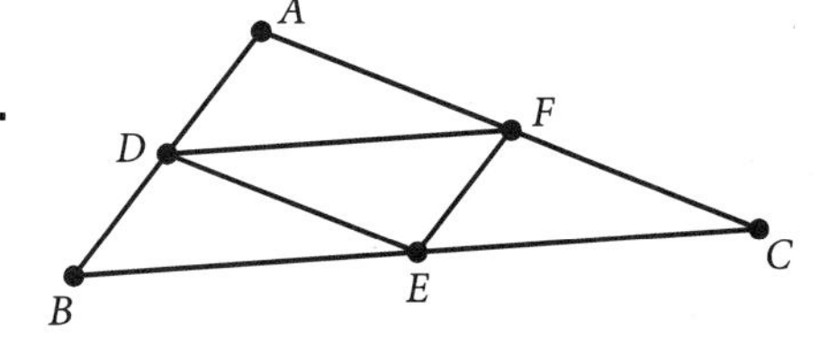

Given: $AB = 6x - 2$, $FE = 17 - 6x$, $CA = 5y - 7$, $DE = 2y - 1$, $CB = 6y - 3x$

2. x ____________ 3. y ____________ 4. AB ____________ 5. FE ____________

6. CA ____________ 7. DE ____________ 8. CB ____________ 9. DF ____________

Use the conjectures from your text and the figures below to find the indicated values.

10. DC ____________ 11. KJ ____________ 12. GF ____________

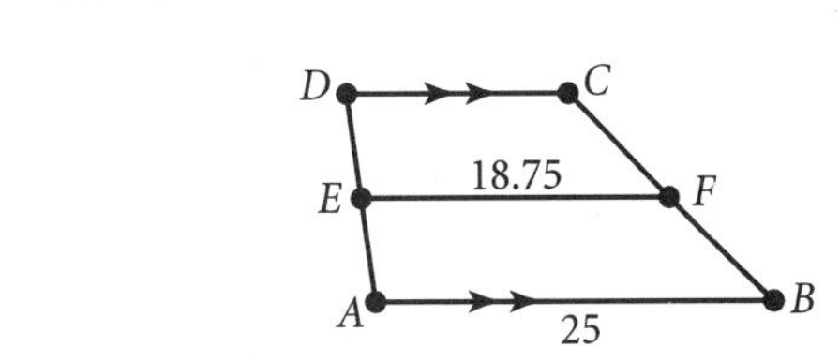

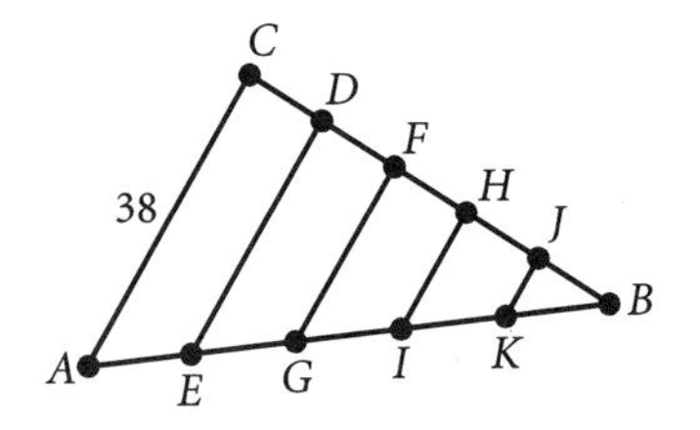

Figure *ABCD* at the right is a rhombus. *E, F, G,* and *H* are midpoints. In Exercises 13 and 14, what type of quadrilateral is formed by the indicated vertices? Explain your reasoning.

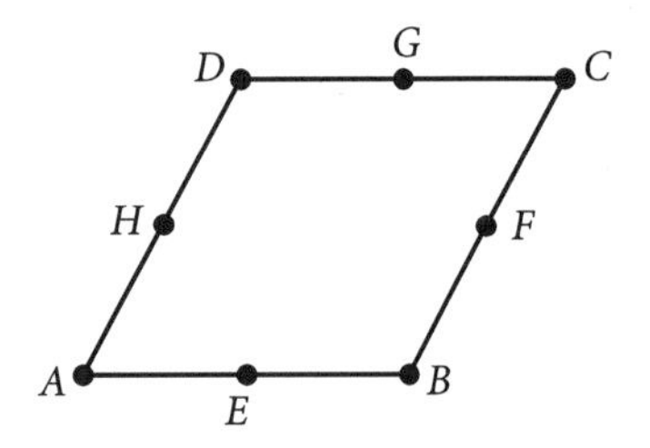

13. $EFGH$ ____

14. $EHDB$ ____

NAME ______________________ CLASS ____________ DATE ____________

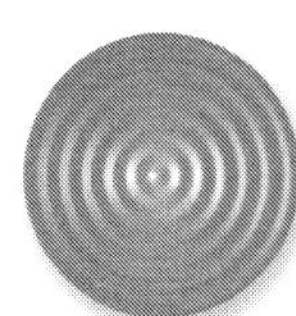

Practice Masters Level C

3.7 Midsegments of Triangles and Trapezoids

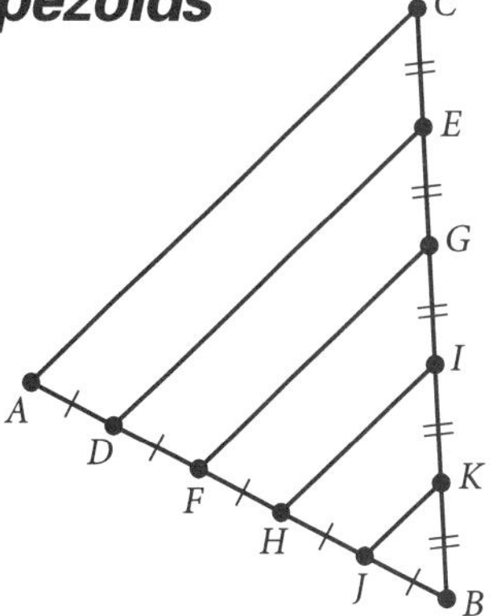

In the figure at the right, $ED = 15$, and $IH = 7.5$. Find the indicated measures.

1. AC ______________________

2. GF ______________________

3. KJ ______________________

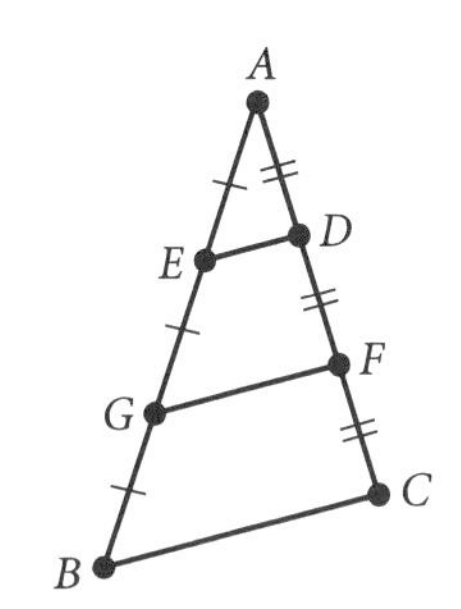

In the figure at the right, $BC = x^2 + x$, and $ED = x + 5$. Find the indicated measures.

4. ED ______________________

5. GF ______________________

6. BC ______________________

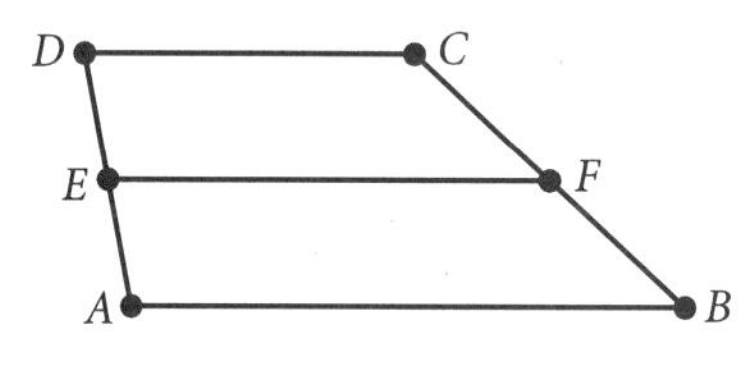

Figure $ABCD$ at the right is a trapezoid with $\overline{DC} \parallel \overline{AB}$. E and F are midpoints. $DC = x^2 - x - 3$, $AB = x^2 + 2x + 2$, and $EF = 2x^2 - 46$. Find the indicated measures.

7. x ____________ 8. DC ____________

9. EF ____________ 10. AB ____________

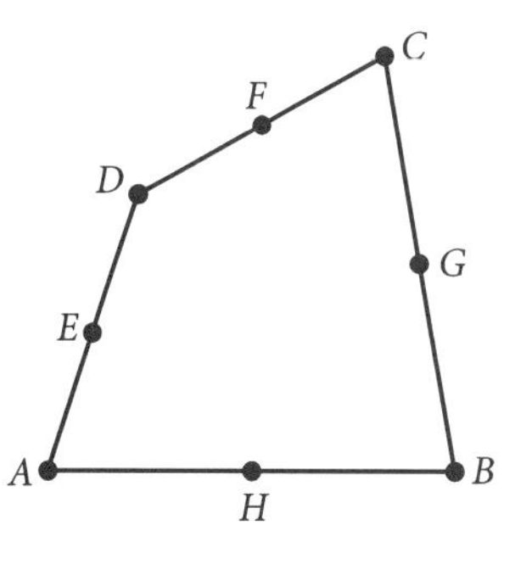

11. In the figure at the right, $AD = DC$ and $CB = AB$. E, F, G, and H are midpoints. What shape is formed by $EFGH$? Write a paragraph proof to justify your answer.

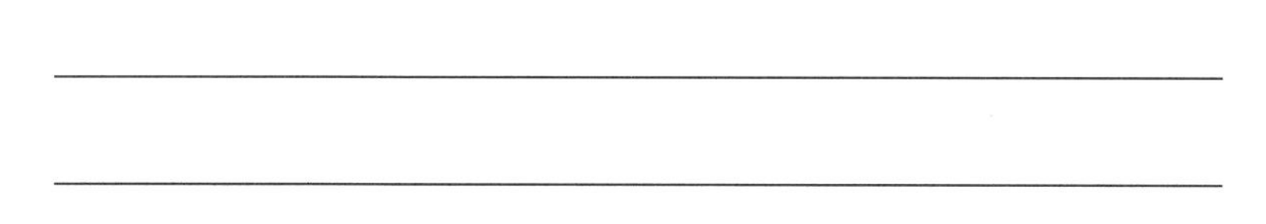

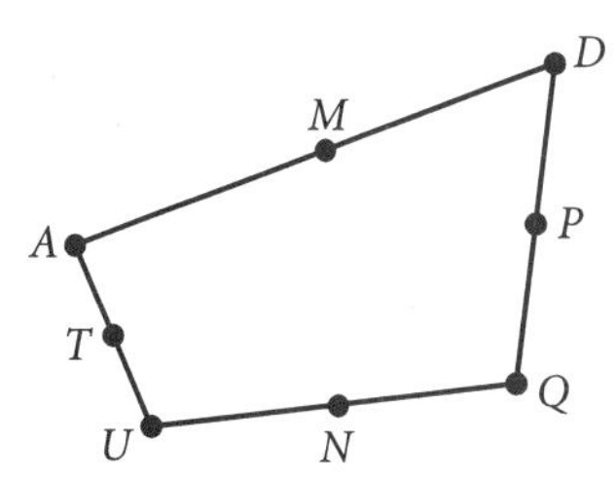

12. M, P, N, and T are the midpoints of quadrilateral $QUAD$ shown at the right. What shape is formed by $MPNT$? Explain your answer.

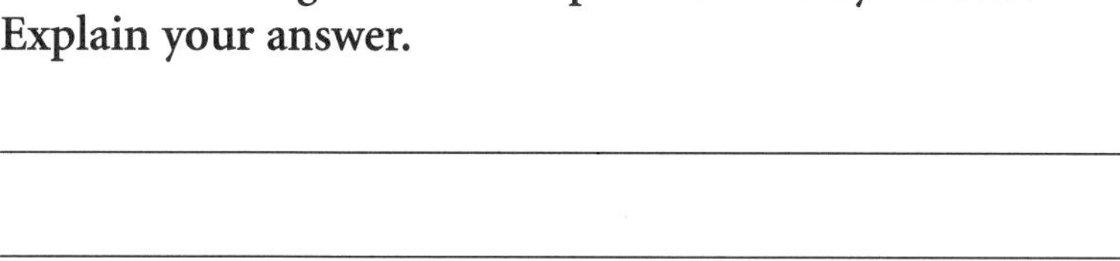

NAME ______________________ CLASS __________ DATE __________

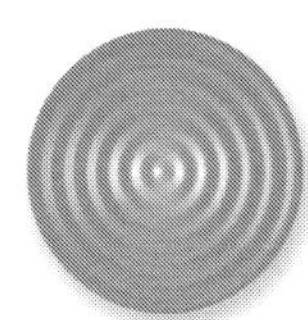

Practice Masters Level A

3.8 Analyzing Polygons with Coordinates

In Exercises 1–3, the endpoints of a segment are given. Determine the slope and midpoint of the segment.

1. $(4, 3)$ and $(-3, -4)$ ______________________
2. $(5, 8)$ and $(-7, 6)$ ______________________
3. $(-3, -8)$ and $(-5, -4)$ ______________________

In Exercises 4–6, the endpoints of two segments are given. State whether the segments are parallel, perpendicular, or neither. Justify your answer.

4. $(-2, 1)$ and $(3, 7)$; $(1, 1)$ and $(-3, 11)$ ______________________
5. $(2, -1)$ and $(6, 11)$; $(-3, -7)$ and $(-1, -1)$ ______________________
6. $(-4, 0)$ and $(2, 3)$; $(-2, 1)$ and $(4, -11)$ ______________________
7. On the grid provided, graph quadrilateral $ABCD$. What type of quadrilateral is this? Justify your answer.

   $A(3, 2), B(1, -2), C(2, -5), D(4, -1)$

   ____

   ____

   ____

   ____

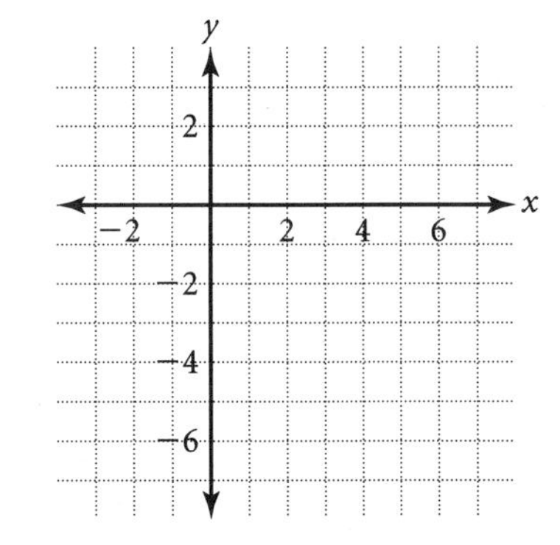

8. The endpoints of two segments are given. Draw each segment on the grid provided, then connect the endpoints to each other. What type of quadrilateral do you think this is? Explain your answer.

   $\overline{AC}$ has endpoints $(-3, 11)$ and $(2, -4)$.

   $\overline{BD}$ has endpoints $(-6, 5)$ and $(3, 8)$.

   ____

   ____

   ____

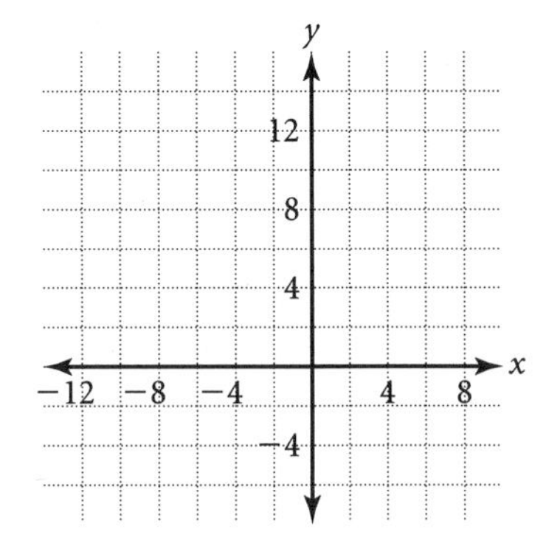

NAME ______________________ CLASS ______________ DATE ______________

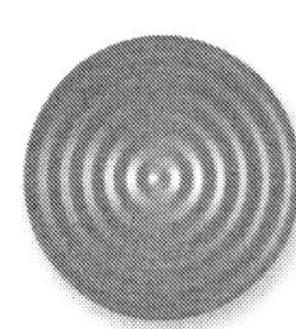

Practice Masters Level B

3.8 *Analyzing Polygons with Coordinates*

Use the grid at the right for Exercises 1–3.

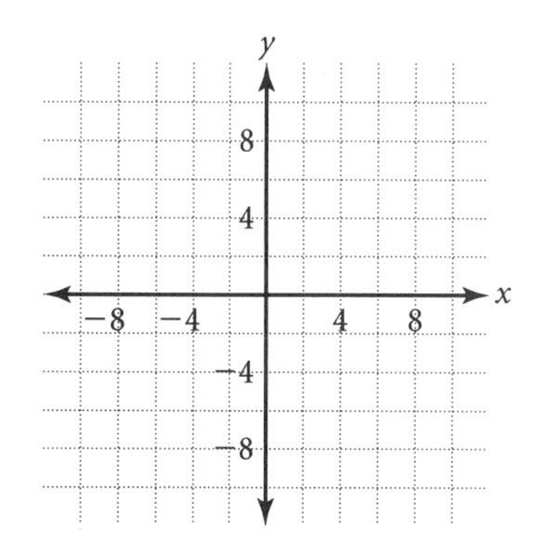

1. Graph $\triangle ABC$ with vertices $A(-5, 4)$, $B(1, 1)$, and $C(-7, 2)$.

2. $\triangle ABC$ is translated using the rule $T(x, y) = (x + 6, y - 5)$. List and graph the coordinates of the image, $\triangle A'B'C'$.

____

3. What is the shape determined by $CC'B'B$? Justify your answer.

____

Use the grid at the right for Exercises 4 and 5. Points *A*, *B*, and *C* are three vertices of a parallelogram.

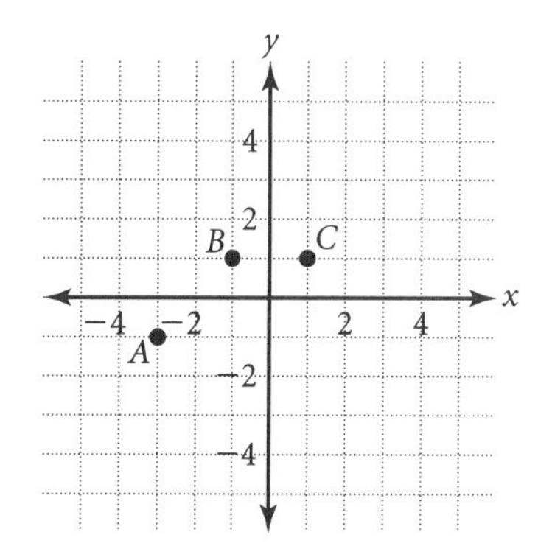

4. How many parallelograms can be formed using these three points? Explain your answer.

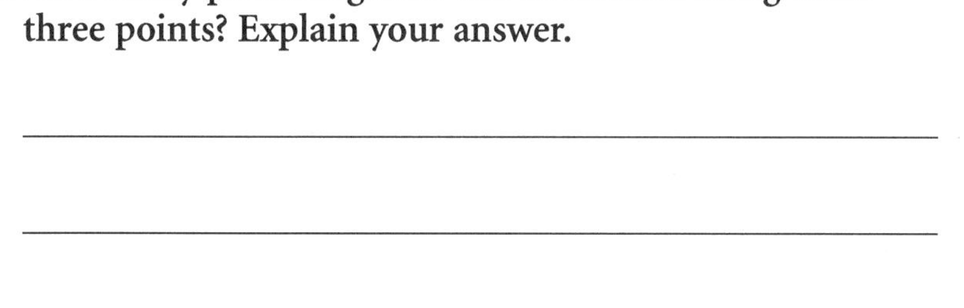

____

____

____

5. Give the coordinates of the fourth vertex of the other parallelograms. ______________

For Exercises 6 and 7, one endpoint of $\overline{AB}$ is $(-2, -7)$. The midpoint of the segment is $(1, -1)$.

6. Find the coordinates of *B*. ______________

7. If you drew a line perpendicular to $\overline{AB}$ through the midpoint, what would be the slope of that line? ______________

8. A segment has slope $\frac{3}{4}$. One endpoint of the segment has coordinates $(-2, 7)$. Find the coordinates of the other endpoint. ______________

9. A triangle has vertices $(-3, 4)$, $(4, 6)$, and $(-7, 18)$. Use slopes to determine whether the triangle is a right triangle. Justify your answer.

____

NAME ______________________ CLASS ______________ DATE ______________

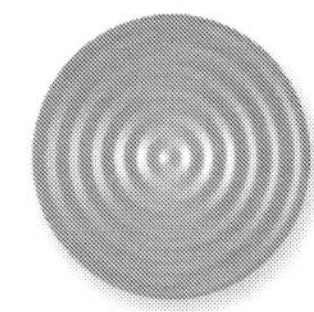

Practice Masters Level C

3.8 Analyzing Polygons with Coordinates

1. One endpoint of a line segment is $(-7, -15)$. The midpoint of the segment is $(-5, -4)$. Find the other endpoint. ______________

2. What is the slope of a line drawn through the midpoint, perpendicular to the segment in Exercise 1? ______________

For Exercises 3–5, a quadrilateral has vertices $(2, -5)$, $(-8, -5)$, $(-2, 10)$, and $(6, 5)$.

3. Graph the quadrilateral on the grid provided.

4. What type of quadrilateral is this? Justify your answer.

____

____

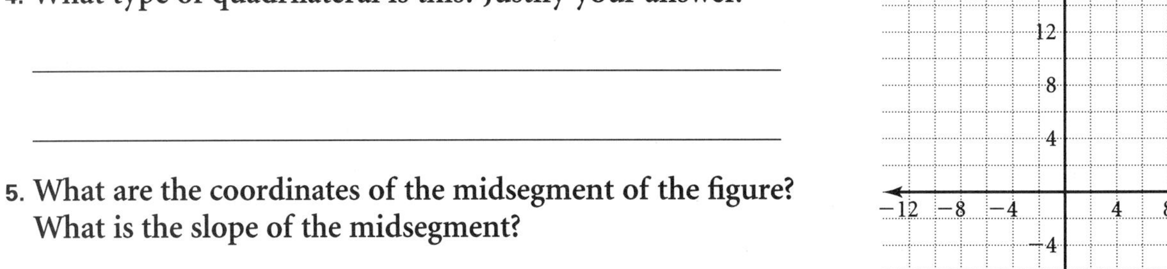

5. What are the coordinates of the midsegment of the figure? What is the slope of the midsegment?

____

For Exercises 6–8, vertex *A* of square *ABCD* has coordinates $(-1, 1)$. The coordinates of the intersection of the diagonals is $(2, 4)$.

6. Find the coordinates of vertices *B*, *C*, and *D*. ______________________________

7. $m\angle CAB =$ ______________ 8. slope of $\overline{CA} =$ ______________

Use the grid at the right for Exercises 9 and 10. Points *A*, *B*, and *C* are three vertices of a parallelogram.

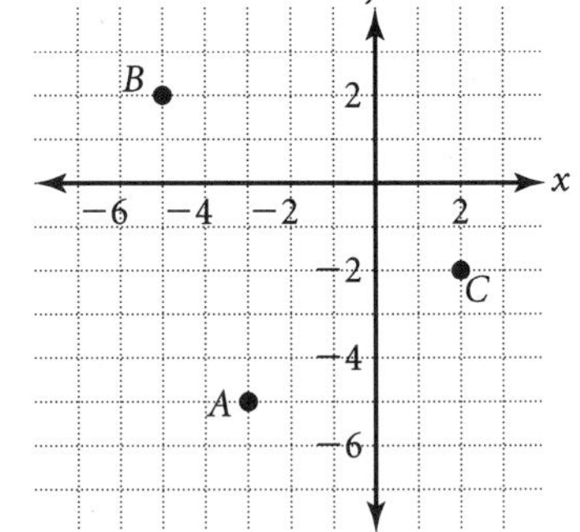

9. How many parallelograms can be formed using these three points? Explain your answer.

____

____

____

10. Give the coordinates of the fourth vertex of the other parallelograms.

____

NAME ______________________ CLASS ______________ DATE ______________

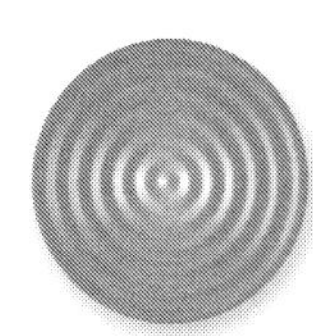

Practice Masters Level A

4.1 Congruent Polygons

1. Write three different congruence statements about the figures below.

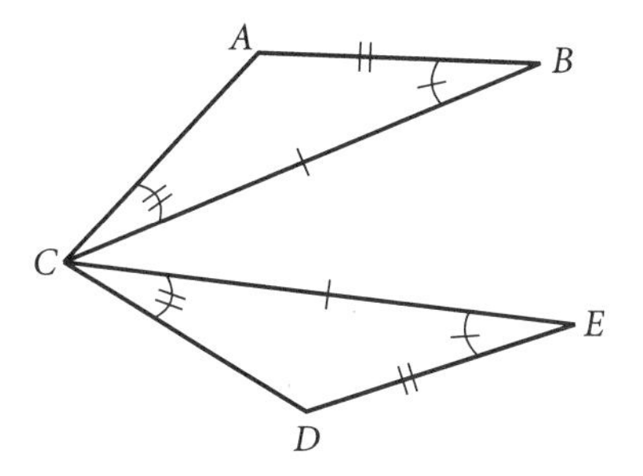

____

____

____

2. Write a congruence statement for the figure at the right.

____

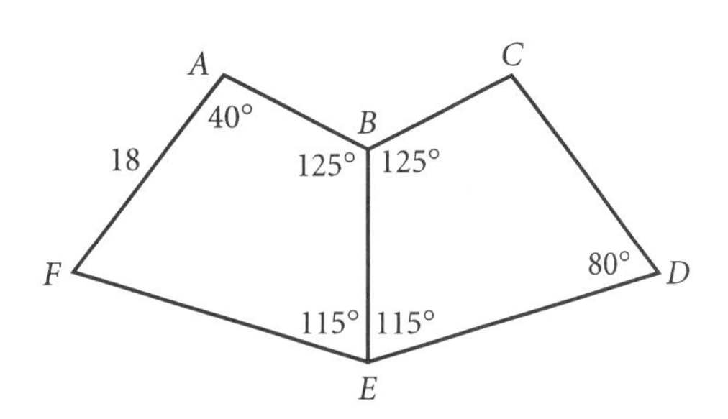

3. Given the following information, name three pairs of congruent triangles.

$\angle ABC \cong \angle BAD$,
$\angle ADB \cong \angle BCA$,
$\angle DAC \cong \angle CBD$,
$\angle ADC \cong \angle BCD$,
$\overline{AD} \cong \overline{BC}$,
$\overline{AE} \cong \overline{BE}$, and
$\overline{ED} \cong \overline{EC}$

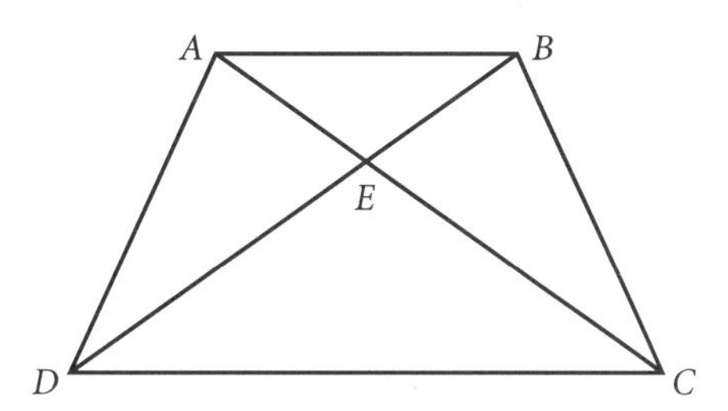

____

____

____

Given that pentagon *QRSTU* ≅ pentagon *JKLMN*, complete the statements.

4. $\angle S \cong$ ________ 5. $\angle T \cong$ ________ 6. $\angle K \cong$ ________ 7. $\angle J \cong$ ________

8. $\angle N \cong$ ________ 9. $\overline{ST} \cong$ ________ 10. $\overline{MN} \cong$ ________ 11. $\overline{RQ} \cong$ ________

Suppose that △*ABD* ≅ △*FEC*. Find the measure of each.

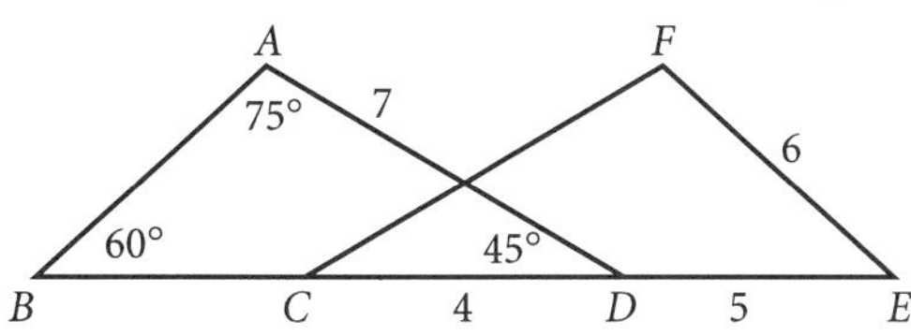

12. $m\angle E =$ ________ 13. $m\angle F =$ ________

14. $m\angle FCE =$ ________ 15. $BD =$ ________

16. $AB =$ ________ 17. $FC =$ ________

18. $m\angle CGD \cong$ ________ 19. $m\angle AGC =$ ________

NAME ______________________ CLASS ______________ DATE ______________

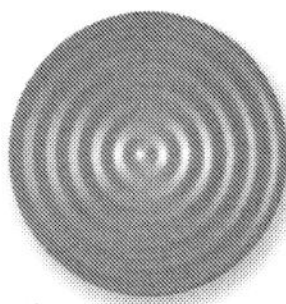

Practice Masters Level B

4.1 Congruent Polygons

1. Write three different congruence statements about the figure below.

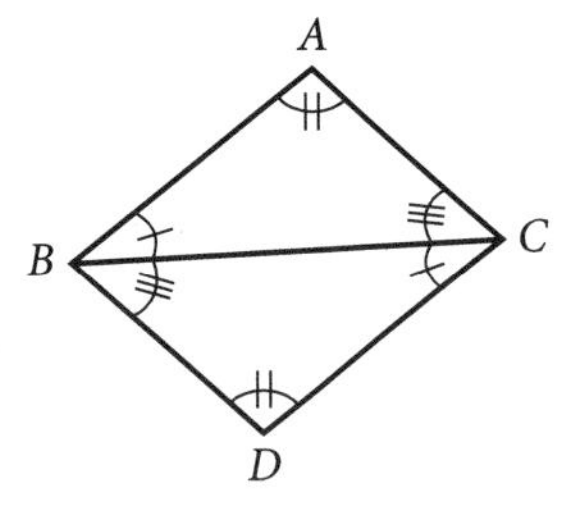

____

____

____

2. Write a congruence statement for the figures at the right.

____

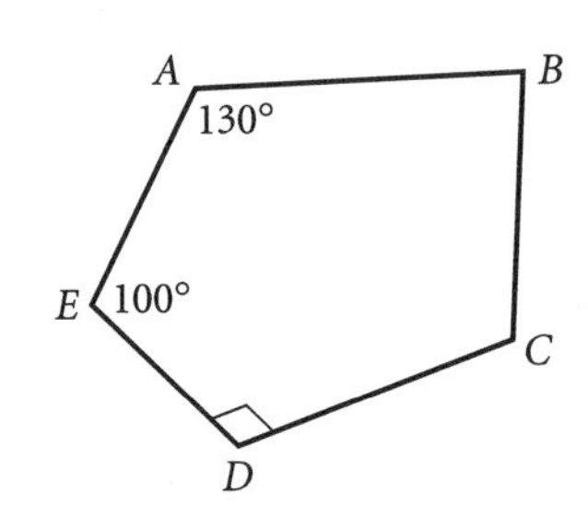

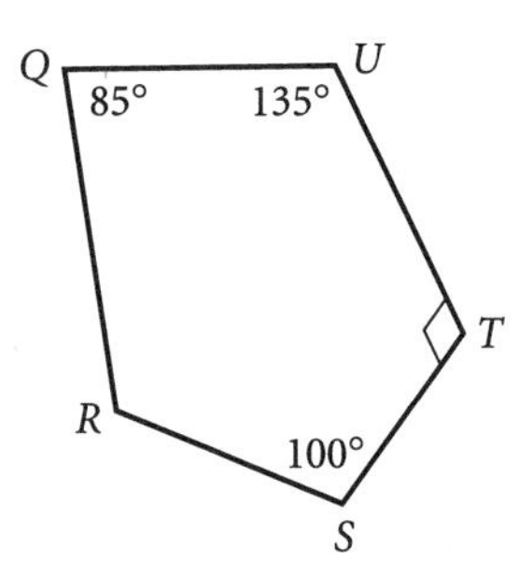

Use the given information for Exercises 3–9.
Given: $\triangle AFG \cong \triangle BFE$ and $\triangle AEG \cong \triangle BGE$

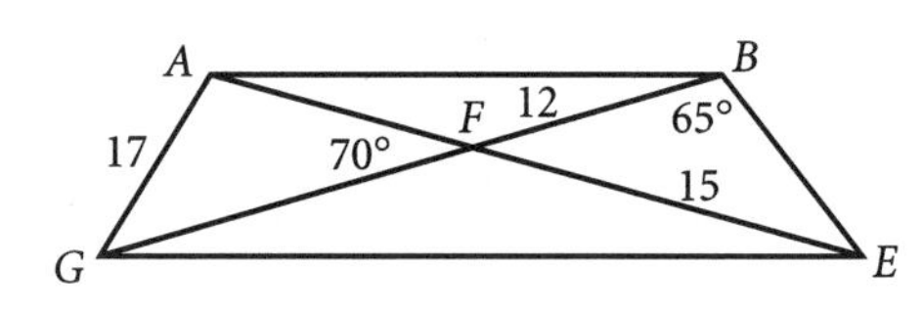

3. $m\angle BFE =$ ______________
4. $m\angle AEG =$ ______________
5. $m\angle AGF =$ ______________
6. $BG =$ ______________
7. $FG =$ ______________
8. $EB =$ ______________
9. Are $\overline{AB}$ and $\overline{GE}$ parallel? Support your answer.

____

____

Given that polygon *BCDEFG* ≅ polygon *LMNPQR* ≅ polygon *TVWXYZ*, complete the statements.

10. $\angle B \cong$ ______________ $\cong$ ______________
11. $\overline{XY} \cong$ ______________ $\cong$ ______________
12. $\overline{VT} \cong$ ______________ $\cong$ ______________
13. $\overline{NM} \cong$ ______________ $\cong$ ______________
14. $\angle Q \cong$ ______________ $\cong$ ______________
15. $\angle W \cong$ ______________ $\cong$ ______________
16. $\overline{RQ} \cong$ ______________ $\cong$ ______________
17. $\angle M \cong$ ______________ $\cong$ ______________

NAME ______________________ CLASS ____________ DATE ____________

Practice Masters Level C

4.1 Congruent Polygons

For Exercises 1–4, use the figures below and complete the statements.

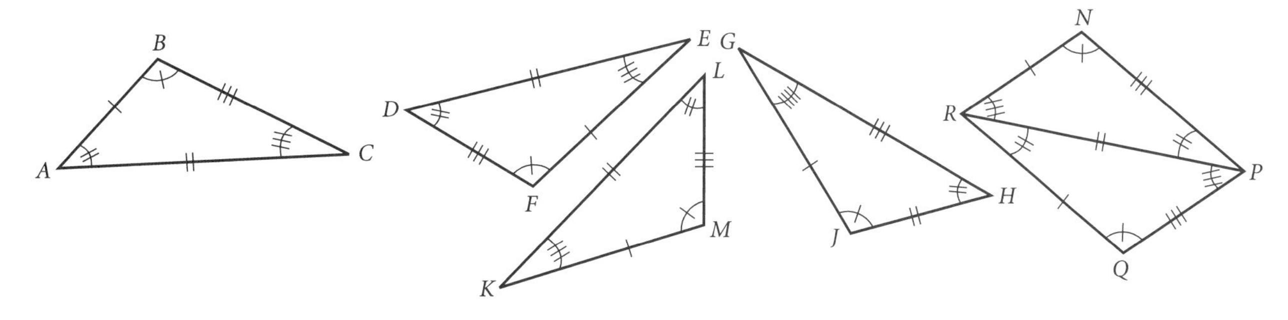

1. $\triangle ABC \cong$ ______________

2. $\triangle KLM \cong$ ______________

3. $\triangle RNP \cong$ ______________

4. $\triangle GHJ \cong$ ______________

5. Given that $\overline{BG} \parallel \overline{CE}$, $\angle A \cong \angle D$, $m\angle GBF = 50°$, $\overline{GF} \cong \overline{EF}$, and $\overline{CD} \cong \overline{BA}$, name four pairs of congruent figures.

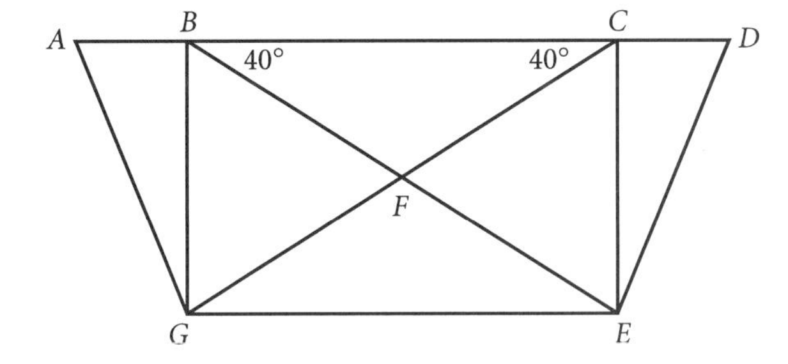

____

____

____

____

For Exercises 6–8, the vertices of $\triangle ABC$ and $\triangle EDC$ are $A(7, 9)$, $B(11, 5)$, $C(3, 3)$, $D(-5, 1)$, and $E(-1, -3)$.

6. Graph $\triangle ABC$ and $\triangle EDC$ on the coordinate grid.

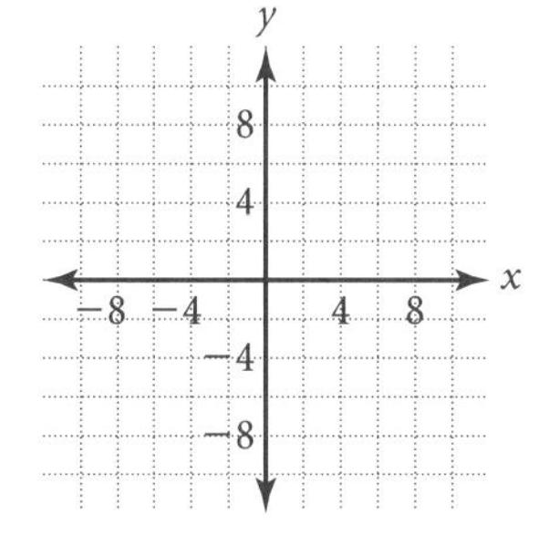

7. Are $\triangle ABC$ and $\triangle EDC$ congruent? Explain.

____

____

8. What postulate, theorem or definition justifies your answer to Exercise 7?

____

NAME ______________________ CLASS ____________ DATE ____________

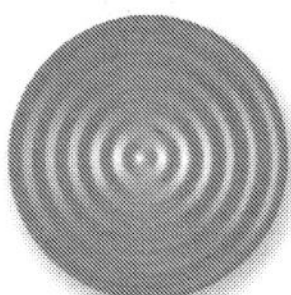

Practice Masters Level A

4.2 Triangle Congruence

Determine whether each pair of triangles can be proven congruent by using the SSS, SAS, or ASA Congruence Postulate. If so, identify which postulate is used.

1. ______________________

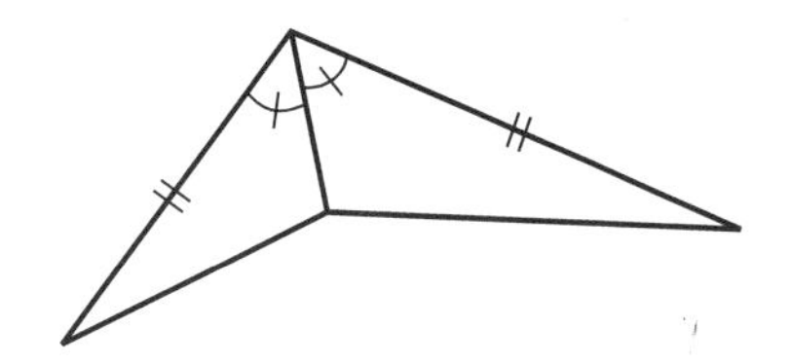

2. ______________________

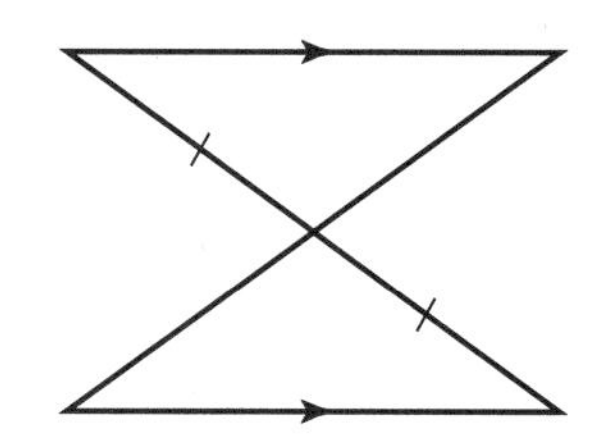

3. ______________________

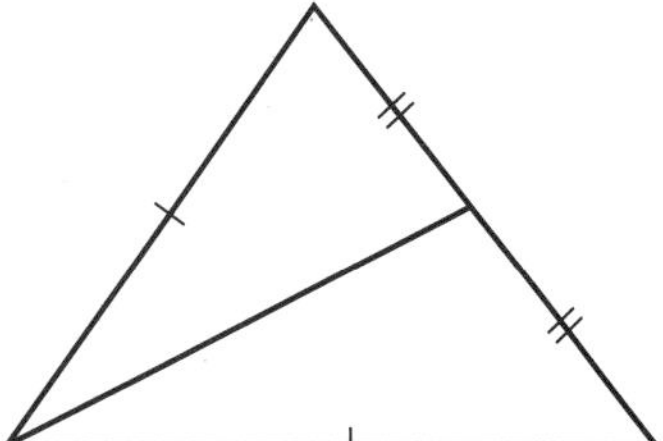

4. ______________________

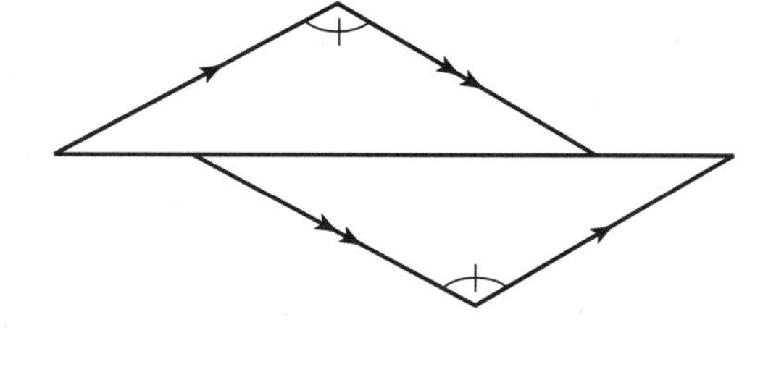

For each postulate or theorem stated below, give the other sides or angles that must be congruent to prove $\triangle ABC \cong \triangle CDA$.

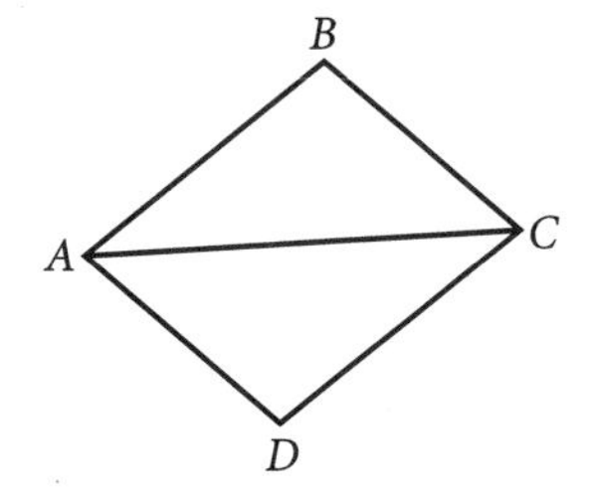

5. ASA ______________________

6. SAS ______________________

7. SSS ______________________

For Exercises 8–14, some triangle measures are given. Is there exactly one triangle that can be constructed with the given measurements? If so, identify the postulate that justifies the answer.

8. $\triangle ABC$; $m\angle A = 37°$, $AB = 8$, and $AC = 10$ ______________________

9. $\triangle FGH$; $m\angle G = 85°$, $m\angle F = 60°$, and $GF = 12$ ______________________

10. $\triangle JKL$; $JK = 5$, $LJ = 7$ and $KL = 5$ ______________________

11. $\triangle NOP$; $m\angle P = 51°$, $NO = 7$, and $NP = 9$ ______________________

12. $\triangle RST$; $m\angle S = 62°$, $m\angle T = 41°$, and $RS = 10$ ______________________

13. $\triangle UVW$; $m\angle U = 30°$, $VU = 8.2$, and $UW = 5.7$ ______________________

14. $\triangle XYZ$; $m\angle Y = 61°$, $m\angle X = 36°$, and $m\angle Z = 83°$ ______________________

NAME ______________________ CLASS ____________ DATE ____________

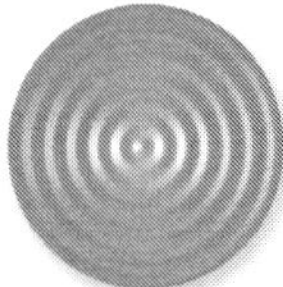

Practice Masters Level B

4.2 Triangle Congruence

Determine whether each pair of triangles can be proven congruent by using the SSS, SAS, or ASA congruence postulate. If so, identify which postulate is used.

1. ____________

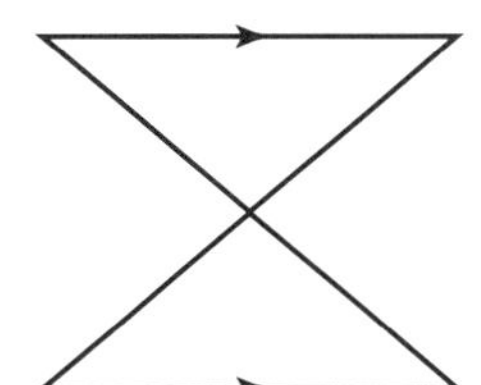

2. ____________

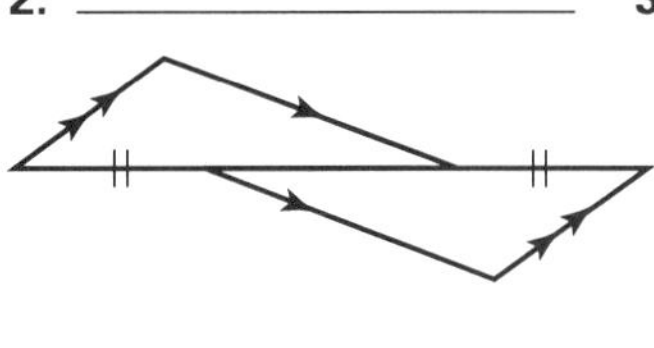

3. ____________

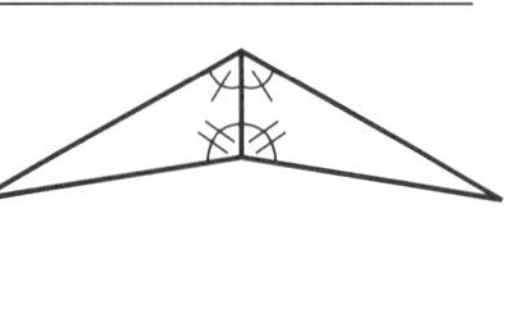

4. ____________

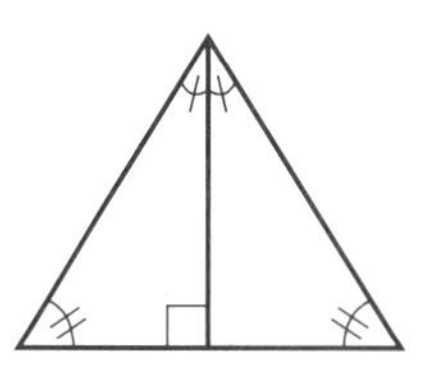

For each postulate or theorem stated below, give the other sides or angles that must be congruent to prove $\triangle ABC \cong \triangle DEF$.

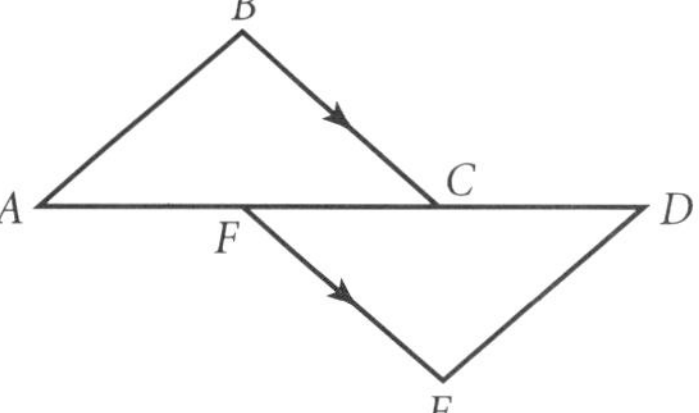

5. ASA ______________________

6. SAS ______________________

For Exercises 7–9, some triangle measures are given. Is there exactly one triangle that can be constructed with the given measurements? If so, identify the postulate that justifies the answer.

7. $\triangle ABC$; $m\angle A = 90°$, $AB = 7$, and $BC = 11$ ______________________

8. $\triangle DEF$; $m\angle D = 46°$, $m\angle E = 44°$, and $m\angle F = 90°$ ______________________

9. $\triangle JKL$; $m\angle K = 39°$, $JL = 8$, and $KL = 13$ ______________________

If $\triangle BAD \cong \triangle CDA$, $m\angle BAD = 75°$, $m\angle CDA = (4x - 13)°$, $AD = 11$, $BA = 9$ and $CD = \frac{2}{3}y - 3$, find the following:

10. $x =$ ______________________

11. $y =$ ______________________

Complete the proof.

Given: $\overline{BD}$ bisects $\angle ABC$; $\overline{AB} \cong \overline{CB}$

Prove: $\triangle ABD \cong \triangle CBD$

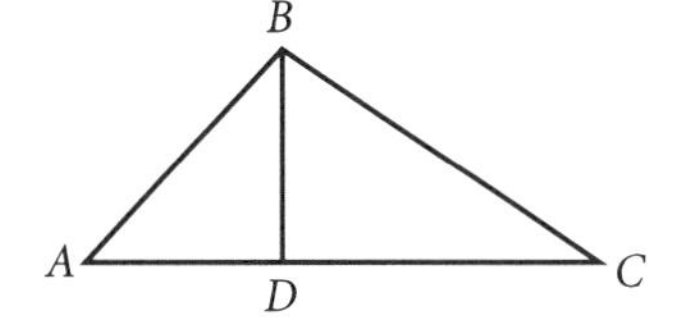

Statements	Reasons
$\overline{BD}$ bisects $\angle ABC$.	Given
12.	Definition of angle bisector
$\overline{AB} \cong \overline{CB}$	Given
$\overline{BD} \cong \overline{BD}$	**13.**
14.	**15.**

NAME ______________________ CLASS ____________ DATE ____________

Practice Masters Level C

4.2 Triangle Congruence

Determine whether each pair of triangles can be proven congruent by using the SSS, SAS, or ASA congruence postulate. If so, identify what postulate is used.

1. ______________ **2.** ______________ **3.** ______________ **4.** ______________

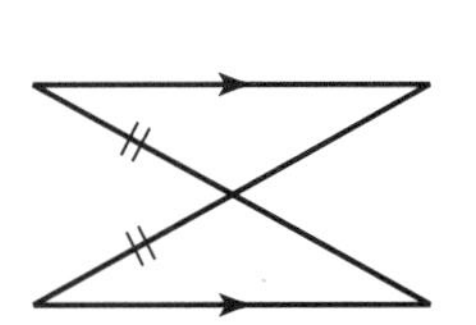

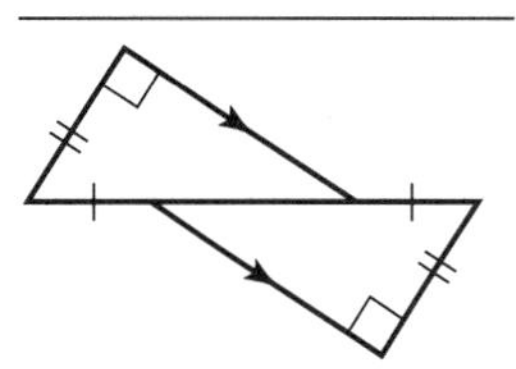

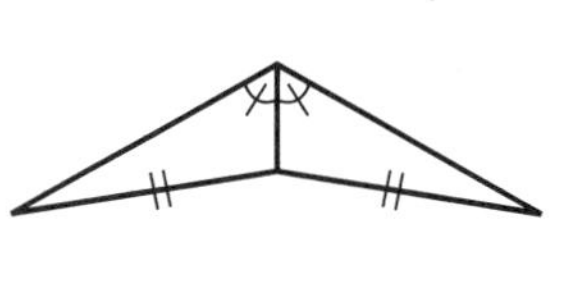

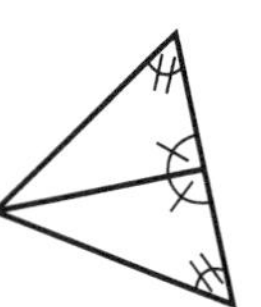

Decide if the statements are *always true, sometimes true,* or *never true.* Give the postulate that supports the answer.

5. Two right triangles are congruent if their legs are congruent. ______________
6. Two triangles are congruent if their corresponding congruent parts include two sides and an angle. ______________
7. Two triangles are congruent if their corresponding congruent parts include two angles and a side. ______________
8. Two right triangles are congruent of a corresponding leg and an acute angle are congruent. ______________
9. Two triangles are congruent if all three corresponding angles are congruent. ______________

For Exercises 10 and 11, $\triangle CAB \cong \triangle FED$, $m\angle F = (3x - 11y)°$, $m\angle D = (x + y + 7)°$.

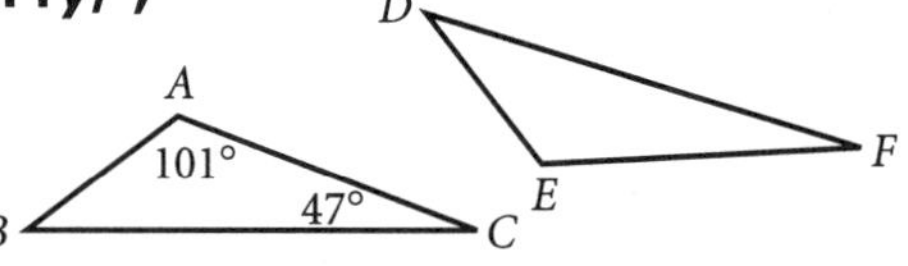

10. $x =$ ______________ **11.** $y =$ ______________

Complete the following proof:

Given: $\triangle AED$, $\overline{AB} \cong \overline{CD}$, $\overline{AE} \cong \overline{DE}$
Prove: $\triangle CEA \cong \triangle BED$

Statements	Reasons
$\overline{AB} \cong \overline{CD}$	Given
$\overline{BC} \cong \overline{BC}$	**12.**
$AB + BC = AC$	**13.**
$CD + BC = BD$	
$AB + BC = CD + BC$	**14.**
$\overline{AC} \cong \overline{BD}$	**15.**
$\overline{AE} \cong \overline{DE}$	Given
$\angle A \cong \angle D$	**16.**
$\triangle CEA \cong \triangle BED$	**17.**

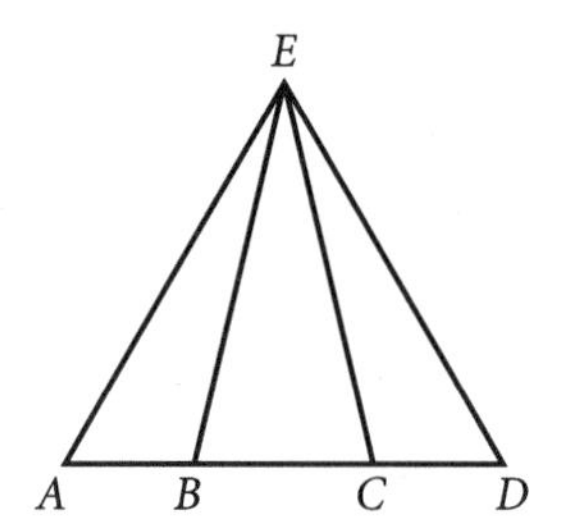

Practice Masters Level A

4.3 Analyzing Triangle Congruence

Provide an example or a counterexample to explain your answer for Exercises 1 and 2.

1. If $\angle A \cong \angle Q$, $m\angle R = m\angle B$, and $m\angle C = m\angle S$, is $\triangle ABC \cong \triangle QRS$? ______________

____

2. If $m\angle A = m\angle Q$, $m\angle B = m\angle R$, and $\overline{AC} \cong \overline{QS}$ is $\triangle ABC \cong \triangle QRS$? ______________

____

In Exercises 3–7, determine whether the pairs of triangles can be proven congruent. If so, write a congruence statement and name the postulate or theorem used.

3. ______________

4. ______________

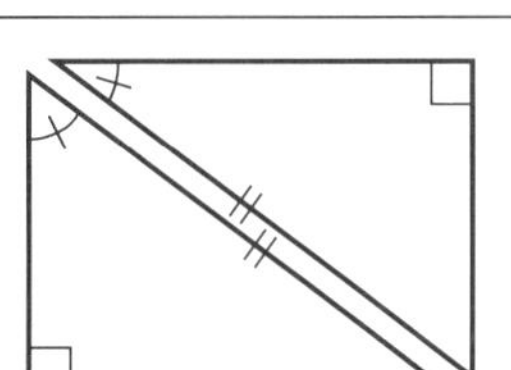

5. ______________

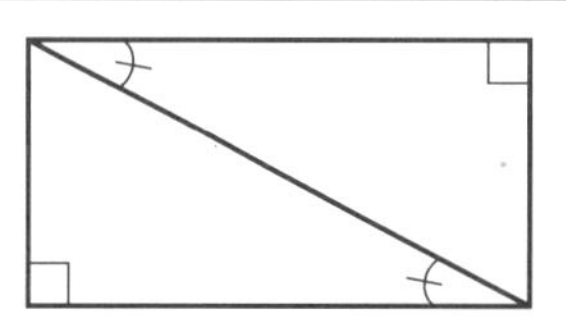

6. ______________

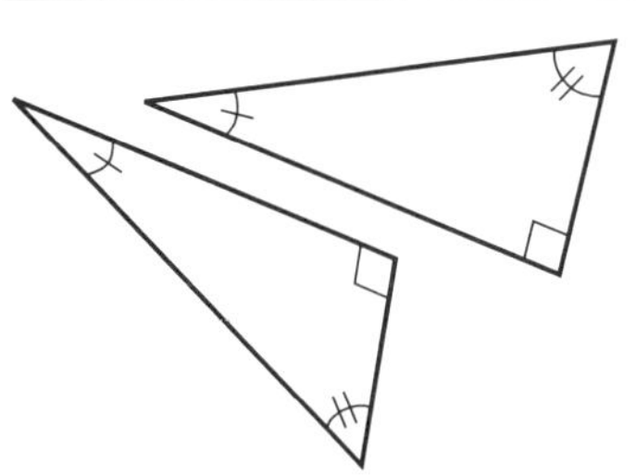

7. ______________

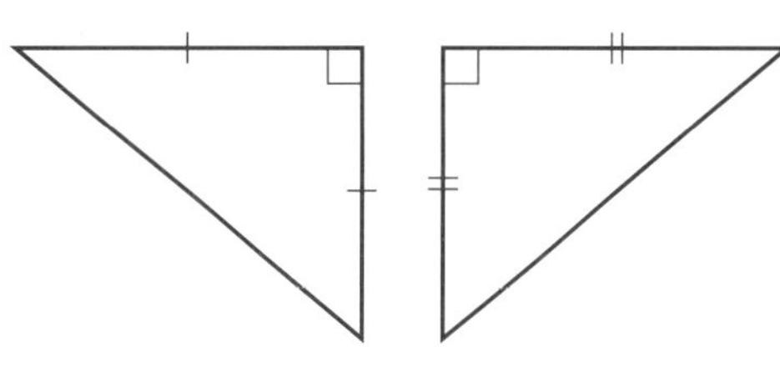

For Exercises 8–12, $\triangle ABC$ and $\triangle DEF$ are right triangles. Given the following information, which theorem or postulate can you use to prove $\triangle ABC \cong \triangle DEF$?

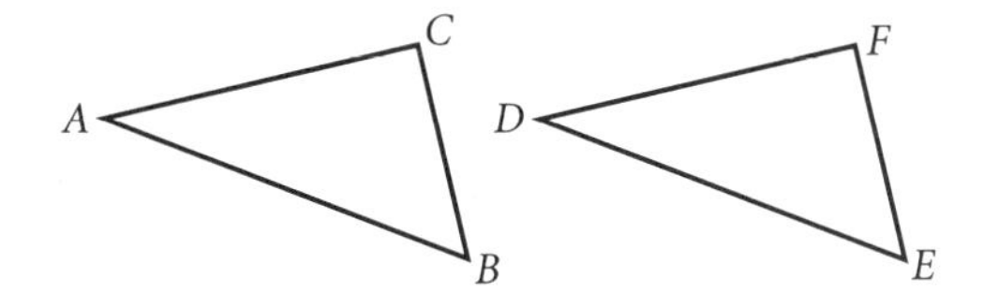

8. $\overline{AC} \cong \overline{DF}$; $\angle A \cong \angle D$
   $\angle C \cong \angle F = 90°$ ______________

9. $\overline{AB} \cong \overline{DE}$; $\angle A \cong \angle D$
   $\angle C \cong \angle F = 90°$ ______________

10. $\angle C \cong \angle F$; $\angle B \cong \angle E$
    $\overline{BC} \cong \overline{EF}$ ______________

11. $\angle B \cong \angle E$; $\overline{AB} \cong \overline{DE}$
    $\overline{CB} \cong \overline{FE}$ ______________

12. $\angle C \cong \angle F$; $\angle B \cong \angle E$
    $\overline{AC} \cong \overline{DF}$ ______________

NAME ______________________ CLASS __________ DATE __________

Practice Masters Level B

4.3 Analyzing Triangle Congruence

For Exercises 1–6, determine whether the given combination of angles and sides determines a unique triangle. If so, identify the theorem or postulate that supports the answer. If not, give a counter example.

1. $\triangle ABC$; $m\angle B = 41°$; $m\angle C = 68°$; $m\angle A = 51°$ ______________

2. $\triangle DEF$; $m\angle E = 90°$; $DE = 16$; $EF = 12$ ______________

3. $\triangle JKL$; $m\angle J = 38°$; $LK = 4$; $JK = 7$ ______________

4. $\triangle MNO$; $m\angle O = 90°$; $MN = 8$; $OM = 5$ ______________

5. $\triangle PQR$; $PR = 7$; $RQ = 8$; $PQ = 13$ ______________

6. $\triangle VWX$; $m\angle V = 36°$; $VW = 8$; $WX = 6$ ______________

Decide whether the given information is enough to say that $\triangle ABC \cong \triangle DEF$. Identify the theorem or postulate that supports your decision.

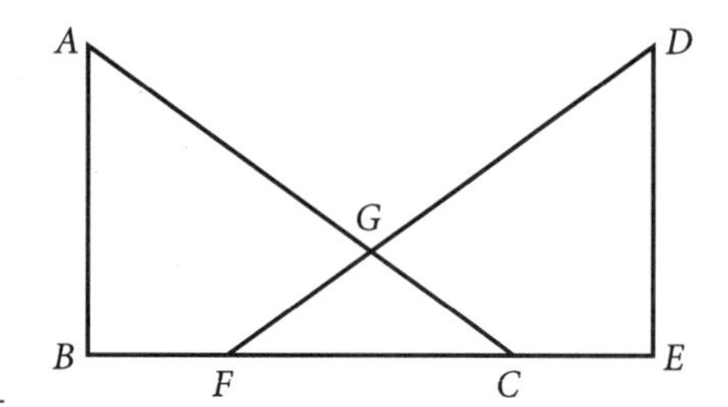

7. $\overline{AB} \cong \overline{DE}$; $\overline{BF} \cong \overline{CE}$; $\overline{AB} \perp \overline{BE}$; $\overline{DE} \perp \overline{BE}$ ______________

8. $\angle DFC \cong \angle ACF$; $\angle A \cong \angle D$; $\overline{AB} \cong \overline{DE}$ ______________

9. $\overline{AC} \cong \overline{DF}$; $\overline{AB} \cong \overline{DE}$; $\angle B$ and $\angle E = 90°$ ______________

10. $\angle AGF \cong \angle DGC$; $\angle A \cong \angle D$; $\overline{AB} \cong \overline{DE}$ ______________

11. $AB \parallel DE$; $\overline{AB} \perp \overline{BE}$; $\overline{AB} \cong \overline{DE}$; $\angle GFC \cong \angle GCF$ ______________

Complete the following proof.

Given: $\overline{XQ} \perp \overline{RQ}$; $\overline{XT} \perp \overline{ST}$; $\overline{RQ} \cong \overline{ST}$

Prove: $\triangle RXQ \cong \triangle SXT$

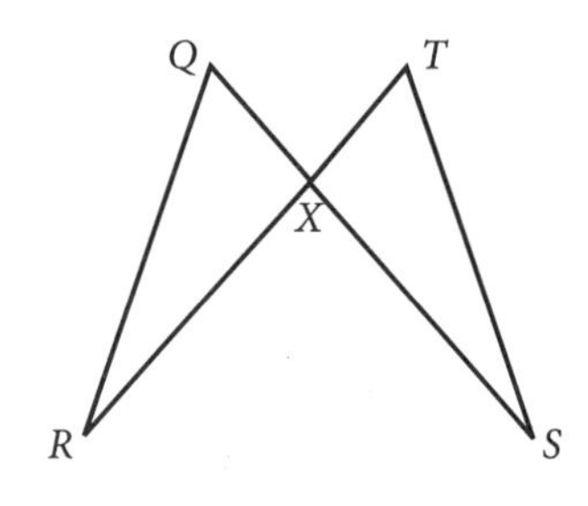

Statements	Reasons
$\overline{XQ} \perp \overline{RQ}$; $\overline{XT} \perp \overline{ST}$; $\angle RQX$, $\angle STX$ are right angles.	Given **12.**
13.	**14.**
$\overline{RQ} \cong \overline{ST}$	**15.**
$\angle QXR \cong \angle TXS$	**16.**
17.	**18.**

NAME ______________________ CLASS __________ DATE __________

Practice Masters Level C

4.3 Analyzing Triangle Congruence

Use the given information to decide which triangles are congruent. Identify the postulate or theorem that justifies your answer.

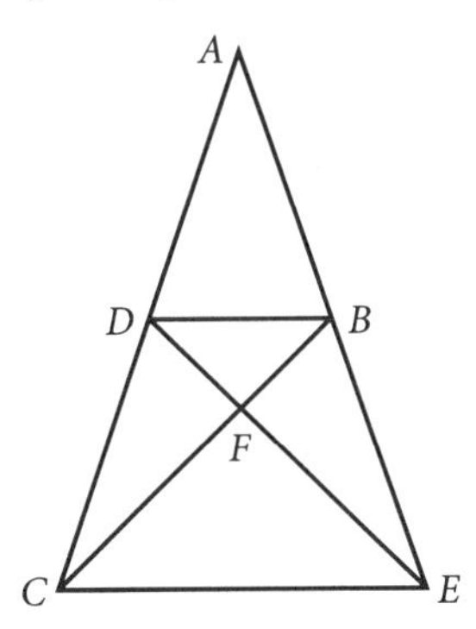

1. $\overline{AC} \cong \overline{AE}$
   $\angle ACB \cong \angle AED$
   $\triangle$ ________ $\cong \triangle$ ________
   by ____________

2. $\overline{DB} \perp \overline{AE}$
   $\angle DAE \cong \angle DEA$
   $\triangle$ ________ $\cong \triangle$ ________
   by ____________

3. Point F is the midpoint of $\overline{BC}$ and $\overline{DE}$.
   $\triangle$ ________ $\cong \triangle$ ________
   by ____________

4. $\overline{BC} \cong \overline{DE}$
   $\overline{CD} \cong \overline{BE}$
   $\triangle$ ________ $\cong \triangle$ ________
   by ____________

Decide whether the following statements are *always true, sometimes true,* or *never true.* If it is sometimes or never true, provide a counter example explaining why it is not always true.

5. Congruent triangles are co-planar. ____________

6. A quadrilateral can be congruent to a pentagon. ____________

7. Two regular triangles are congruent. ____________

8. If an edge of one cube is congruent to the edge to a second cube, the faces of the cubes are congruent. ____________

9. Isosceles right triangles are congruent. ____________

Complete the following proof.

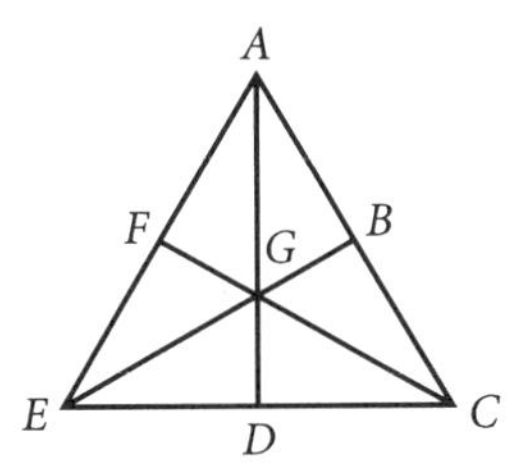

Given: $\overline{AD}$ bisects $\angle EAC$;
$\overline{AE} \cong \overline{AC}$;
F is the midpoint of $\overline{AE}$;
B is the midpoint of $\overline{AC}$.

Prove: $\triangle GAF \cong \triangle GAB$

Statements	Reasons
$\overline{AD}$ bisects $\angle EAC$.	Given
10.	**11.**
$\overline{AE} \cong \overline{AC}$ F is the midpoint of $\overline{AE}$.	Given
B is the midpoint of $\overline{AC}$.	Given
12.	**13.**
14.	**15.**
16.	**17.**
18.	**19.**

NAME ______________________ CLASS ____________ DATE ____________

Practice Masters Level A

4.4 Using Triangle Congruence

Find each indicated measure.

1.

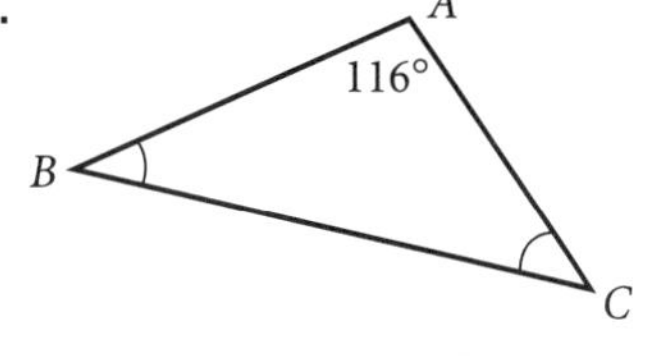

$m\angle C$ ______________

2.

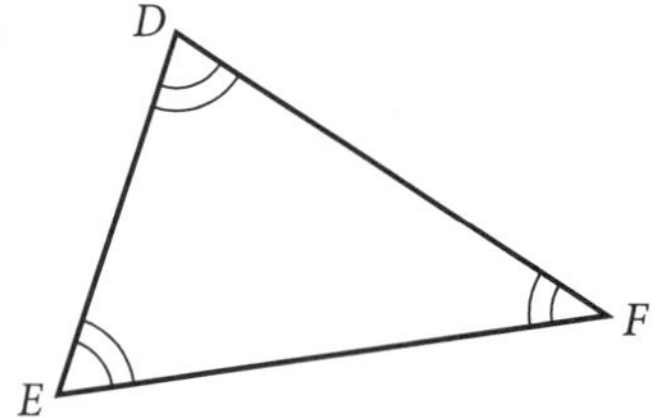

$m\angle F$ ______________

3.

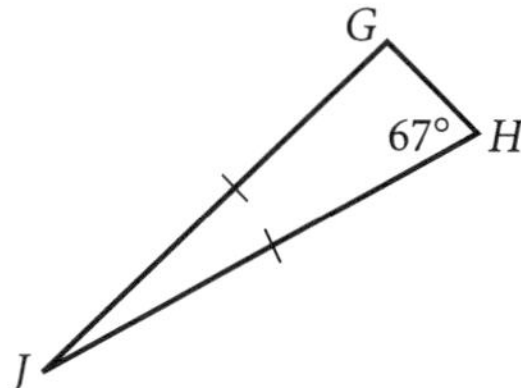

$m\angle J$ ______________

4.

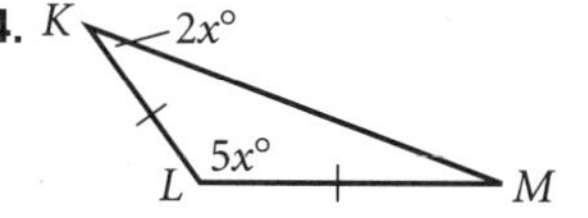

$m\angle L$ ______________

5.

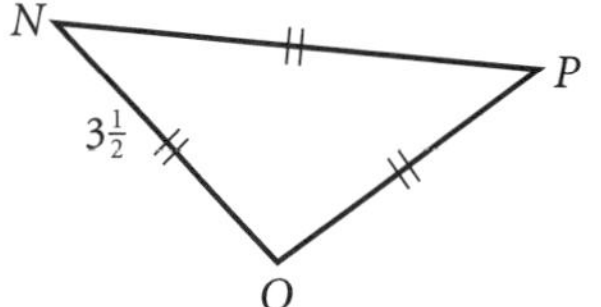

NP ______________

6.

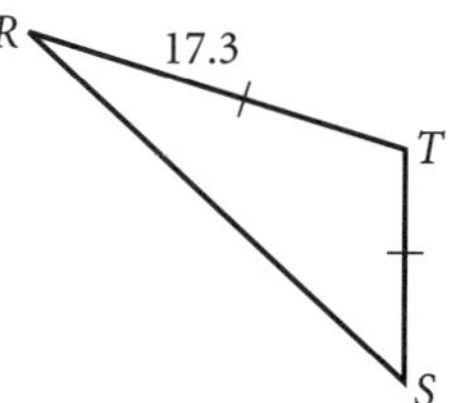

ST ______________

Given: $\triangle WVZ \cong \triangle XZV$, $\overline{VY} \cong \overline{ZY}$, $VX = 12$, $m\angle WVY = 98°$, and $m\angle YVZ = 27°$. Find each indicated measure.

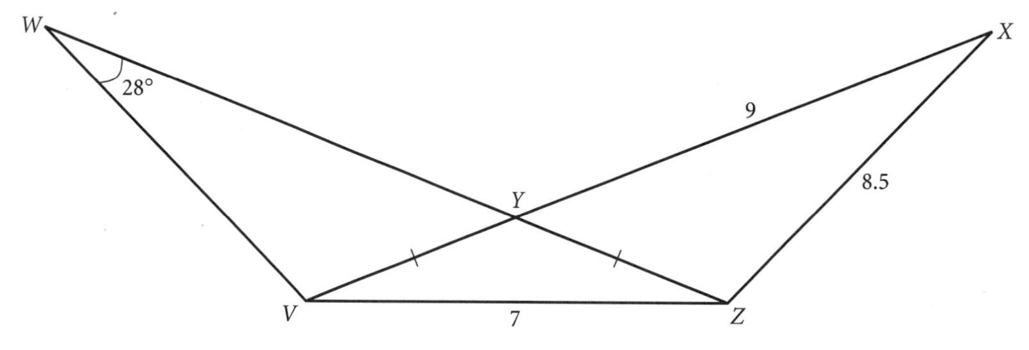

7. VY ______________

8. WZ ______________

9. VW ______________

10. $m\angle WYV$ ______________

11. $m\angle XZV$ ______________

12. $m\angle XYZ$ ______________

13. $m\angle XZY$ ______________

14. $m\angle YZV$ ______________

15. $m\angle VYZ$ ______________

16. Suppose that $\triangle ABC \cong \triangle DEF$. If $AB = 13$, how long is DE? Justify your answer.

____

____

NAME ______________________ CLASS ____________ DATE ____________

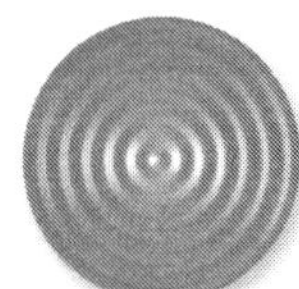

Practice Masters Level B

4.4 Using Triangle Congruence

Find each indicated measure.

1.

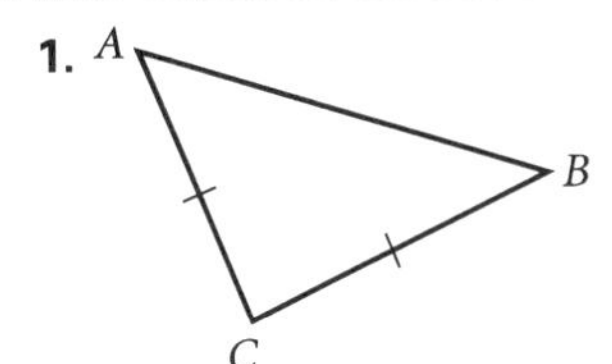

$m\angle A = m\angle B = 2x°; m\angle C = 40°$

$m\angle B$ ____________

2.

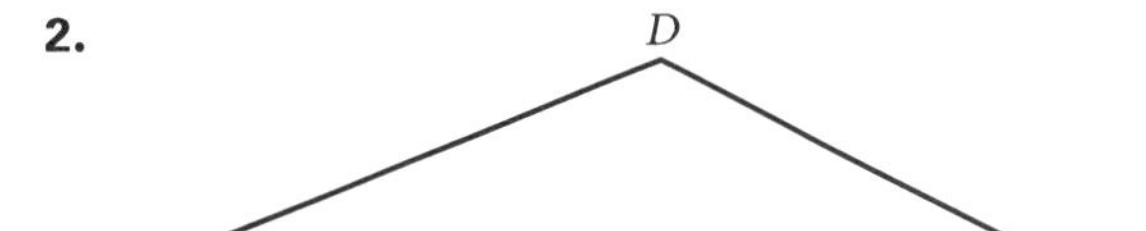

$m\angle E = 32°; m\angle D = 8x°; EF = 2x$

EF ____________

3.

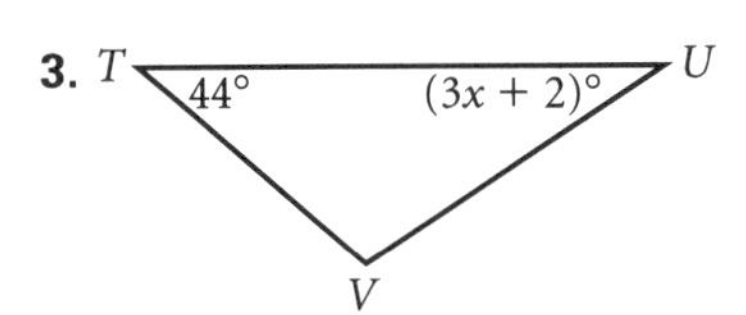

If $VU = 2x - 9$, find TV. ____________

4.

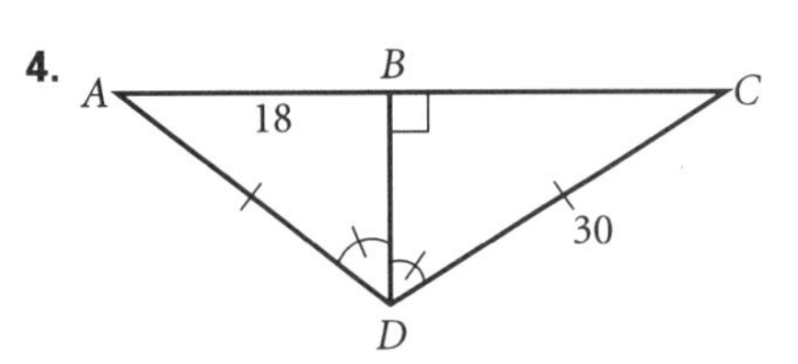

AC ____________

Find each indicated measure.

5. x ________ y ________

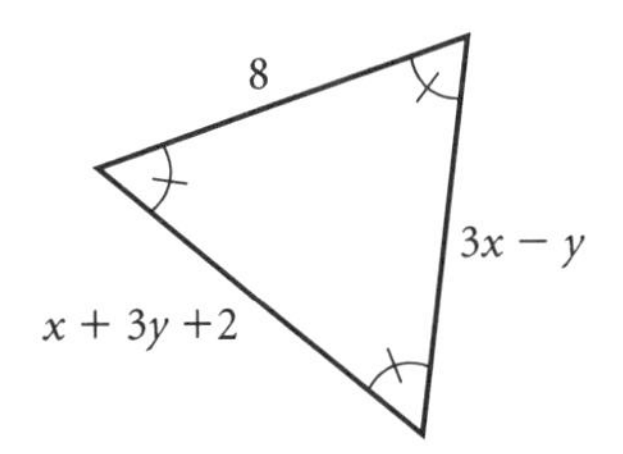

6. $m\angle N$ ____________

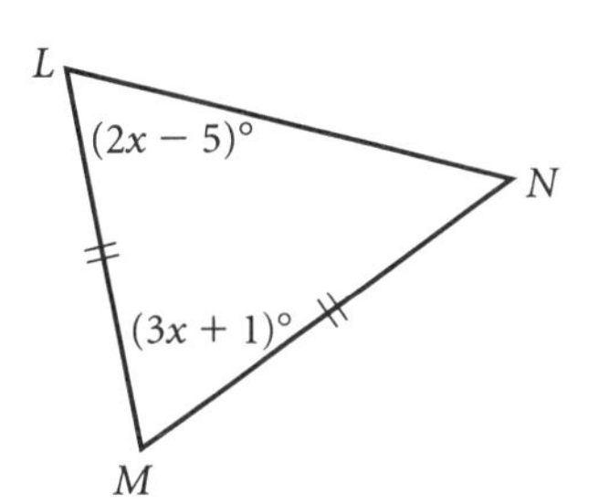

Use the diagram to complete a flowchart proof.

Given: $\triangle XYZ$ and $\triangle XDC$ are isosceles;
Y and Z trisect $\overline{DC}$.

Prove: $\angle XYD \cong \angle XZC$

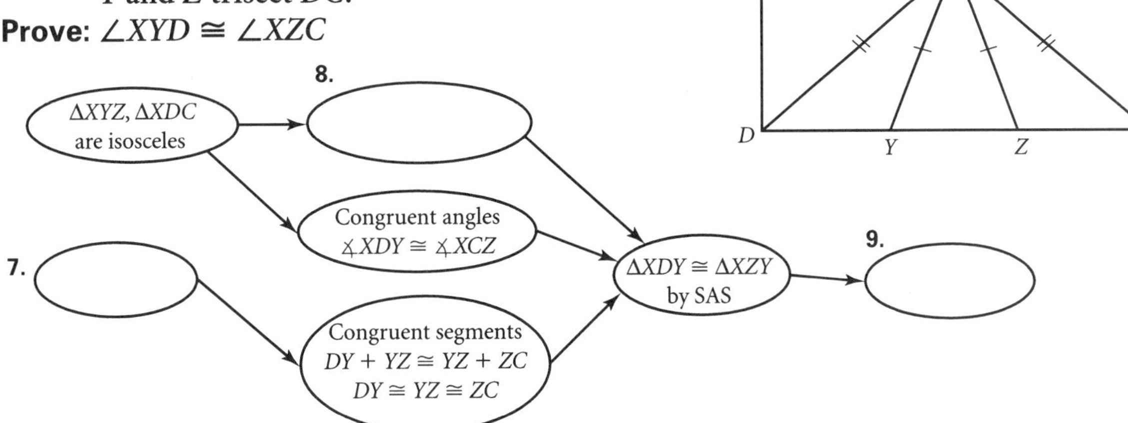

NAME ______________________ CLASS ______________ DATE ______________

Practice Masters Level C

4.4 Using Triangle Congruency

Decide whether the following statements are *always true, sometimes true,* or *never true.*

1. An altitude of an isosceles triangle divides it into two congruent triangles. ______________

2. An altitude of an equilateral triangle divides it into two congruent triangles. ______________

3. If two sides of a triangle are congruent, the angles opposite those sides are congruent. ______________

4. If two triangles are congruent, angles and sides have the same measure. ______________

5. Find the length of each side.

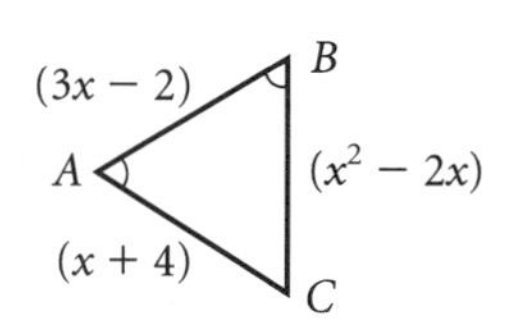

$AB =$ ________

$BC =$ ________

$AC =$ ________

6. In $\triangle MNL$, $m\angle M = (11x - 28)°$,

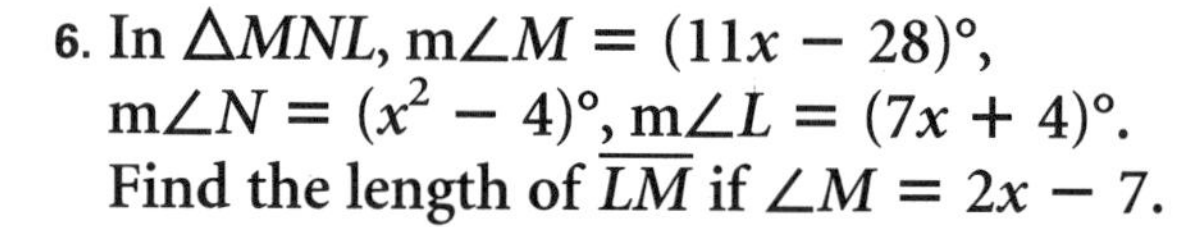

$m\angle N = (x^2 - 4)°$, $m\angle L = (7x + 4)°$.
Find the length of $\overline{LM}$ if $\angle M = 2x - 7$.

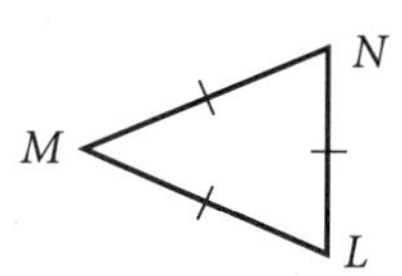

____

7. In $\triangle ABC$, $m\angle A = (2x + 5)°$, and $m\angle ABE = \frac{1}{4} m\angle AEB$.
Find $m\angle EBD$ if $\angle A \cong \angle C$.

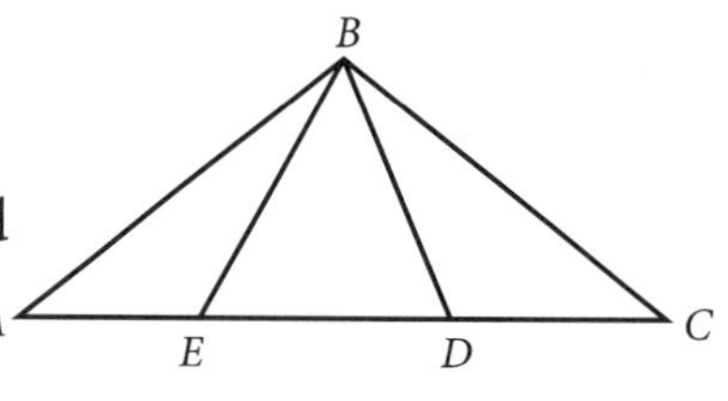

____

Write a paragraph proof.

8. **Given:** BD bisects $\angle ABC$; $\overline{AB} \cong \overline{CB}$
Prove: $\triangle ADC$ is isosceles.

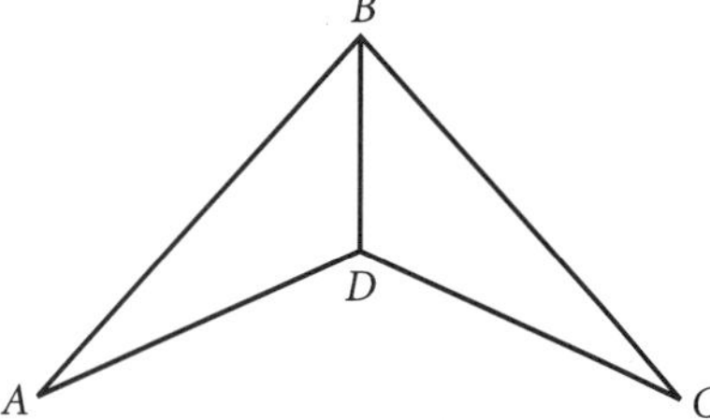

____

____

____

____

____

____

____

NAME ______________________ CLASS __________ DATE __________

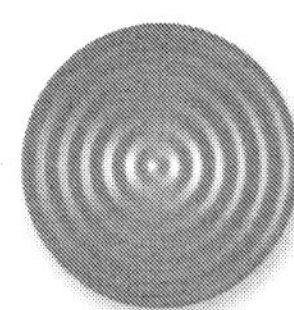

Practice Masters Level A

4.5 Proving Quadrilateral Properties

In Exercises 1–6, find the indicated measures for each parallelogram.

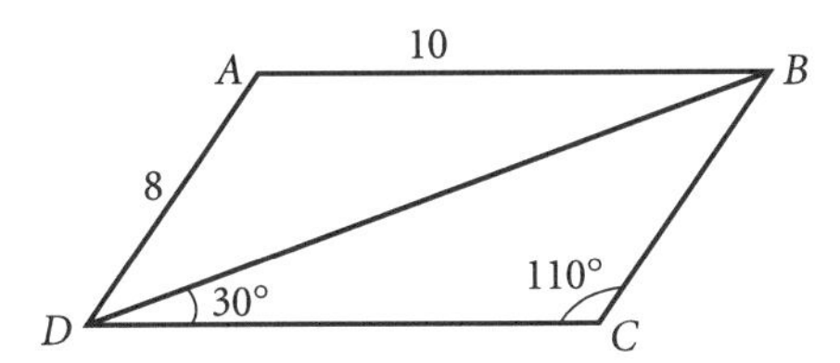

1. m∠*DBC* ________
2. m∠*BDA* ________
3. m∠*A* ________
4. *BC* ________
5. *DC* ________
6. m∠*ADC* ________

***ABCD* is a parallelogram with diagonals *AC* and *BD*.**

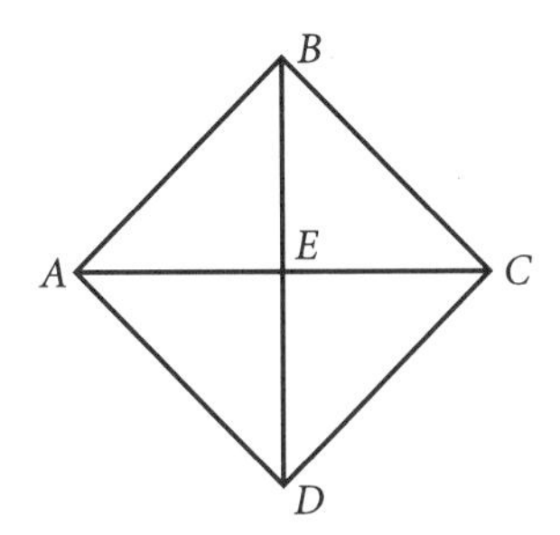

7. If $EC = 2x - 7$ and $AE = x + 2$, how long is $\overline{AC}$?

____

8. $\overline{BD}$ divides ∠*ADC* into two angles that measure 43° and 26°. Find m∠*ABC*.

____

***BDEF* is a rhombus. Points *D, E,* and *F* are midpoints of $\overline{AB}$, $\overline{AC}$ and $\overline{AB}$, respectively, and m∠*DEB* = 37°. Find each measure.**

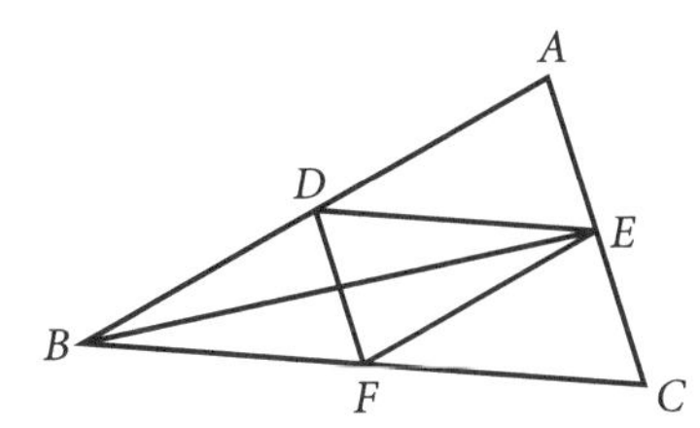

9. m∠*FEB* ________
10. m∠*DBF* ________
11. m∠*EDB* ________
12. m∠*EBF* ________
13. m∠*EFC* ________
14. m∠*BDF* ________

***ABDE* is a parallelogram with *BC* ≅ *BD*.**

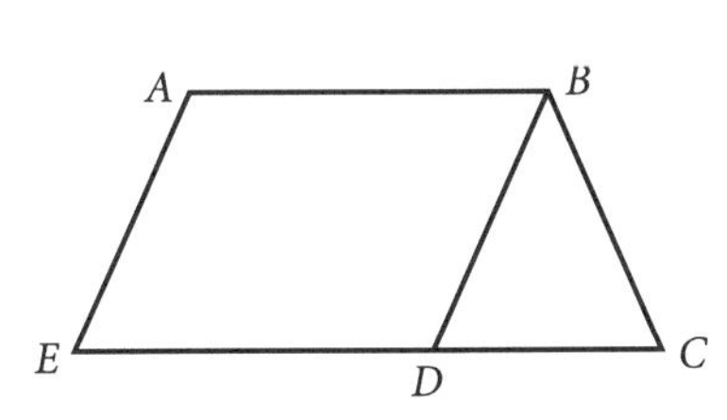

15. If m∠*E* = 71°, find m∠*EAB*. ________
16. If m∠*BDC* = 58°, find m∠*EAB*. ________
17. If m∠*A* = $(2x + 11)°$, m∠*B* = 77°, find *x*. ________

18. In parallelogram *ABCD*, $\overline{BD}$ is a diagonal, $\overline{AE} \perp \overline{BD}$, and $\overline{CF} \perp \overline{BD}$. List all congruent triangles.

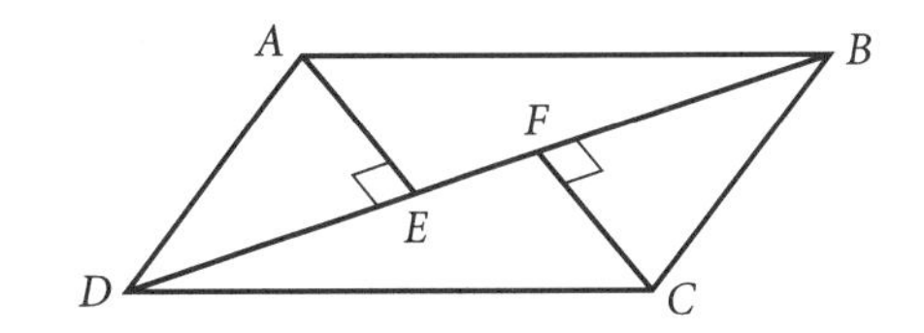

____

NAME ______________________ CLASS ____________ DATE ____________

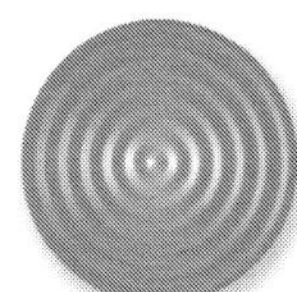

Practice Masters Level B

4.5 Proving Quadrilateral Properties

***ABCD* is a parallelogram with diagonals $\overline{BD}$ and $\overline{AC}$.**

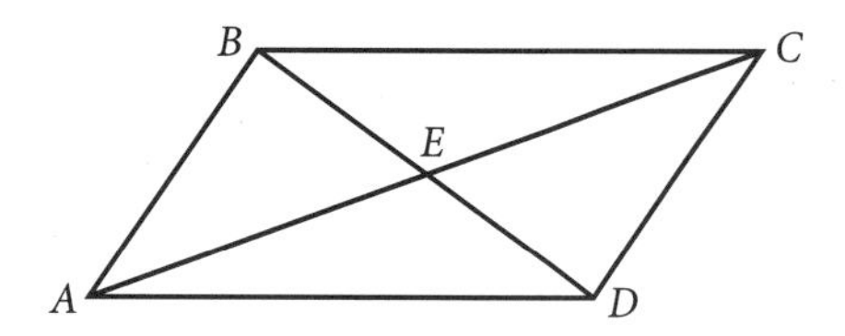

1. If m$\angle CBD$ = 26°, m$\angle DCA$ = 72°, and m$\angle DEC$ = 81°. Find m$\angle BAD$. ____________

2. If $AC = 3x + 5y$, $EC = 2x + y$, $BC = 3x + y$, and $AD = 5$. Find the length of AC. ____________

***BDEF* is a rhombus. Points *D*, *E*, and *F* are midpoints of $\overline{AB}$, $\overline{AC}$ and $\overline{BC}$, respectively.**

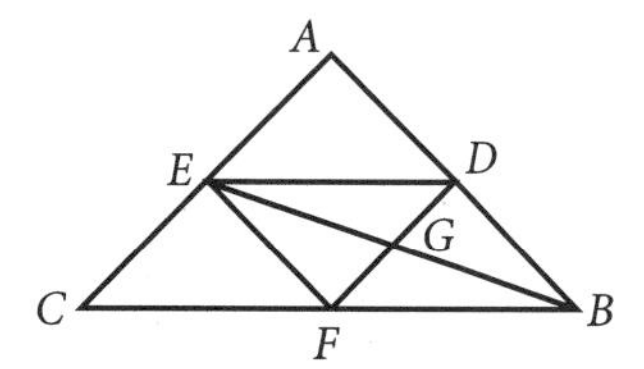

3. If $GE = x + 6$, $BE = 16$, $DF = 30$ and $GF = 2y - x$. Find x and y. ____________

4. If $GE^2 + GD^2 = ED^2$, find the perimeter of $BDEF$. ____________

***ABDE* is a parallelogram with $\overline{BC} \cong \overline{BD}$. Use the diagram for Exercises 5–8.**

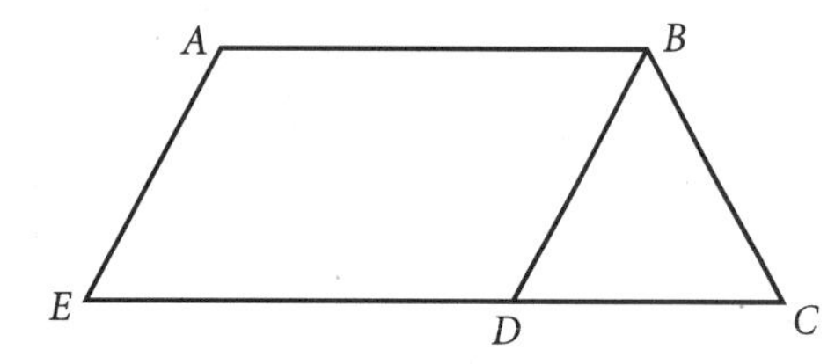

5. If m$\angle BDC$ = 58°, find m$\angle EAB$. ____________

6. If m$\angle DBC = 3x$, m$\angle BCD = 6x$, find m$\angle EAB$.

   ____

7. If m$\angle DBC = 3x$, m$\angle BCD = 6x$, find m$\angle ABD$.

   ____

8. m$\angle DCB = (3x + 10)°$. Express m$\angle AED$ in terms of x. ____________

Complete the proof.

Given: Parallelogram $AECF$;
$\overline{ED} \cong \overline{FB}$

Prove: $\triangle ABF \cong \triangle CDE$

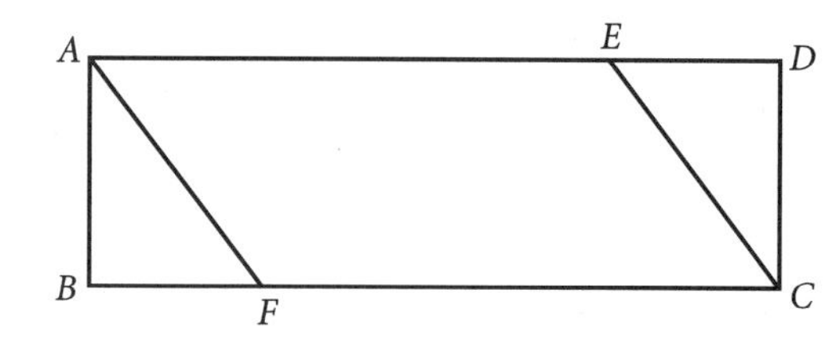

Statements	Reasons
$AECF$ is a parallelogram.	Given
$AF \parallel EC$	**9.**
$\angle FAE \cong \angle CED$	**10.**
$\angle AFB \cong \angle ECF$	**11.**
$\angle FAE \cong \angle ECF$	**12.**
$\angle CED \cong \angle AFB$	**13.**
$\overline{AF} \cong \overline{EC}$	**14.**
$\overline{ED} \cong \overline{FB}$	**15.**
$\triangle ABF \cong \triangle CDE$	

NAME ______________________ CLASS ____________ DATE ____________

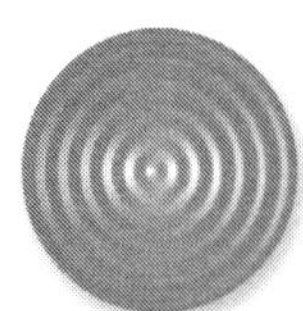

Practice Masters Level C

4.5 Proving Quadrilateral Properties

For Exercises 1–5, use parallelogram *ABDE* where $\overline{BC} \cong \overline{BD}$.

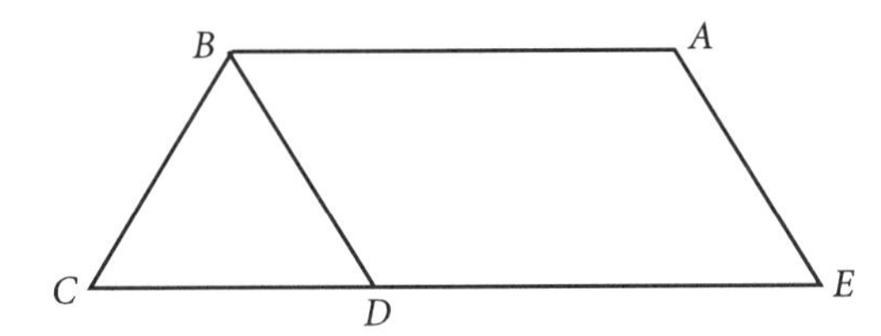

1. If $m\angle BAE = 112°$, find $m\angle BCD$. ________

If $m\angle AED = (5x - 24)°$ and $m\angle BDE = (4x + 15)°$ find:

2. $m\angle EAB$ ________
3. $m\angle ABD$ ________
4. $m\angle BCD$ ________
5. $m\angle DBC$ ________

Decide if the following statements are *always true, sometimes true,* or *never true.*

6. A rectangle is a rhombus. ________
7. A square is a rhombus. ________
8. A rhombus is a rectangle. ________
9. A rhombus is a square. ________
10. A kite is a parallelogram. ________
11. A kite is a rhombus. ________
12. A rhombus is a kite. ________

Quadrilateral *ABCD* is a parallelogram.

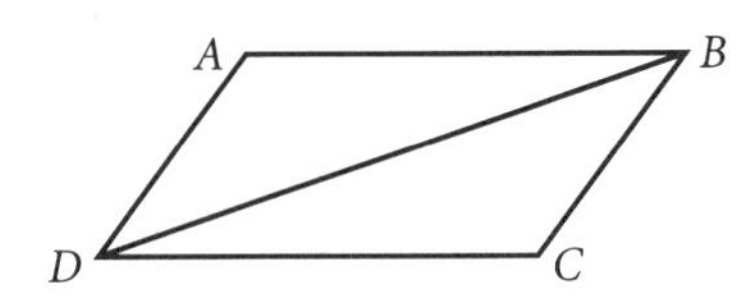

13. If $m\angle CDB = 24°$, $m\angle A = (6x + 9)°$ and $m\angle BDA = 33°$. Find x. ________

14. The perimeter of $ABCD$ is 56. Find the dimensions if $AB = 3x + 7$ and $DA = x - 3$. ________

Complete the following proof.

Given: Parallelogram $ABCD$ with diagonals $\overline{AC}$ and $\overline{BD}$

Prove: $EG \cong FG$

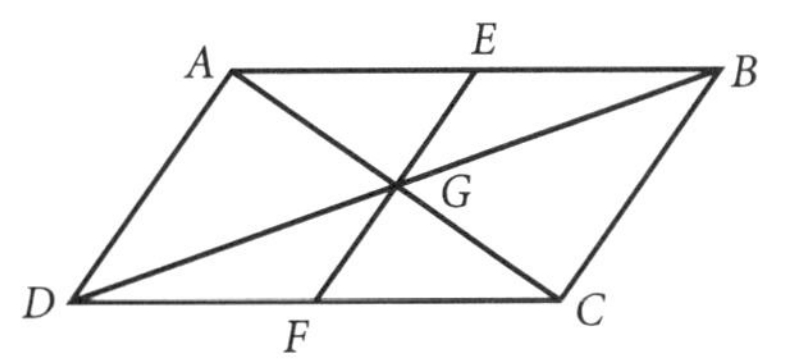

Statements	Reasons
15.	Given
$\angle BAC \cong DCA$	**16.**
$\overline{AG} \cong \overline{CG}$	**17.**
18.	Vertical angles are congruent.
19.	ASA $\cong$ ASA
20.	**21.**

NAME ______________________ CLASS __________ DATE __________

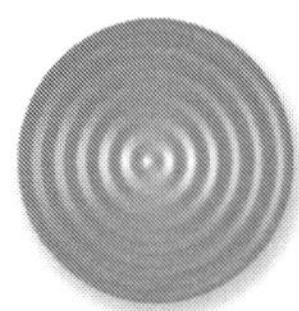

Practice Masters Level A

4.6 Conditions for Special Quadrilaterals

For Exercises 1–3, *ABCD* is a square.

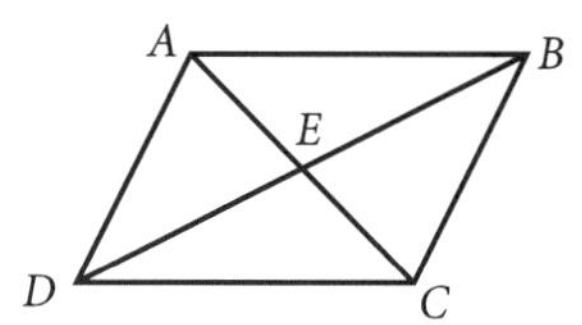

1. List all of the right angles. ______________
2. List all of the 45° angles. ______________
3. List all of the congruent segments. ______________

For Exercises 4–6, *ABCD* is a kite.

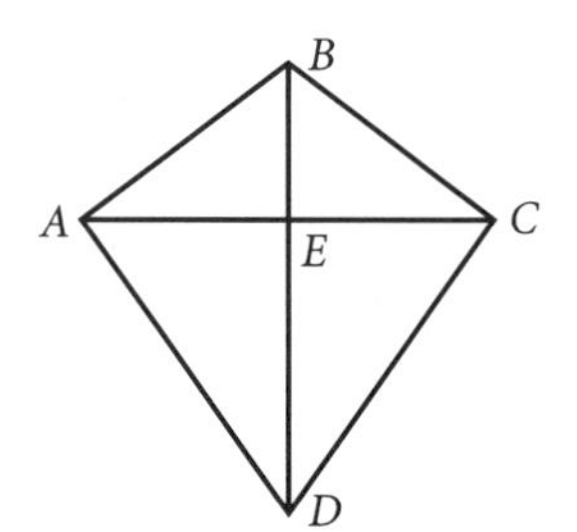

4. List all of the congruent angles. ______________
5. List all of the congruent segments. ______________
6. List all of the right angles. ______________

For Exercises 7 and 8, *ABCD* is a rhombus.

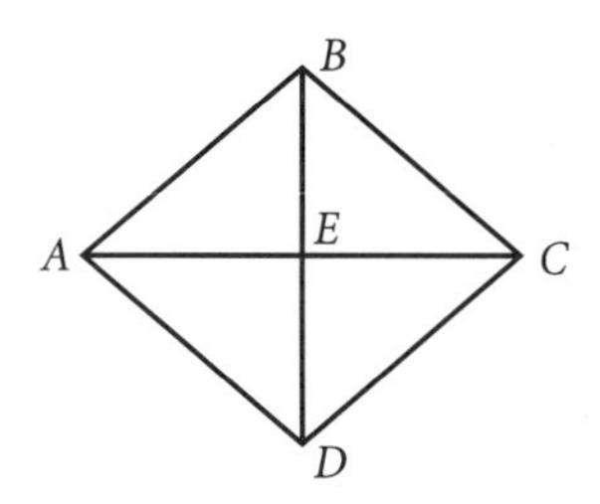

7. List all of the right angles. ______________
8. List all of the congruent segments. ______________

Quadrilateral *ABCD* has diagonals $\overline{AC}$ and $\overline{DB}$. For the conditions given below, state whether the quadrilateral is a rhombus, rectangle, parallelogram, or neither. Give the theorem or postulate that justifies the conclusion.

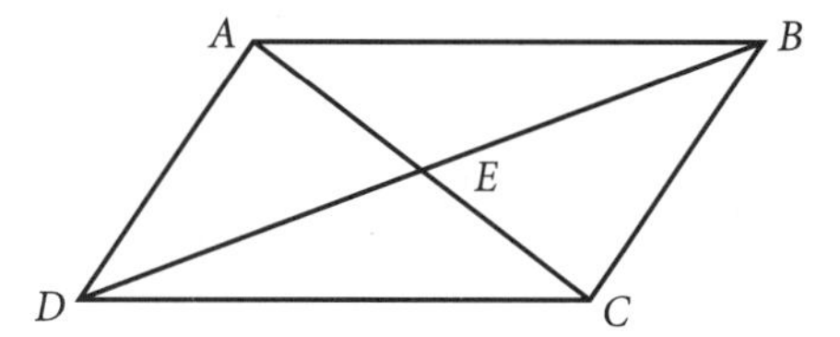

9. $\angle BAC \cong \angle DCA, \angle DAC \cong BCA$ ______________

10. $\angle AEB \cong \angle CED, \angle BEC \cong DEA$ ______________

11. $\overline{AC} \cong \overline{BD}, \overline{AC} \perp \overline{BD}$ ______________

12. $\overline{AB} \cong \overline{CD}, \overline{AD} \cong \overline{BC}$ ______________

NAME ______________________ CLASS ______________ DATE ______________

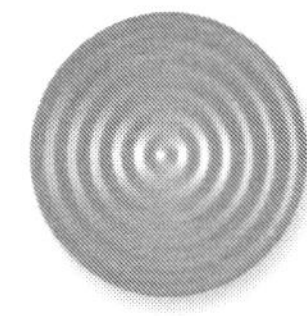

Practice Masters Level B

4.6 Conditions for Special Quadrilaterals

Quadrilateral *ABCD* has diagonals $\overline{AC}$ and $\overline{BD}$ intersecting at *E*. For the conditions given below, state whether the quadrilateral is a rhombus, rectangle, parallelogram, or neither. Then give the theorem or postulate that justifies the conclusion.

1. $\overline{BC} \cong \overline{AD}, \overline{BE} \cong \overline{ED}$ ______________________
2. E is the midpoint of $\overline{BD}$ and $\overline{AC}$. ______________________
3. $\triangle ABC \cong \triangle DCB$ ______________________
4. $\angle ABD \cong \angle CDB, \overline{AD} \parallel \overline{BC}$ ______________________
5. $\overline{CB} \perp \overline{BA}$; $\angle ABC$ is supplementary to $\angle BCD$; $\overline{AB} \cong \overline{DC}$ ______________________
6. $\triangle BEC \cong \triangle DEA$ ______________________
7. $\overline{AB} \cong \overline{BC}, \overline{AD} \cong \overline{DC}$ ______________________

For Exercises 8–10, refer to the diagram of parallelogram *ABCD*. State whether each set of conditions below is sufficient to prove that *ABCD* is a square. Explain your reasoning.

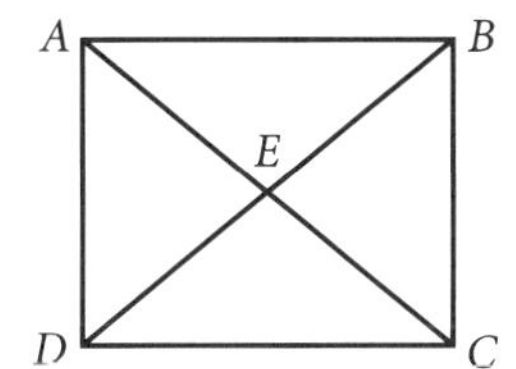

8. $\overline{AD} \perp \overline{DC}; \overline{AD} \cong \overline{BC}$ ______________________
9. $\overline{AD} \perp \overline{DC}; \overline{AD} \cong \overline{DC}$ ______________________
10. $\overline{AC} \perp \overline{BD}$ ______________________

Complete the following proof.

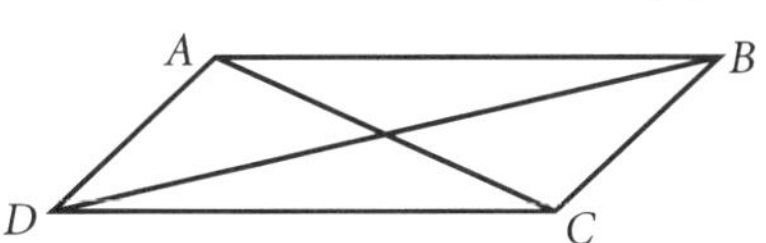

Given: *ABCD* is a rhombus;
$\triangle ABC \cong \triangle DCB$

Prove: *ABCD* is a rectangle.

Statements	Reasons
Rhombus *ABCD*	11.
$\overline{AG} \parallel \overline{CG}$	12.
13.	Given
14.	15.
$\angle ABC$ is supplementary to $\angle DCB$.	16.
17.	Definition of right angle
ABCD is a rectangle.	18.

NAME ______________________ CLASS ______________ DATE ______________

Practice Masters Level C

4.6 Conditions for Special Quadrilaterals

Quadrilateral *ABCD* has diagonals $\overline{AC}$ and $\overline{BD}$ intersecting at *E*. For the conditions given below, state whether the quadrilateral is a rhombus, rectangle, parallelogram, or neither. Then give the theorem or postulate that justifies the conclusion.

1. $\overline{BD} \perp \overline{AC}$, $\overline{AB} \cong \overline{BC}$ ______________________
2. $\angle ABD \cong \angle CBD$, $\angle ADB \cong \angle CDB$ ______________________
3. $\angle DAC \cong \angle BCA$, $\overline{AB} \cong \overline{DC}$ ______________________
4. $\triangle ABC \cong \triangle ADC$, $\triangle ABD \cong \triangle CDB$ ______________________
5. $\triangle ABD$ and $\triangle CDB$ are isosceles with vertex angles at *A* and *C*, respectively. ______________

***ABCD* is a parallelogram. $AB = 5x$, $BC = 3x - 2.2$, $BE = 4y - 5.5$, $ED = \frac{1}{2}y + 5$, and $AC = 5y - 2$.**

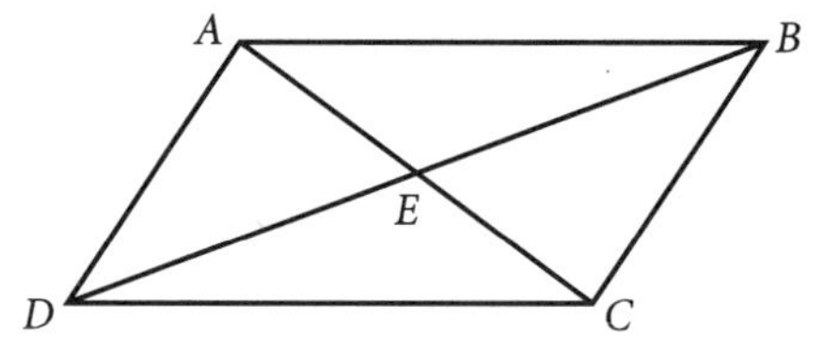

6. Find the perimeter of $\triangle DEC$. ______________
7. Find the area of *ABCD*. ______________
8. What type of special quadrilateral is *ABCD*? ______________

In $\triangle ABC$, points *D*, *E*, and *F* are midpoints of $\overline{AB}$, $\overline{BC}$, and $\overline{AC}$, respectively; $DE = 2x + 7$, $AB = 6x - 1$, $EC = 4x - 8$, and $\angle A \cong \angle C$. Find the following:

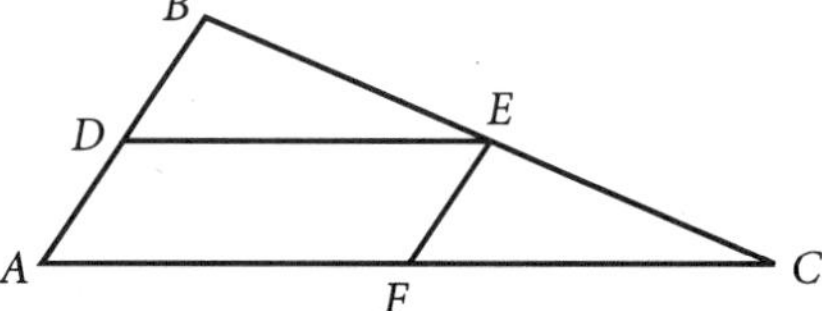

9. the perimeter of *DEFA* ______________
10. the perimeter of *AFEB* ______________

In $\triangle ABC$, $\overline{DE} \parallel \overline{CA}$ and points *D* and *F* are midpoints of $\overline{BC}$ and $\overline{AC}$, respectively.

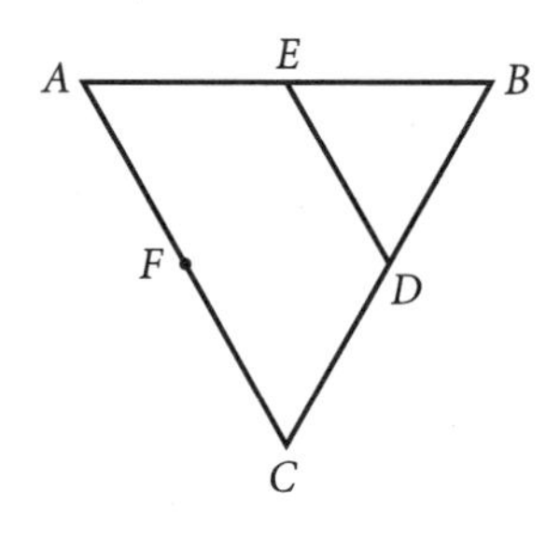

11. If $DE = 13.5$, find the length of $\overline{CA}$. ______________
12. If $\triangle ABC$ is isosceles with $\angle A \cong \angle B$, $AB = 113\frac{2}{5}$, and $DE = 43\frac{1}{2}$, find the perimeter of $\triangle ABC$. ______________
13. If $\triangle ABC$ is equilateral, what kind of quadrilateral is *AEDF*? Justify your answer. ______________

Practice Masters Level A

4.7 Compass and Straightedge Constructions

Given lengths *a* and *b*, use a compass and straightedge to construct the following constructions in the space provided.

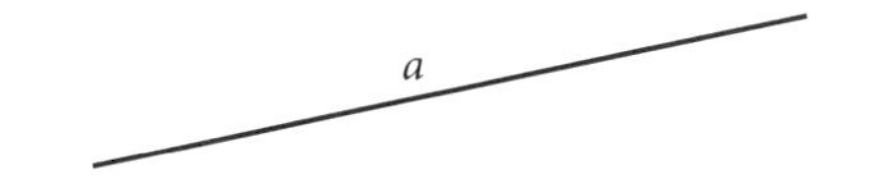

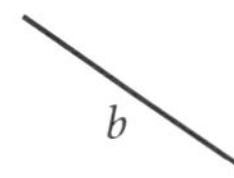

1. Construct a.

2. Construct b.

3. Construct $b + a$.

4. Construct $a - b$.

Given angles *c* and *d*, use a compass and straightedge to complete the constructions.

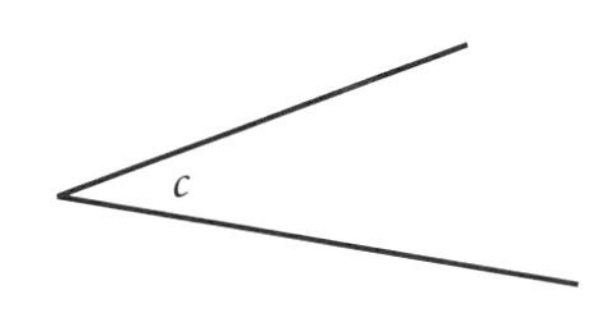

d

5. Construct $\angle c$.

6. Construct $\angle d$.

7. Construct $\angle d - \angle c$.

8. Construct $\frac{1}{2}\text{m}\angle d$.

9. Construct an equilateral triangle with side b.

10. Construct the angle bisector of $\angle d$.

NAME ______________________ CLASS __________ DATE __________

Practice Masters Level B

4.7 Compass and Straightedge Constructions

Given lengths *a* and *b* and angles *c* and *d*, use a compass and straightedge to construct the following constructions in the space provided.

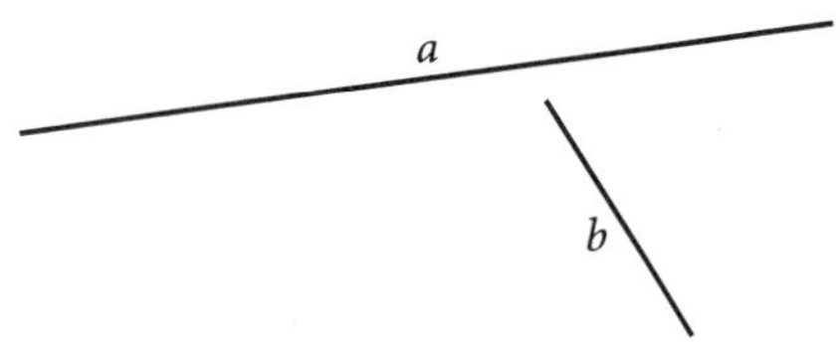

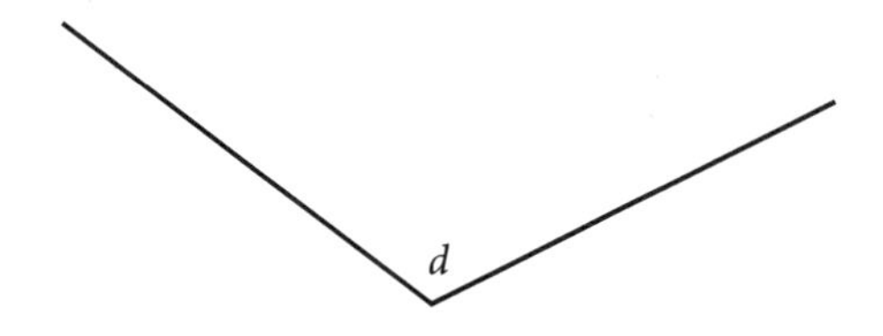

1. Construct an isosceles triangle with base a and altitude b.
2. Construct a square with sides b.
3. Construct a right triangle with leg b and acute angle c.
4. Construct parallel lines that are b distance apart.
5. Construct a rectangle with one side length b and diagonal of length a.
6. Construct a rhombus with one angle d and sides of length b.
7. Construct a kite with one pair of sides of length a and the other pair of sides of length c.
8. Construct an obtuse triangle using a and b for two sides, and angles c and d for two angles.

NAME ______________________ CLASS ____________ DATE ____________

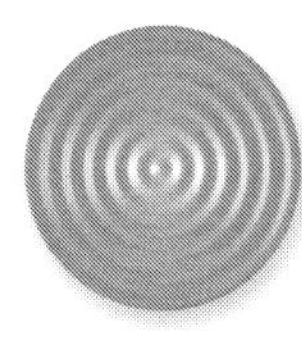

Practice Masters Level C

4.7 Compass and Straightedge Constructions

Given lengths *a*, *b*, and *c*, and angles *d*, *e*, and *f*, use a compass and straightedge to construct the following constructions in the space provided.

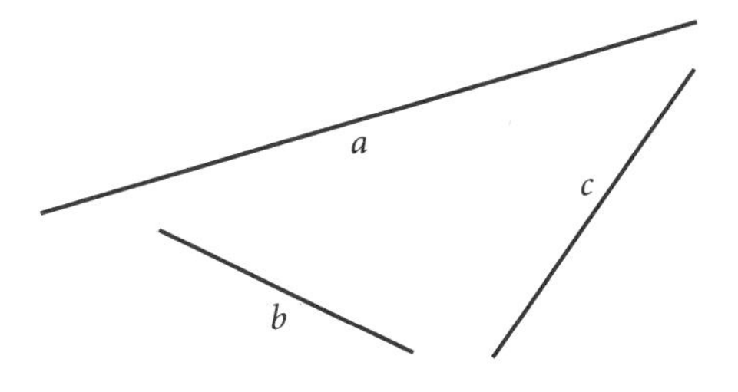

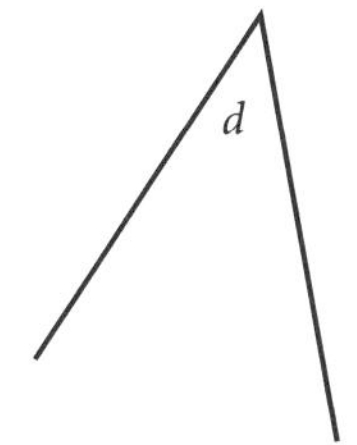

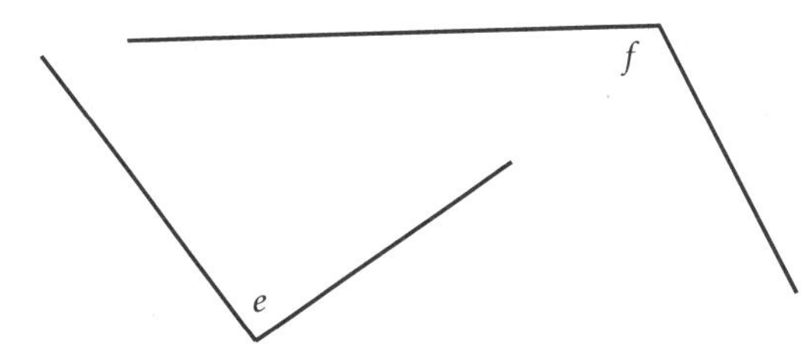

1. Construct a right triangle with hypotenuse *a* and one leg *c*.

2. Construct an isosceles triangle with base of length *a* and base angles that measure *c*.

3. Construct a parallelogram with sides of length *a* and *b*, and one pair of angles that measure *d*.

4. Construct parallel lines so that alternate interior angles measure *c*.

5. Construct a right triangle with a smaller leg of length *a*. Construct the circumscribed circle.

6. Construct an obtuse triangle with one angle of measure *d* and a smallest side of length *a*. Construct the inscribed circle.

NAME ______________________ CLASS __________ DATE __________

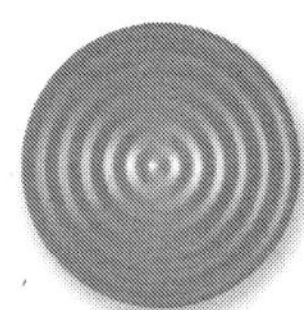

Practice Masters Level A

4.8 *Constructing Transformations*

Translate each figure below by the direction and distance of the given translation vector.

1. 

2. 

3. 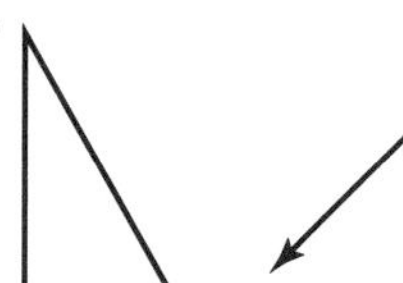

Reflect the figure over the given line.

4. 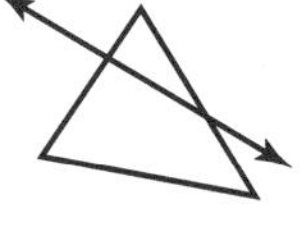

5. 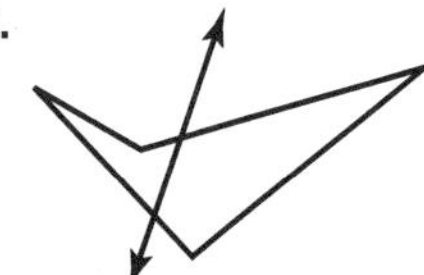

6. 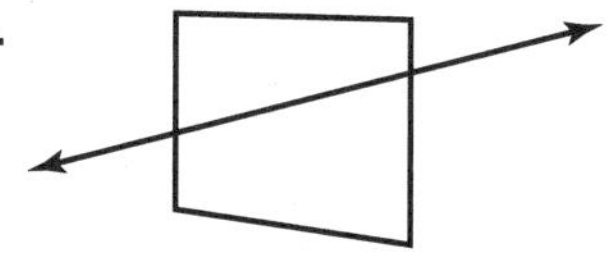

State whether each triangle described below is possible. Explain the reason for your answer.

7. $AB = 11, BC = 8, CA = 6$ ______________________

8. $GH = 21, HJ = 5, GJ = 16$ ______________________

9. $KL = 9, LM = 4, MK = 6$ ______________________

10. $WX = 11, XY = 4, WY = 7$ ______________________

11. $MN = 24, MO = 12, NO = 10$ ______________________

12. $FG = 3, GH = 4, FH = 5$ ______________________

13. $JK = 31, JL = 17, KL = 4$ ______________________

NAME ______________________ CLASS ____________ DATE ____________

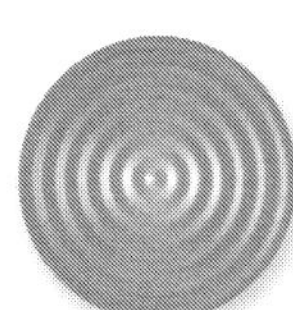

Practice Masters Level B

4.8 Constructing Transformations

Reflect each figure below across the given line.

1.

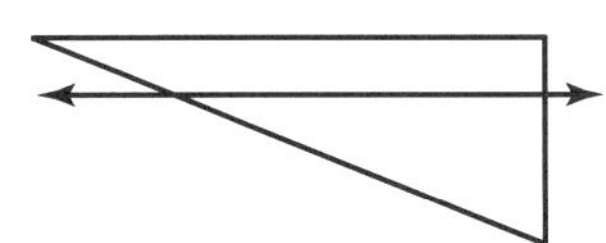

2.

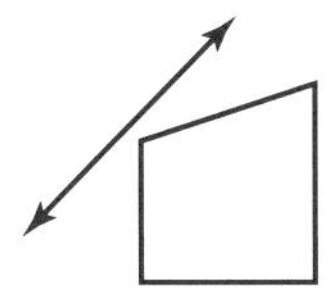

3.

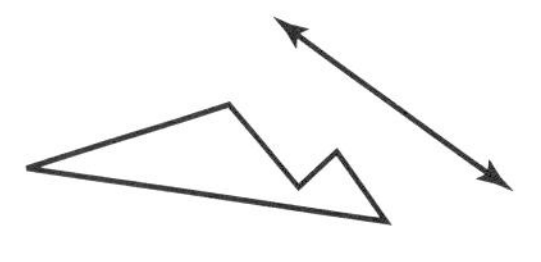

Rotate the segment about the given point by the angle below it.

4.

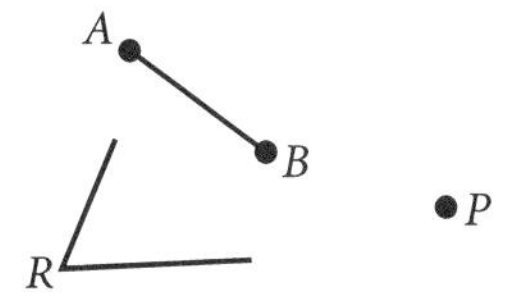

5.

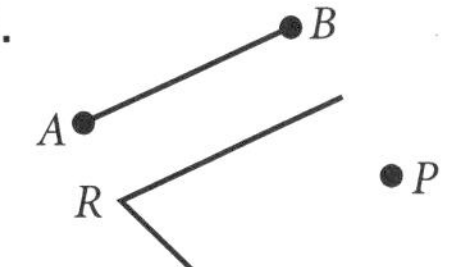

6.

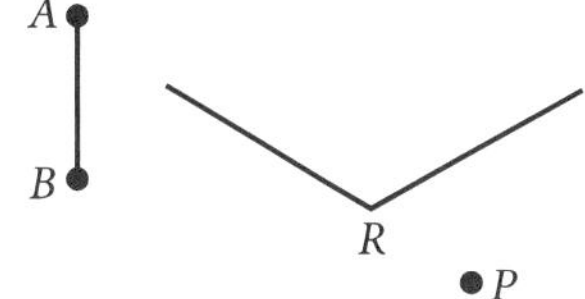

For Exercises 7–12, state the two values the length of the third side of the triangle must be between, to make the triangle possible.

7. $AB = 5, BC = 9, AC =$ ______________________

8. $WX = 132, XY = 417, WY =$ ______________________

9. $DE = 11.4, EF = 4.2, DF =$ ______________________

10. $GH = 3\frac{3}{4}, HJ = 9\frac{3}{8}, GJ =$ ______________________

11. $LM = \sqrt{27}, NL = 5\sqrt{3}, MN =$ ______________________

12. $PQ = 0.03, RP = 0.11, PR =$ ______________________

13. In $\triangle PQR$ shown, what values are possible for PQ?

____

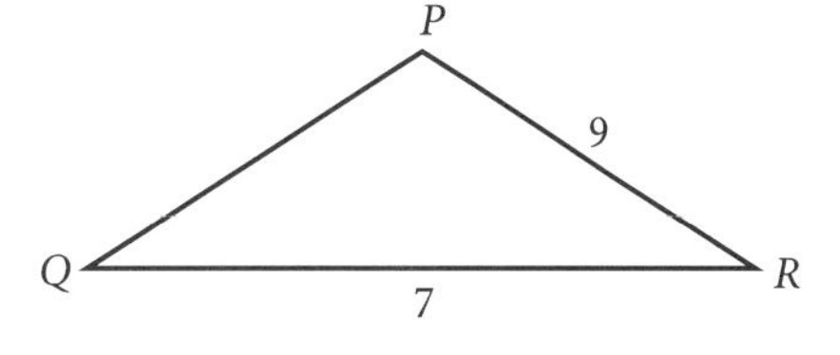

14. Write a paragraph proof.

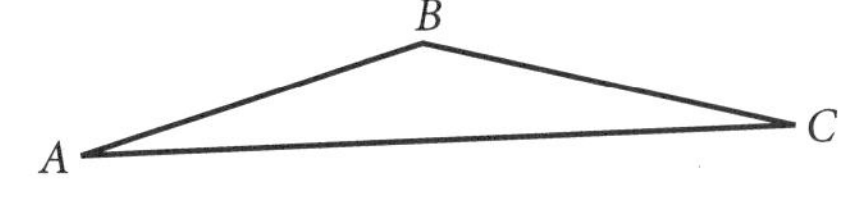

Given: $\triangle ABC$, with $AB + BC = AC$
Prove: ABC is not a triangle.

____

____

____

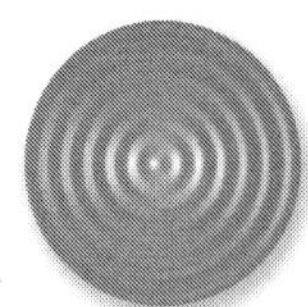

Practice Masters Level C

4.8 Constructing Transformations

Translate the figure, the direction and distance of the given translation vector. Then rotate the figure 180° around *P*.

1.

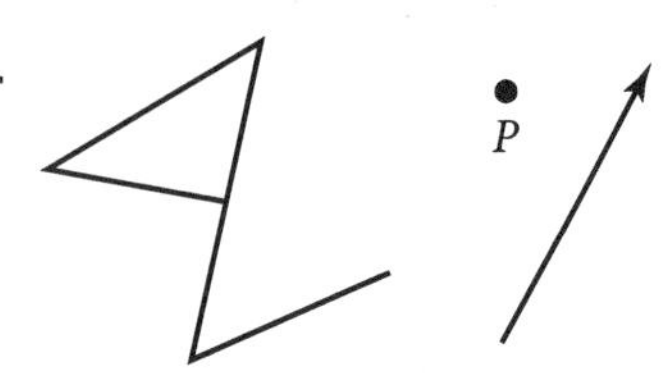

2.

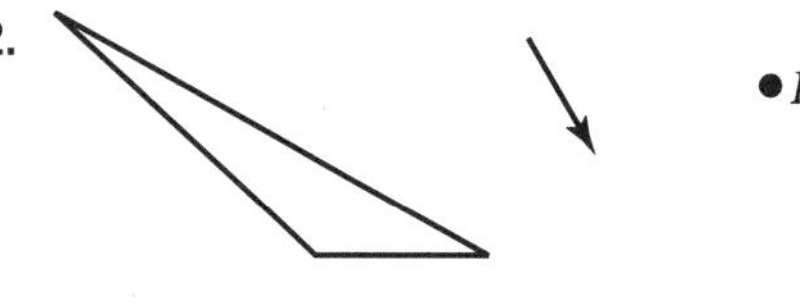

Rotate the figure 180° around *P*. Then translate the figure, the direction and distance of the given translation vector.

3.

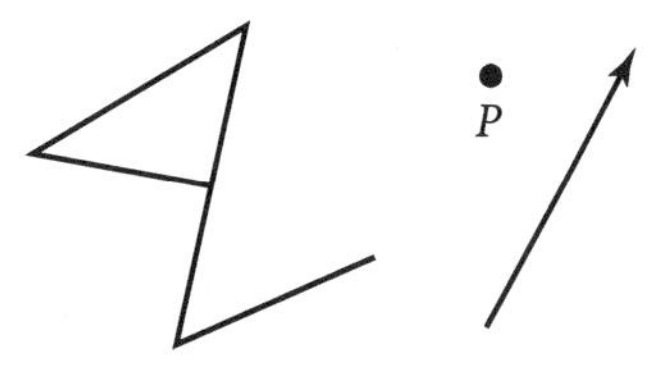

4.

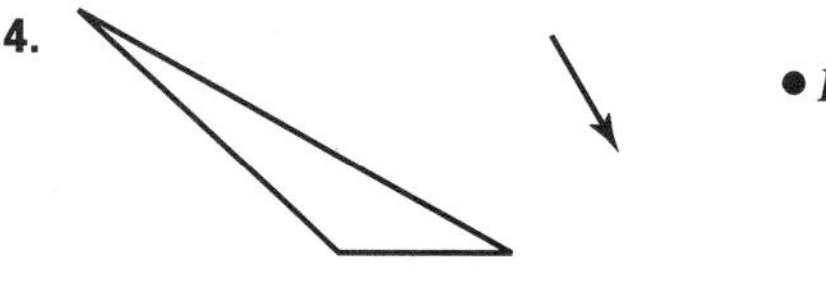

5. Based on Exercises 1–4, are translations and rotations commutative? ____________

For each triangle below, determine the largest and smallest possible values for *x*.

6. $AB = 7, BC = x, AC = 11$ ____________
7. $DE = 2x, DF = 3, EF = 5$ ____________
8. $GH = 17, HJ = 2x, GJ = x$ ____________
9. $KL = 2x, LM = 8x, KM = 12$ ____________

In each figure below, decide between what two values *AB* must fall.

10. $AB =$ ____________

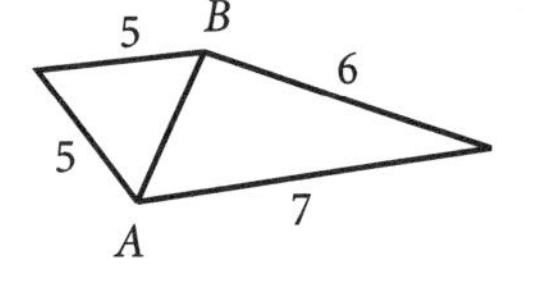

11. $AB =$ ____________

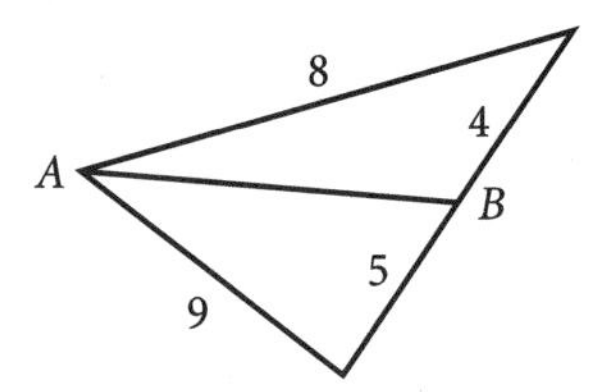

12. $AB =$ ____________

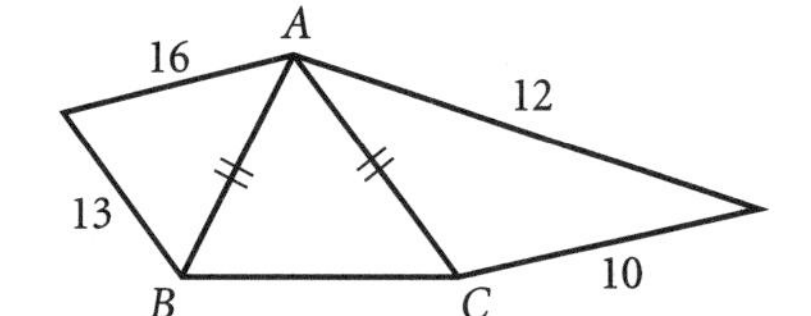

13. $AB =$ ____________

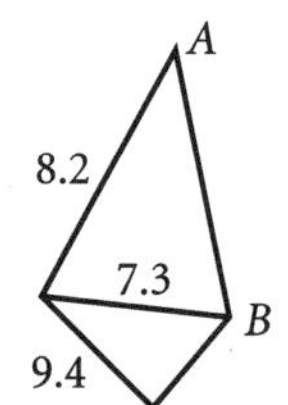

Practice Masters Level A

5.1 Perimeter and Area

For each figure below, determine the measure of the perimeter and area.

1. Perimeter __________ Area __________

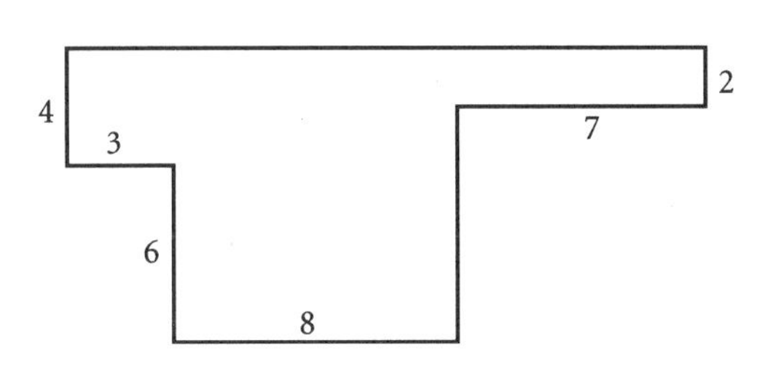

2. Perimeter __________ Area __________

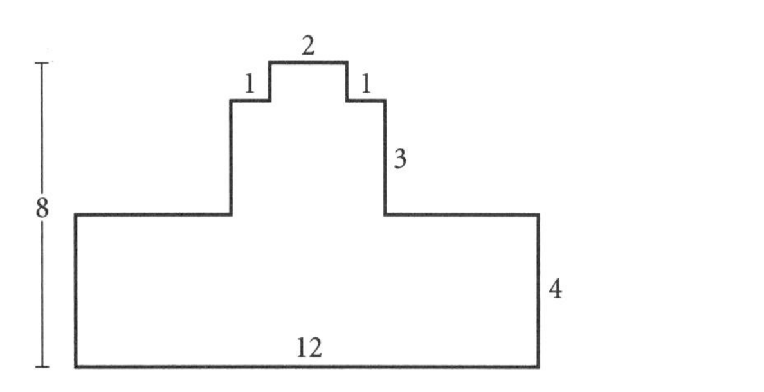

3. Perimeter __________ Area __________

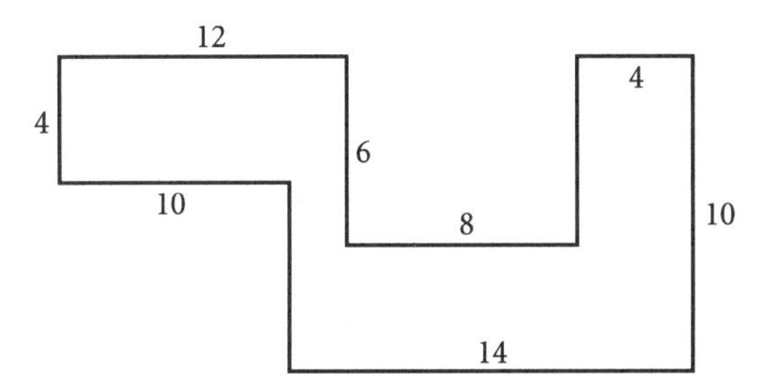

4. Perimeter __________ Area __________

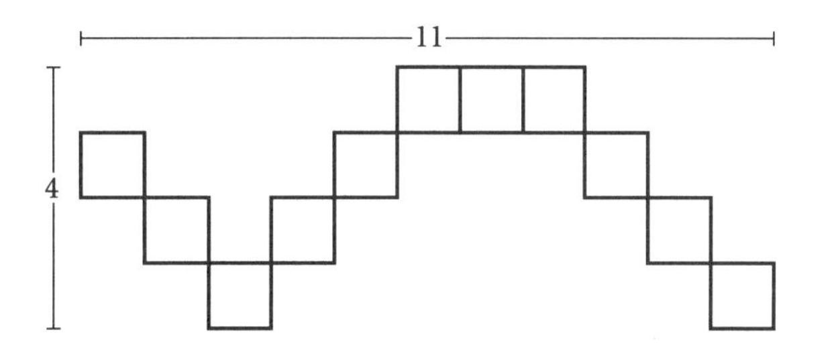

For Exercises 5–8, use the figure to find the indicated perimeters and areas. *ADKQ* measures 20-inch-by-20-inch.

5. the perimeter of *ADKQ* __________
6. the area of *ADKQ* __________
7. the perimeter of *NGJRLM* __________
8. the area of *NGJRLM* __________

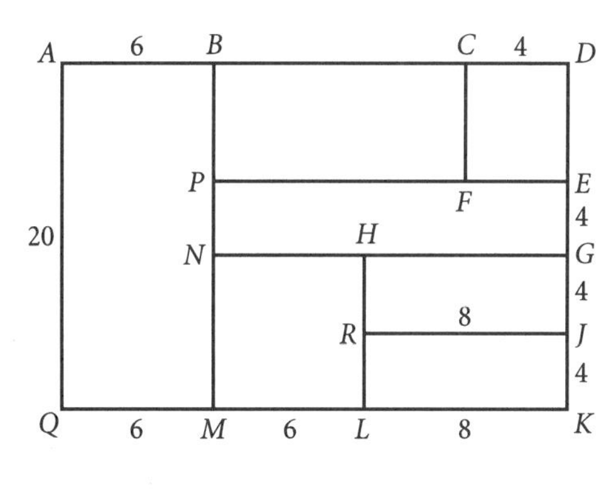

The perimeter of a rectangle is 12 meters. If the side lengths are given by integers, find all possible dimensions, and determine the area of each.

	Length	Width	Perimeter	Area
9.	__________	__________	12 meters	__________
10.	__________	__________	12 meters	__________
11.	__________	__________	12 meters	__________

NAME ______________________ CLASS __________ DATE __________

Practice Masters Level B

5.1 Perimeter and Area

For each figure below, determine the measure of the perimeter and area.

1. Perimeter = ________ Area = ________

2. Perimeter = ________ Area = ________

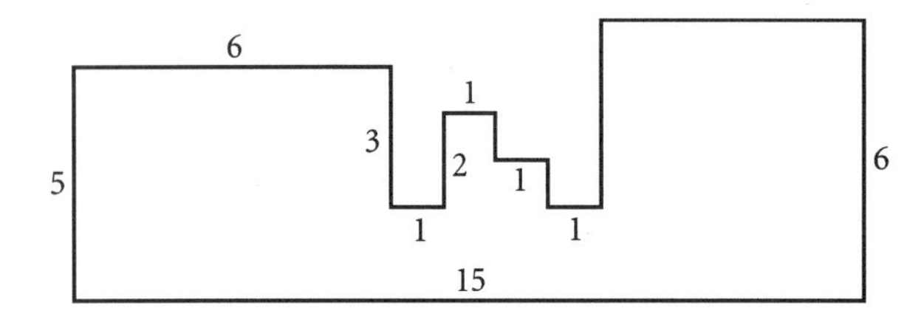

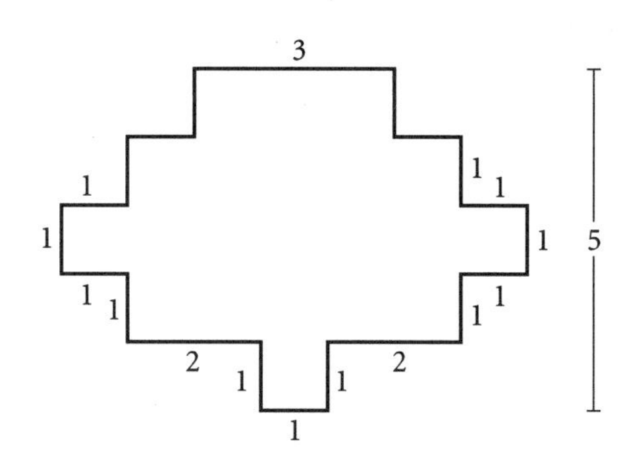

For Exercises 3–10, use the figure and measures to find the indicated perimeters and areas. In *ACDM, AB* = 27, *BC* = *CD* = 18, *AP* = *PN* = 6, *MH* = 9, *TR* = 3 and *HT* = 15.

3. the perimeter of *ACDM* ________
4. the area of *ACDM* ________
5. the perimeter of *PABETV* ________
6. the area of *PABETV* ________
7. the area of *GBAMTRJL* ________
8. the perimeter of *GBAMTRJL* ________
9. Draw $\triangle BPF$. What is its area? ________
10. Draw pentagon *BNREC*. What is its area? ________

Complete the table to determine the possible perimeters that produce a rectangle with an area of 24 square units. The side lengths are all given by integers.

	Length	Width	Area	Perimeter
11.	________	________	24 sq. units	________
12.	________	________	24 sq. units	________
13.	________	________	24 sq. units	________
14.	________	________	24 sq. units	________

NAME ______________________ CLASS ____________ DATE ____________

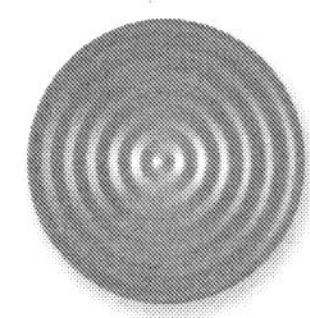

Practice Masters Level C

5.1 Perimeter and Area

For Exercises 1–12, use the figure and measures to find the indicated perimeters and areas. In *ADERT*, *AB* = *RL* = *FE* = 8, *BK* = *PN* = *HG* = 4, *BD* = 16, *DE* = 14, and *HD* = *AT* = 11. Find the indicated measure.

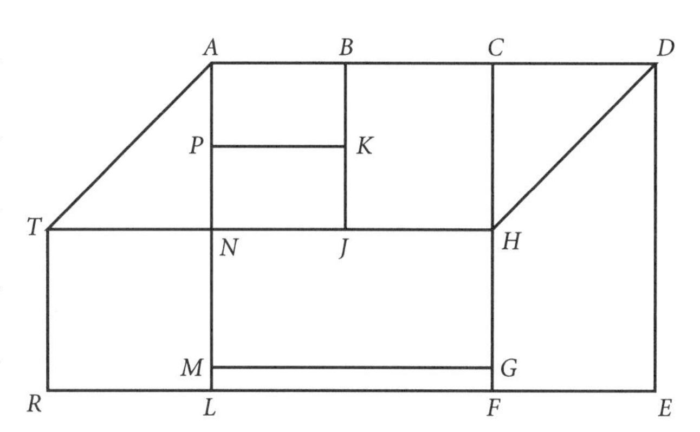

1. the perimeter of *ABJHFL* ____________
2. the area of *ABJHFL* ____________
3. the perimeter of *ERTHD* ____________
4. the area of *ERTHD* ____________
5. the perimeter of *JTREDHCB* ____________
6. the area of *JTREDHCB* ____________
7. the area of quadrilateral *RTAL* ____________
8. Find two figures that each have an area of 32 square units. ____________ and ____________
9. Find a figure whose perimeter is 48 units. ____________
10. Find a figure whose area is 48 square units. ____________
11. Find a figure whose perimeter is 71 units. ____________
12. Find a figure whose area is 80 square units. ____________

Complete the table to determine the possible perimeters that would produce a rectangle with an area of 48 square feet. The side lengths are integers.

	Length	Width	Perimeter	Area
13.	____	____	____	48 square feet
14.	____	____	____	48 square feet
15.	____	____	____	48 square feet
16.	____	____	____	48 square feet
17.	____	____	____	48 square feet

NAME ______________________ CLASS ____________ DATE ______________

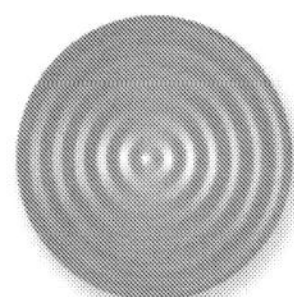

Practice Masters Level A

5.2 Areas of Triangles, Parallelograms, and Trapezoids

Find the area of each figure. Give the formula that you used to find the area.

1.

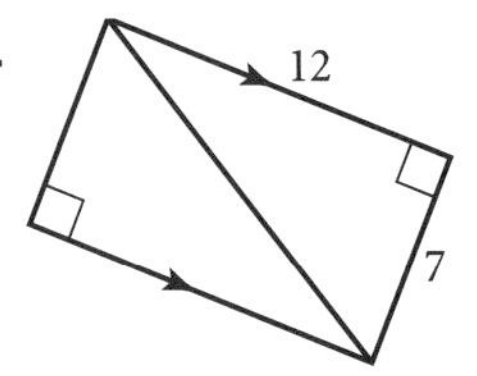

Formula ______________

Area ______________

2.

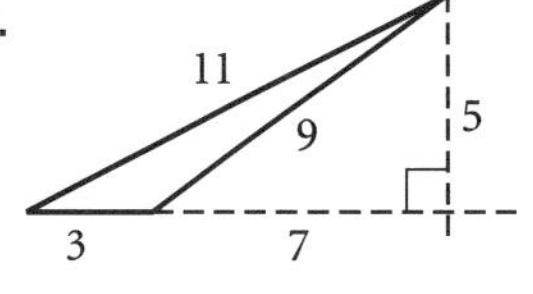

Formula ______________

Area ______________

3.

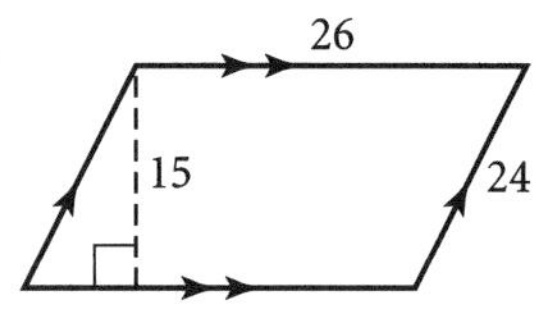

Formula ______________

Area ______________

4.

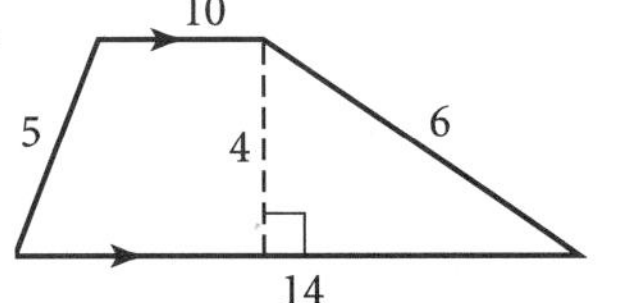

Formula ______________

Area ______________

5.

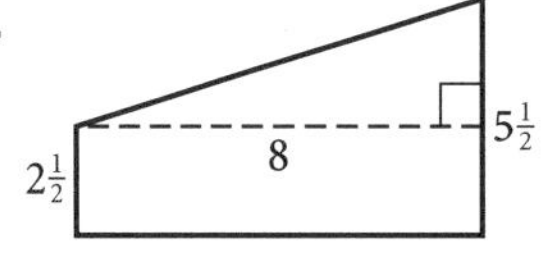

Formula ______________

Area ______________

6.

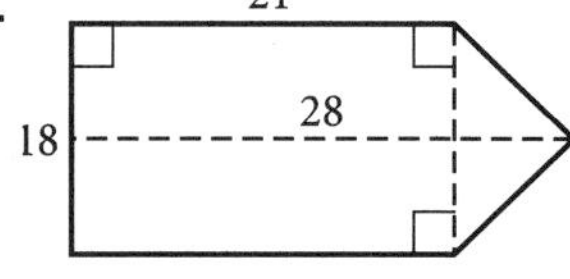

Formula ______________

Area ______________

7. ΔABC is a right triangle with one leg 12, one leg 16 and a hypotenuse of 20. Find the area of the triangle. ______________

8. Parallelogram *FGHJ* has sides of 7.3 and 4.6. If its height to the shorter side is 3, what is the area? ______________

9. A trapezoid has bases of 15 and 8, legs of 7 and 9, and a height of 6. What is the area of the trapezoid? ______________

10. In ΔWXY, the base is 11 and the area is 55. What is its height? ______________

11. The area of a parallelogram is 15. If the height is 2.5, what is the length of one of the sides? ______________

Find the indicated measures in trapezoid *ABDC*.

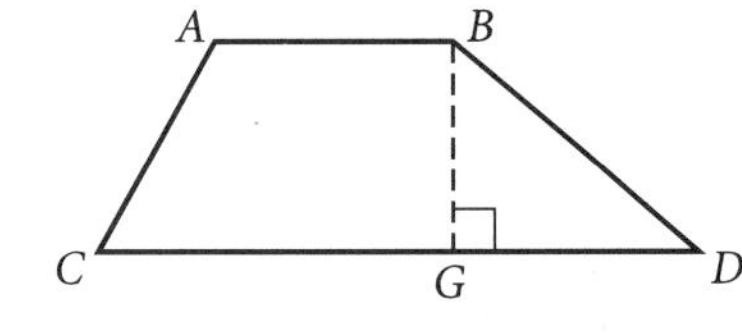

12. $AB = 17$
$BG = 10$
Area $ABDC = 200$ units^2
Find CD. ______________

13. $AB = 3.3$
$CD = 8.1$
Area $ABCD = 42.75$ units^2
Find BG. ______________

NAME ______________________ CLASS ______________ DATE ______________

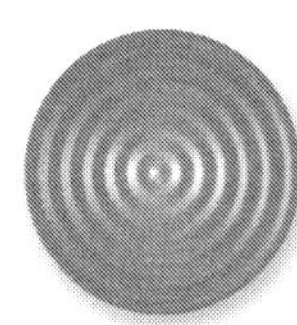

Practice Masters Level B

5.2 Areas of Triangles, Parallelograms, and Trapezoids

Find the area of each figure. Give the formula that you used to find the area.

1.

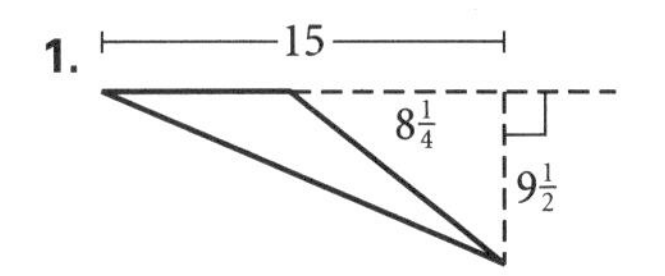

Formula ______________

Area ______________

2.

Formula ______________

Area ______________

3.

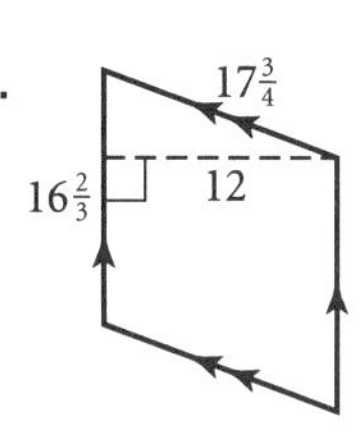

Formula ______________

Area ______________

4. The area of $\triangle GHJ$ is 24 m^2. If the height is 8 m, what is the length of the base? ______________

5. The area of a parallelogram is 12 m^2. If one side measures $3\frac{1}{2}$ m, what is the height? ______________

6. Trapezoid *ABCD* has an area of 36 m^2, a smaller base of 4 m and a height of 6 m. How long is the other base? ______________

In *ABCD, BE* = *BC, FB* = 11, *AD* = 12, *FD* = 10, *EC* = $5\frac{1}{2}$, and *DC* = $13\frac{1}{2}$. Find the area of each figure.

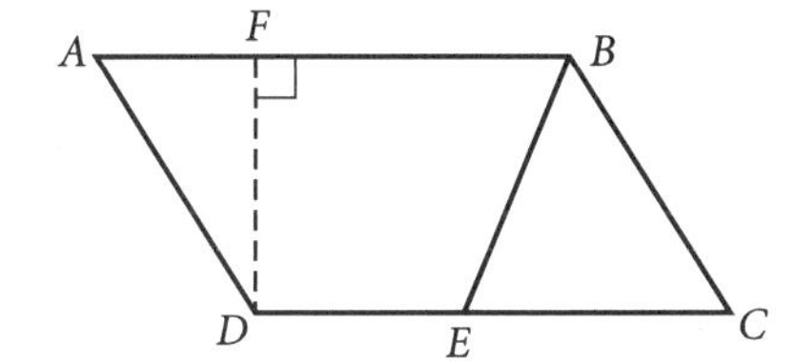

7. *ABCD* ______________

8. $\triangle FDA$ ______________

9. *FBED* ______________

10. $\triangle BEC$ ______________

11. *FBCD* ______________

12. *ABED* ______________

13. In $\triangle ABC$ with altitude *BD*, the area is 91 ft^2, the altitude is $3x + 7$ ft, and the base measures 14 ft. Find the length of the altitude. ______________

In quadrilateral *ABED*, $AB \parallel DC$, *AB* = 24, *BE* = 30, and *BC* = 18. Find the measures and areas for each problem below.

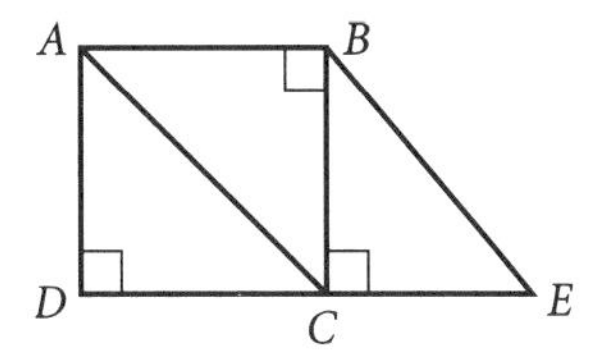

14. area of $\triangle ADC$ ______________

15. area of $\triangle BCE$ ______________

16. area of *ABCD* ______________

17. area of *ABED* ______________

18. area of *ABEC* ______________

NAME ______________________ CLASS ______________ DATE ______________

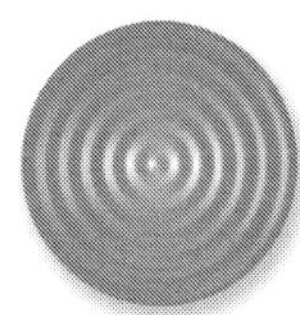

Practice Masters Level C

5.2 Areas of Triangles, Parallelograms, and Trapezoids

In figure *ABCD*, the ratio of *AB*:*BC*:*CD*:*DA* = 20:15:9:16. The perimeter of the figure is 180 units.

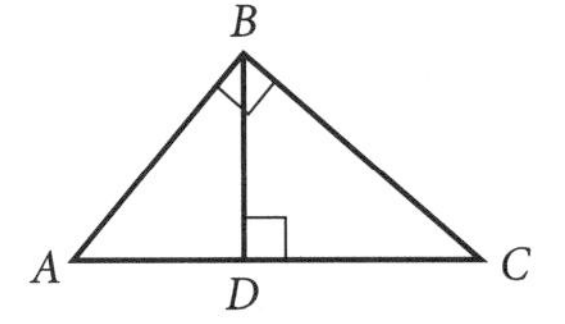

1. Find *BD*. ______________
2. Find the area of $\triangle ABC$. ______________
3. Find the area of $\triangle BDC$. ______________

In trapezoid *GHJK*, $KJ = x^2 + 2$, $GH = 3x + 1$, $GK = 6$, and the area of *GHJK* = 93 square meters.

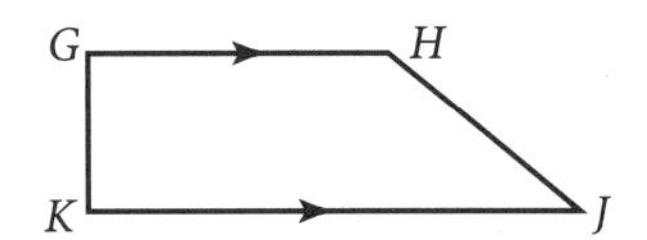

4. Find *GH*. ______________
5. Find *KJ*. ______________

In *ABCD*, $AB = 4x - 1$, $BC = 2x + 3$, $AE = 7$, and the perimeter of *ABCD* is 40 units.

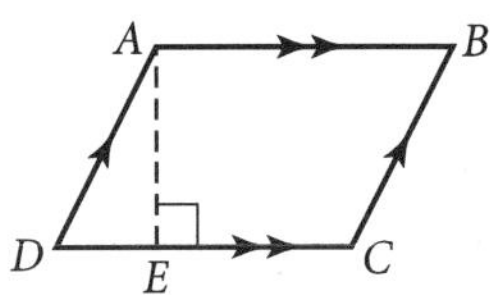

6. Find the dimensions of *ABCD*. ______________
7. Find the area of *ABCD*. ______________

In trapezoid *WXYZ*, altitude $WV = x$, base $WX = 2x + 1$, base $ZY = 6x - 5$, and the area of *WXYZ* equals 30 square units.

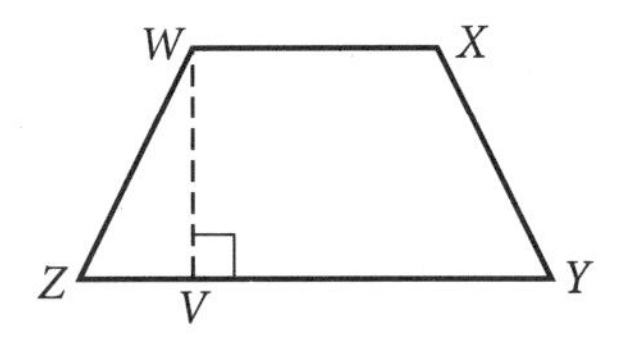

8. Find *WX*. ______________
9. Find *ZY*. ______________
10. Find altitude *WV*. ______________

The area of $\triangle ADE$ = 48 square units, $AD = 10$, $DE = 12$, and the ratio of $DE:BC = 1:3$.

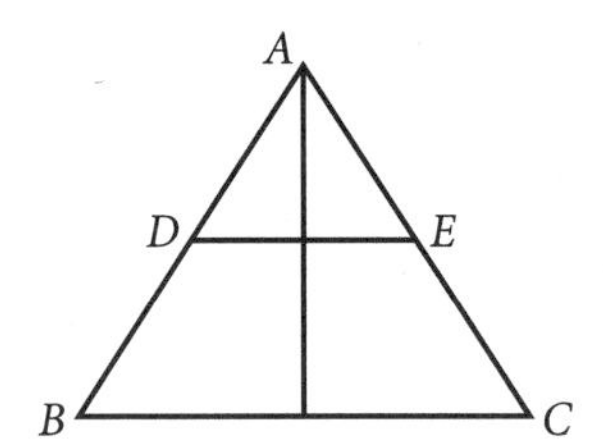

11. Find the altitude of $\triangle ABC$. ______________
12. Find the area of $\triangle ABC$. ______________
13. Find the area of quadrilateral *DECB*. ______________

NAME ______________________ CLASS ____________ DATE ____________

Practice Masters Level A

5.3 *Circumferences and Areas of Circles*

Find the radius of the circle with the given measurements. Give your answers exactly, in terms of π, and rounded to the nearest tenth.

1. $C = 34$ ______________

2. $A = 25\pi$ ______________

3. $C = 6\frac{3}{4}$ ______________

4. $A = 23.78$ ______________

5. $A = 157$ ______________

6. $A = 314$ ______________

7. What happens to the radius of a circle whose area is halved? See Exercises 5 and 6 to help you with your answer. ______________

In Exercises 8–10, find the circumference and area of each circle. Use 3.14 for π.

8. $r = 7$ ______________

9. $r = 12$ ______________

10. $r = 6$ ______________

11. What happens to the area of a circle whose radius is halved? Compare Exercises 9 and 10 to help you with your answer. ______________

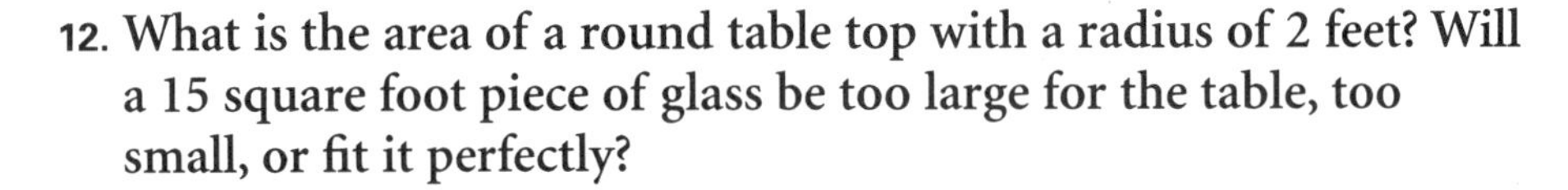

12. What is the area of a round table top with a radius of 2 feet? Will a 15 square foot piece of glass be too large for the table, too small, or fit it perfectly? ______________

For Exercises 13–15, find the exact area of the shaded region.

13. ______________

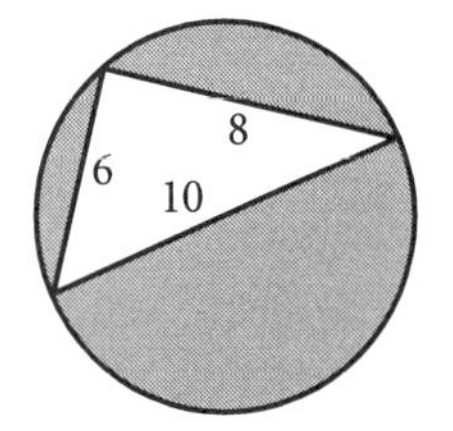

14. ______________

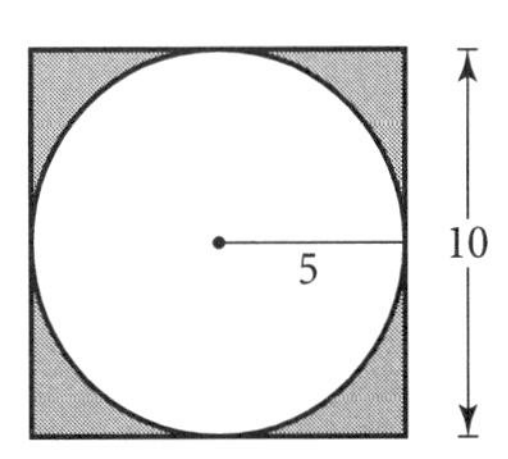

15. ______________

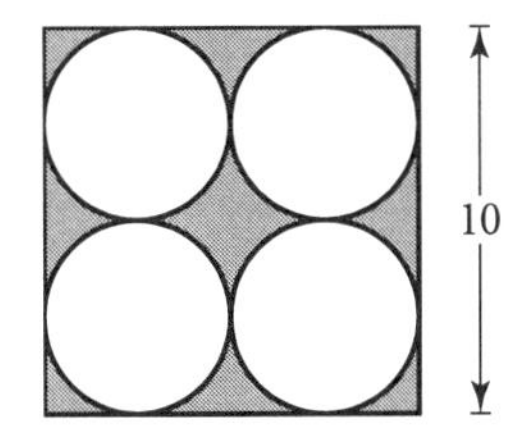

16. Which shaded area is greater, the area inscribed in the square from Exercise 14 or the shaded area of the four circles from Exercise 15?

____

Practice Masters Level B

5.3 Circumferences and Areas of Circles

1. Find the area of the circle with a circumference of 48 units. __________

2. Joe has 225 feet of fencing. He plans to enclose part of his yard for his dog. What dimensions and shape should he use to give his dog the greatest area in which to run? Complete the following chart to help you with your answer. __________

Shape	Dimensions	Area
Triangle		
Rectangle		
Square		
Circle		

3. Parallelogram *ABCD* has an area of 30 square meters. Altitude $BG = 4$ meters. Find *BC*.

____

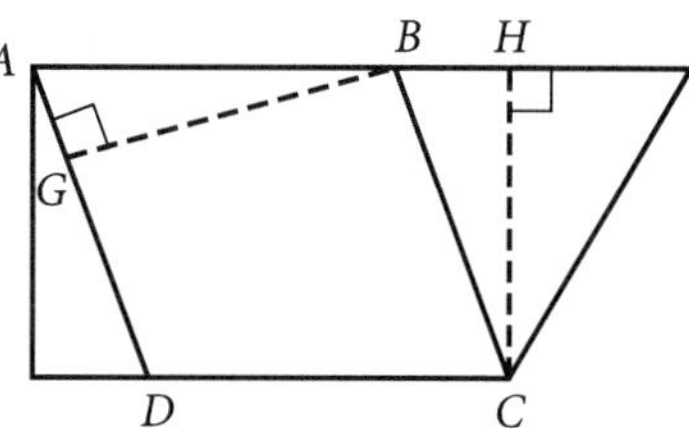

4. George used 450 feet of fencing to enclose a circular pool. He wants to leave a 3-foot walkway around the first fence. How many feet of fencing will he need for the second fence? __________

5. The round school track is 0.25 miles long as measured by the inner edge. If the track itself is 6 feet wide, how much farther does the outside person run than the person on the inside, if both runners start and stop at the same place on the track? __________

Find the shaded area.

6. __________ 7. __________ 8. __________

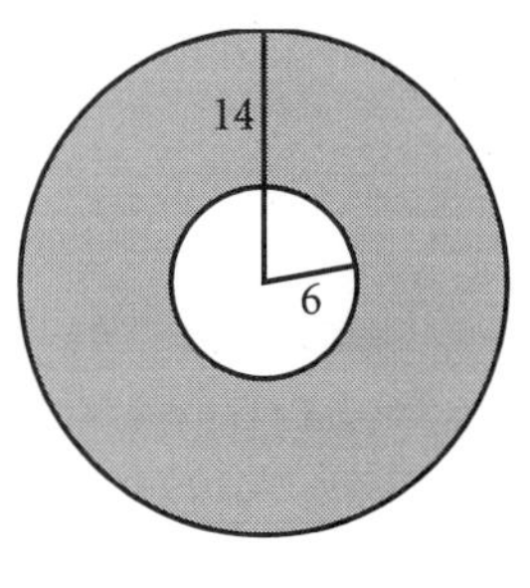

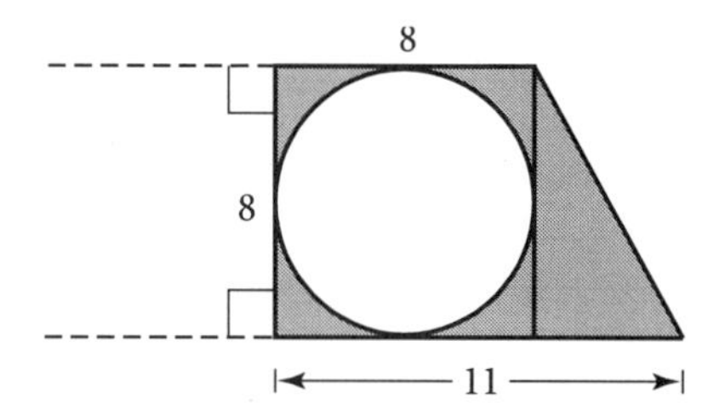

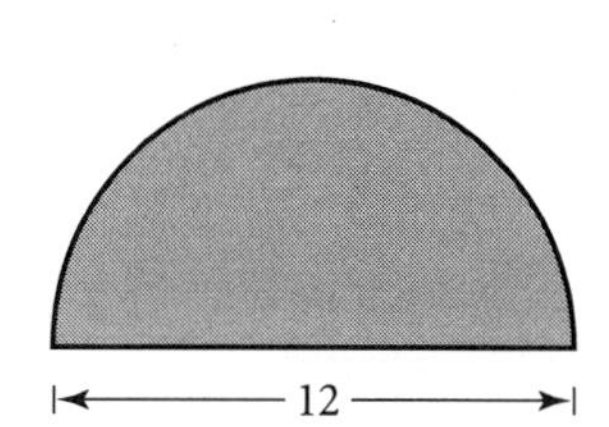

9. Find the circumference of the circle that circumscribes a square whose perimeter is 16 units. __________

NAME ______________________ CLASS ____________ DATE ____________

Practice Masters Level C

5.3 Circumferences and Areas of Circles

In the oval shaped figure, the area is 198.5 square feet and its perimeter is 55.4 feet.

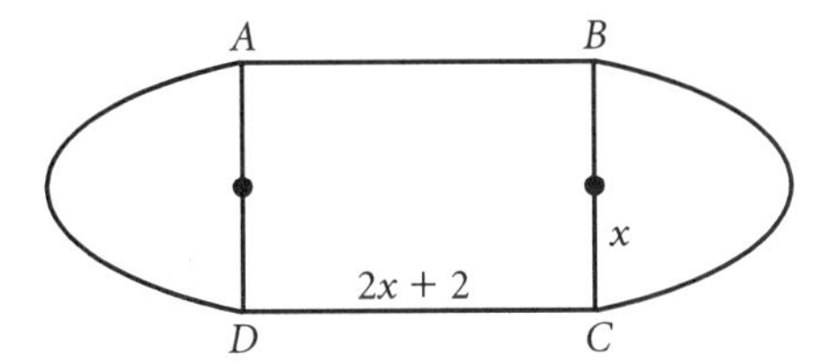

1. Find BC. ______________

2. Find DC. ______________

3. John's car tires have a radius of 15.5 inches. If his car travels 1150 revolutions per minute, how fast is he traveling, in miles per hour?

____

4. How many revolutions per minute should his tires make in order to travel 65 miles per hour?

____

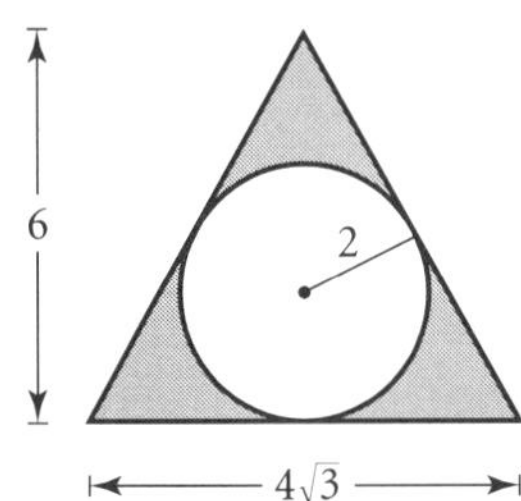

5. Find the area of the shaded region in the figure at the right.

____

6. A builder is constructing houses on a cul-de-sac. A street leading into the cul-de-sac needs to be 30 feet. If the frontage of each lot needs to be at least 35 feet, and the builder plans to build 5 houses on the cul-de-sac, what should be the radius of the cul-de-sac?

____

7. Find the area of the shaded part of the figure at the right if the radius of the largest circle is 12, the middle circle 8, and the smallest circle 4.

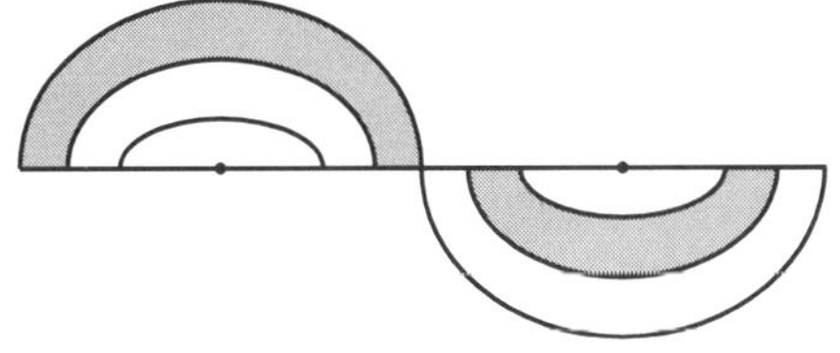

____

8. Sharon and Matt are making sugar cookies. If the dough rolls out into a 9 inch by 13 inch rectangle, how many cookies with a radius of 2 inches can they make?

____

9. Find the area of the rectangle that circumscribes three 4-inch circles. ______________

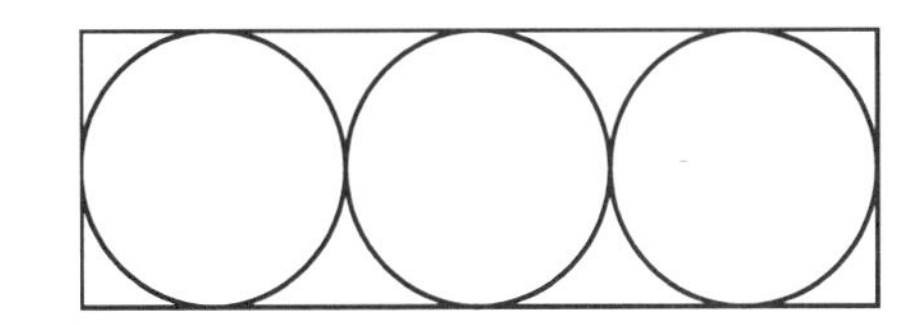

NAME ______________________ CLASS __________ DATE __________

Practice Masters Level A

5.4 The Pythagorean Theorem

For Exercises 1–5, two sides of a right triangle are given. Find the missing side length. Round answers to the nearest tenth.

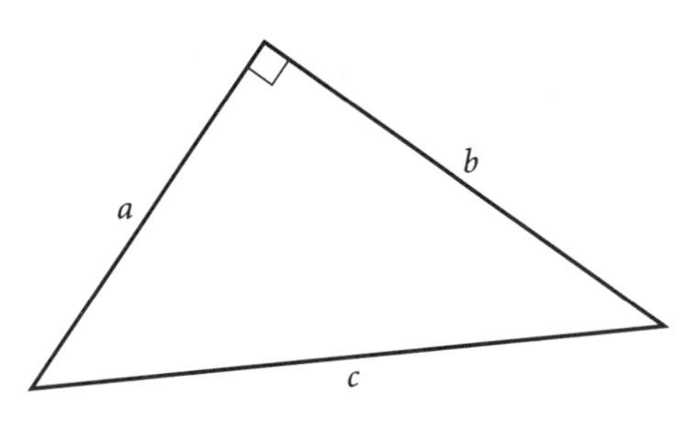

1. $a = 7$ $b = 12$ $c =$ __________
2. $a = 8$ $b =$ __________ $c = 13$
3. $a =$ __________ $b = 11\frac{1}{2}$ $c = 20$
4. $a = 6\frac{1}{2}$ $b = 8\frac{3}{4}$ $c =$ __________
5. $a =$ __________ $b = 4.6$ $c = 11.1$

Each of the following triples represents the side lengths of a triangle. Determine whether the triangle is right, acute, or obtuse.

6. 4, 5, 6 __________
7. 7, 7, 10 __________
8. $3, 1\frac{1}{4}, 3\frac{1}{4}$ __________
9. 2.7, 7.3, 9 __________

Solve.

10. If the diagonal of a square is 9 units long, how long is each side? __________
11. If the side of a square is 7 units long, how long is the diagonal? __________
12. A rectangular suitcase measures 2 feet by 3 feet. Can an umbrella that is 42 inches long be packed lying flat in the suitcase? __________

A rectangular box has a length of 15 inches, a width of 9 inches, and a height of 6 inches.

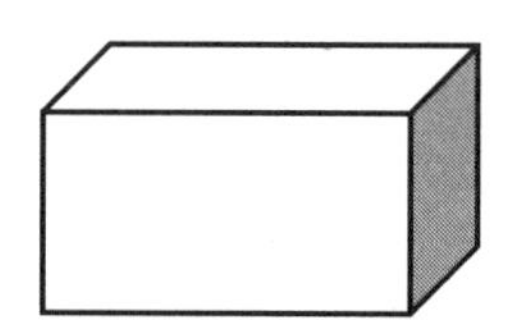

13. What is the length of the diagonal of the bottom of the box?

____

14. What is the length of the diagonal of the box from the corner of the top to an opposite corner of the bottom?

____

NAME ______________________ CLASS ____________ DATE ____________

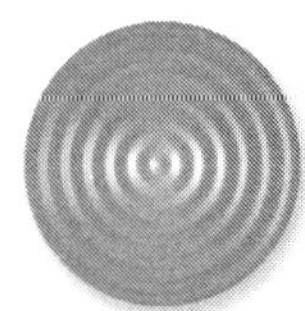

Practice Masters Level B

5.4 The Pythagorean Theorem

Two sides of a right triangle are given. Find the missing side lengths. Give answers to the nearest tenth.

1. $a = 9$ $\quad b =$ ________ $\quad c = 12$

2. $a = 3\sqrt{3}$ $\quad b =$ ________ $\quad c = 6$

3. $a =$ ________ $\quad b = \frac{3}{4}$ $\quad c = 1$

4. $a =$ ________ $\quad b = \sqrt{2}$ $\quad c = \sqrt{6}$

5. $a{:}b = 1{:}3$ $\quad c = 5$ $\quad a =$ ________ $\quad b =$ ________

6. $a{:}b = 1{:}2$ $\quad c = 4\sqrt{3}$ $\quad a =$ ________ $\quad b =$ ________

Decide whether each set of numbers can represent the side lengths of a right triangle.

7. 6, 27, 30 ________ 8. 50, 50, 60 ________ 9. 14, 48, 52 ________

Solve.

10. A rectangle has a perimeter of 20 and a width of 4. Find the length of its diagonal. ________

11. The altitude of an equilateral triangle is 12. What is its perimeter? ________

12. The diagonals of a rhombus are in a 2:1 ratio. If the perimeter of the rhombus is 40, find the length of each diagonal. ________

For Exercises 13–14, refer to $\triangle ABC$ with altitude $\overline{BD}$.

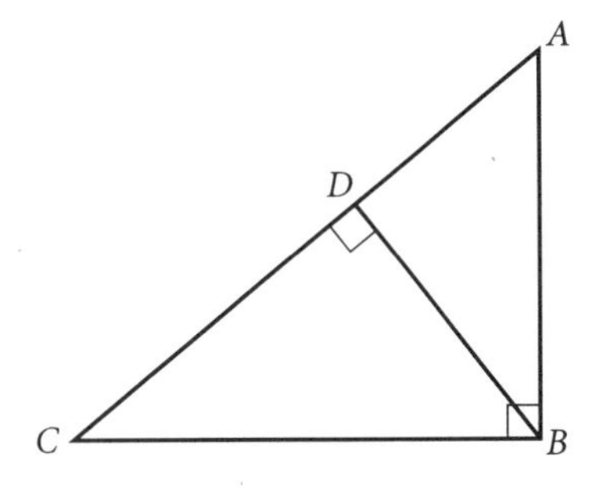

13. If $AB = x + 3$, $BC = 3x$, and $AC = 17$. Find the area of $\triangle ABC$.

____

14. If the ratio of $BC{:}AB$ is 3:4 and $AC = 10\sqrt{3}$, find the area of $\triangle ABC$.

____

NAME ______________________ CLASS ____________ DATE ____________

Practice Masters Level C

5.4 The Pythagorean Theorem

1. The side of a square equals the diagonal of a second square. What is the ratio of the perimeter of the larger square to that of the smaller square? ____________

2. The ratio of the sides of a triangle are 5:13:12. If the perimeter is 15, what is the area of the triangle? ____________

3. What kind of triangle, right, acute, or obtuse, has sides of 4, $4\sqrt{2}$, and 6? ____________

4. Find the length of a leg of a right triangle if the second leg measures $\frac{1}{2}$ and the hypotenuse is $\frac{5}{8}$. ____________

One leg of a right triangle, $\overline{AB}$, measures $x + 1$, the other leg measures $4x$, and the hypotenuse is $4x + 1$. Find the indicated measure.

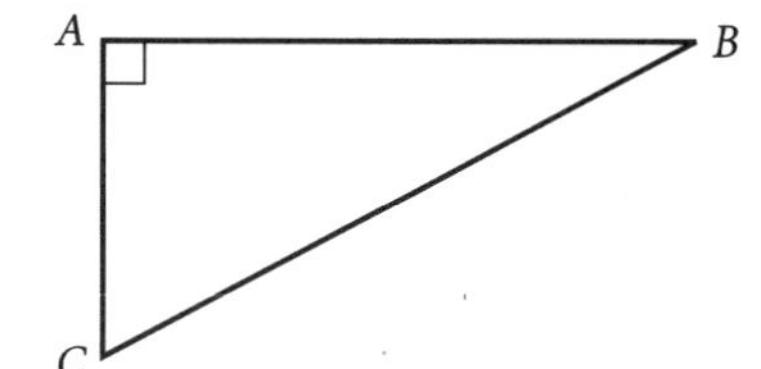

5. AB ____________
6. BC ____________
7. Area of $\triangle ABC$ ____________
8. AC ____________

In $\triangle WXY$ with altitude YZ, $XY = 6$, and $XZ = 2$. Find the indicated measure.

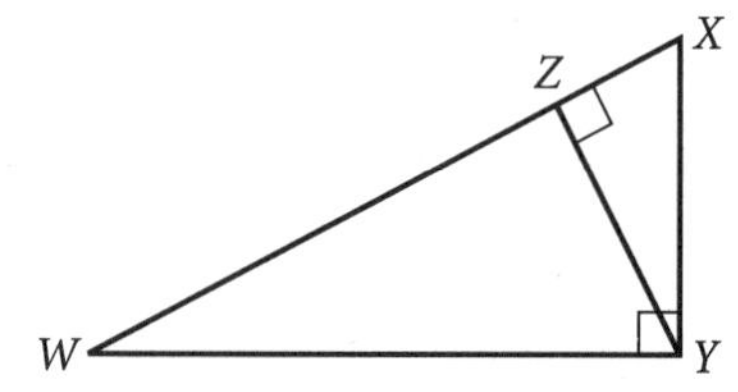

9. YZ ____________
10. WY ____________
11. WZ ____________
12. WX ____________

In isosceles trapezoid $ABCD$, base DC measures 35, the height is 12, and $AD = BC = 15$.

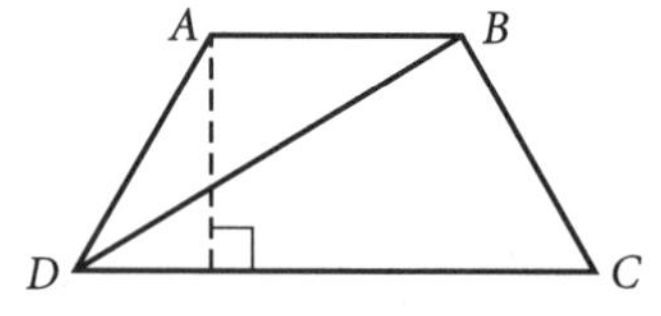

13. Find the area of $ABCD$. ____________

14. Find the area of $\triangle DBC$. ____________

15. A man travels 5 miles north, then 2 miles east, followed by 1 mile north and then 4 miles east. How far is he from his starting point? ____________

16. Find the length of each side of a cube if the diagonal of one face is 8. ____________

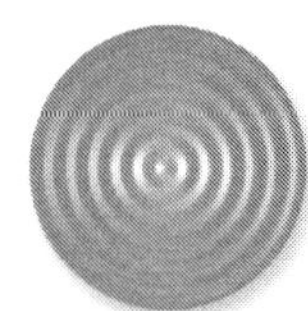

Practice Masters Level A

5.5 Special Triangles and Areas of Regular Polygons

For each given length, find the remaining two lengths. Give your answers in simplest radical form.

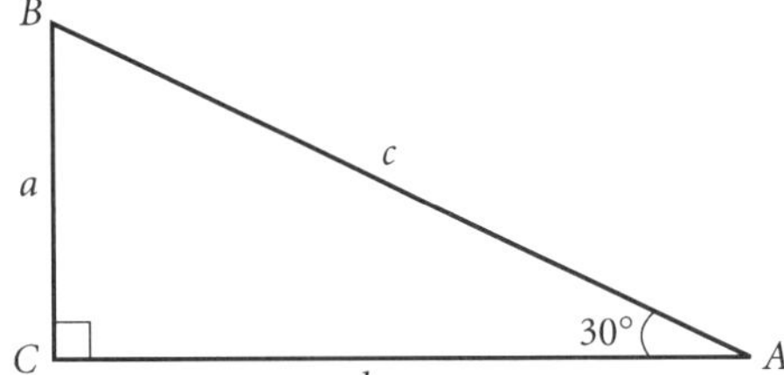

1. $a = 6$ Find: b ________ c ________

2. $b = 6$ Find: a ________ c ________

3. $c = 6$ Find: a ________ b ________

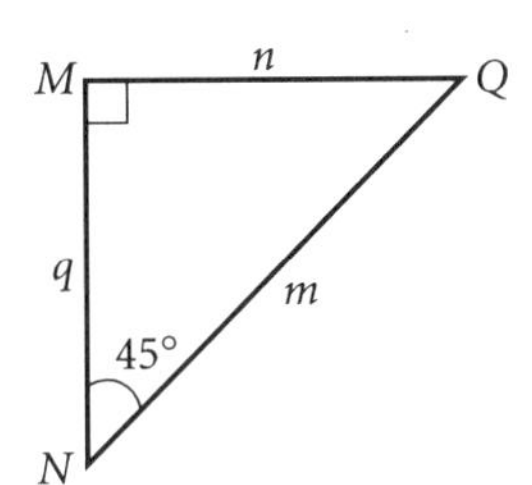

4. $q = 7\sqrt{2}$ Find: n ________ m ________

5. $m = 4\sqrt{2}$ Find: q ________ n ________

6. $s = 3\sqrt{3}$ Find: t ________ r ________

7. $t = 3\sqrt{3}$ Find: s ________ r ________

8. $r = 3\sqrt{3}$ Find: s ________ t ________

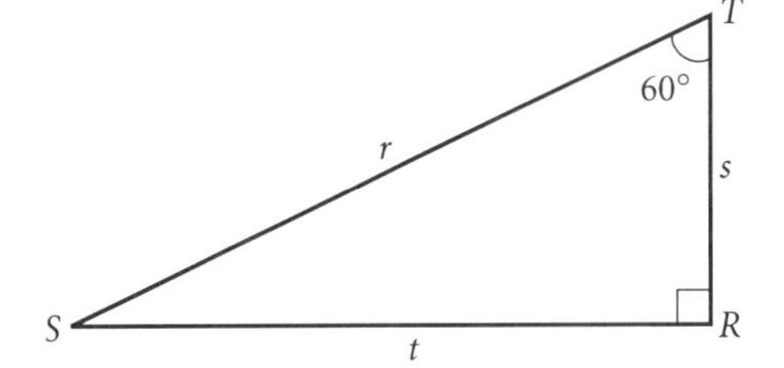

Refer to the square for Exercises 9 and 10.

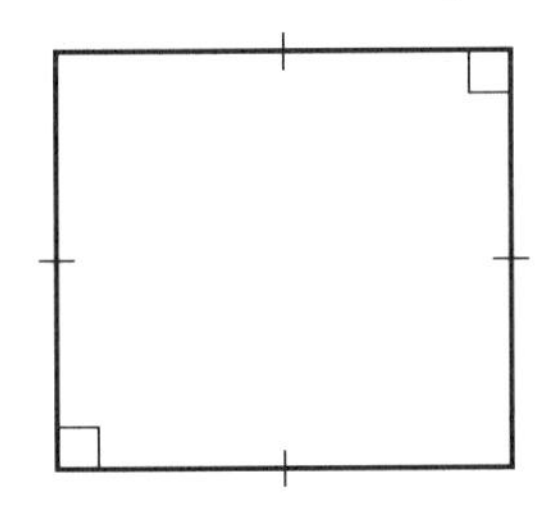

9. The diagonal of a square is $\sqrt{30}$. Find its perimeter. ________

10. Find the area of the square with a diagonal of $\sqrt{30}$. ________

$\triangle ABC$ is an equilateral triangle with a side of 6.

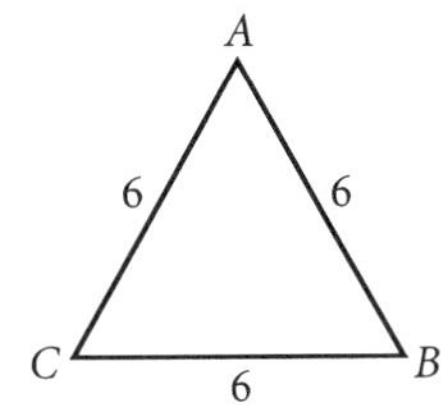

11. Find an altitude to one side. ________

12. Find the area of the triangle. ________

NAME ______________________ CLASS ____________ DATE ____________

Practice Masters Level B

5.5 Special Triangles and Areas of Regular Polygons

In $\triangle ABC$, $\overline{AC} \perp \overline{BC}$, $\overline{CD}$ is the altitude to AB. Use the figure to find the missing measures in Exercises 1–6.

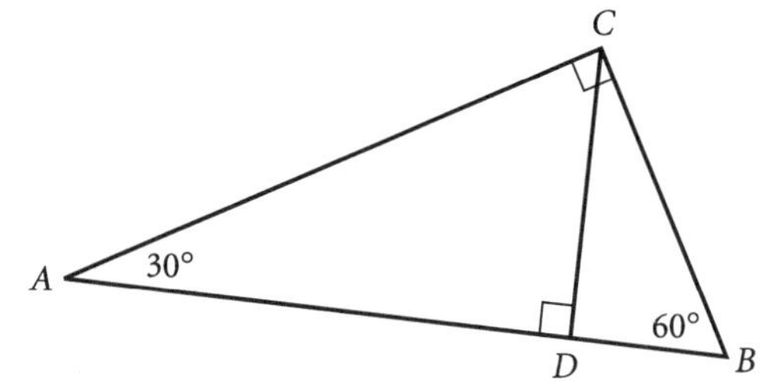

	AB	BC	CD	AD	DB	AC
1.	8	____	____	____	____	____
2.	____	2	____	____	____	____
3.	____	____	4	____	____	____
4.	____	____	____	9	____	____
5.	____	____	____	____	10	____
6.	____	____	____	____	____	12

For Exercises 7–9, refer to the regular hexagon, *ABCDEF*.

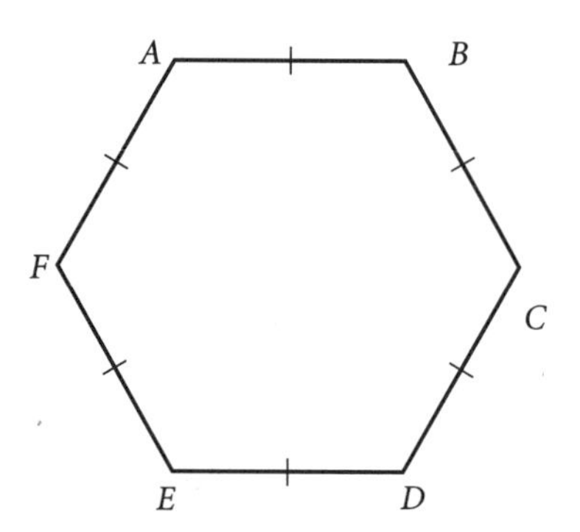

7. If the area of *ABCDEF* is 841.8 square units, find the length of each side. ____________

8. If the area of *ABCDEF* is 841.8 square units, find the length of the apothem. ____________

9. If the apothem equals 4, what is the area?

____

For Exercises 10 and 11, refer to trapezoid *TQRS*.

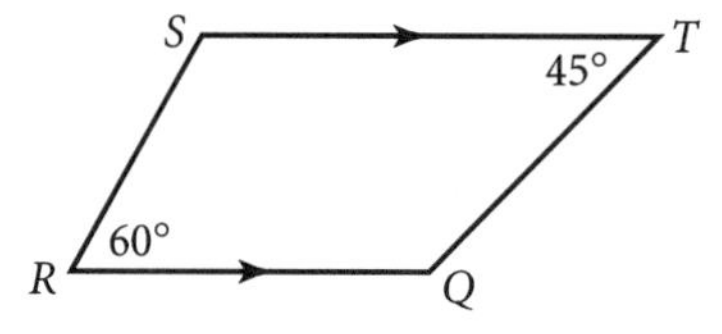

10. Find the perimeter of *TQRS*. ____________

11. Find the area of *TQRS*. ____________

In the figure at the right, $m\angle BAC = 45°$ and $m\angle D = 30°$.

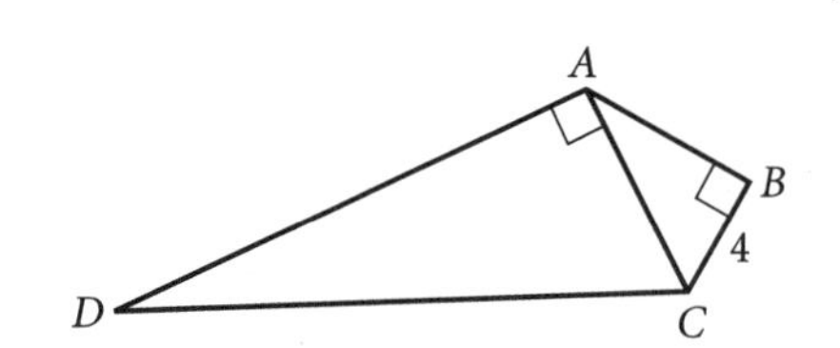

12. Find *AC*. ____________

13. Find *AD*. ____________

14. Find *CD*. ____________

NAME ________________ CLASS ________ DATE ________

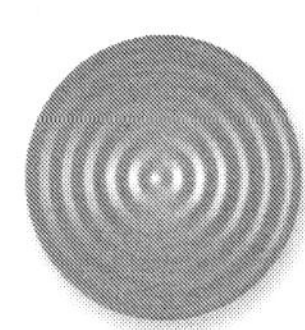

Practice Masters Level C

5.5 Special Triangles and Areas of Regular Polygons

In $\triangle ABC$, $m\angle ABC = 90°$ and $AC = 16$.

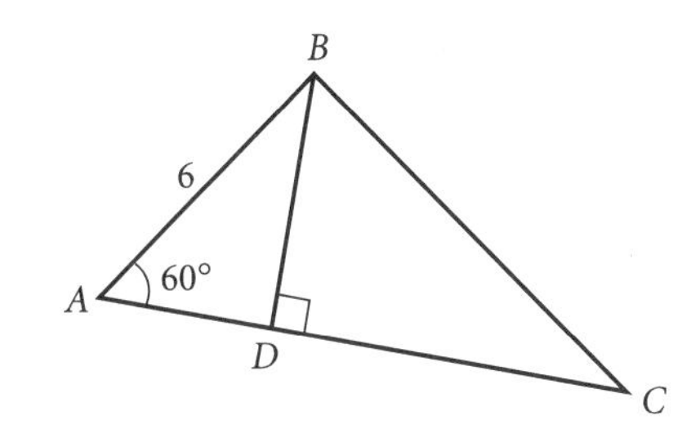

1. Find BC. ________
2. Find BD. ________
3. Find DC. ________

$\triangle PQR$ and $\triangle PSR$ are right triangles, that are also perpendicular to each other. $QR = 10$, $m\angle QRP = 45°$ and $m\angle SRP = 60°$.

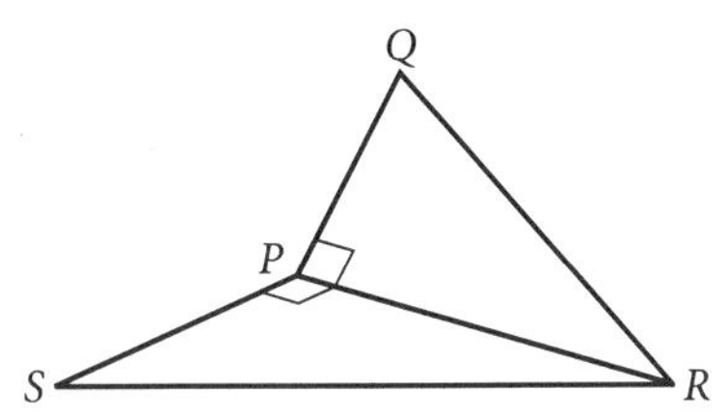

4. $SP =$ ________
5. $SR =$ ________
6. $QP =$ ________
7. $PR =$ ________

8. Find the area of a regular hexagon with a radius of $7\sqrt{3}$ units. ________

$\triangle ABC$ has altitude $\overline{BD}$. Find the following measures in terms of x.

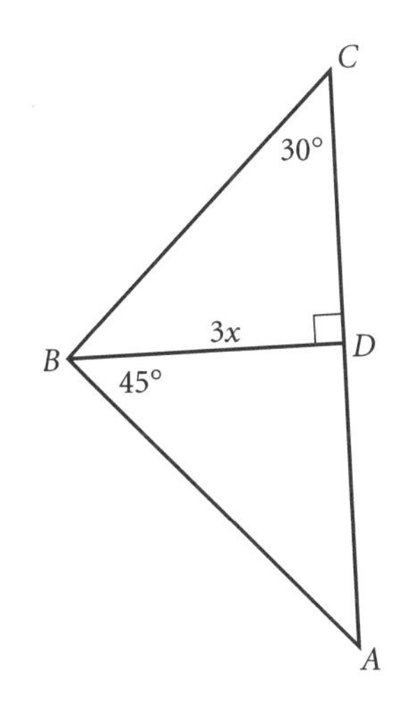

9. AB ________
10. AC ________
11. AD ________
12. BC ________
13. area of $\triangle BDC$ ________
14. area of $\triangle BDA$ ________
15. area of $\triangle ABC$ ________

The figure below is a rectangular box. For Exercises 16–18, leave answers in simplest radical form.

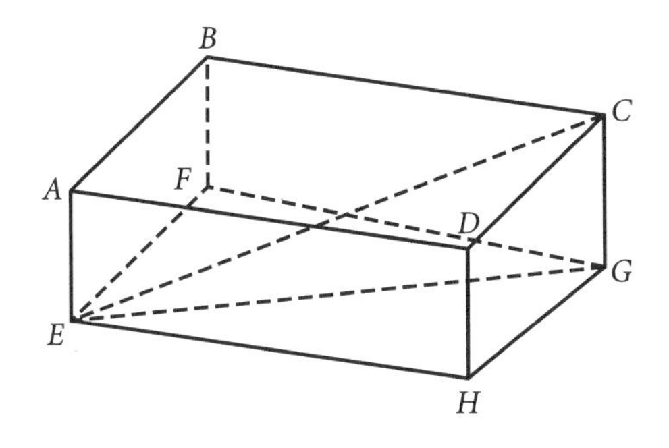

16. If $EH = 6$, $GH = 2$, $DH = 4$, find the length of diagonal $\overline{CE}$. ________
17. If $EH:HG:DH$ is 3:4:12 and $CE = 26$, find:
    a. EH ________
    b. HG ________
    c. DH ________
    d. GE ________
18. If $EH = GH = DH = 4$, find GE. ________

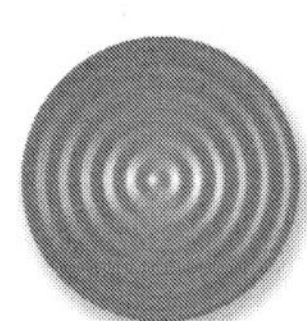

Practice Masters Level A

5.6 The Distance Formula and the Method of Quadrature

Find the distance between each pair of points. Round your answers to the nearest hundredth.

1. (3, 6) and (5, 7) ______________
2. (−6, 4) and (2, 7) ______________
3. (4, −3) and (−9, −6) ______________
4. (−8, −1) and (−1, −8) ______________
5. $\left(\frac{1}{2}, 2\frac{1}{2}\right)$ and $\left(3\frac{1}{2}, 6\right)$ ______________
6. (0, 0) and (10, 16) ______________

Graph each set of points. Join the points in a smooth curve. Use quadrature to estimate the area between the curve and the *x*-axis.

7. (−1, 0), (0, 1), (1, 2), (2, 3), $\left(3, 3\frac{1}{2}\right)$, (4, 3), (5, 2), (6, 1) and (7, 0)

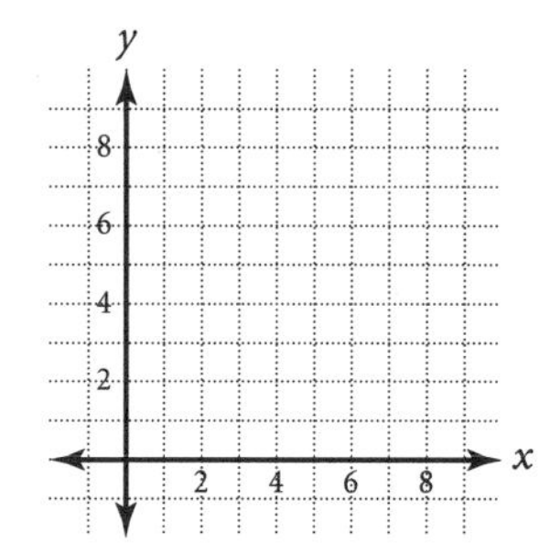

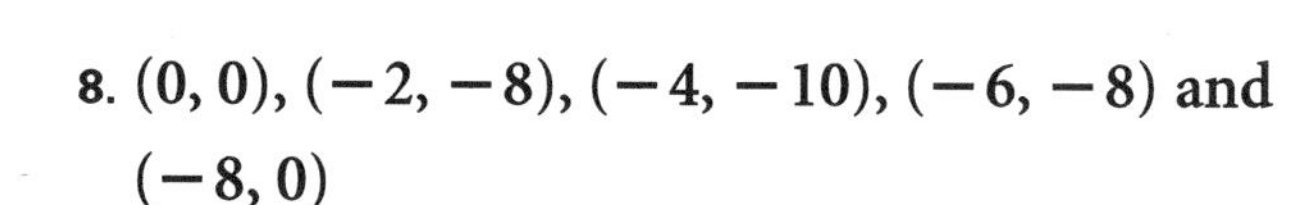

8. (0, 0), (−2, −8), (−4, −10), (−6, −8) and (−8, 0)

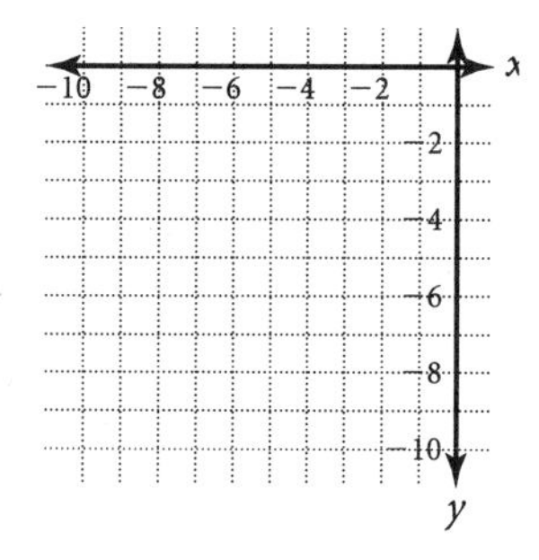

For Exercises 9–12, refer to △*ABC*.

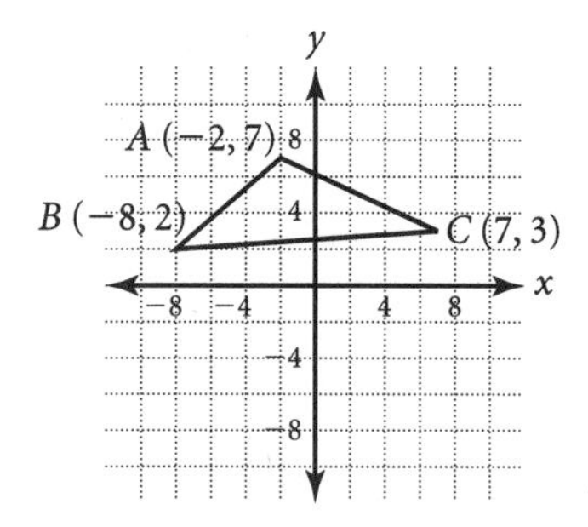

9. Find *AB*. ______________
10. Find *AC*. ______________
11. Find *BC*. ______________
12. What kind of triangle is *ABC*? ______________

13. The coordinates of a triangle are (−2, 1), (1, 3), and (3, 0). Classify the triangle as right, isosceles, or equilateral.

____

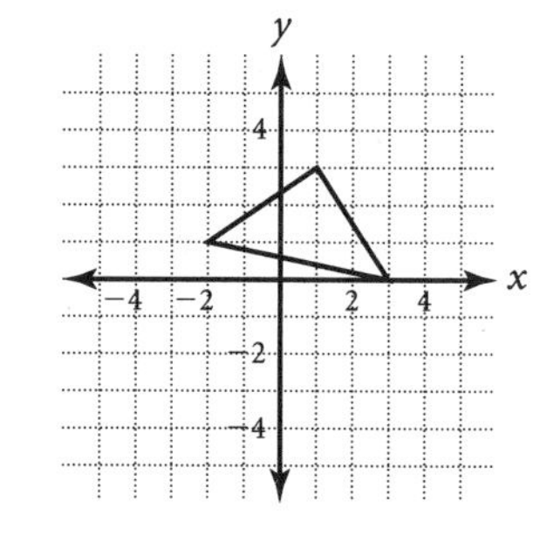

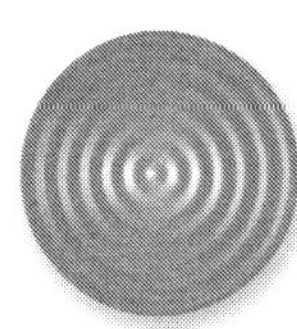

Practice Masters Level B

5.6 The Distance Formula and the Method of Quadrature

For Exercises 1–8, refer to the diagram. Leave answers in simplest radical form.

1. Find the length of $\overline{AE}$. ______________________

2. Find the length of $\overline{EC}$. ______________________

3. Find the length of $\overline{AC}$. ______________________

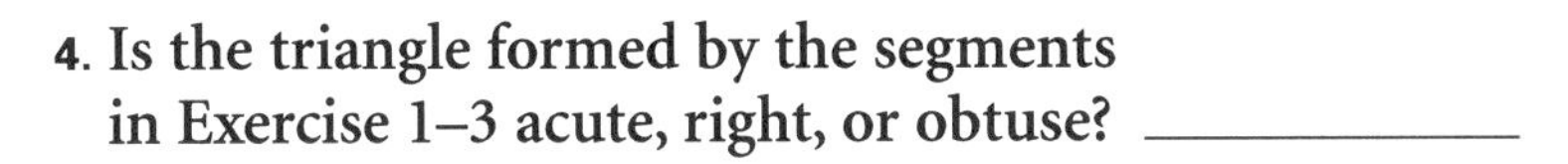

4. Is the triangle formed by the segments in Exercise 1–3 acute, right, or obtuse? ____________

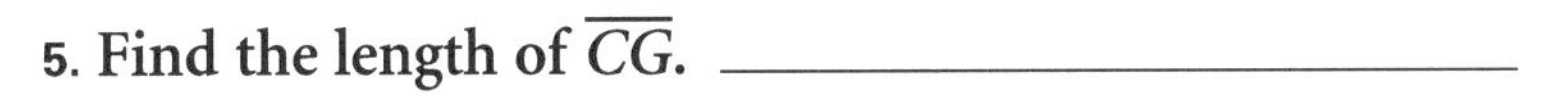

5. Find the length of $\overline{CG}$. ______________________

6. Find the length of $\overline{AG}$. ______________________

7. Is $\triangle ACG$ acute, right, or obtuse? ______________________

8. What kind of quadrilateral is *FHCD*? Use the distance formula to support your answer. ______________________

____

Graph each set of points. Join the points in a smooth curve. Use quadrature to estimate the area between the curve and the *x*-axis.

9. $(-6, 0)$, $\left(-5\frac{1}{2}, 2\right)$, $(-4, 3)$ $(-3, 2)$ $(-2, 1)$ $\left(-\frac{1}{2}, 2\right)$, $(1, 4)$, $(2, 5)$, $(4, 3)$, and $(5, 0)$

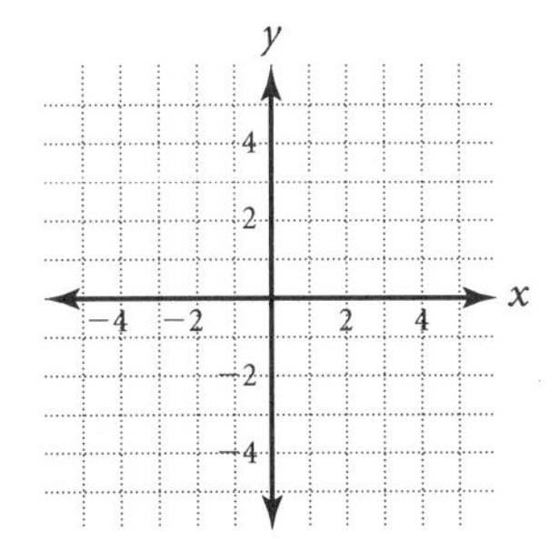

10. $(-4, 0)$, $(-3, -1)$, $(-2, 0)$, $(-1, 1)$, $(0, 3)$, $(1, 4)$, $(2, 3)$, $(3, 1)$, $(4, 0)$, $(5, -1)$, and $(6, 0)$

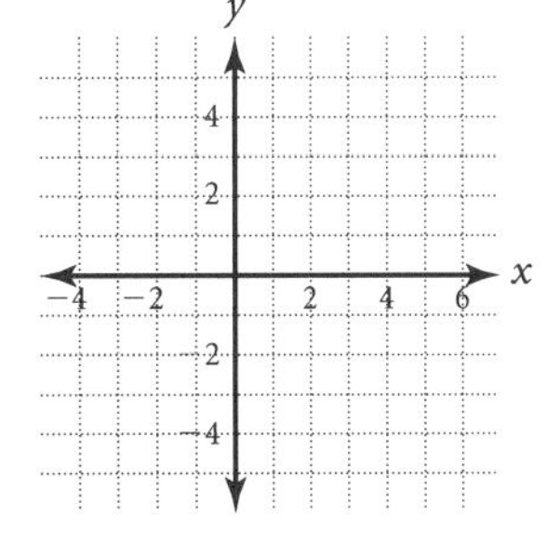

11. If $\overline{AB}$ has endpoints of $(2, -5)$ and $(8, -1)$, and $\overline{CD}$ has endpoints of $(-4, -6)$ and $(10, 8)$, how long is the segment joining the midpoints of $\overline{AB}$ and $\overline{CD}$? ______________________

Practice Masters Level C

5.6 The Distance Formula and the Method of Quadrature

1. Suppose $\overline{AC}$ has a length of 5. If A has the coordinates (5, 1) and C has the coordinates (8, y), find y. ______________

2. Suppose $\overline{WZ}$ has a length of 5. If Z has the coordinates (2, 6), and W has the coordinates (x, 3), find x. ______________

3. The midpoint of $\overline{HJ}$ is (6, –1). If H has coordinates (2, −3), what are the coordinates of J? ______________

4. The midpoint of $\overline{PQ}$ is $(-3\frac{1}{2}, -2)$. If P has coordinates (4, 4), what are the coordinates of Q? ______________

5. One leg of an isosceles triangle has coordinates (2, 4) at the vertex and (−2, 1) at the base. Find the coordinates of the third point if it lies on the x-axis. ______________

For Exercises 6–11, refer to the graph.

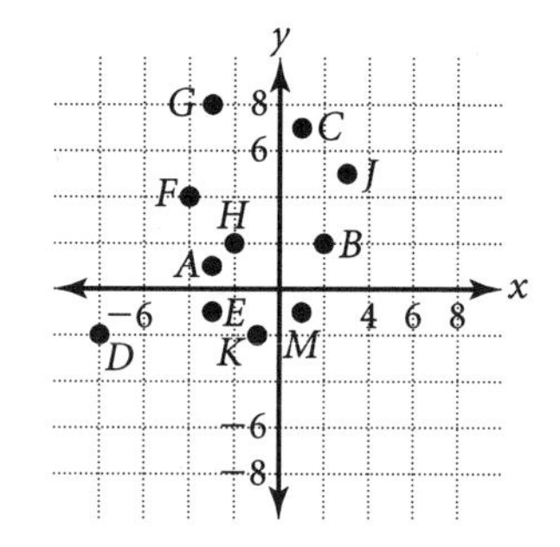

6. Find the area of $\triangle ABC$. ______________

7. Find the area of $\triangle DEF$. ______________

8. What do you notice about the triangles in Exercises 6 and 7?

____

9. Find the area of $\triangle GHJ$. ______________

10. Find the area of $\triangle HMK$. ______________

11. How do $\triangle GHJ$ and $\triangle HMK$ compare? ______________

____

12. Use the integral x-values from −2 to 6 to find and graph points for $y = \frac{1}{4}(x - 2)^2 + 6$. Then find the area from the curve to the x-axis between −2 and 6. ______________

x	−2	−1	0	1	2	3	4	5	6
y									

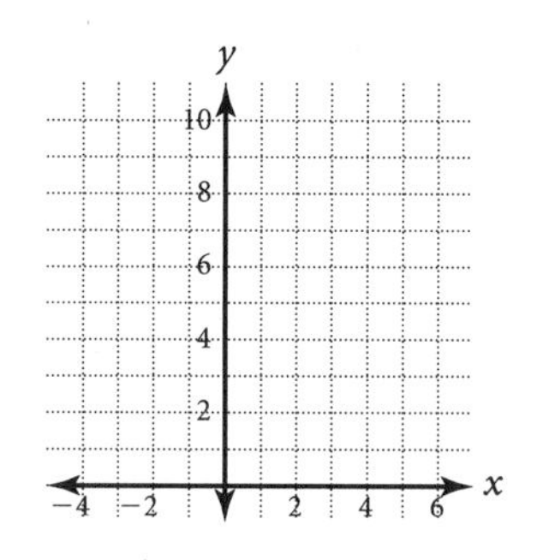

NAME ______________________ CLASS ____________ DATE ____________

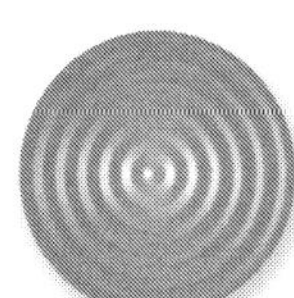

Practice Masters Level A

5.7 Proofs Using Coordinate Geometry

Determine the coordinates of the unknown vertex or vertices of each figure below.

1. isosceles triangle GJH ______________

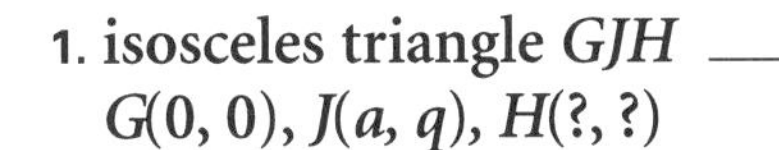

$G(0, 0)$, $J(a, q)$, $H(?, ?)$

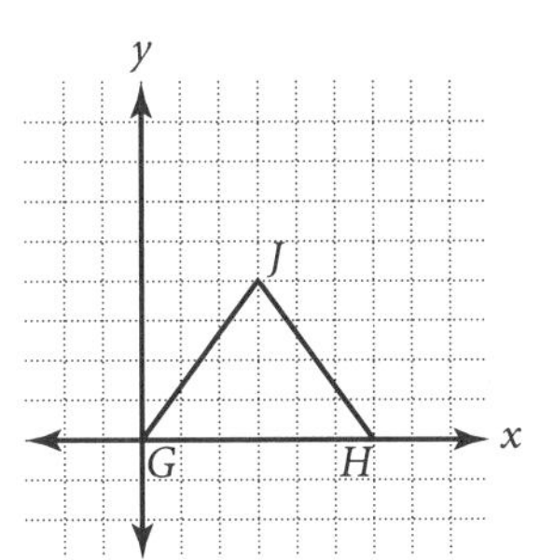

2. rhombus $LMNP$ ______________
$L(-q, 3a)$, $M(0, a)$, $N(?, ?)$, $P(?, ?)$

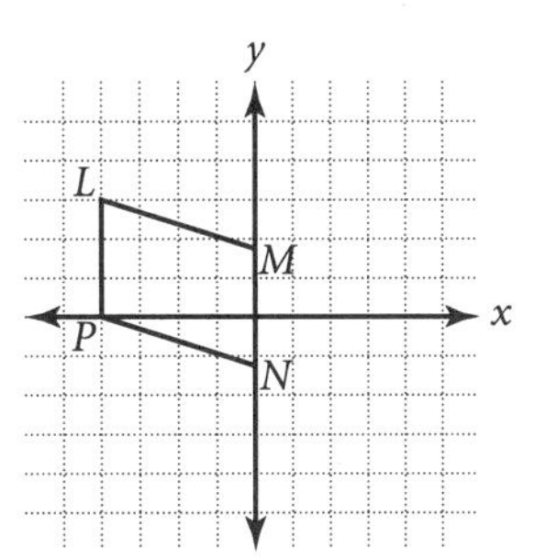

3. parallelogram $ABCD$ ______________
$A(0, 0)$, $B(?, ?)$, $C(a + b, c)$

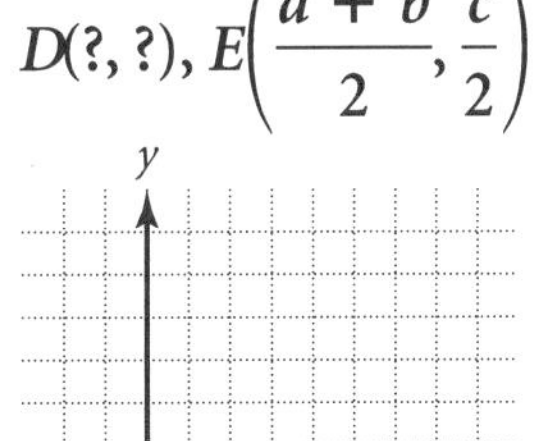

$D(?, ?)$, $E\left(\frac{a + b}{2}, \frac{c}{2}\right)$

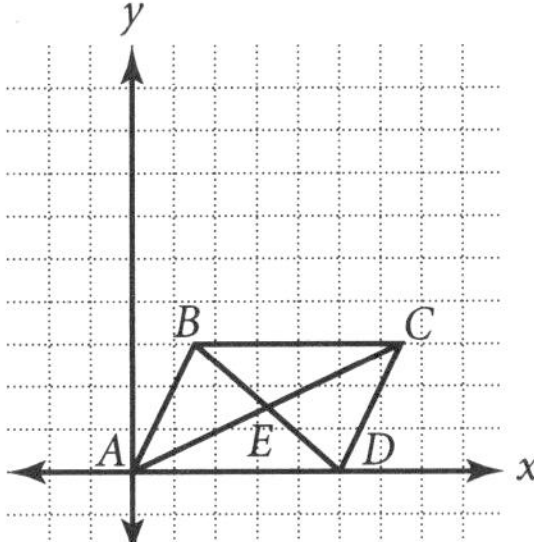

4. trapezoid $TQRW$ ______________

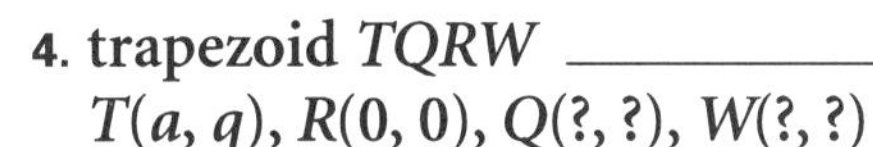

$T(a, q)$, $R(0, 0)$, $Q(?, ?)$, $W(?, ?)$

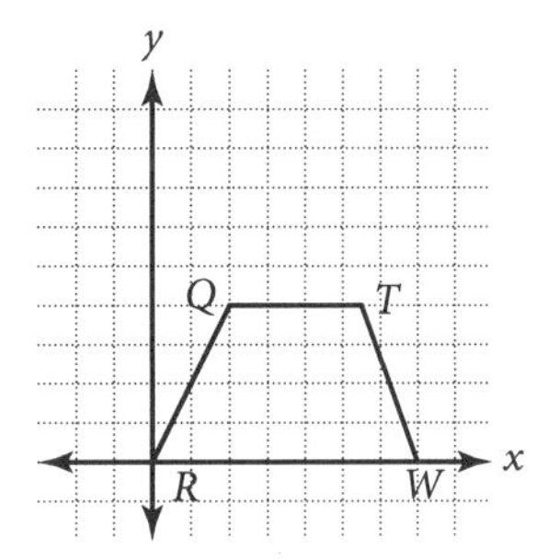

5. In Exercise 3, what is the relationship between $\overline{DB}$ and $\overline{AC}$? What can you conclude about the diagonals of a parallelogram? ______________

____

Use the diagram for Exercises 6 and 7.

y
4
2
−4 −2 2 4 x
−2
−4
P_1 P_2 P_3 P_4

6. If point P_1 is the reflection of P_2 across the line $y = x$, what are the coordinates of P_1? ______________

7. If the coordinates of P_3 are the reflection of P_4 across the line $y = x$, what are the coordinates of P_3? ______________

Fill in the blank.

8. To reflect a point across the line $y = x$, you ______________ the x- and y-coordinates of the point.

NAME ______________________ CLASS ____________ DATE ____________

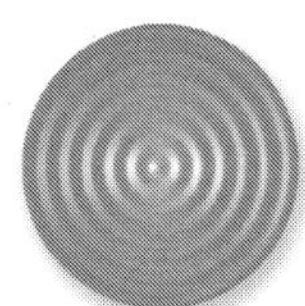

Practice Masters Level B

5.7 Proofs Using Coordinate Geometry

Refer to the diagram for Exercises 1–8.

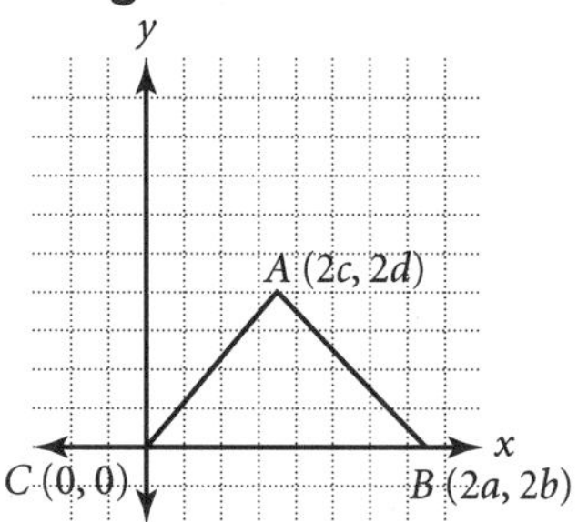

1. Find the midpoint of $\overline{AB}$. ____________

2. Find the midpoint of $\overline{AC}$. ____________

3. Find the slope of the segment joining the midpoints of $\overline{AB}$ and $\overline{AC}$. ____________

4. Find the slope of $\overline{CB}$. ____________

5. What can be concluded about the segment joining the midpoints of these two sides of the triangle? ____________

6. Find the length of the segment joining the midpoints of $\overline{AB}$ and $\overline{CB}$. ____________

7. How does this compare to the length of $\overline{CB}$? ____________

8. Generalize about the segment joining any midpoints of two sides of a triangle. ____________

9. *EFGH* is a parallelogram with diagonal $\overline{EG}$. Use distance and/or slope formulas to show that the diagonal of a parallelogram divides it into two congruent triangles.

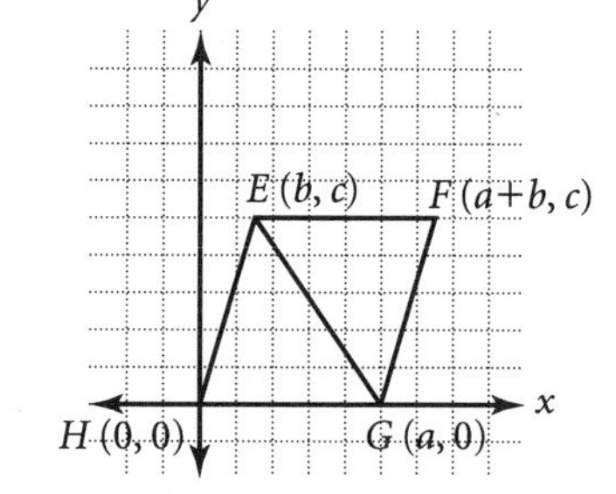

____

____

____

***JKLM* is a square with diagonals $\overline{JL}$ and $\overline{KM}$. Use distance and/or slope formulas to prove each statement.**

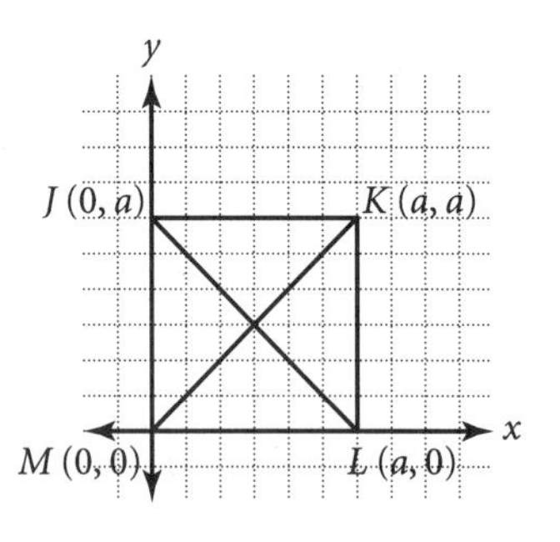

10. Diagonals are congruent. ____________

____

11. Diagonals bisect each other. ____________

____

12. Diagonals are perpendicular to each other. ____________

____

13. To reflect point *A* across the line $y = x$, you can reverse the *x*- and *y*-coordinates of point *A* to obtain *A*'. The line joining points *A* and *A*' is ____________ with a slope of ____________.

NAME ______________________ CLASS ____________ DATE ____________

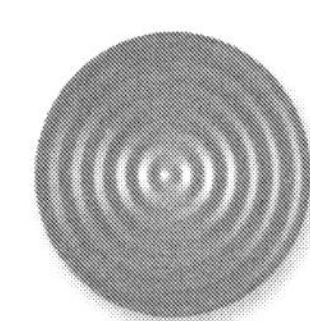

Practice Masters Level C

5.7 Proofs Using Coordinate Geometry

1. Determine the coordinates of the endpoints of midsegment $\overline{DE}$ that joins $\overline{AB}$ and $\overline{BC}$. Show that the midsegment is parallel to $\overline{AC}$ and half its length.

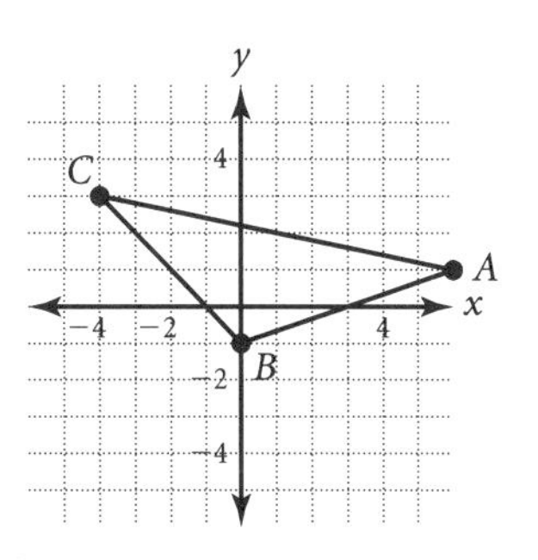

Refer to the diagram for Exercises 2–13.

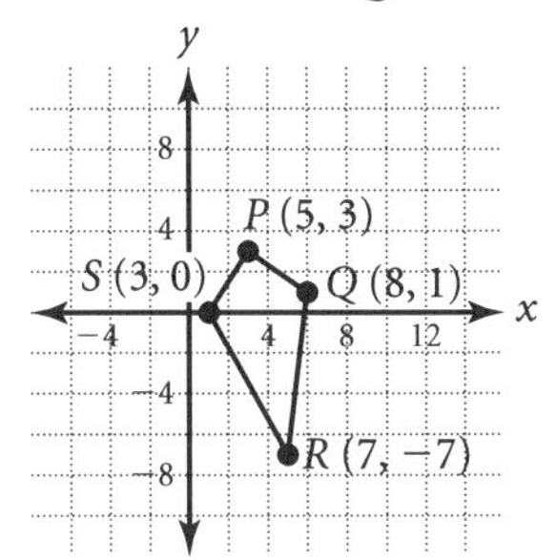

2. Find the length of $\overline{PQ}$. ____________

3. Find the length of $\overline{QR}$. ____________

4. Find the length of $\overline{RS}$. ____________

5. Find the length of $\overline{SP}$. ____________

6. What kind of quadrilateral is *PQRS*? ______

7. What is true about the diagonals of the figure? Show that this is true using coordinate geometry. ____________

8. Reflect *SPQR* across the line $y = x$. What are the new coordinates? ____________

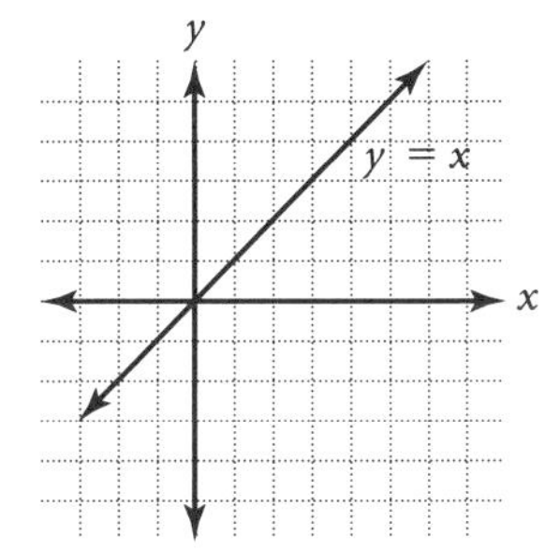

9. Find the length of $\overline{P'Q'}$. ____________

10. Find the length of $\overline{Q'R'}$. ____________

11. Find the length of $\overline{R'S'}$. ____________

12. Find the length of $\overline{S'P'}$. ____________

13. What kind of quadrilateral is $P'Q'R'S'$?

____

14. Use the distance/slope formulas to show that the diagonals of a rhombus divide it into four congruent triangles.

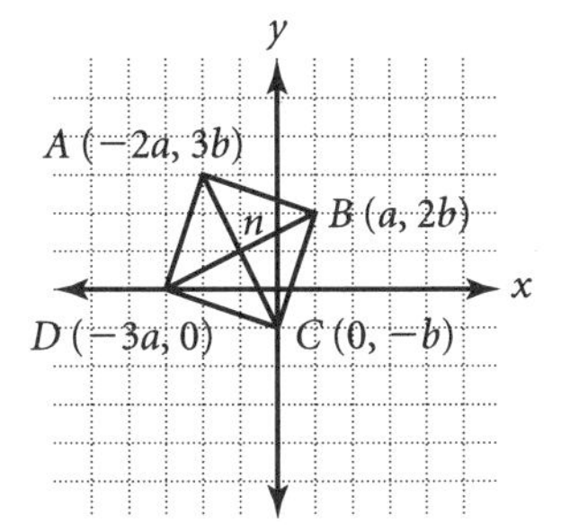

____

____

____

NAME ______________________ CLASS ____________ DATE ____________

Practice Masters Level A

5.8 Geometric Probability

Use the points on the number line for Exercises 1–3.

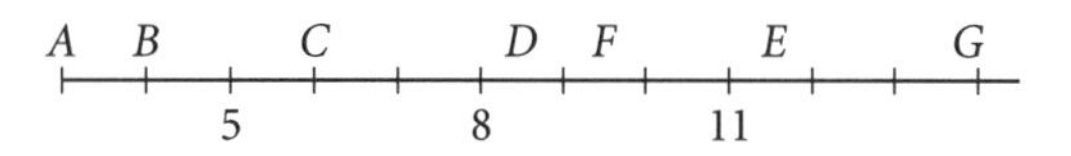

1. Find the probability that a point on $\overline{AG}$ lies between C and F. ____________

2. Find the probability that a point on $\overline{AG}$ lies between G and B. ____________

3. Find the probability that a point on $\overline{BF}$ lies between E and G. ____________

Refer to the rectangular field for Exercises 4–8.

4. A rectangular field measures 27 feet by 15 feet. Find the area of the field. ____________

5. A small shed is on the field. Its dimensions are 8 feet by 10 feet. What is its area? ____________

6. What is the probability that a single drop of rain that lands in the field would hit the shed? ____________

7. What is the probability that a single drop of rain that lands in the field would *not* hit the shed? ____________

8. There is a large oak tree in one corner whose branches have a diameter of 20 feet. What is the probability that a single drop of rain that lands in the field would miss both the shed and the tree? (Assume the shed is not under the tree.) ____________

The box below has been divided into rectangles of equal area. Refer to the box for Exercises 9–10.

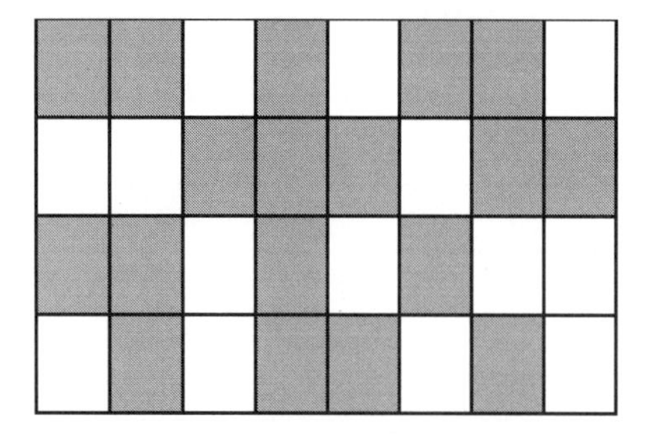

9. a. If only the first column is darkened, what fraction of the entire box has been darkened? ____________

   b. What percentage is this? ____________

   c. What decimal portion is this? ____________

10. Find the probability that a small grain of rice, randomly tossed onto the grid, will land in a clear box? ____________

For each number below, change to a decimal probability.

11. 65% ____________ 12. $\frac{3}{8}$ ____________ 13. 0.25% ____________ 14. $\frac{5}{12}$ ____________

For each number below, change to a fractional probability.

15. 0.45 ____________ 16. 87.5% ____________ 17. 17.5% ____________ 18. 0.125 ____________

Practice Masters Level B

5.8 Geometric Probability

Use the points on the number line for Exercises 1–7.

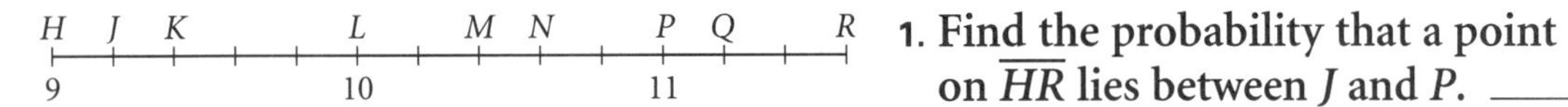

1. Find the probability that a point on $\overline{HR}$ lies between J and P. ____________

2. Find the probability that a point on $\overline{HR}$ lies between N and K. ____________

3. Find the probability that a point on $\overline{MN}$ lies between J and K. ____________

4. Find the probability that a point on $\overline{LQ}$ lies between M and P. ____________

5. The probability that a point on $\overline{JR}$ lies between ________ and ________ is 25%.

6. The probability that a point on $\overline{JQ}$ lies between ________ and ________ is 0.50.

7. The probability that a point on $\overline{KP}$ lies between ________ and ________ is $\frac{3}{4}$.

A dartboard is made up of concentric circles with the following radii:

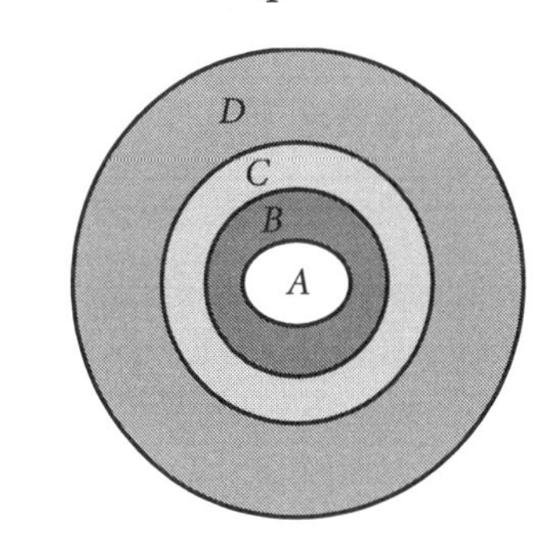

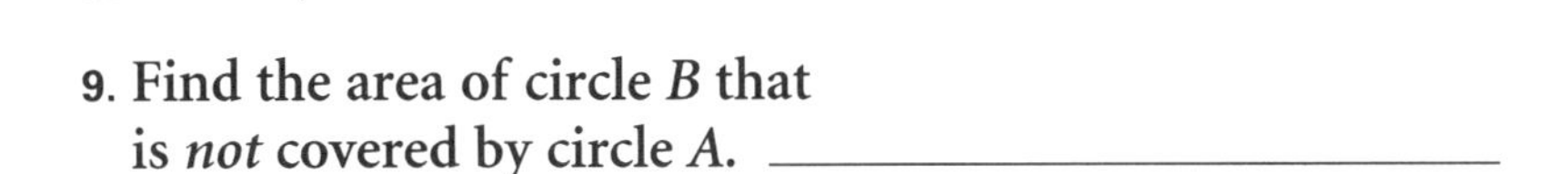

Circle A: $r = 2$ inches Circle B: $r = 4$ inches

Circle C: $r = 6$ inches Circle D: $r = 10$ inches

8. Find the area of circle A. ____________

9. Find the area of circle B that is *not* covered by circle A. ____________

10. Find the area of circle C that is *not* covered by circle A or B. ____________

11. Find the area of the dartboard that is *not* covered by circles A, B, or C. ____________

The circles on the dartboard are painted on a rectangular piece of corkboard that is 2 feet by 30 inches. Find the probability of each event, assuming the dart lands on the corkboard.

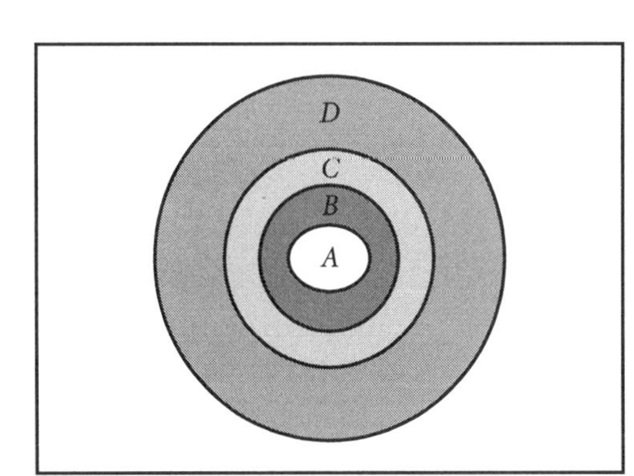

12. A random dart lands on one of the circles.

____

13. A random dart lands on circle C or D.

____

14. A random dart will make a bull's-eye. ____________

15. A random dart falls only on circle C. ____________

NAME ______________________ CLASS ______________ DATE ______________

Practice Masters Level C

5.8 Geometric Probability

A circular spinner is divided into 8 equal sections. The radius of the circle is 12 inches. Find each probability.

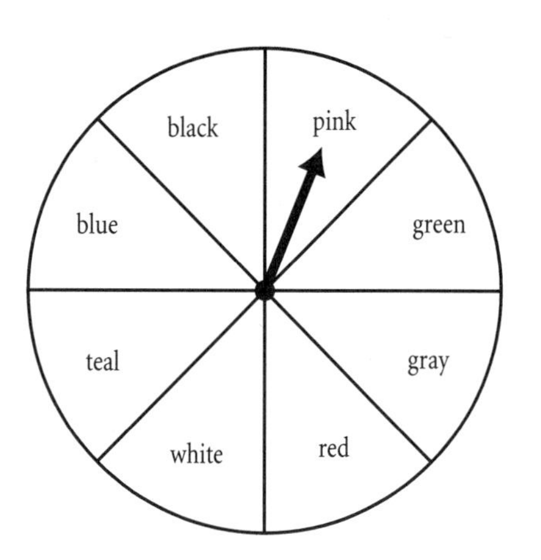

1. P (green)
2. P (not gray or teal)
3. P (not blue)
4. P (teal, white, green, blue)
5. P (green, red or black)
6. P (black, red)
7. P (brown)
8. P (all but pink)

For Exercises 9–12, triangle *ABC* is inscribed in a rectangle, which is inscribed in a circle, which is inscribed in a square. Express each probability as a decimal to the nearest hundredth.

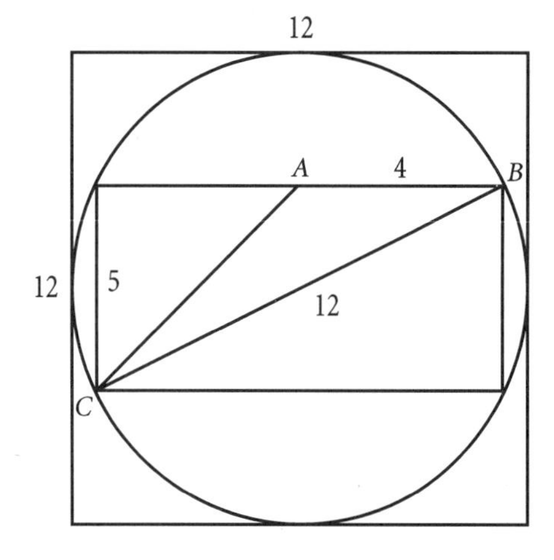

9. What is the probability that a pebble dropped on the figure will land *only* in triangle ABC?

____

10. What is the probability that the pebble will land in the rectangle, but *not* in triangle ABC?

____

11. What is the probability that the pebble will land in the circle, but *not* in the rectangle or triangle ABC?

____

12. What is the probability that the pebble will land in the square, but *not* the circle, rectangle or triangle ABC?

____

A regular hexagon with a side of 12 inches is randomly thrown onto an isosceles trapezoid whose bases measure 30 inches and 45 inches, and whose height is 24 inches.

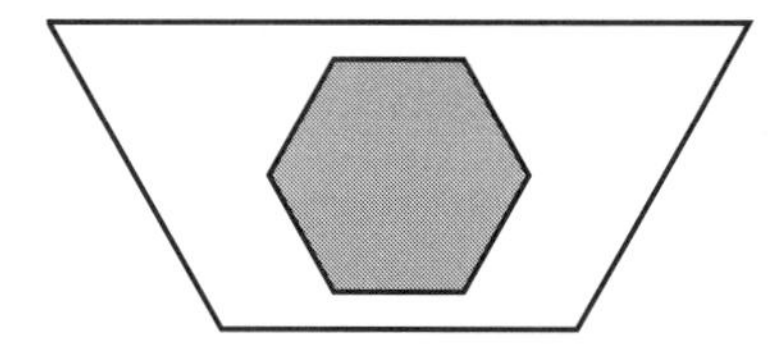

13. What is the probability that a fly will land on the regular hexagon? ______________________
14. What is the probability that the fly will land on the trapezoid but *not* the hexagon? ______________________

NAME ______________________ CLASS ____________ DATE ____________

Practice Masters Level A

6.1 Solid Shapes

For Exercises 1–4, refer to the isometric drawing at the right. Assume that no cubes are hidden from view.

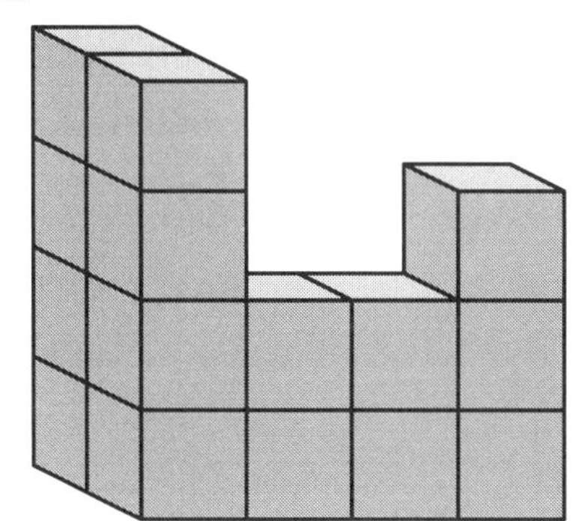

1. Give the volume in cubic units.

   ____

2. Give the surface area in square units.

   ____

3. Draw six orthographic views of the solid. Consider the edge with a length of 4 to be the front of the figure.

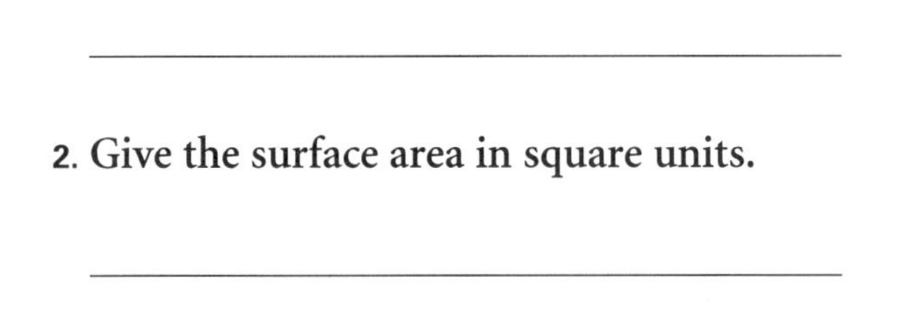

4. On the isometric dot paper provided, draw the solid from a different view.

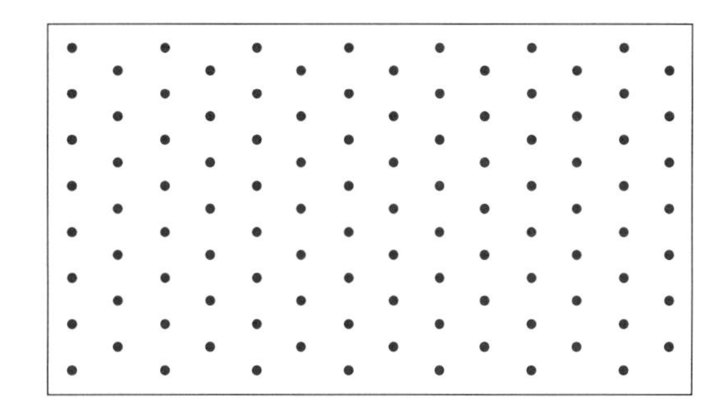

5. Each of the four solids at the right has a volume of 8 cubic units. Draw two other solids with a volume of 8 cubic units that are completely different than the solids shown.

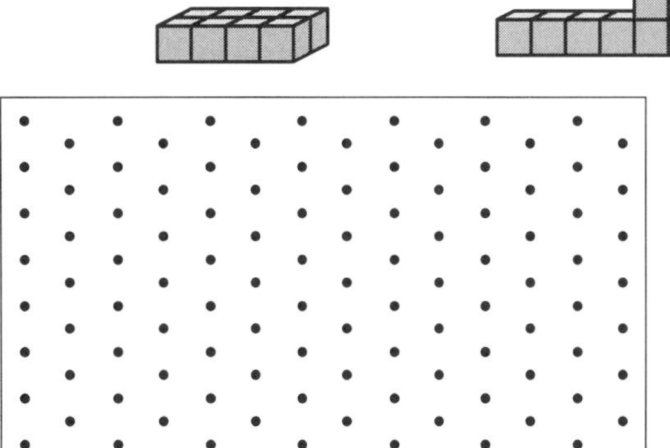

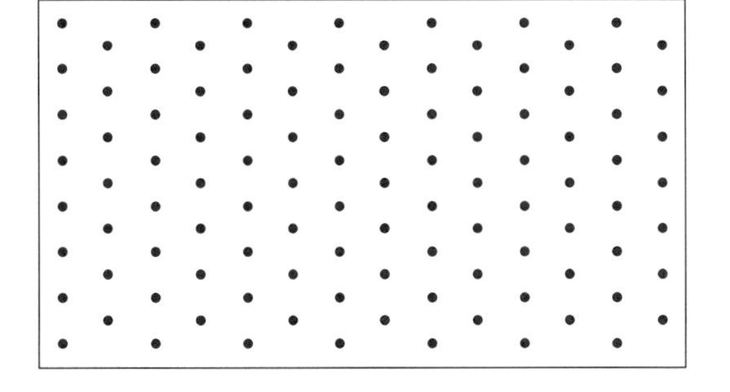

NAME ______________________ CLASS ____________ DATE ____________

Practice Masters Level B

6.1 Solid Shapes

For Exercises 1–3, refer to the isometric drawing at the right. Assume that no cubes are hidden from view.

1. Give the volume in cubic units.

____

2. Give the surface area in square units.

____

3. Draw six orthographic views of the solid. Consider the edge with a length of 5 to be the front of the figure.

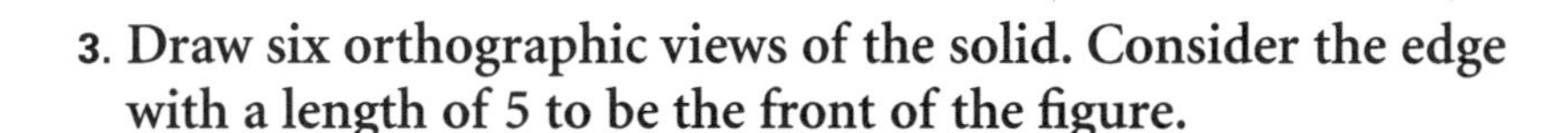

__________ __________ __________

__________ __________ __________

4. On the isometric dot paper provided, create a solid with a volume of 10 cubic units that includes one hidden cube.

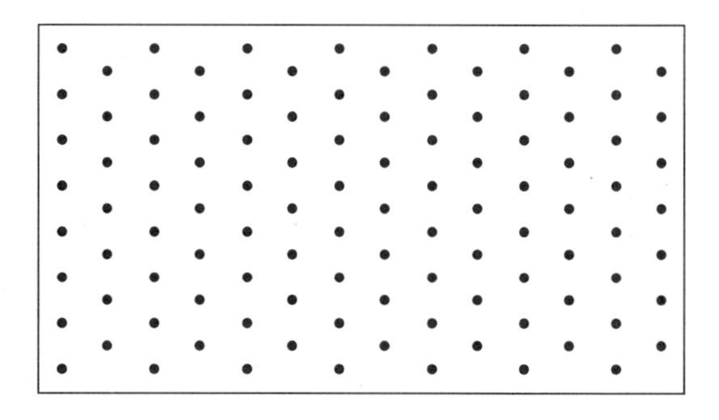

5. Draw six orthographic views of your solid from Exercise 4. Label the front of the solid in each view.

__________ __________ __________

__________ __________ __________

Practice Masters Level C

6.1 Solid Shapes

Refer to the isometric drawing at the right. Assume that four cubes are hidden from view.

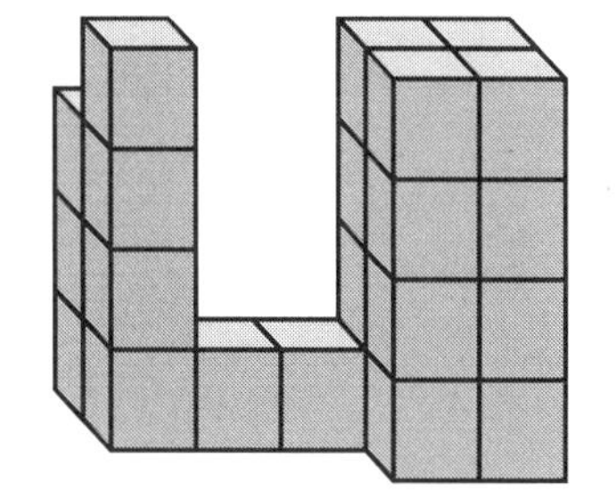

1. Give the volume in cubic units.

____

2. Give the surface area in square units.

____

3. On the isometric dot paper provided, create a solid with a volume of 12 cubic units with at least two stacks of three high and two stacks of two high with no hidden cubes.

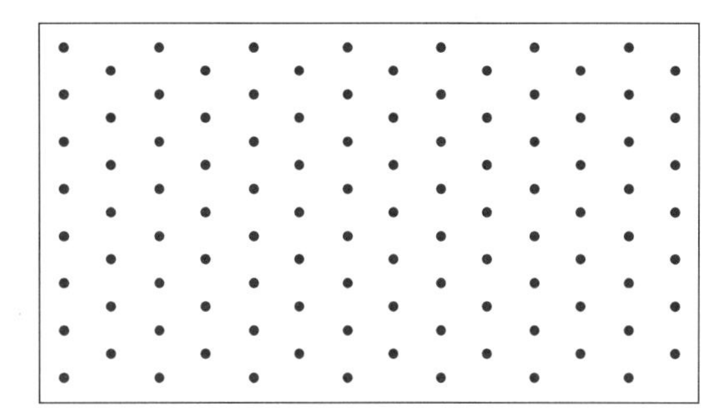

4. On the isometric dot paper provided, create a solid with a volume of 15 cubic units that includes as many hidden cubes as possible. Make sure your figure has at least two stacks that are three units high and two stacks that are two units high. State the number of cubes that are hidden.

____

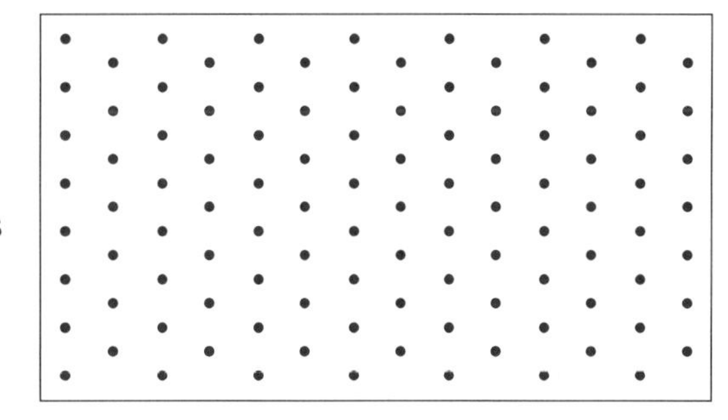

5. Draw six orthographic views of your solid from Exercise 4. Label the front of the solid in each view.

________ ________ ________ ________ ________ ________

NAME ______________________ CLASS ______________ DATE ______________

Practice Masters Level A

6.2 Spatial Relationships

For Exercises 1–4, refer to the figure at the right.

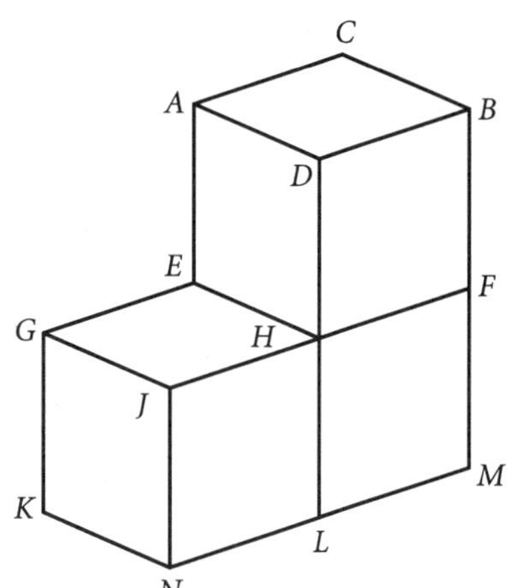

1. Name three pairs of parallel segments.

____

2. Name two pairs of line segments that are skew to each other.

____

3. Name two line segments that are perpendicular to *ADBC*.

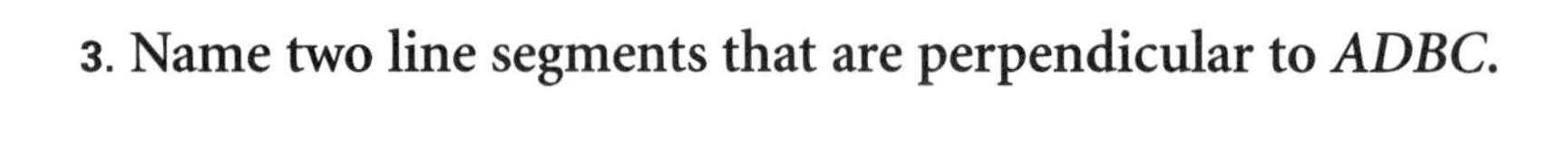

____

4. How many faces are in the polyhedron? ______________

Decide whether each statement below is *always true, sometimes true,* or *never true.*

5. If plane $\mathcal{P}$ is parallel to plane $\mathcal{Q}$, then the lines on $\mathcal{P}$ are parallel to the lines on $\mathcal{Q}$. ______________

6. If two lines are parallel, then the planes that contain the lines are parallel. ______________

7. If two planes are perpendicular, then the lines on the planes are perpendicular to each other. ______________

Complete the following proof: Given: $\overline{AD} \perp$ plane $\mathcal{M}$, $\overline{AC} \cong \overline{BC}$ **Prove:** $\triangle ADC \cong \triangle BDC$

Statements	Reasons
$\overline{AD} \perp$ plane $\mathcal{M}$ $\overline{AC} \cong \overline{BC}$	Given
8. $\overline{AD} \perp$ ________ $\overline{BD} \perp$ ________	Lines perpendicular to a plane are perpendicular to all lines in the plane passing through the same point.
$\angle ADC$ and $\angle BDC$ are right angles.	**9.**
$\triangle ADC$ and $\triangle BDC$ are right triangles.	**10.**
11. $\overline{DC} \cong$ ________	Reflexive Property
$\triangle ADC \cong \triangle BDC$	**12.**

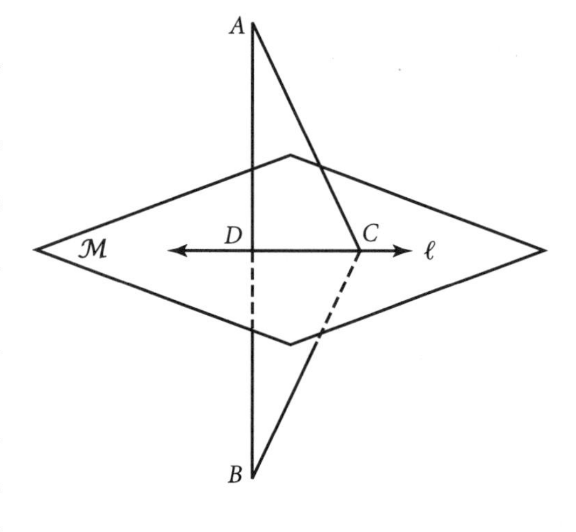

Practice Masters Level B

6.2 Spatial Relationships

For Exercises 1–3, refer to the solid made with cubes at the right.

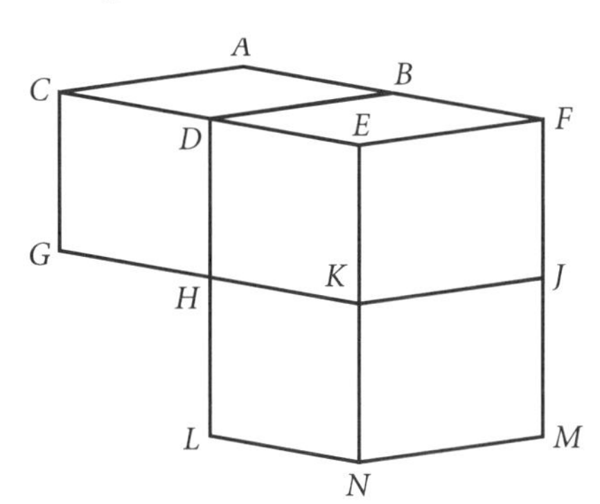

1. Name two pairs of parallel planes.

____

2. Name three segments that are parallel to $\overline{DH}$.

____

3. Name three segments perpendicular to $\overline{DC}$.

____

Decide whether the following statements are *always true, sometimes true,* or *never true*.

4. Parallel planes intersect in a plane. _______________

5. Perpendicular planes intersect in a plane. _______________

6. The intersection of a line and a plane is a point. _______________

Complete the following proof: Given: $ADHE \cong BCGF$, $\overline{CD} \perp \overline{BC}$ and $\overline{BA} \perp \overline{BC}$
Prove: $ABCD$ is a parallelogram.

Statements	Reasons
7. $ADHE \cong BCGF$, $\overline{CD} \perp \overline{BC}$ and $\overline{BA} \perp \overline{BC}$	Given
$\overline{AD} \cong$ ____________ $\overline{EH} \cong$ ____________	Polygon Congruence Postulate
$\overline{CD} \perp \overline{BC}$ and $\overline{BA} \perp \overline{BC}$	8.
$\overline{AD} \parallel \overline{BC}$	9.
$ABCD$ is a parallelogram.	10.

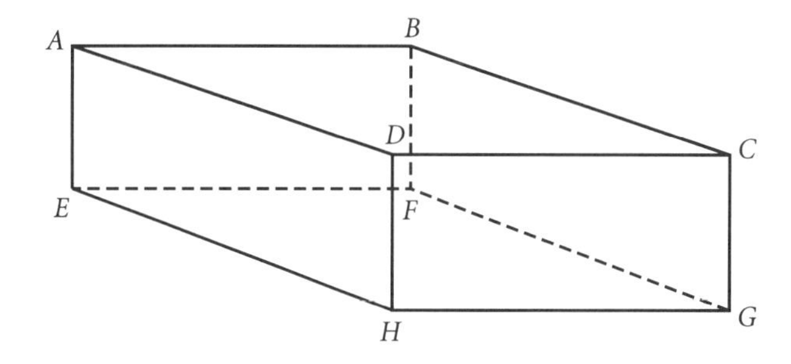

NAME ______________________ CLASS __________ DATE __________

Practice Masters Level C

6.2 *Spatial Relationships*

For Exercises 1–4, refer to the regular figure at the right.

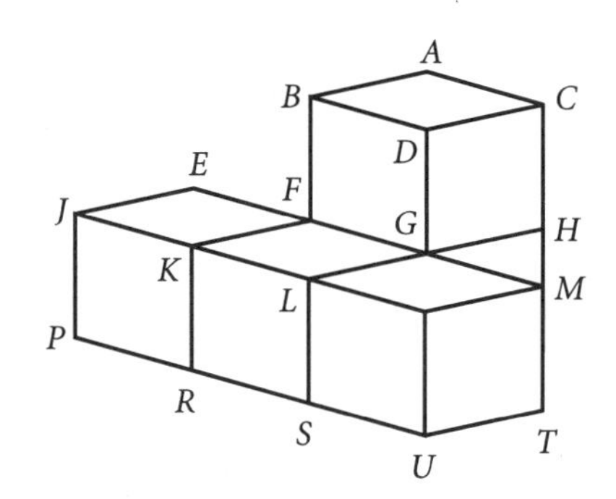

1. List two pairs of parallel planes.

____

2. List three pairs of parallel lines.

____

3. Give two pairs of skew lines. How do you know they are skew?

____

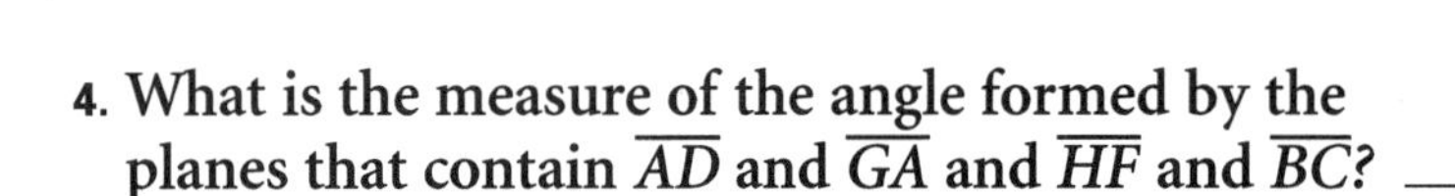

4. What is the measure of the angle formed by the planes that contain $\overline{AD}$ and $\overline{GA}$ and $\overline{HF}$ and $\overline{BC}$? _______________

For Exercises 5–8, refer to the figure at the right.

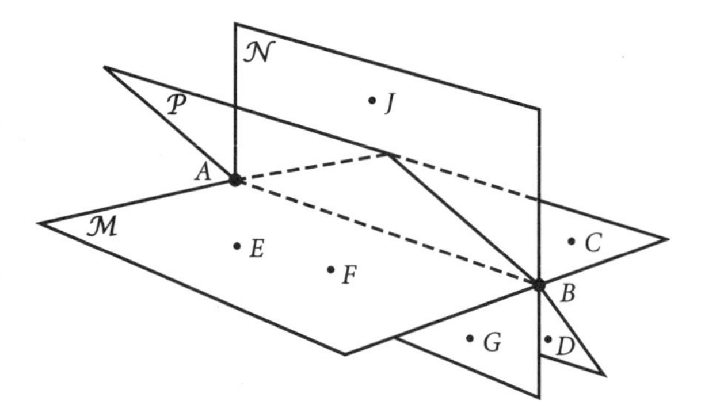

5. Name the intersection of planes $\mathcal{N}$, $\mathcal{P}$, $\mathcal{M}$.

____

6. Name a plane that contains points *J* and *G*.

____

7. How many planes can contain point *B*?

____

8. Is there a plane that contains $\overline{AB}$? How many?

____

Complete each statement with *sometimes, always,* or *never*.

9. Polyhedrons __________ have at least one pair of parallel planes.

10. Polyhedrons __________ have at least one pair of skew lines.

NAME ______________________ CLASS ____________ DATE ____________

Practice Masters Level A

6.3 Prisms

For Exercises 1–4, refer to the right trapezoidal prism at the right.

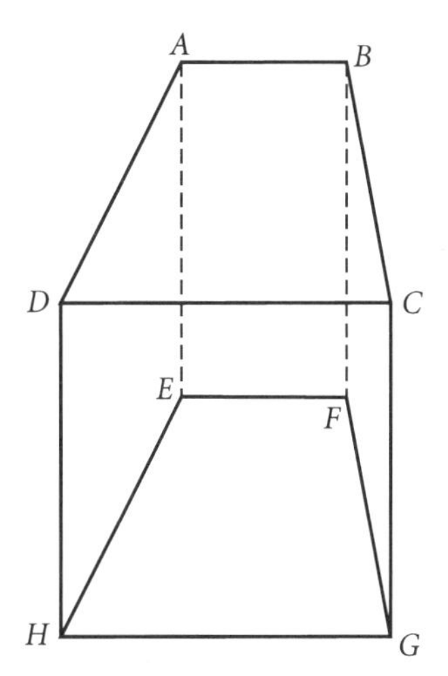

1. Name the bases of the prism.

____

2. Name the lateral faces of the prism.

____

3. Is the figure a regular right prism? Why or why not?

____

4. What geometric figure makes up the faces?

____

For Exercises 5–12, refer to the drawing of the right rectangular prism. Complete the following table. Round all answers to the nearest tenth.

	Length, l	Width, w	Height, h	Diagonal, d
5.	5	4	3	______
6.	7	6	11	______
7.	12	10	8	______
8.	$4\frac{1}{2}$	$2\frac{1}{2}$	$3\frac{1}{2}$	______
9.	5	7	6	______
10.	______	4	8	12
11.	2	______	3	18
12.	4	6	______	12

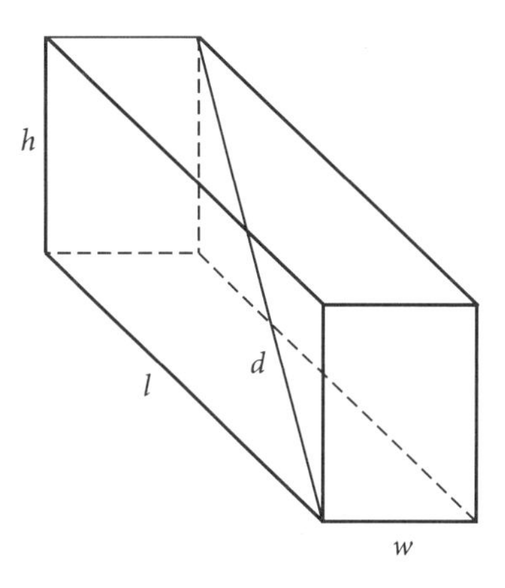

13. Can you use the same formula for finding the length of a diagonal for all right rectangular prisms? Why or why not? ______________________

____

NAME ______________________ CLASS ____________ DATE ____________

Practice Masters Level B

6.3 Prisms

For Exercises 1–6, refer to the right triangular prism at the right.

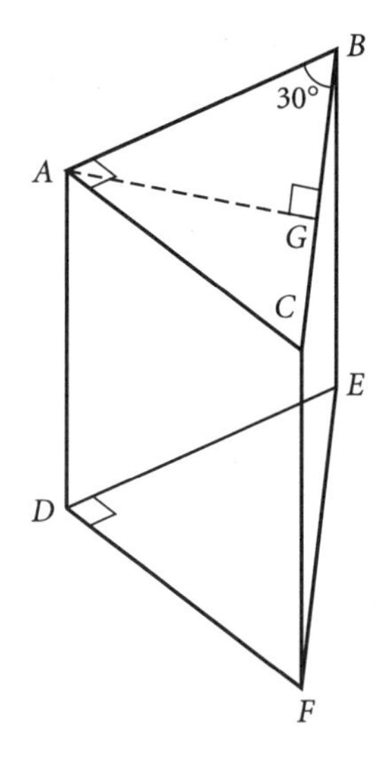

1. Name the bases using exact geometric language.

   ____

2. Name the lateral faces. ______________________

3. What geometric shape describes the faces? ____________

4. If AG is $6\sqrt{3}$ feet, what is the perimeter of the base? ______

5. Are any of the faces parallel? ______________________

6. Find as many lines skew to $\overline{AG}$ as possible. ____________

For Exercises 7–8, refer to the drawing of the rectangular prism. Complete the following table. Round answers to the nearest tenth.

	Length, l	Width, w	Height, h	Diagonal, d
7.	4.1	7.8	9.3	________
8.	$42\frac{1}{3}$	$16\frac{1}{4}$	$18\frac{1}{2}$	________

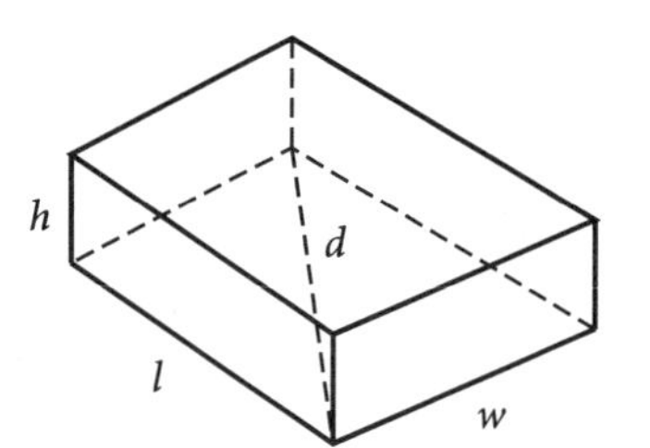

For Exercises 9–12, refer to the drawing of the regular hexagonal prism. Complete the following table. Use the formula $d = \sqrt{4l^2 + h^2}$. Round answers to the nearest tenth.

	Length, l	Height, h	Diagonal, d
9.	9	6	________
10.	4	6	________
11.	________	8	$2\sqrt{137}$
12.	$\sqrt{26}$	________	15

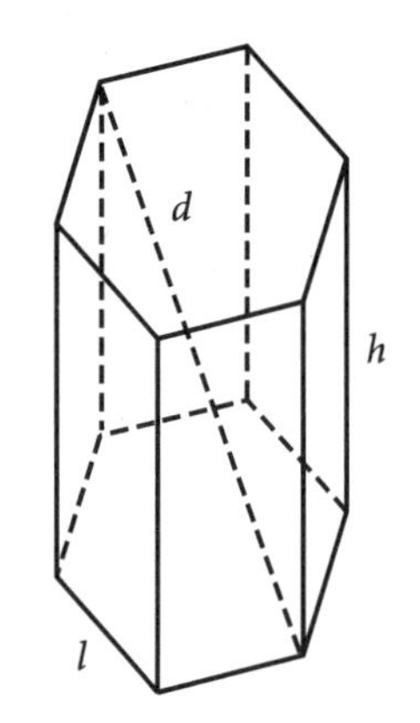

13. Can you use the same formula for finding the length of a diagonal for rectangular prisms, for all prisms? Why or why not? ______________________

Practice Masters Level C

6.3 Prisms

For Exercises 1–4, refer to the regular hexagonal prism at the right.

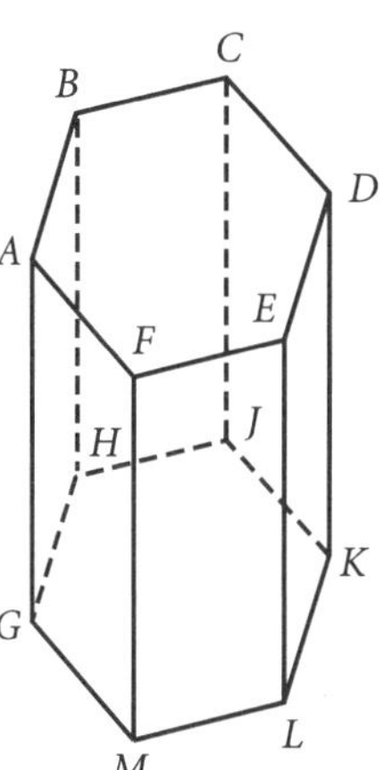

1. Name the bases and give the precise geometric name for it.

____

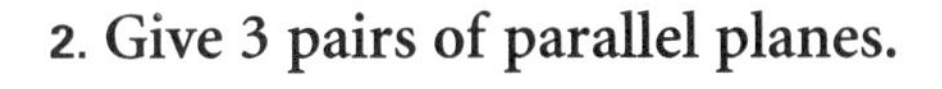

2. Give 3 pairs of parallel planes.

____

3. If the apothem is $4\sqrt{3}$ inches, find the perimeter of the base.

____

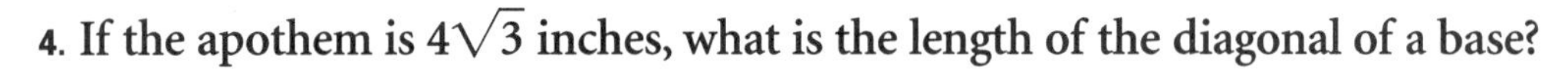

4. If the apothem is $4\sqrt{3}$ inches, what is the length of the diagonal of a base?

____

For Exercises 5–6, refer to the drawing of the rectangular prism. Complete the following table. Round all answers to the nearest tenth.

	Length, l	Width, w	Height, h	Diagonal, d
5.	$\sqrt{24}$	$\sqrt{18}$	3	______
6.	$4\sqrt{2}$	$3\sqrt{3}$	$7\sqrt{6}$	______

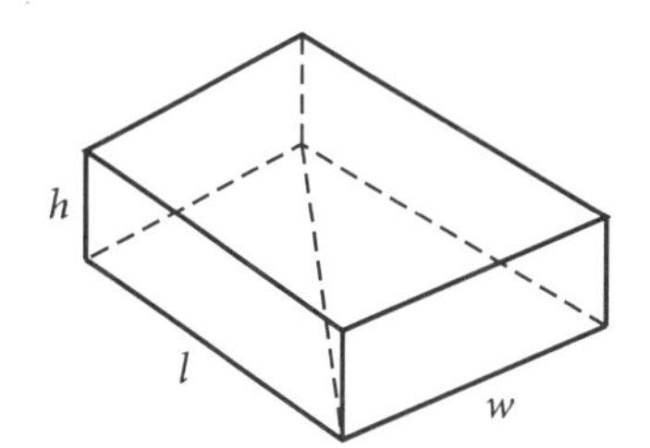

For Exercises 7–10, refer to the drawing of the right triangular prism. Complete the following table. Round all answers to the nearest tenth.

	Side, s	Altitude, a	Height, h	Diagonal, d
7.	$6\sqrt{3}$	______	6	______
8.	______	$\sqrt{3}$	______	$\sqrt{15}$
9.	8	______	12	______
10.	______	6	______	$6\sqrt{2}$

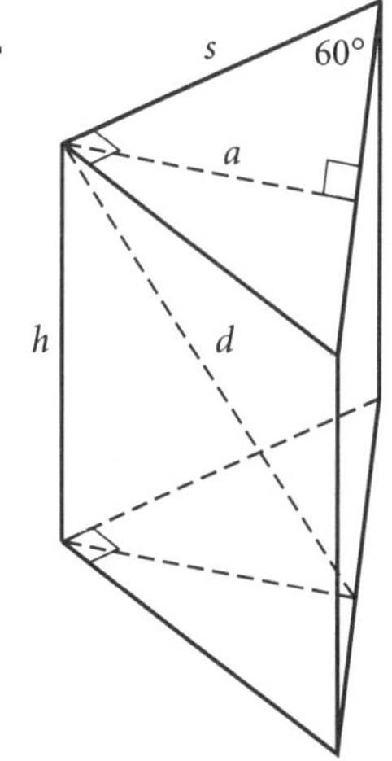

11. Give a formula for the diagonal of a regular hexagon given the length of a side and the apothem. Explain why the formula is correct.

____

____

NAME ______________________ CLASS ____________ DATE ____________

Practice Masters Level A

6.4 Coordinates in Three Dimensions

Name the octant, coordinate plane, or axis in which each point is located.

1. $(2, 4, -3)$ ______________________
2. $(-2, 4, 3)$ ______________________
3. $(2, -4, 3)$ ______________________
4. $(-2, -4, 3)$ ______________________
5. $(2, -4, -3)$ ______________________
6. $(-2, 4, -3)$ ______________________
7. $(2, 4, 3)$ ______________________
8. $(-2, -4, -3)$ ______________________

For Exercises 9–12, label the coordinate axes and locate each pair of points in a three-dimensional coordinate system. Find the distance between the points, and find the midpoint of the segment connecting them.

9. $(2, 1, 5)$ and $(4, 7, 7)$

____

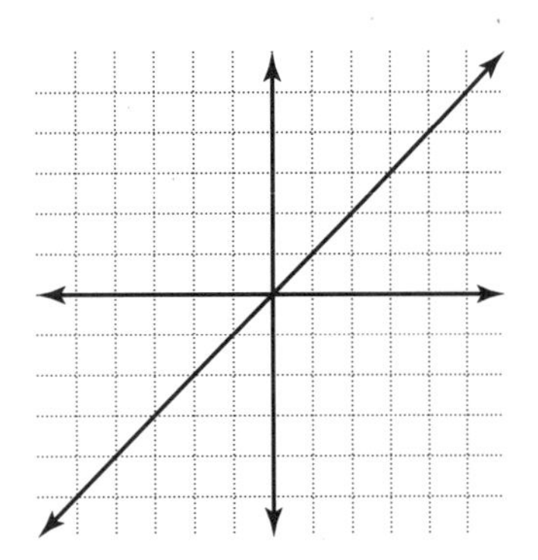

10. $(6, -1, 3)$ and $(9, 3, 5)$

____

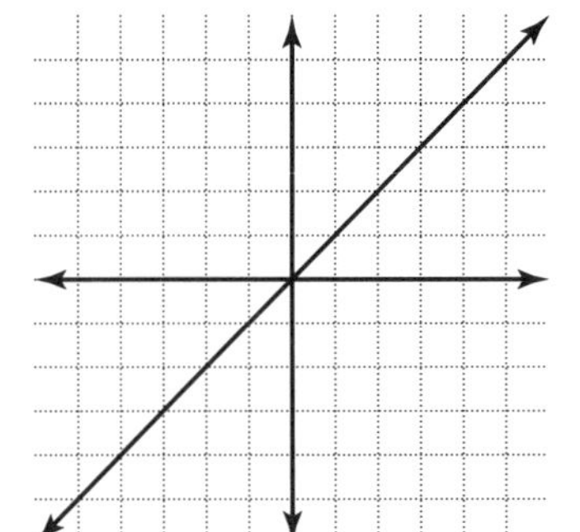

11. $(-3, 2, -5)$ and $(4, -2, 2)$

____

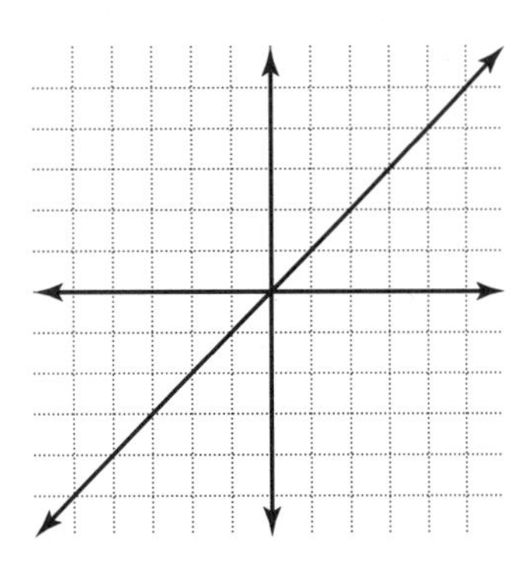

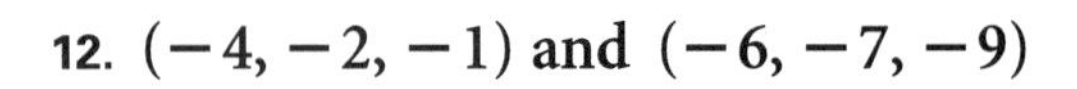

12. $(-4, -2, -1)$ and $(-6, -7, -9)$

____

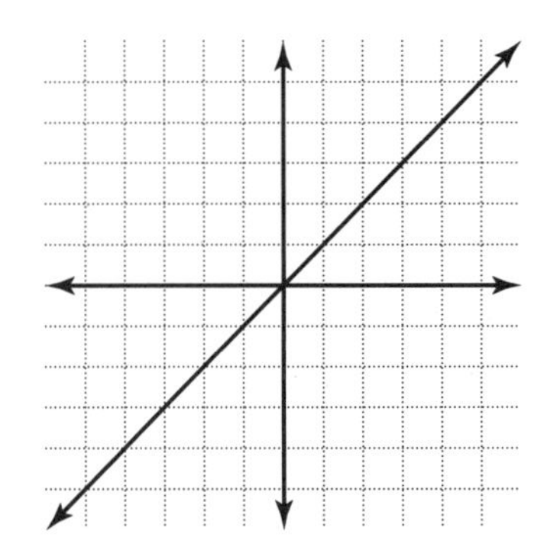

NAME ______________________ CLASS __________ DATE __________

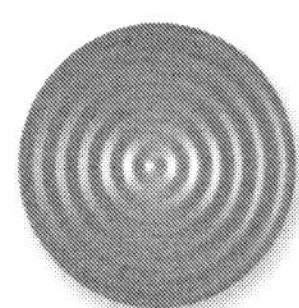

Practice Masters Level B

6.4 Coordinates in Three Dimensions

Name the octant, coordinate plane, or axis in which each point is located.

1. $(-5, 7, 2)$ ______________
2. $(0, 7, -8)$ ______________
3. $(4, -11, 6)$ ______________
4. $(-8, -10, 4)$ ______________
5. $(6, -16, -3)$ ______________
6. $(-9, 0, 0)$ ______________
7. $(-2, 8, -6)$ ______________
8. $(-5, -9, 0)$ ______________

For Exercises 9–10, label the coordinate axes and locate each pair of points in a three-dimensional coordinate system. Find the distance between the points, and find the midpoint of the segment connecting them.

9. $(3, -4, 5)$ and $(-4, 5, 3)$

____

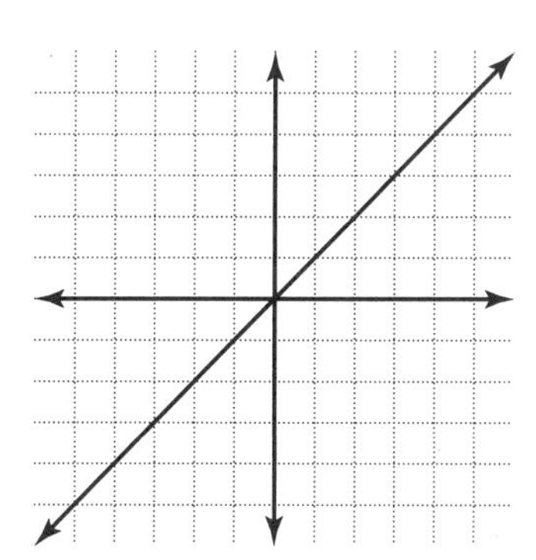

10. $(7, -8, 4)$ and $(2, 4, 9)$

____

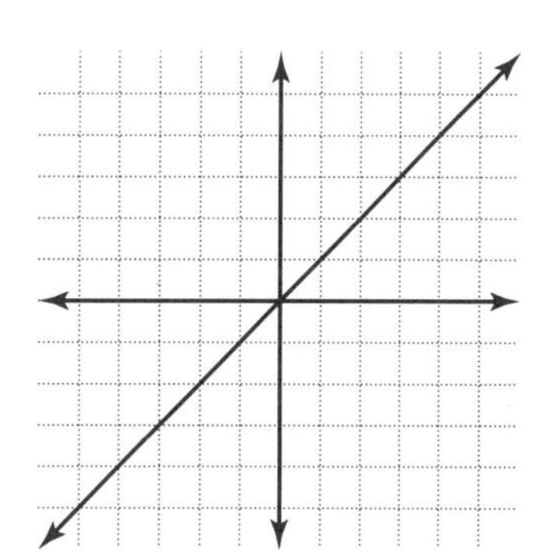

For Exercises 11–21, refer to the right rectangular prism below. Determine the coordinates of each point.

11. point H __________
12. point G __________
13. point B __________
14. point F __________

Find each measure.

15. AD __________
16. AB __________
17. EC __________
18. HD __________
19. FA __________
20. GB __________
21. area of $FGCB$ __________

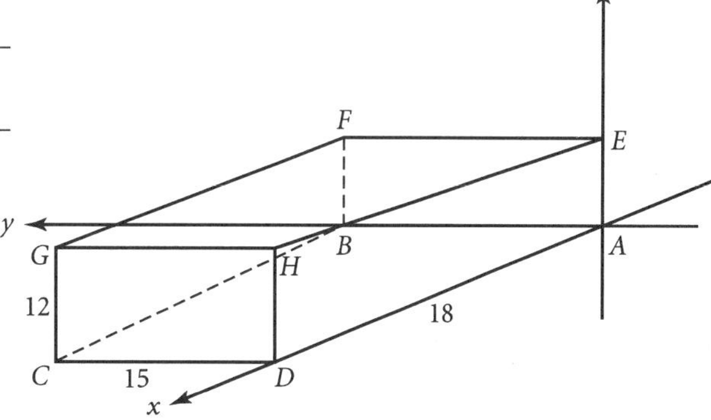

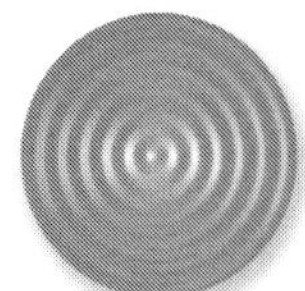

Practice Masters Level C

6.4 Coordinates in Three Dimensions

Name the octant, coordinate plane, or axis in which each point is located.

1. $(0, -3, 12)$ ____________ 2. $(17, -11, 31)$ ____________

3. $(-8, 0, 0)$ ____________ 4. $(-9, 4, 0)$ ____________

5. $(5, 9, 21)$ ____________ 6. $(0, 0, 43)$ ____________

For Exercises 7–8, label the coordinate axes and locate each pair of points in a three-dimensional coordinate system. Find the distance between the points and find the midpoint of the segment connecting them.

7. $(12, -15, 21)$ and $(8, 5, 12)$

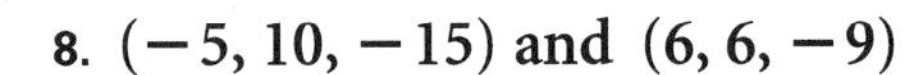

8. $(-5, 10, -15)$ and $(6, 6, -9)$

____________ ____________

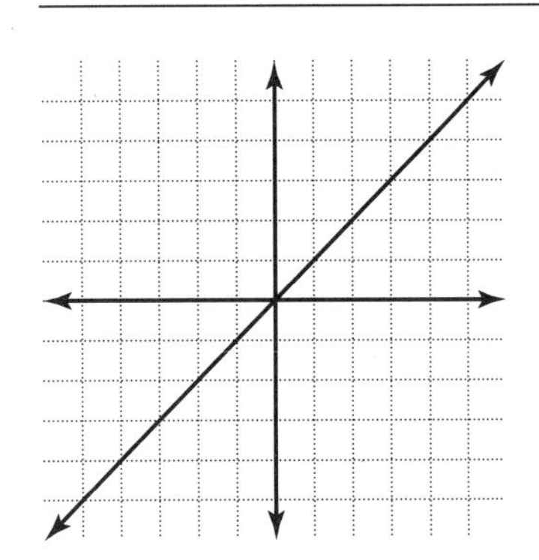

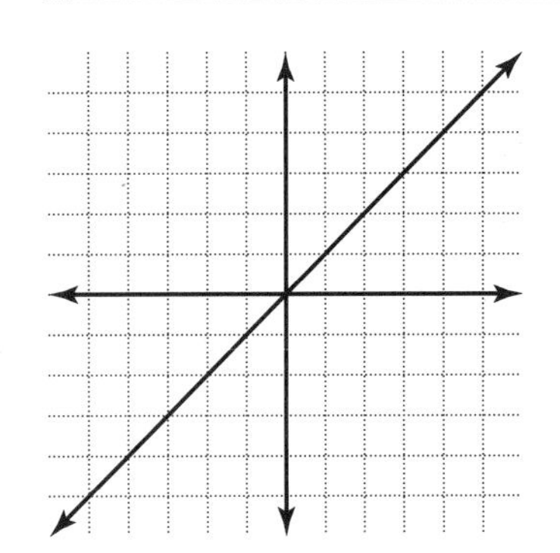

Given the midpoint, *M*, and one endpoint $\overline{AB}$, find the following:

	Other endpoint of $\overline{AB}$	***AB***
9. $M(-6, 9, 4)$ and $A(-2, 6, 7)$	____________	____________
10. $M(-3, 6, 0)$ and $B(0, -7, 4)$	____________	____________
11. $M(22, 12, -8)$ and $A(3\frac{1}{2}, -6, -12)$	____________	____________

For Exercises 12–17, refer to the diagram of the right rectangular prism. Determine the following:

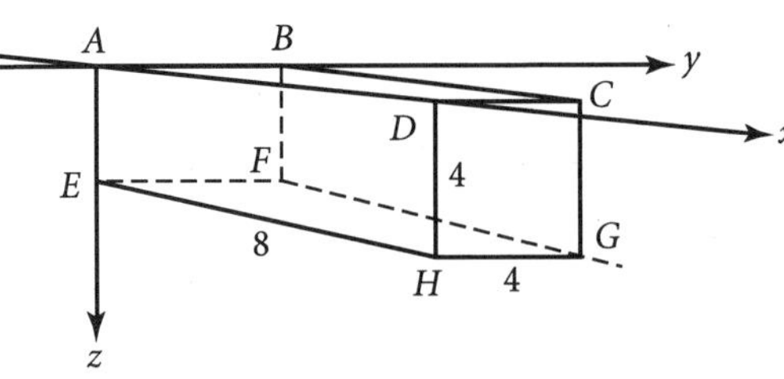

12. point B ________ 13. point F ________

14. point H ________ 15. GH ________

16. AG ________ 17. CF ________

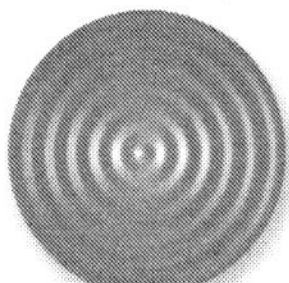

Practice Masters Level A

6.5 *Lines and Planes in Space*

Find the *x*-, *y*-, and *z*-intercepts for each equation.

1. $5x + 2y - 3z = 9$ ______________________

2. $-2x - 5y + z = 12$ ______________________

3. $6x - 4y + 2z = 16$ ______________________

4. $-x + 4y - 4z = 12$ ______________________

5. What is the difference in the graphs of the equations $3x + 5y = 9$ and $3x + 5y + z = 9$? ______________________

____

In the coordinate plane provided, label your axes and plot the line defined by the parametric equations.

6. $x = t - 1$
   $y = 3 - t$

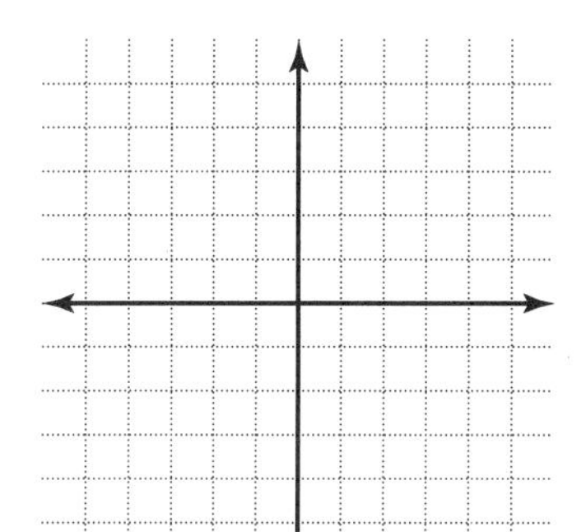

7. $x = 4t$
   $y = t - 2$

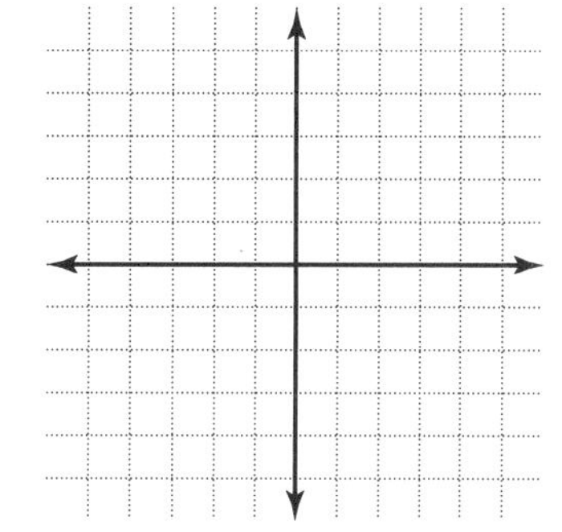

8. $x = 3$
   $y = 2t - 5$
   $z = t + 1$

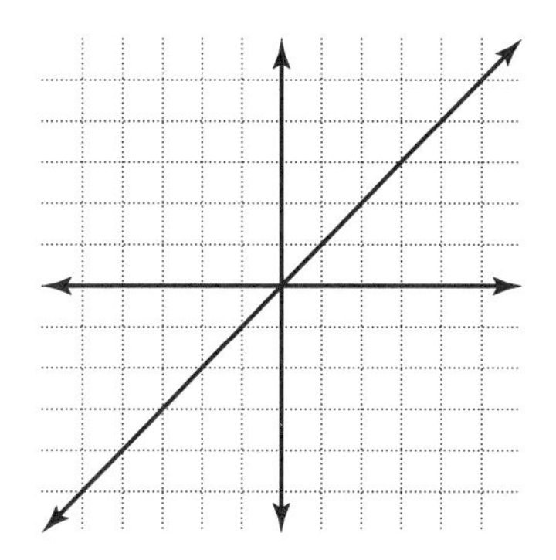

9. $x = -t + 2$
   $y = t - 2$
   $z = 3t$

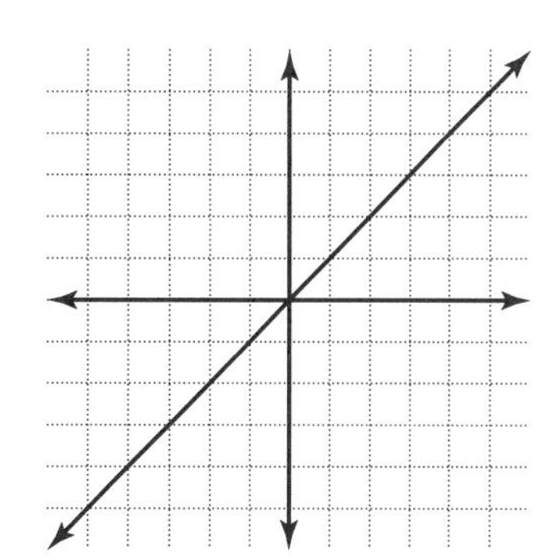

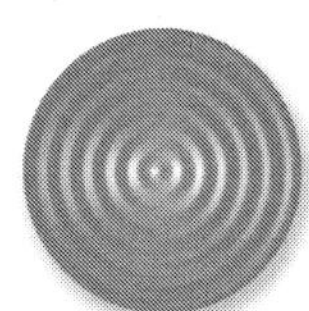

Practice Masters Level B

6.5 *Lines and Planes in Space*

Use the intercepts to sketch the plane defined by each equation below.

1. $2x + 3y - z = 12$

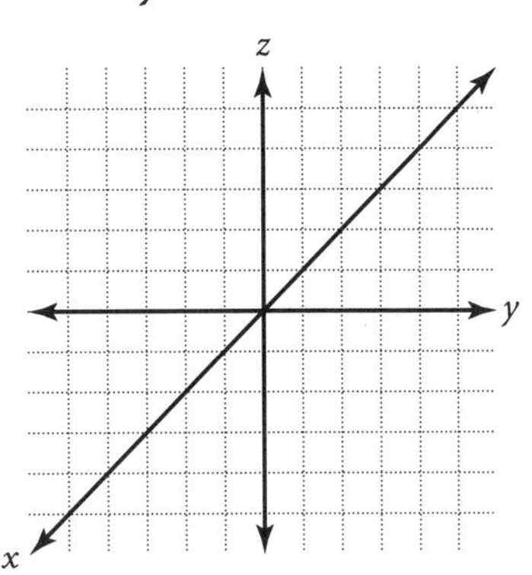

2. $-3x + y - 3z = 9$

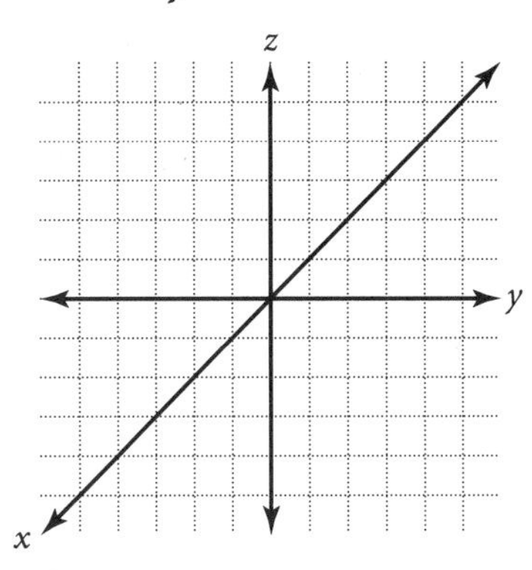

In the coordinate plane provided, label your axes and plot the line defined by the parametric equations.

3. $x = 4t - 2$
   $y = 3 - t$

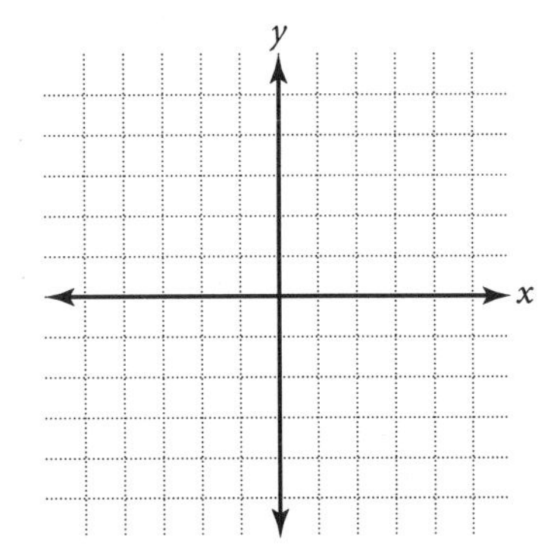

4. $x = \frac{1}{2}t - 4$
   $y = 2t + 6$

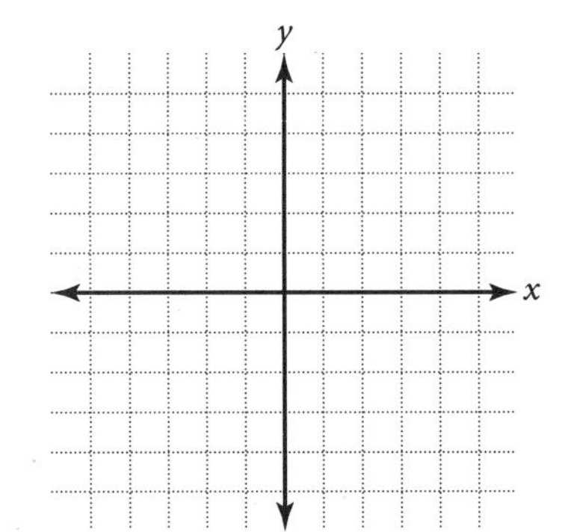

5. $x = -t$
   $y = 2t + 1$
   $z = t - 2$

6. $x = t - 4$
   $y = 3 - t$
   $z = \frac{1}{2}t + 3$

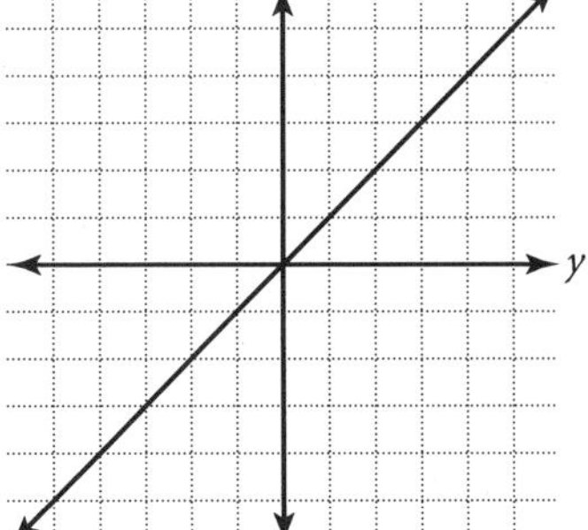

Practice Masters Level C

6.5 Lines and Planes in Space

Use the intercepts to sketch the plane defined by each equation below.

1. $6x - 4y + 3z = 12$

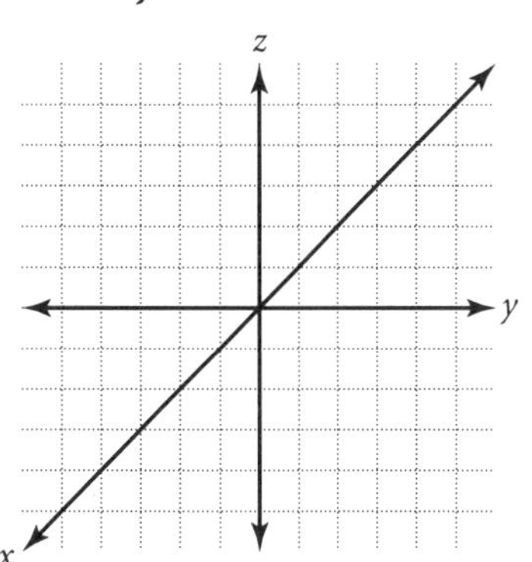

2. $-3x - 2y + z = 6$

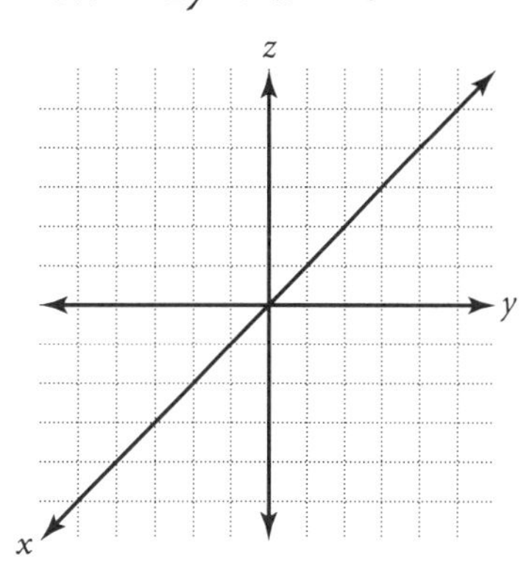

In the coordinate plane provided, label your axes and plot the line defined by the parametric equations.

3. $x = -4t - 3$
   $y = 2 - t$

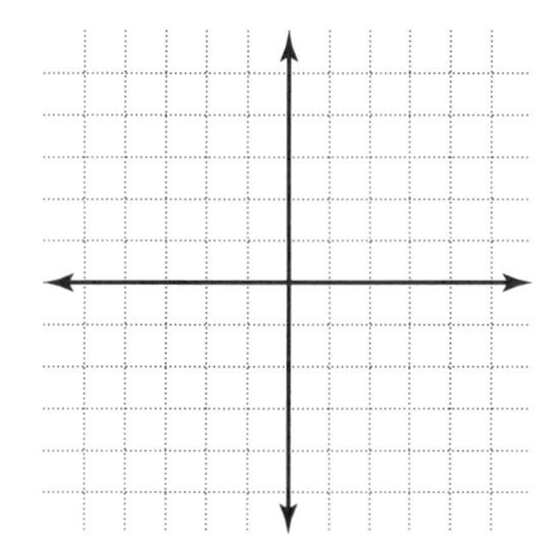

4. $x = 2t + 3$
   $y = 3t$
   $z = 4 - t$

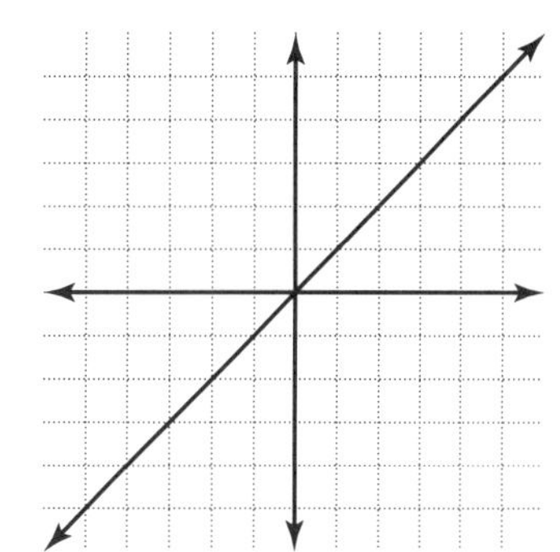

Write each set of symmetric equations in parametric form.

5. $\dfrac{x+6}{-4} = \dfrac{y-4}{3} = \dfrac{z+2}{2}$

   ________ ________ ________

6. $\dfrac{\frac{1}{2}x + 3}{6} = \dfrac{y-1}{-3} = \dfrac{z-5}{-5}$

   ________ ________ ________

Write each set of parametric equations in symmetric form.

7. $x = 4t + 3$
   $y = 3 - t$
   $z = 3t + 1$ ____________________

8. $x = -t - 4$
   $y = 2t$
   $z = 3 - t$ ____________________

NAME ______________________ CLASS __________ DATE __________

Practice Masters Level A

6.6 *Perspective Drawing*

In Exercises 1–4, locate the vanishing point for the figure and draw the horizon line.

1.

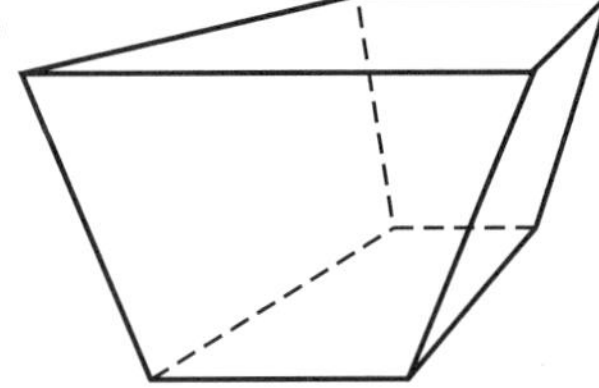

2.

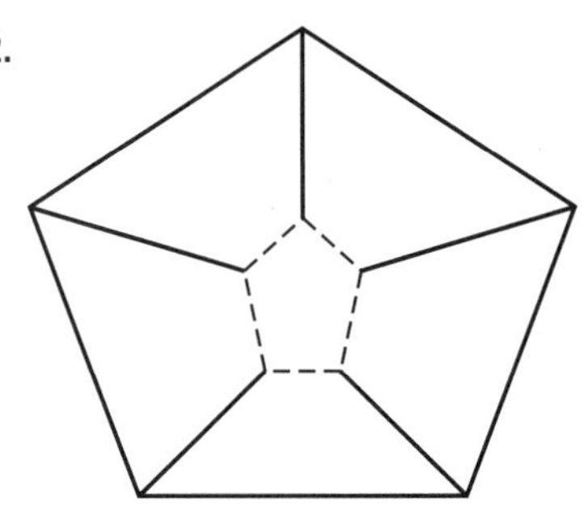

3.

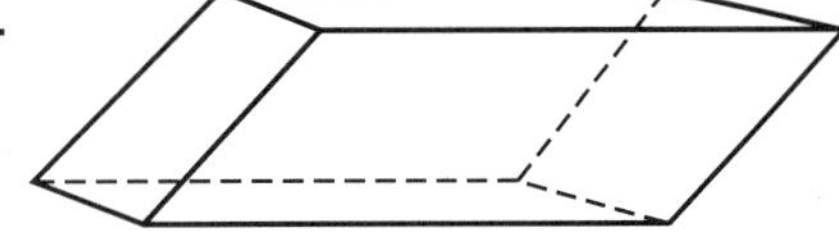

4.

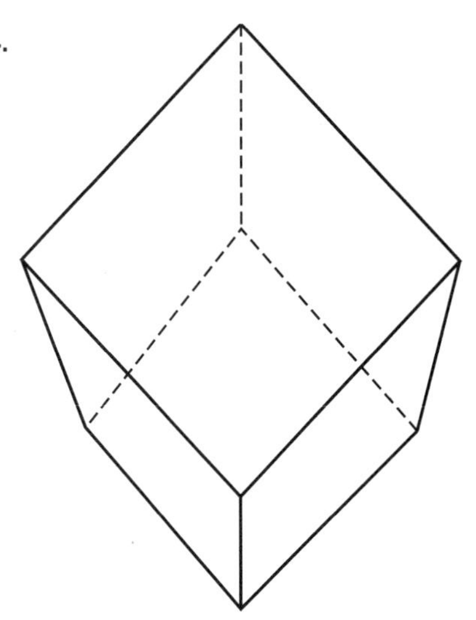

5. In the space provided, make a one-point perspective drawing of a rectangular solid. Place the vanishing point above the solid.

____

6. In the space below, make a two-point perspective drawing of a rectangular solid. Place the vanishing points above the solid.

____

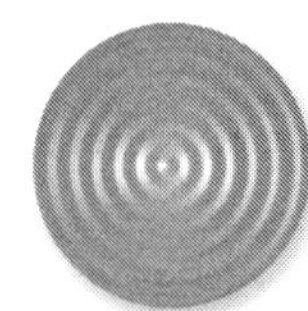

Practice Masters Level B

6.6 *Perspective Drawing*

In Exercises 1–4, locate the vanishing point for the figure and draw the horizon line.

1.

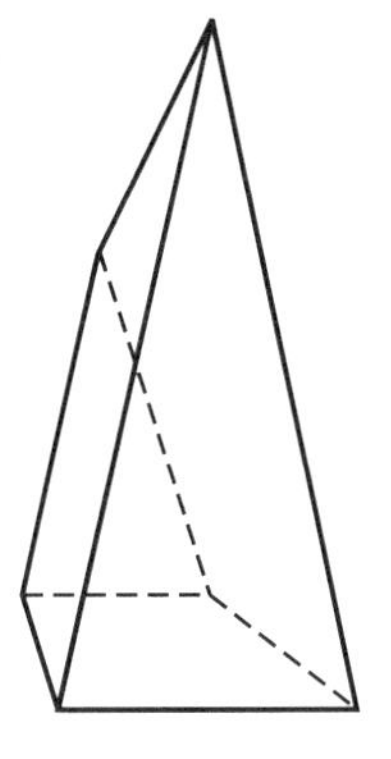

2.

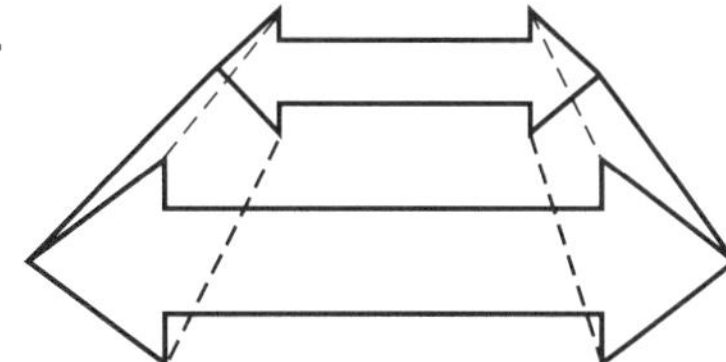

3.

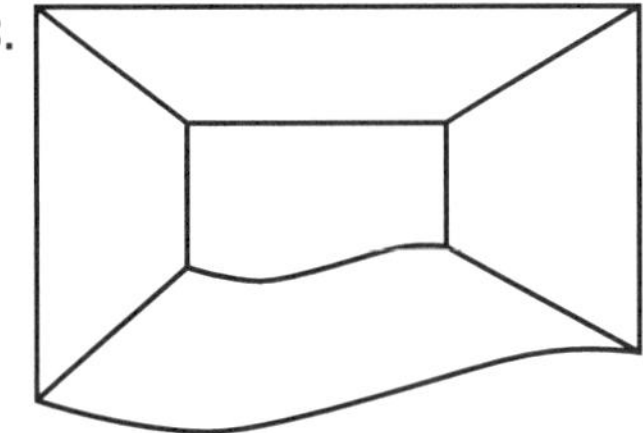

4.

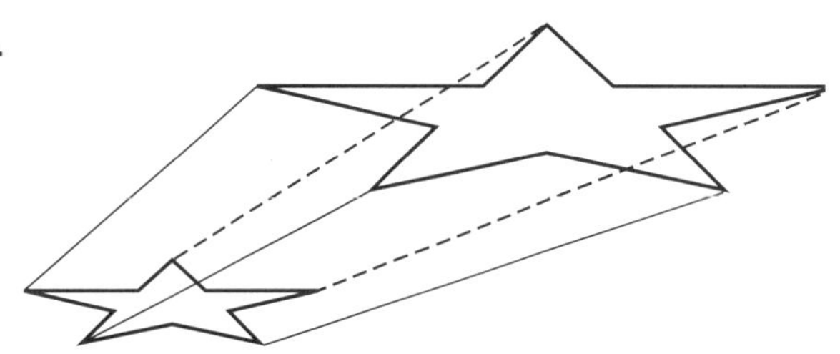

5. In the space provided, make a two-point perspective drawing of a rectangular solid. Place the vanishing points to the right of the solid.

6. Create a perspective drawing of the word MATH.

Practice Masters Level C

6.6 Perspective Drawing

In Exercises 1–4, locate the vanishing point for the figure and draw the horizon line.

1.

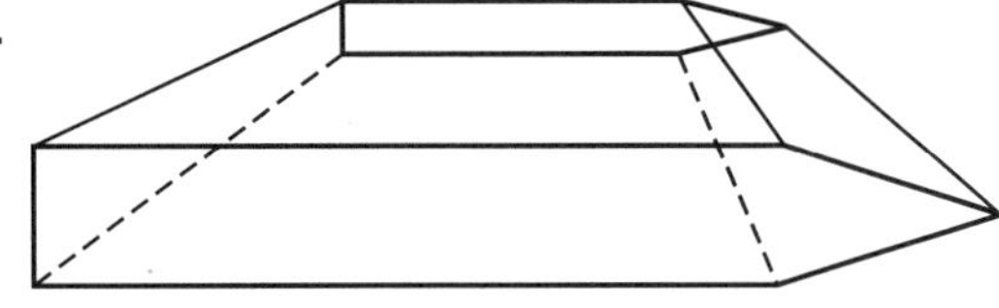

2.

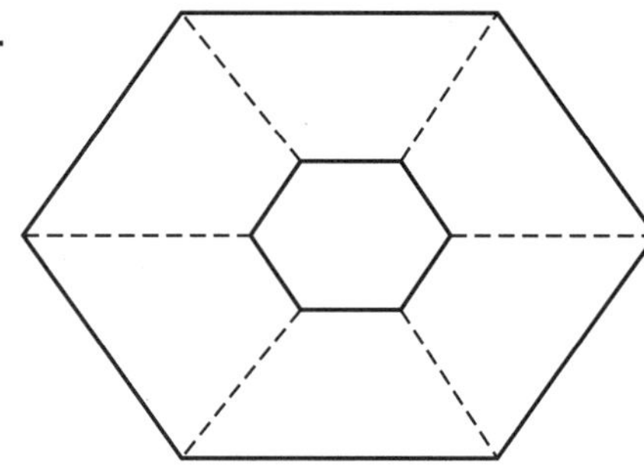

3.

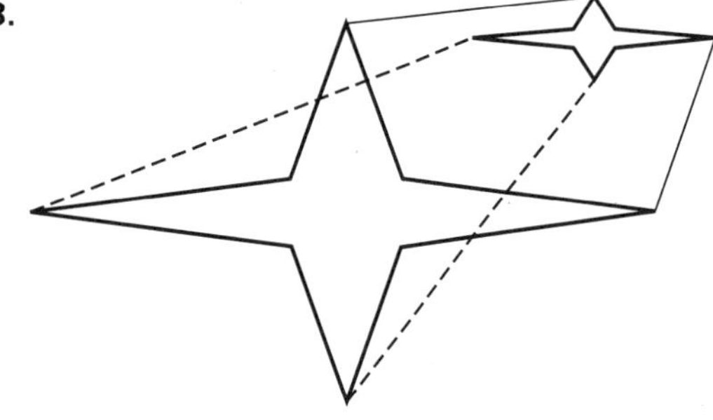

4.

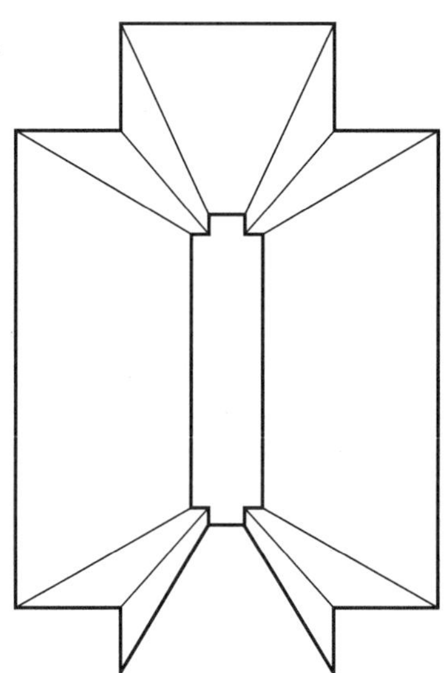

5. Make a two-point perspective drawing of a rectangular solid. Place the vanishing points to the left of the solid.

6. Use one-point or two-point perspective to draw a perspective view of the following:

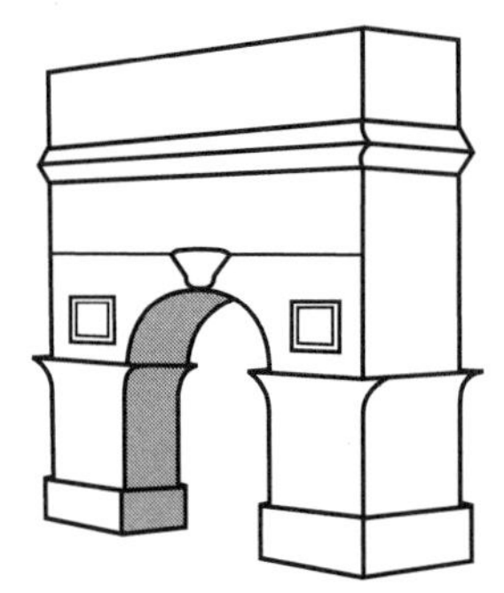

NAME ______________________ CLASS ____________ DATE ____________

Practice Masters Level A

7.1 Surface Area and Volume

Find the surface area and volume for each rectangular prism having the given dimensions.

1. $1 \times 1 \times 1$ ______________
2. $1 \times 4 \times 5$ ______________
3. $1 \times 2 \times 7$ ______________
4. $3 \times 4 \times 5$ ______________
5. $2 \times 2 \times 2$ ______________
6. $2 \times 2 \times 5$ ______________
7. $2 \times 5 \times 5$ ______________
8. $5 \times 5 \times 5$ ______________
9. $3 \times 3 \times 3$ ______________
10. $3 \times 3 \times 4$ ______________
11. $3 \times 4 \times 3$ ______________
12. $4 \times 3 \times 4$ ______________
13. $3 \times 4 \times 4$ ______________
14. $4 \times 4 \times 4$ ______________

Determine the surface-area-to-volume ratio for a rectangular prism with the given dimensions. Show all of your steps.

15. $1 \times 1 \times 1$ ______________
16. $2 \times 2 \times 2$ ______________
17. $3 \times 3 \times 3$ ______________
18. $4 \times 4 \times 4$ ______________
19. $5 \times 5 \times 5$ ______________
20. $6 \times 6 \times 6$ ______________
21. The side of a cube is 3 inches. Find the surface-area-to-volume ratio. ______________
22. The side of a cube is 12 centimeters. Find the surface-area-to-volume ratio. ______________
23. To make an open box, a square is cut from each corner of a 10-inch-by-10-inch cardboard. What is the whole-number length for the side of the square that will create a box having the greatest volume? ______________

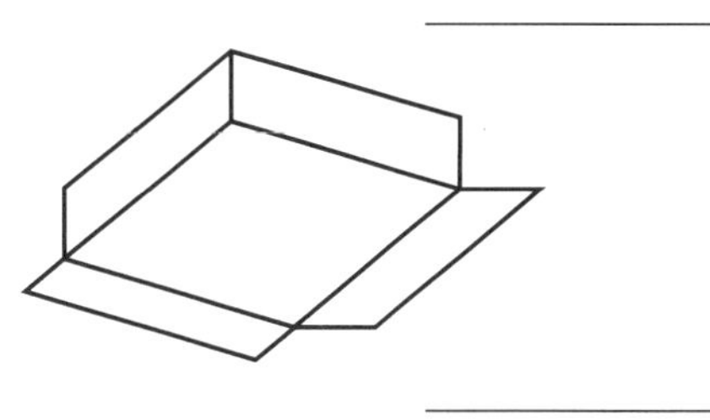

24. The dimensions of Box A are 3 inches by 9 inches by 8 inches. The dimensions of Box B are 2 inches by 12 inches by 9 inches. The volumes are the same. Which has the smaller surface area? ______________

NAME ______________________ CLASS __________ DATE __________

Practice Masters Level B

7.1 Surface Area and Volume

Determine the surface-area-to-volume ratio for a rectangular prism with the given dimensions.

1. $6 \times 10 \times 10$ ____________
2. $4 \times 4 \times 8$ ____________
3. $10 \times 11 \times 20$ ____________
4. $2 \times 3 \times 4$ ____________
5. $3.2 \times 5.1 \times 7$ ____________
6. $5 \times 12 \times 15$ ____________

For Exercises 7–10, determine whether you should maximize the volume or minimize the surface area. Explain your reasoning.

7. designing baby food jars that will hold 4 ounces of fruit ____________
8. building a rabbit pen with a limited amount of fencing ____________
9. building a silo that cannot be taller than 20 feet high ____________
10. designing a box whose length is two times smaller than the width ____________

Solve.

11. The side of a cube is 18 inches. Find the surface-area-to-volume ratio. ____________
12. The volume of a rectangular prism is 2000 cubic centimeters. Two of the sides are 8 centimeters and 10 centimeters. Find the length of the missing side. ____________
13. The surface-area-to-volume ratio of a cube is 1 to 2. Find the smallest possible length of its side. ____________

The dimensions of Box A are 2 inches by 6 inches by 10 inches. The dimensions of Box B are 3 inches by 5 inches by 8 inches. Use this information for Exercises 14–16.

14. Find the difference in volumes. ____________
15. Find the difference in surface areas. ____________
16. Find the ratio of surface area of Box A to that of Box B. ____________

NAME ______________________ CLASS __________ DATE __________

Practice Masters Level C

7.1 Surface Area and Volume

For Exercises 1 and 2, determine whether you should maximize the volume or minimize the surface area. Explain your reasoning.

1. designing a child's cup that will hold no more than 6 ounces of juice

____

2. constructing a sand box with a limited amount of lumber

____

3. Compare the surface-area-to-volume ratio of a $s \times s \times 2$ rectangular prism with that of a $s \times s \times s$ rectangular prism as s decreases. __________

4. The volume of a rectangular prism is 135 cubic inches. Two of its sides are 3 inches and 9 inches. Find the surface area. __________

5. The surface area of a rectangular prism is 504 cubic inches. Two of its sides are 6 inches and 12 inches. Find the volume. __________

6. The volume of a rectangular prism is 500 cubic centimeters. Two of its sides are 5 centimeters and 10 centimeters. Find the surface-area-to-volume ratio. __________

7. The volume of a rectangular prism is 216 cubic inches. Two of its sides are 6 inches and 12 inches. Find the surface-area-to-volume ratio. __________

8. The surface-area-to-volume ratio of a rectangular prism is 4 to 5. Two of the sides are 10 centimeters and 20 centimeters. Find the length of the missing side. __________

9. The surface-area-to-volume ratio of a rectangular prism is 13 to 18. Two of the sides are 9 centimeters and 12 centimeters. Find the length of the missing side. __________

10. The surface-area-to-volume ratio of a cube is 2 to 5. Find the length of the side. __________

11. The surface-area-to-volume ratio of a cube is 10 to 3. Find the length of the side. __________

12. The surface-area-to-volume ratio of a cube is 3 to 2. Find the surface area. __________

13. The surface-area-to-volume ratio of a cube is 3 to 4. Find the volume. __________

NAME ______________________ CLASS ____________ DATE ____________

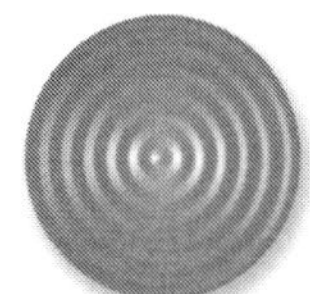

Practice Masters Level A

7.2 *Surface Area and Volume of Prisms*

Find the volume of a prism with the given dimensions.

1. $B = 20 \text{ in.}^2$, $h = 5$ in. ____________
2. $B = 8 \text{ cm}^2$, $h = 9$ cm ____________
3. $B = 12 \text{ cm}^2$, $h = 4$ cm ____________
4. $B = 15 \text{ in.}^2$, $h = 15$ in. ____________
5. $B = 20 \text{ in.}^2$, $h = 6$ in. ____________
6. $B = 6 \text{ in.}^2$, $h = 20$ in. ____________
7. $B = 11 \text{ cm}^2$, $h = 3$ cm ____________
8. $B = 3 \text{ cm}^2$, $h = 11$ cm ____________
9. The bases of a right rectangular prism are two congruent triangles, each with a height of 7 centimeters and a base of 2 centimeters. The height of the prism is 10 centimeters. What is the volume? ____________

Use the given dimensions to find the volume of each prism with rectangular bases.

10. $l = 4, w = 4, h = 9$ ____________
11. $l = 2, w = 1, h = 6$ ____________
12. $l = 3, w = 2, h = 1$ ____________
13. $l = 9, w = 5, h = 1$ ____________
14. $l = 4, w = 3, h = 3$ ____________
15. $l = 16, w = 10, h = 5$ ____________
16. $l = 12, w = 10, h = 2$ ____________
17. $l = 15, w = 6, h = 7$ ____________

Find the surface area of a right rectangular prism with the given dimensions.

18. $l = 7, w = 8, h = 12$ ____________
19. $l = 11, w = 10, h = 3$ ____________
20. $l = 6, w = 5, h = 2$ ____________
21. $l = 4, w = 10, h = 3$ ____________
22. $l = 3, w = 4, h = 5$ ____________
23. $l = 5, w = 3, h = 4$ ____________
24. $l = 20, w = 15, h = 6$ ____________
25. $l = 15, w = 6, h = 20$ ____________
26. A right prism has two congruent squares for its bases. The sides of the squares measure 5 inches. The height of the prism is 10 inches. Find the surface area of the prism. ____________
27. The following statement is an example of what principle? If two solids have equal heights and the cross sections formed by every plane parallel to the bases of both solids have equal areas, then the two solids have equal volumes. ____________

NAME ______________________ CLASS ____________ DATE ____________

Practice Masters Level B

7.2 *Surface Area and Volume of Prisms*

1. If you know the base area and the volume of a right prism, explain how you can find the height. ______________________

2. Explain how to determine the surface area of a right prism if you are given a net of the prism. ______________________

Find the surface area and volume of a right prism with the given base shape, base dimensions, and prism height, *h*. Round to the nearest tenth, if necessary.

3. square base whose sides measure 3 meters; $h = 14$ meters ____________

4. equilateral triangle base whose sides measure 6 inches; $h = 8$ inches ____________

5. regular hexagon base whose sides measure 10 centimeters; $h = 4$ centimeters ____________

6. regular octagon base whose apothem is 2.8 meters and perimeter is 51.2 meters; $h = 8.1$ meters ____________

7. regular pentagon base whose apothem is 2 centimeters and perimeter is 25 centimeters; $h = 2$ centimeters ____________

8. rectangular base whose length is 4.2 units and width is 2.5 units; $h = 1.5$ units ____________

9. a right triangle base whose hypotenuse is 17 inches and one leg is 15 inches; $h = 5$ inches ____________

10. a regular hexagon whose apothem is 3 feet; $h = 9$ feet ____________

11. A container shaped like an oblique prism can hold 22 ounces of mustard. Another container, in the shape of a right prism, has the same height as the oblique prism. The areas of the bases of each prism are equal. How much mustard can the right prism hold? Explain. ______________________

    ____

12. The volume of a right prism is 1297 square centimeters. The base is an equilateral triangle whose sides are each 24 centimeters. Find the height of the prism. ____________

NAME ______________________ CLASS ____________ DATE ____________

Practice Masters Level C

7.2 *Surface Area and Volume of Prisms*

Use the figure below for Exercises 1–5. The hole in the center of the figure is in the shape of a triangular prism and goes all the way through the cube.

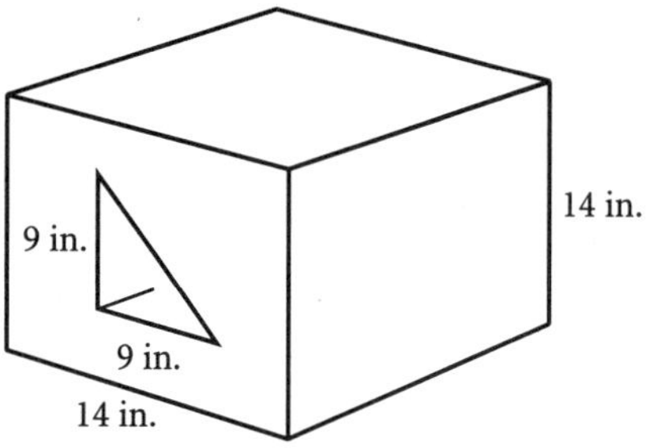

1. Find the surface area of the hole to the nearest tenth.

   ____

2. Explain how you would find the surface area of the entire figure. ______________________

   ____

   ____

3. Find the surface area of the figure, including that of the hole. Round to the nearest tenth. ____________

4. Find the volume of the hole. ____________

5. Find the volume of the figure with the hole. ____________

6. The surface area of a right rectangular prism is 112 square units. The height is twice the width. The length is 4 units more than the height. Find the volume. ____________

7. The surface area of a right rectangular prism is 486.08 square meters. The length is 2.5 times the width. The height is 1.5 times the length. Find the volume. ____________

8. The ratio of the area of the base of a right prism to the area of its lateral sides, is 2 to 3. The total surface area is 35 square meters. The height of the prism is 3 meters. Find the volume. ____________

9. A right triangular prism has an isosceles right triangle for a base. The height of the prism is 10 meters. The surface area is 441.42 square meters. Find the volume. ____________

10. If two prisms have equal volumes and the cross sections formed by every plane are parallel to the bases of both solids, what must be true about the heights of each prism? ____________

NAME ______________________ CLASS ______________ DATE ______________

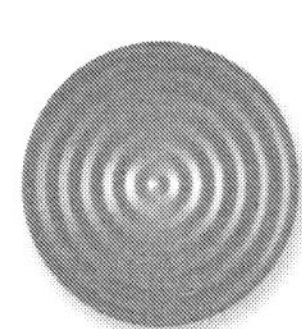

Practice Masters Level A

7.3 Surface Area and Volume of Pyramids

Use the pyramid at the right for Exercises 1–7.

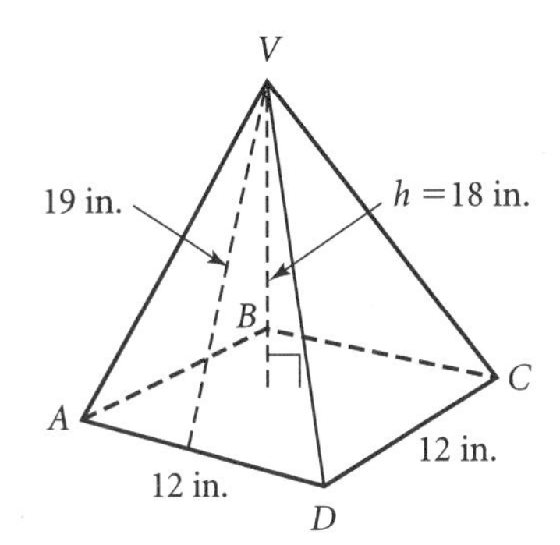

1. Find the area of $\triangle VAB$. ____________
2. Find the area of $\triangle VBC$. ____________
3. Find the area of $\triangle VCD$. ____________
4. Find the area of $\triangle VDA$. ____________
5. Find the area of $ABCD$. ____________
6. Find the surface area of the pyramid. ____________
7. Find the volume of the pyramid. ____________

Use the pyramid with an equilateral triangular base for Exercises 8–11.

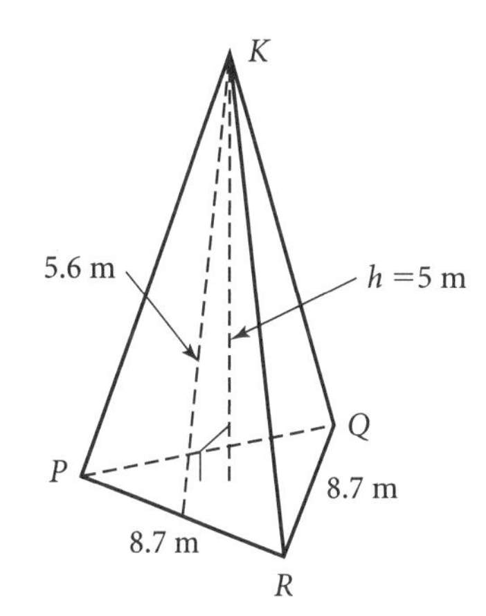

8. Find the area of each lateral face. ____________
9. The height of $\triangle PQR$ is 7.5. Find the area of $\triangle PQR$. ____________
10. Find the surface area of the pyramid. ____________
11. Find the volume of the pyramid. ____________
12. The area of the base of a pyramid is 24 square units. The height of the pyramid is 8 units. Find the volume. ____________
13. The perimeter of the square base of a pyramid is 24 units. The slant height is 10 units. Find the surface area of the pyramid. ____________
14. If the height of the pyramid in Exercise 13 is 8 units, what is the volume? ____________

Find the surface area of each regular pyramid with the given side length, *s*, and slant height *l*. The number of sides of the base is given by *n*.

15. $s = 3, l = 4, n = 4$ ____________
16. $s = 12, l = 14, n = 4$ ____________
17. $s = 15, l = 17, n = 3$ ____________

NAME ______________________ CLASS __________ DATE __________

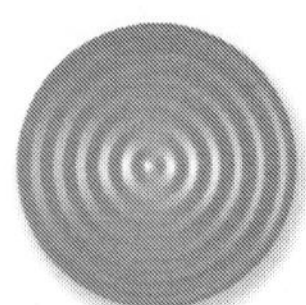

Practice Masters Level B

7.3 Surface Area and Volume of Pyramids

Use the pyramid at the right for Exercises 1–6. The base of the pyramid is an equilateral triangle whose perimeter measures 120 inches. The volume of the pyramid is 3464.1 cubic inches.

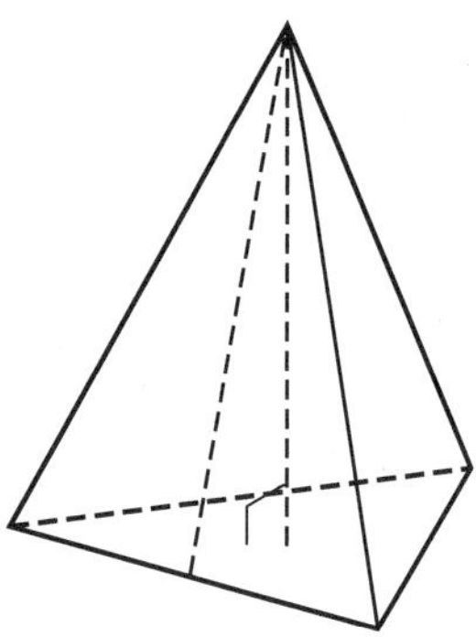

1. Find the length of a side of the base. __________
2. Find the area of the base of the pyramid to the nearest hundredth. __________
3. Find the height of the pyramid. __________
4. The apothem of the triangular base is 11.547 inches. Find the slant height of the pyramid to the nearest hundredth. __________
5. Find the lateral area of the pyramid. __________
6. Find the total surface area of the pyramid. __________

Use the pyramid with a square base for Exercises 7–13. The area of the base is 81 square meters and its volume is 108 cubic meters.

7. Find the height of the pyramid. __________
8. Find the length of a side of the base. __________
9. Find the length of the apothem in the base. __________
10. Find the slant height of a lateral face. __________
11. Find the perimeter of the base. __________
12. Find the lateral area of the pyramid. __________
13. Find the total surface area of the pyramid. __________

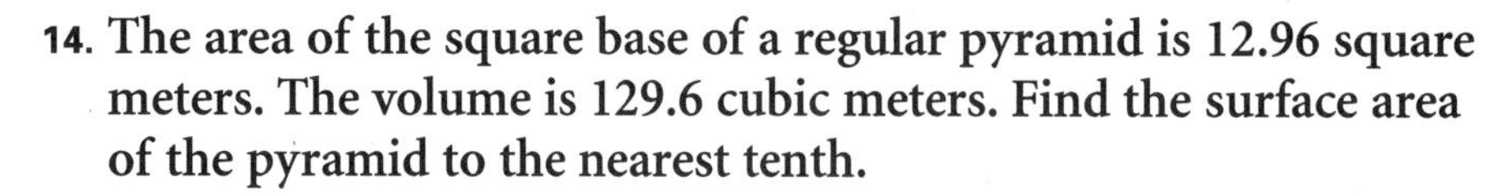

14. The area of the square base of a regular pyramid is 12.96 square meters. The volume is 129.6 cubic meters. Find the surface area of the pyramid to the nearest tenth. __________

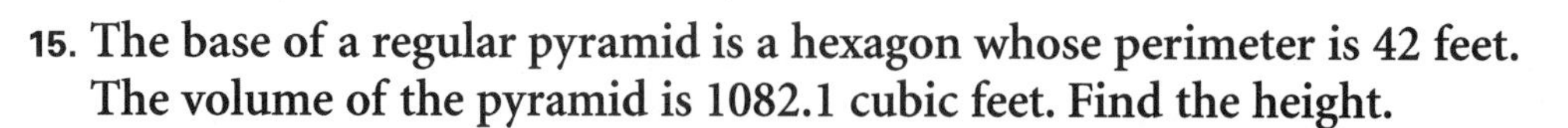

15. The base of a regular pyramid is a hexagon whose perimeter is 42 feet. The volume of the pyramid is 1082.1 cubic feet. Find the height. __________

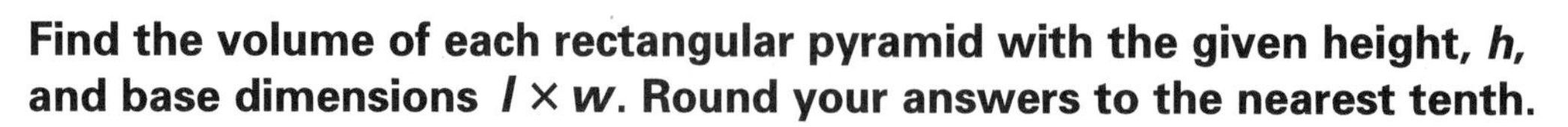

Find the volume of each rectangular pyramid with the given height, *h*, and base dimensions $l \times w$. Round your answers to the nearest tenth.

16. $h = 6, l = 10, w = 12$ __________
17. $h = 20, l = 8, w = 2$ __________
18. $h = 5, l = 7, w = 3$ __________

Practice Masters Level C

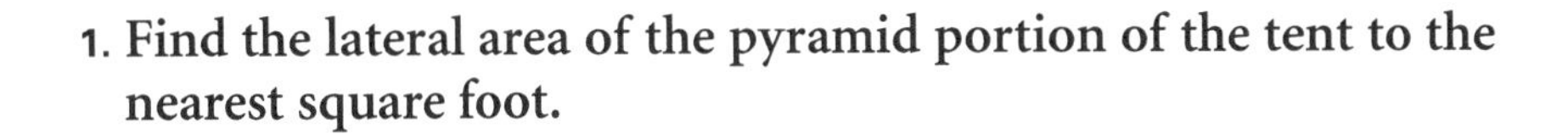

7.3 *Surface Area and Volume of Pyramids*

A tent consists of a pyramid atop a rectangular prism as shown in the figure at the right. The total height of the tent is 12 feet. Use this information for Exercises 1–4.

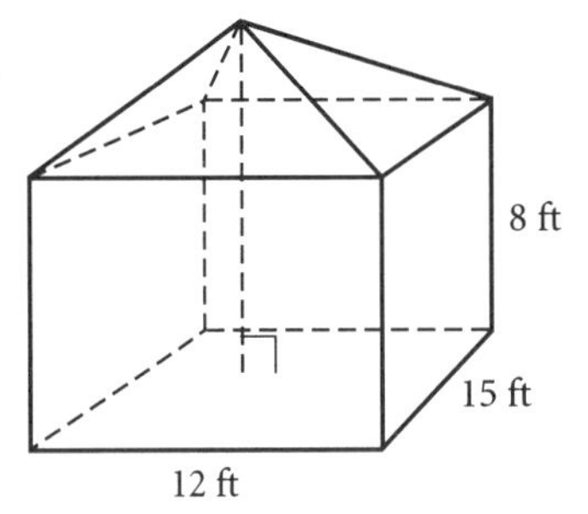

1. Find the lateral area of the pyramid portion of the tent to the nearest square foot. ______________________

2. Find the lateral area of the prism portion of the tent to the nearest square foot. ______________________

3. Find the total surface area of the tent to the nearest square foot. ______________________

4. Find the volume of the tent to the nearest cubic foot. ______________________

5. A tent has a square base and a total height of 12 feet. The height of the pyramid top is 4 feet. The volume of the tent is the same as the volume of the tent above. How wide is it to the nearest inch? ______________________

6. A tent has a square base and a total height of 6 feet. The height of the pyramid top is 2 feet. The prism base is a cube. The entire tent including the floor, is made of canvas. To the nearest tenth of a square yard, how much canvas is used to make the tent? ______________________

A decorative ornament is made of solid wood. It is comprised of two congruent regular hexagonal pyramids that share the same base. The perimeter of the hexagon is 72 centimeters. The slant height of each of the faces is 16 centimeters. Use this information for Exercises 7 and 8.

7. Find the volume of the ornament to the nearest tenth. ______________________

8. Find the surface area of the ornament. ______________________

9. The base of a pyramid is a regular hexagon whose area is 50.3 square inches. The height is 6.2 inches. The lateral faces are all congruent. Find the slant height of each lateral face to the nearest hundredth. ______________________

10. The surface area of a regular square pyramid is 864 square meters. The slant height is 15 meters and the height is 12 meters. Find the length of each side of the base. ______________________

NAME ______________________ CLASS ____________ DATE ____________

Practice Masters Level A

7.4 Surface Area and Volume of Cylinders

1. Gordan plans to fill gasoline into a cylindrical tank that is 10 feet by 5 feet by 12 feet to 85% of its capacity. What measurement should Gordan calculate to determine this amount? ____________

2. Sjorn needs to shrink-wrap a cylindrical tube. What measurement should Sjorn calculate to determine the exact amount of shrink-wrap needed? ____________

Find the unknown measure for a right cylinder with radius *r*, height *h*, and surface area *S*. Round your answers to the nearest tenth.

3. $r = 4, h = 3, S =$ ____________
4. $r = 2, h = 10, S =$ ____________
5. $r = 1, h = 1, S =$ ____________
6. $r = 2, h = 2, S =$ ____________
7. $r = 3, h = 3, S =$ ____________
8. $r = 4, h = 4, S =$ ____________
9. $r = 1, h = 3, S =$ ____________
10. $r = 3, h = 1, S =$ ____________
11. $r = 5, h =$ ____________, $S = 80\pi$
12. $r = 2, h =$ ____________, $S = 12\pi$
13. $r = 3, h =$ ____________, $S = 60\pi$
14. $r = 4, h =$ ____________, $S = 112\pi$
15. $r =$ ____________, $h = 8, S = 96\pi$
16. $r =$ ____________, $h = 3, S = 20\pi$
17. $r =$ ____________, $h = 1, S = 84\pi$
18. $r =$ ____________, $h = 10, S = 150\pi$

Find the unknown measure for a right cylinder with radius *r*, height *h*, and volume *V*. Give exact answers.

19. $r = 1, h = 1, V =$ ____________
20. $r = 2, h = 2, V =$ ____________
21. $r = 3, h = 3, V =$ ____________
22. $r = 4, h = 4, V =$ ____________
23. $r = 3, h = 4, V =$ ____________
24. $r = 4, h = 3, V =$ ____________
25. $r = 5, h =$ ____________, $V = 125\pi$
26. $r = 3, h =$ ____________, $V = 90\pi$
27. $r = 7, h =$ ____________, $V = 49\pi$
28. $r = 1, h =$ ____________, $V = 8\pi$
29. $r = 2, h =$ ____________, $V = 24\pi$
30. $r = 9, h =$ ____________, $V = 162\pi$
31. $r =$ ____________, $h = 2, V = 72\pi$
32. $r =$ ____________, $h = 8, V = 72\pi$
33. $r =$ ____________, $h = 3, V = 3\pi$
34. $r =$ ____________, $h = 11, V = 1{,}100\pi$

NAME ________________ CLASS ________ DATE ________

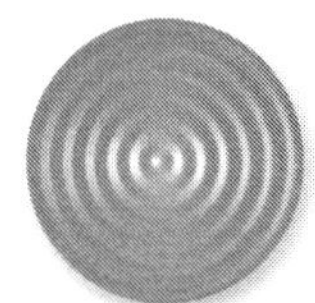

Practice Masters Level B

7.4 Surface Area and Volume of Cylinders

1. Tia and Tyrone both own super-squirt guns. Each water gun has a cylindrical water tank. Tia's water tank measures 12 inches long and has a diameter of 4 inches and Tyrone's water tank measures 10 inches long with a diameter of 6 inches. Write, but do not solve, an equation for finding the difference in the volume of each water tank.

____

Find the unknown measure for a right cylinder with radius *r*, height *h*, and surface area *S*. Round your answers to the nearest tenth.

2. $r = 1, h = 1, S =$ ________
3. $r = 0.5, h = 1, S =$ ________
4. $r = 1, h = 0.5, S =$ ________
5. $r = 0.5, h = 0.5, S =$ ________
6. $r = 1.5, h = 3.4, S =$ ________
7. $r = 3.4, h = 1.5, S =$ ________
8. $r = 7.2, h = 2.4, S =$ ________
9. $r = 2.4, h = 7.2, S =$ ________
10. $r = 4.1, h =$ ________, $S = 48.38\pi$
11. $r = 6.8, h =$ ________, $S = 187.68\pi$
12. $r = 7, h =$ ________, $S = 147\pi$
13. $r = 0.2, h =$ ________, $S = 4.08\pi$
14. $r =$ ________, $h = 5, S = 72\pi$
15. $r =$ ________, $h = 12, S = 56\pi$
16. $r =$ ________, $h = 3, S = 360\pi$
17. $r =$ ________, $h = 2, S = 510\pi$

Find the unknown measure for a right cylinder with radius *r*, height *h*, and volume *V*. Give exact answers.

18. $r = 1, h = 0.2, V =$ ________
19. $r = 0.2, h = 1, V =$ ________
20. $r = 1.2, h = 5, V =$ ________
21. $r = 5, h = 1.2, V =$ ________
22. $r = n, h = an, V =$ ________
23. $r = an, h = n, V =$ ________
24. $r = 2.5, h =$ ________, $V = 18.75\pi$
25. $r = 8, h =$ ________, $V = 96\pi$
26. $r = 0.9, h =$ ________, $V = 16.2\pi$
27. $r = 10.5, h =$ ________, $V = 66.15\pi$
28. $r = 22, h =$ ________, $V = 2371.6\pi$
29. $r = 0.5, h =$ ________, $V = 0.3\pi$
30. $r = 0.6, h =$ ________, $V = 1.98\pi$
31. $r = 40, h =$ ________, $V = 136{,}000\pi$

NAME ______________________ CLASS ______________ DATE ______________

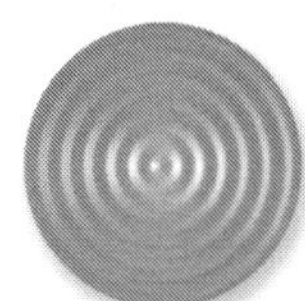

Practice Masters Level C

7.4 Surface Area and Volume of Cylinders

1. The top of the hollow wooden coffee table is a triangular prism. The sides of the triangle measure 36 inches, 48 inches, and 60 inches. The height of the prism is 1.5 inch. Each of the three legs is a cylinder 2 inches wide and 15 inches high. Find the volume to the nearest tenth. ______________

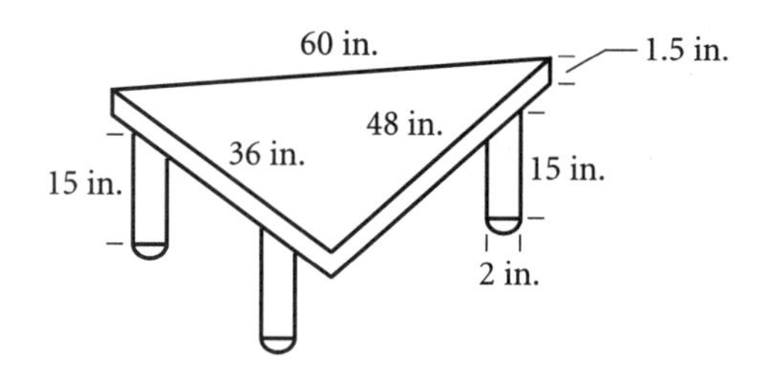

2. The nut is 7 centimeters across its widest width. The circular hole is 2 centimeters across. The nut is 0.25 centimeters deep. Find the volume to the nearest tenth. ______________

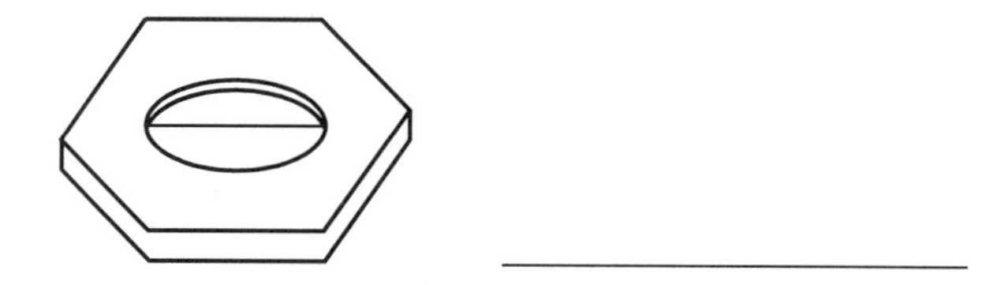

3. The pipe shown is 6 feet across and 22 feet long. The circular hole is 4 feet across. Find the volume to the nearest hundredth. ______________

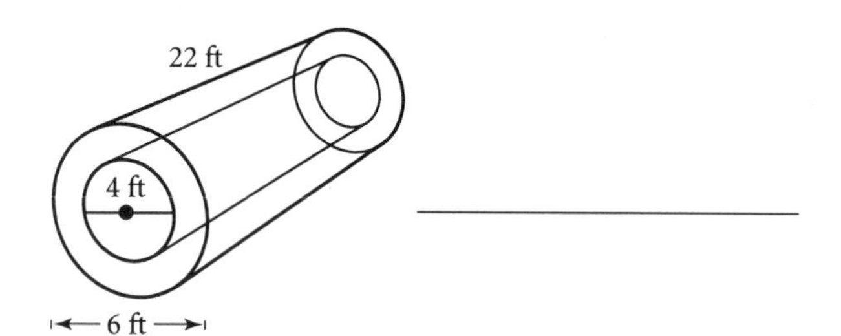

4. The tower consists of 3 stacked right cylinders. The radius of the bottom cylinder is 4 times that of the top cylinder. The radius of the middle cylinder is twice that of the top cylinder. The height of the top cylinder is 4 times that of the bottom cylinder. The height of the middle cylinder is twice that of the bottom cylinder. The radius of the top cylinder is 1 foot. The height of the top cylinder is 8 feet. If the tower is located in a park, what is the surface area of the portion that requires paint? ______________

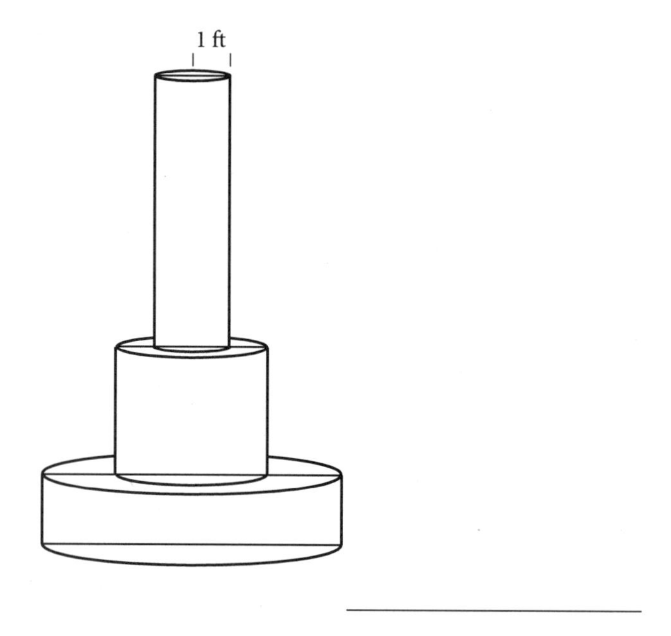

NAME ______________________ CLASS __________ DATE __________

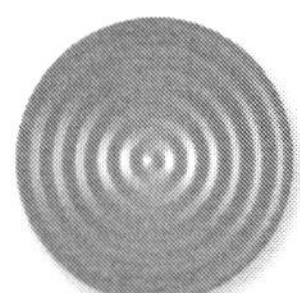

Practice Masters Level A

7.5 Surface Area and Volume of Cones

Find the surface area of each right cone.

1.

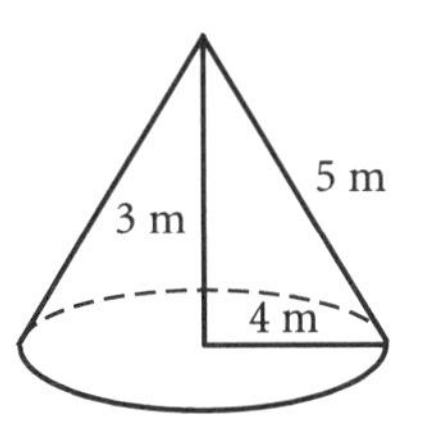

____

2.

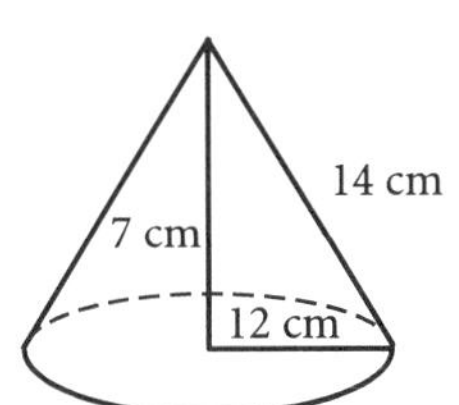

____

3.

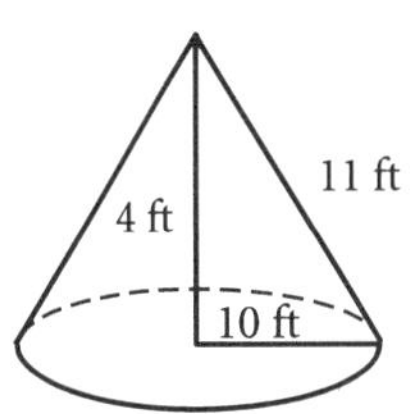

____

4.

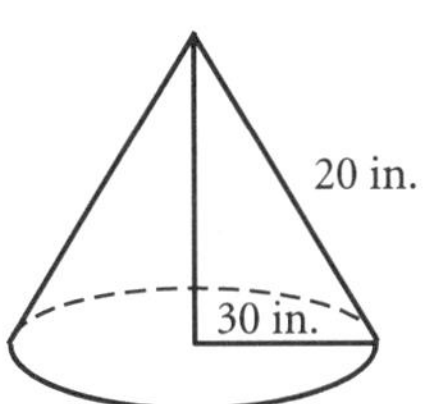

____

5.

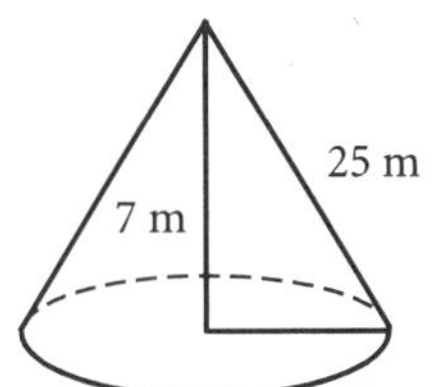

____

6.

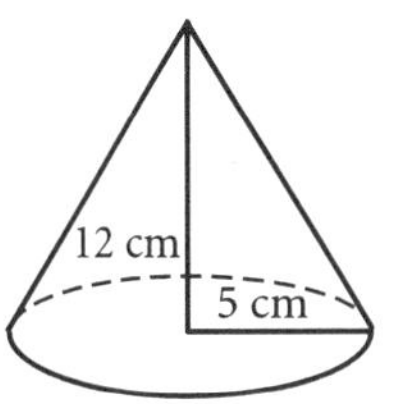

____

Find the volume of each cone. Show exact answers.

7.

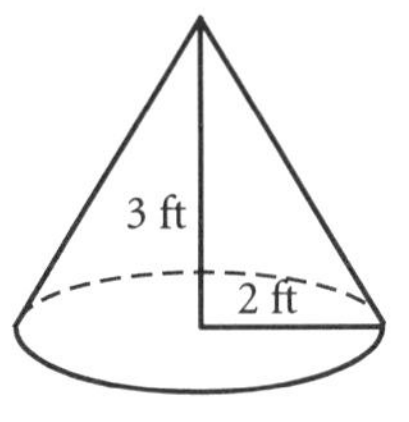

____

8.

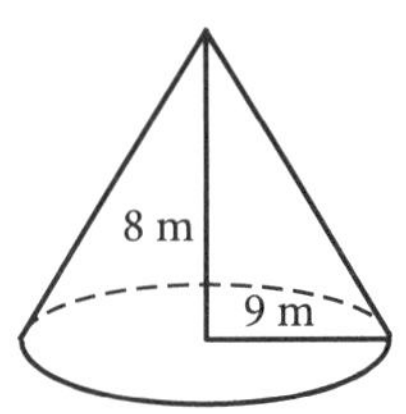

____

9.

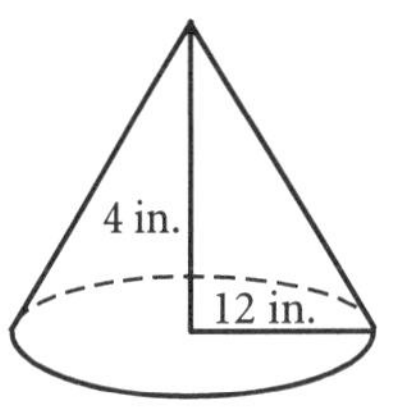

____

10.

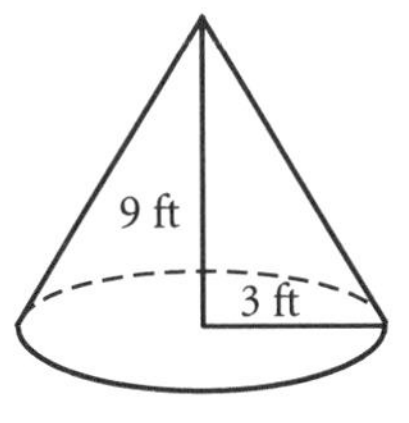

____

11.

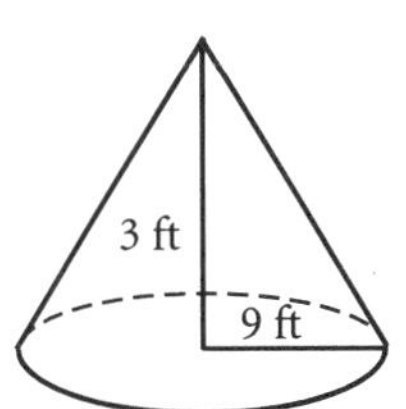

____

12.

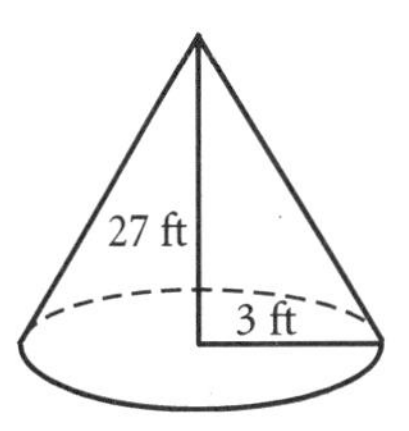

____

13. Write in words the formula for the surface area and the volume of a cone. ______________________

____

NAME ______________________ CLASS ____________ DATE ____________

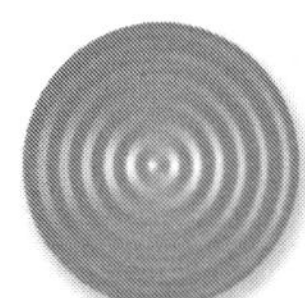

Practice Masters Level B

7.5 *Surface Area and Volume of Cones*

Find the surface area of each right cone to the nearest tenth.

1.

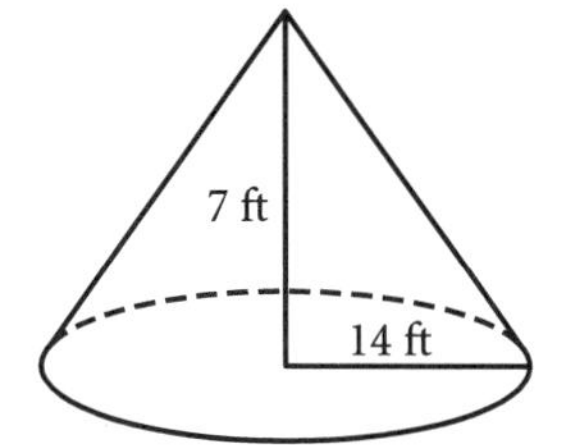

2.

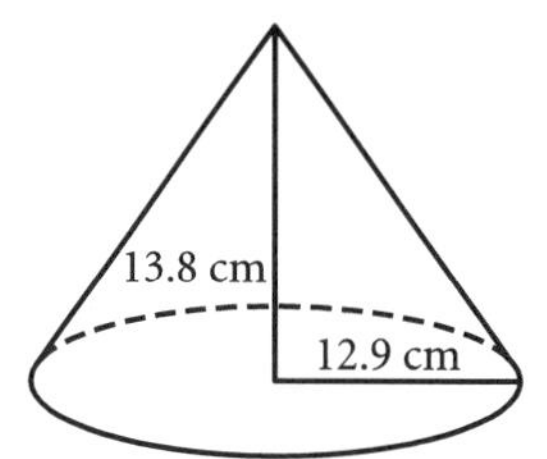

3.

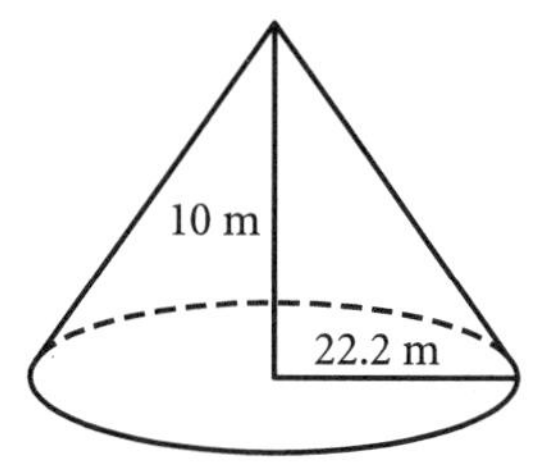

______________ ______________ ______________

4. A right cone has a surface area of 152π square meters. The radius is 8 meters. Write and solve the formula to find the slant height. ______________

5. A right cone has a surface area of 108π square feet. The slant height is twice the radius. Find the radius of the cone. ______________

6. A right cone has a surface area of 525π square meters. The slant height is 5 meters more than the radius. Find the height of the cone to the nearest tenth. ______________

Find the volume of each cone. Show answers to the nearest whole number.

7.

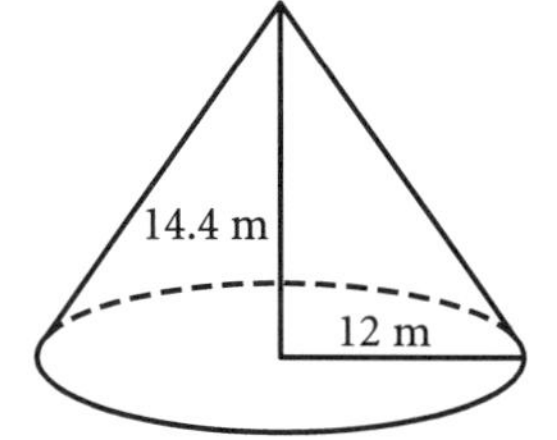

8.

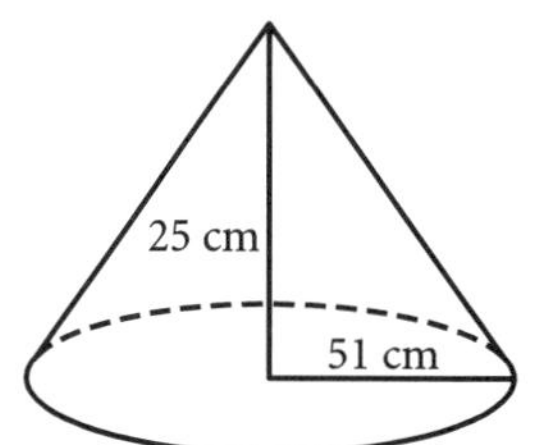

9.

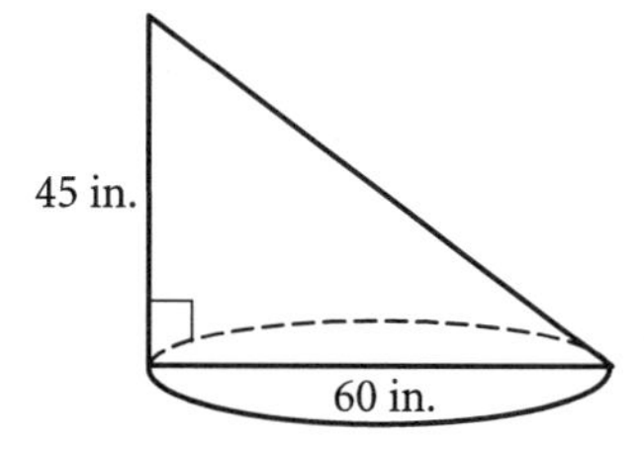

______________ ______________ ______________

10. The volume of a right cone is 27π cubic inches. The height is the same as the radius. Find the surface area of the cone to the nearest hundredth. ______________

11. The heights of a cone and cylinder are equal. They also have the same volume. Find the ratio of the radius of the cylinder to the radius of the cone. ______________

12. The volumes of a cone and cylinder are the same. Their radii are also the same. Find the ratio of the height of the cylinder to the height of the cone. ______________

NAME ______________________ CLASS ____________ DATE ____________

Practice Masters Level C

7.5 *Surface Area and Volume of Cones*

The figure to the right was created by removing the top portion of a cone. Use the figure for Exercises 1–3. Express answers to the nearest whole number.

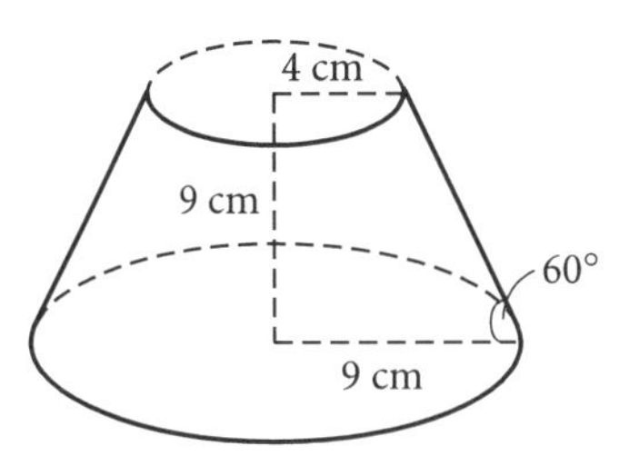

1. Find the volume of the entire cone before the top portion was removed. ____________

2. Find the volume of the missing top portion of the cone. ____________

3. Find the volume of the figure. ____________

4. The surface area and volume of a cone are numerically the same. The radius is 6 centimeters. Find the height. ____________

5. A cone and cylinder have congruent heights and radii. Find the ratio of the volume of the cylinder to the volume of the cone. ____________

6. A cone has a volume of 36π cubic meters. Its height is four times its radius. Find the surface area to the nearest tenth. ____________

The figure to the right was created with two cones that share the same base. Use it for Exercises 7–11. Express answers to the nearest hundredth, if necessary.

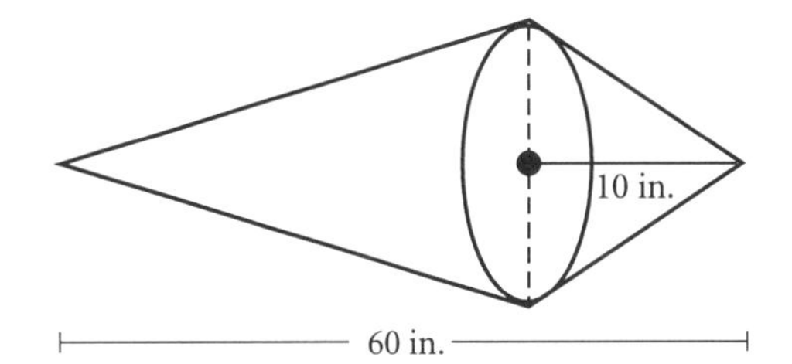

7. The heights of the two cones that comprise this figure are the same. Find the volume of the figure. ____________

8. The heights of the two cones that comprise this figure are the same. Find the surface area of the figure. ____________

9. If the height of one of the cones is twice that of the height of the other cone, find the volume of the figure. ____________

10. The height of one of the cones is twice that of the height of the other cone. Find the surface area of the figure. ____________

11. What conclusion can you draw from Exercises 7–10?

____

NAME ______________________ CLASS ____________ DATE ____________

Practice Masters Level A

7.6 *Surface Area and Volume of Spheres*

Fill in the blank.

1. The set of all points in space that are the same distance, r, for a given center point is known as a ____________.

2. The volume of a cylinder minus the volume of a cone equals the volume of a __________.

3. The formula $4\pi r^2$ can be used to calculate the ____________ of a sphere.

Find the surface area of each sphere, with radius *r* or diameter *d*. Round your answer to the nearest tenth.

4. $d = 6$ ____________
5. $d = 4$ ____________
6. $d = 10$ ____________
7. $r = 7$ ____________
8. $r = 2$ ____________
9. $r = 5$ ____________
10. $d = 9$ ____________
11. $r = 9$ ____________
12. $r = 12$ ____________

Find the surface area of the sphere based on the area, *A*, of a cross section through its center. Express your answer as an exact answer.

13. $A = 19$ ____________
14. $A = 44\pi$ ____________
15. $A = 240$ ____________
16. $A = 50\pi$ ____________
17. $A = 308\pi$ ____________

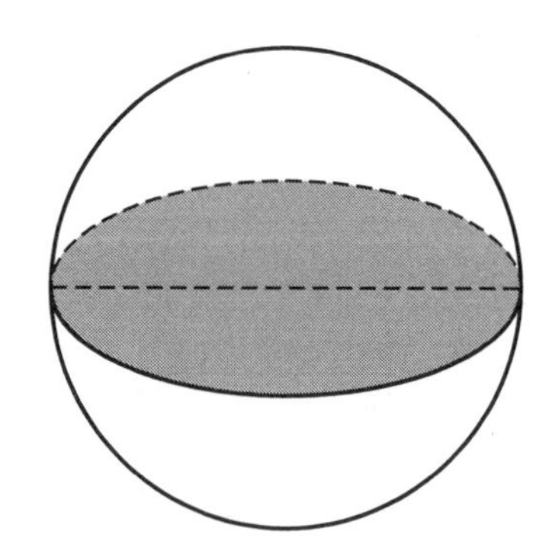

Find the volume of each sphere, with radius *r* or diameter *d*. Round your answer to the nearest tenth.

18. $r = 1$ ____________
19. $d = 4$ ____________
20. $d = 12$ ____________
21. $d = 60$ ____________
22. $r = 5$ ____________
23. $d = 10$ ____________
24. $r = 11$ ____________
25. $d = 22$ ____________
26. $d = 50$ ____________

NAME ______________________ CLASS ______________ DATE ______________

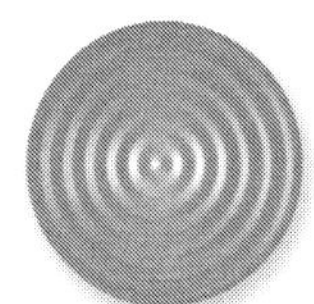

Practice Masters Level B

7.6 Surface Area and Volume of Spheres

Find the surface area of the sphere with the radius *r* or diameter *d*. Express your answers as exact answers in terms of π.

1. $r = 4$ ______________
2. $d = 4$ ______________
3. $d = 10$ ______________
4. $r = 6.4$ ______________
5. $d = 4.2$ ______________
6. $r = 8.7$ ______________

The surface area of a sphere is given. Find the length of the radius to the nearest tenth.

7. 24 ______________
8. 10 ______________
9. 100 ______________
10. 36π ______________
11. 27π ______________
12. 20π ______________

Find the volume of the sphere with radius *r* or diameter *d*. Round your answers to the nearest hundredth.

13. $r = 14$ ______________
14. $d = 6.2$ ______________
15. $r = 2.5$ ______________
16. $r = 50$ ______________
17. $d = 12.9$ ______________
18. $d = 0.54$ ______________
19. $r = 0.1$ ______________
20. $d = 0.1$ ______________
21. $r = 11.1$ ______________

The surface area of a sphere is given. Find the volume to the nearest tenth.

22. 100 ______________
23. 100π ______________
24. 9π ______________
25. 19π ______________
26. 38 ______________
27. 450 ______________
28. 900 ______________
29. 88π ______________
30. 317 ______________

31. Explain what happens to the volume of a sphere when the diameter is doubled. ______________________________

32. Explain what happens to the surface area of a sphere when the diameter is doubled. ______________________________

Find the volume of the sphere based on the area, *A*, of a cross section through its center. Round your answer to the nearest hundredth.

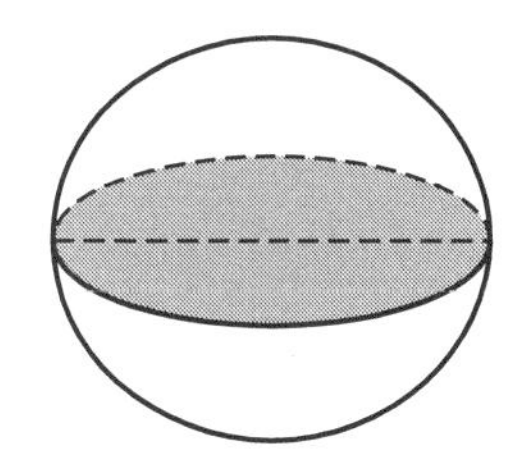

33. $A = 100$ ______________
34. $A = 17\pi$ ______________

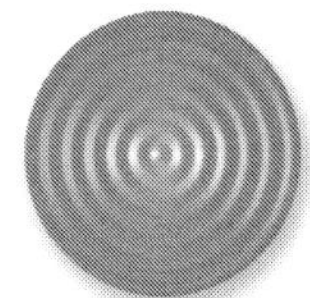

Practice Masters Level C

7.6 Surface Area and Volume of Spheres

1. A right circular cone, sphere, hemisphere, and right cylinder can each fit inside a cube with any flat surface resting completely on the face of the cube. The interior side of the cube measures 30 centimeters. Find the maximum possible volume of each figure. Then determine which figure has the greatest volume.

   cone: __________

   sphere: __________

   hemisphere: __________

   cylinder: __________

   greatest volume: __________

2. The volume-to-surface-area ratio of a sphere is 4 to 1. Find the radius of the sphere. __________

3. The volume-to-surface-area ratio of a sphere is 3 to 5. Find the radius of the sphere. __________

4. Find the exact surface area of a hemisphere whose radius is 8 centimeters. __________

5. The surface area of a sphere and cube are the same. Express the radius r in terms of the side s of the cube. __________

6. The surface area of a sphere is numerically the same as its volume. Find the length of the diameter. __________

This figure is comprised of a cone, cylinder and hemisphere. The heights of the cone, cylinder, and hemisphere are each 30 centimeters. Use this figure for Exercises 7–10.

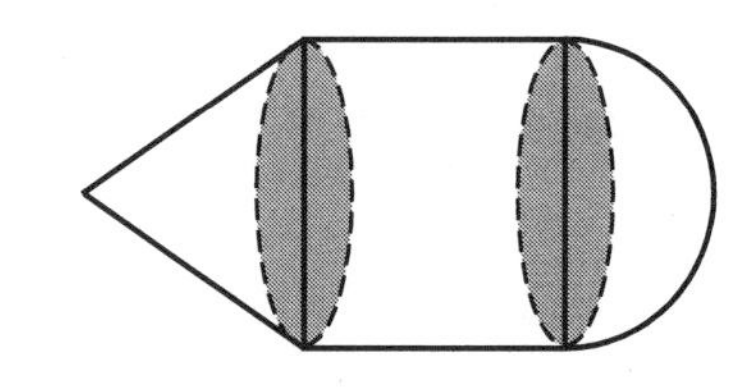

7. Compare the radii of the cone, cylinder, and hemisphere.

   ____

8. Find the total surface area of the figure to the nearest hundredth. __________

9. Find the exact volume of the figure in terms of π. __________

10. Suppose you wanted to cut the figure cross-wise to make two new figures equal in volume, the piece with the cone would be how long? __________

NAME ______________________ CLASS ____________ DATE ____________

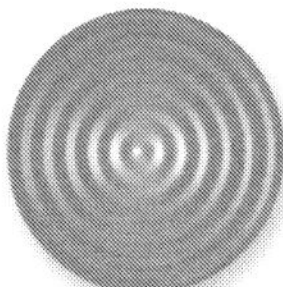

Practice Masters Level A

7.7 Three-Dimensional Symmetry

What are the coordinates of the image if each point below is reflected across the indicated plane in a three-dimensional coordinate system?

1. $(1, 2, 3)$, xy-plane ____________
2. $(2, -5, 6)$, xy-plane ____________
3. $(-1, 7, 2)$, xz-plane ____________
4. $(2, -8, 5)$, xz-plane ____________
5. $(-2, 3, -1)$, yz-plane ____________
6. $(5, -1, -8)$, yz-plane ____________
7. $(9, 9, -8)$, xy-plane ____________
8. $(2, 0, -4)$, yz-plane ____________
9. $(-5, -3, 7)$, xz-plane ____________
10. $(0, -4, 10)$, xy-plane ____________
11. $(2, -4, 1)$, yz-plane ____________
12. $(-11, -8, -6)$, xz-plane ____________

For each half of a pattern given below, describe or name the figure created when it is rotated about the dashed line.

13. 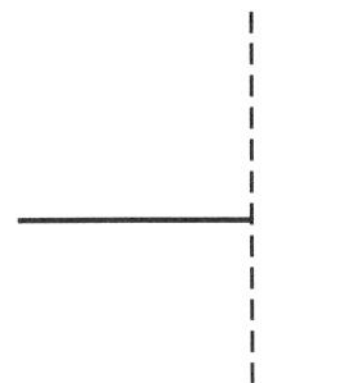

____

14. 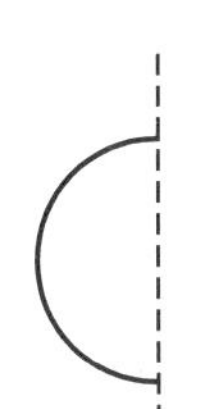

____

15. 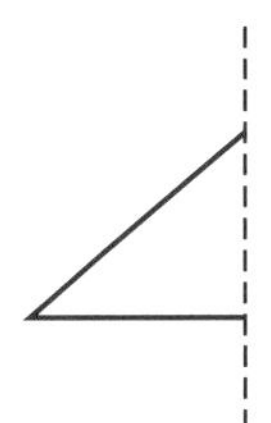

____

16. 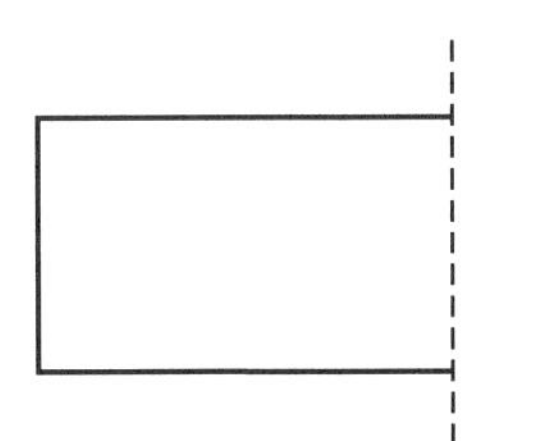

____

17. 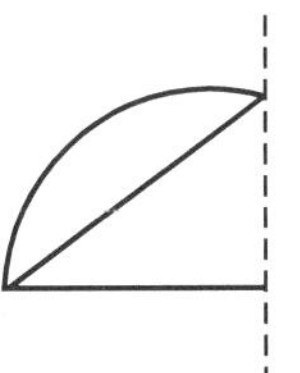

____

18. 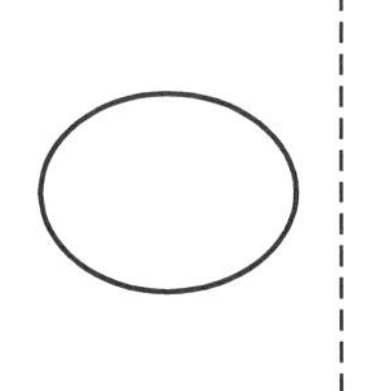

____

The endpoints of segment *AB* are *A*(0, 3, 0) and *B*(0, 3, 3).

19. Which axis is $\overline{AB}$ rotated about to produce a circle with a radius of 3? ____________

20. Which axis is $\overline{AB}$ rotated about to produce the lateral side of a cylinder? ____________

NAME ______________________ CLASS ____________ DATE ____________

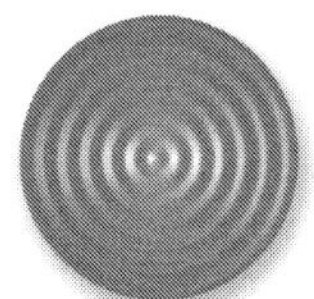

Practice Masters Level B

7.7 Three-Dimensional Symmetry

What are the octants and coordinates of the image if each point below is reflected across the *yz*-plane in a three-dimensional coordinate system? Use the terms *front, back, left, right, top,* and *bottom*.

1. (4, 7, 3)

octants: ______________________

coordinates: ______________________

2. (−2, 5, −9)

octants: ______________________

coordinates: ______________________

3. (−8, 6, 1)

octants: ______________________

coordinates: ______________________

4. (7, −4, 2)

octants: ______________________

coordinates: ______________________

5. (9, −9, 1)

octants: ______________________

coordinates: ______________________

6. (4, 3, −6)

octants: ______________________

coordinates: ______________________

Sketch a pattern that, when rotated about the dashed line, creates the named figure.

7. sphere

8. hemisphere

9. cone

10. cylinder

11. circle

12. donut

NAME ______________________ CLASS ____________ DATE ____________

Practice Masters Level C

7.7 Three-Dimensional Symmetry

Use this graph for Exercises 1–15.
Write answers in terms of π, if necessary.

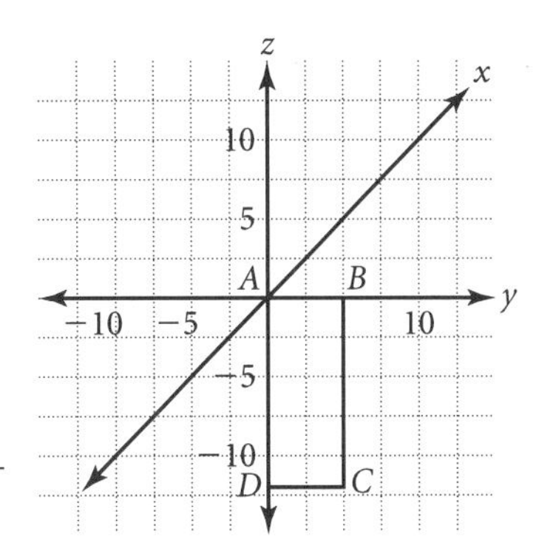

The coordinates below are the image of Point *C* after a reflection across a plane. Name the plane or line of reflection.

1. (0, 5, 12) ______________________

2. (0, −5, −12) ______________________

3. (0, −5, 12) ______________________

Rotate rectangle *ABCD* about the *y*-axis.

4. Name the resulting spatial figure. ____________

5. How long is the radius of the figure? ____________

6. What is the height of the figure? ____________

7. Find the volume of the figure. ____________

8. Find the surface area of the figure. ____________

Rotate rectangle *ABCD* about the *z*-axis.

9. Name the resulting spatial figure. ____________

10. How long is the radius of the figure? ____________

11. Find the volume of the figure. ____________

12. Find the surface area of the figure. ____________

Rotate the diagonal $\overline{AC}$ about the *x*-axis.

13. Name the resulting spatial figure. ____________

14. How long is the radius of the figure? ____________

15. Find the area of the figure. ____________

NAME ______________________ CLASS ______________ DATE ______________

Practice Masters Level A

8.1 Dilations and Scale Factors

In Exercises 1–4, the endpoints of a line segment and a scale factor, *n*, are given. Use the dilation $D(x, y) = (nx, ny)$ to transform each segment, and plot the preimage and the image on the coordinate plane.

1. $(1, 2)$ and $(3, 0)$
   $n = 2$

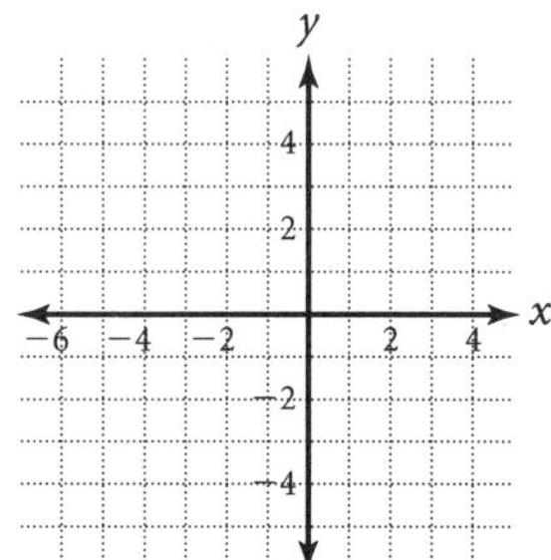

2. $(5, 2)$ and $(3, -1)$
   $n = -1$

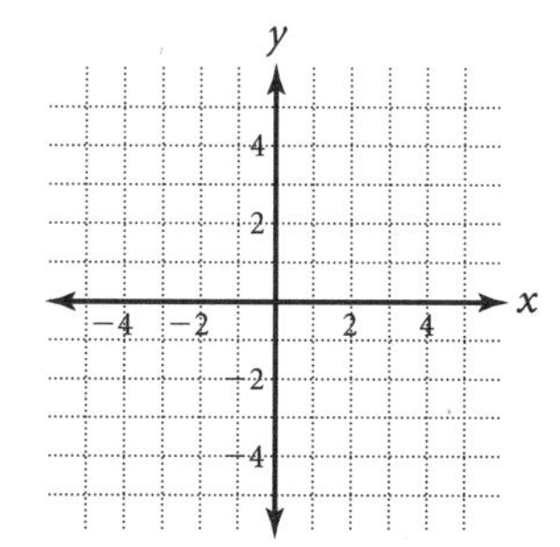

3. $(6, 3)$ and $(0, -3)$
   $n = \frac{2}{3}$

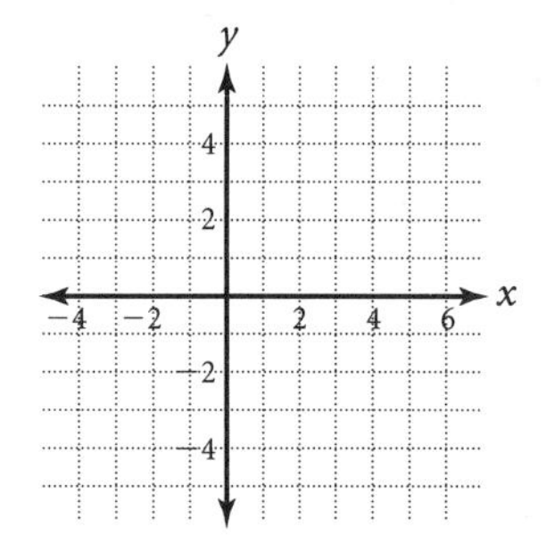

4. $(-4, -2)$ and $(2, 0)$
   $n = \frac{3}{2}$

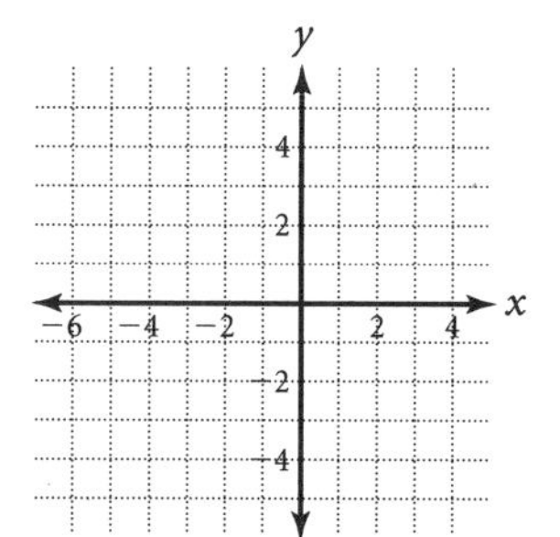

In the space provided, draw the dilation of each figure, using the given scale factor, *n*, and the given point as a center.

5. $n = \frac{1}{2}$

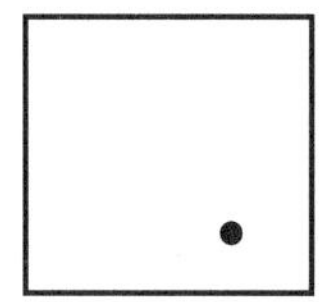

6. $n = 3$

Practice Masters Level B

8.1 Dilations and Scale Factors

Draw the dilation of each figure, using the given scale factor, *n*, and the given point as the center.

1. $n = -1$

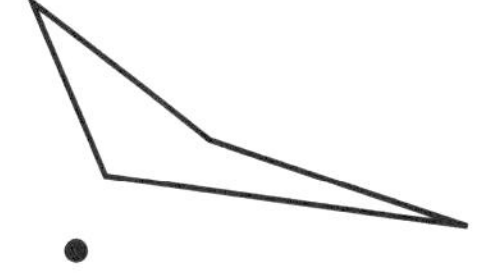

2. $n = 2$

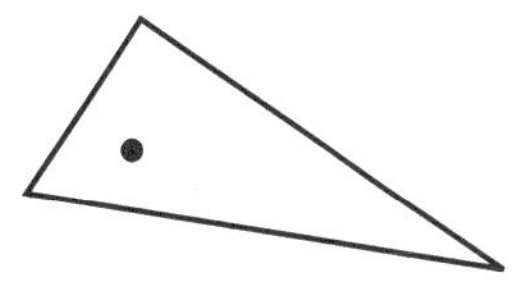

In Exercises 3 and 4, the endpoints of a line segment and a scale factor, *n*, are given. Use the dilation $D(x, y) = (nx, ny)$ to transform each segment, and plot the preimage and the image on the coordinate plane.

3. $(3, 0)$ and $(-1, 2)$

$n = -2$

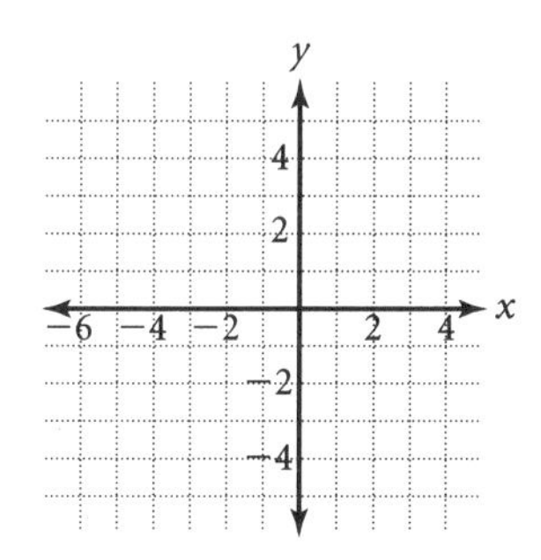

4. $(-2, 2)$ and $(8, 6)$

$n = \frac{1}{2}$

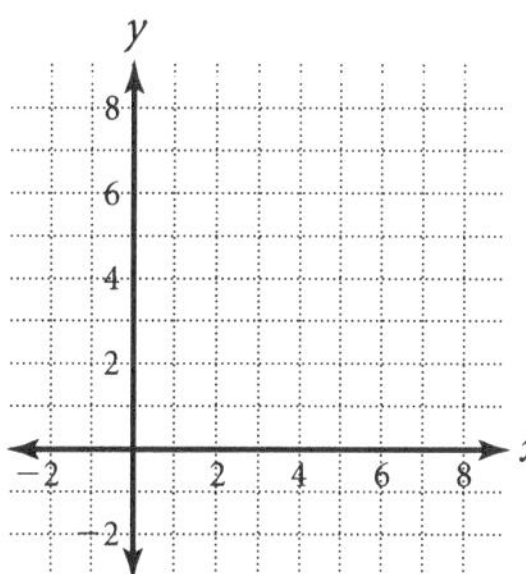

5. Draw the lines joining each image point to its preimage in Exercises 3 and 4.
   a. What seems to be true about the lines in Exercise 3? ______
   b. What seems to be true about the lines in Exercise 4? ______
   c. Complete this statement:
      The center of a dilation $D(x, y) = (nx, ny)$ is located: ______.

In Exercises 6 and 7, the dashed figures represent the preimages of dilations and the solid figures represent the images. Find the scale factor of each dilation.

6.

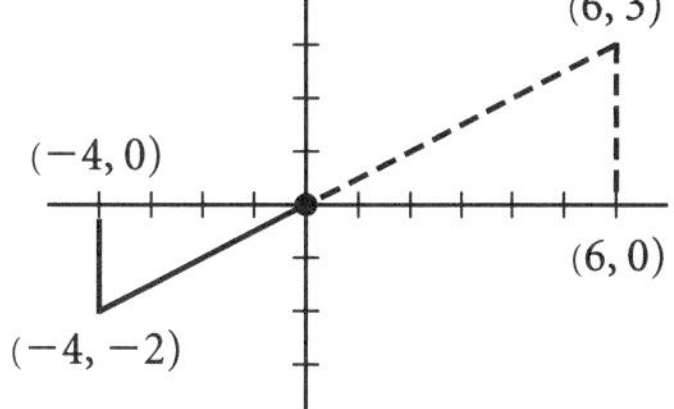

______

7.

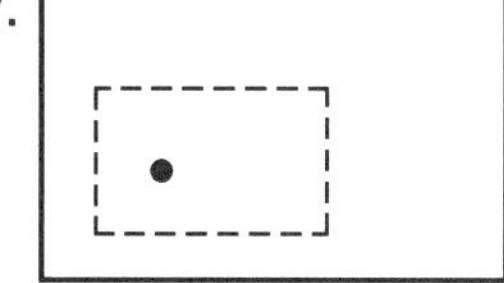

______

NAME ______________________ CLASS __________ DATE __________

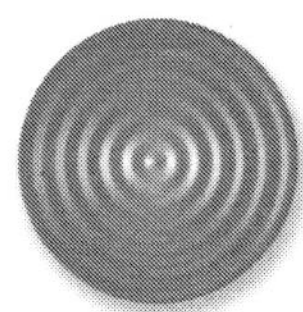

Practice Masters Level C

8.1 Dilations and Scale Factors

In Exercises 1 and 2 the dashed figures represent the preimages of dilations and the solid figures represent the images. Locate the center of each dilation.

1. ______________________

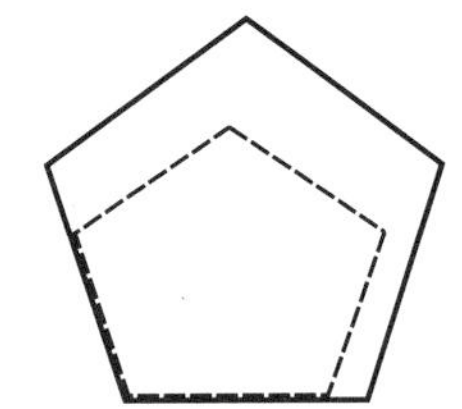

2. ______________________

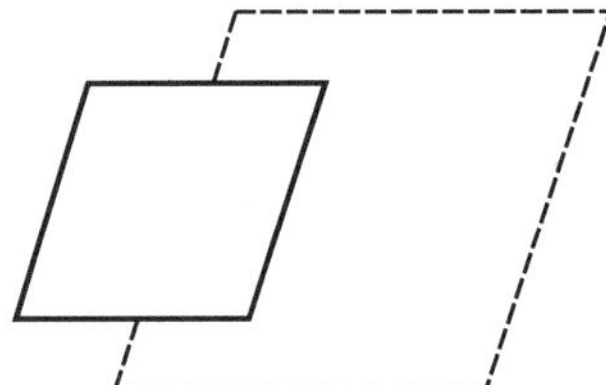

3. Draw the image of triangle ABC under the dilation with center at P and scale factor $n = 3$.

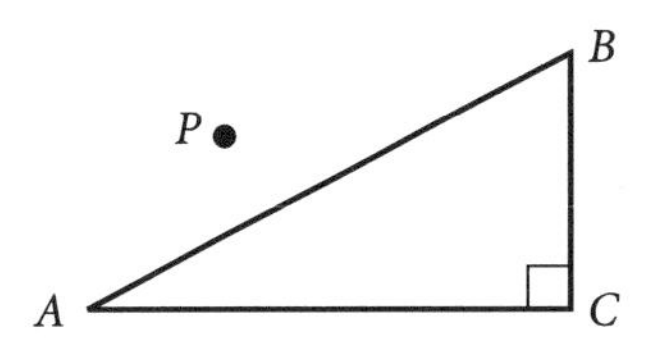

In Exercises 4–6, refer to $\triangle ABC$ and its image in Exercise 3.

4. What is the ratio of the length of a side of the image to the length of the corresponding side of the preimage? ______________

5. Find the area of $\triangle ABC$ and of its image triangle. What is the ratio of the area of the image to the area of $\triangle ABC$?

____

6. Find the measures of $\angle A$, $\angle B$, and $\angle C$ and the images of those angles. What is the ratio of the measure of each image angle to its preimage?

____

NAME ______________________ CLASS ____________ DATE ____________

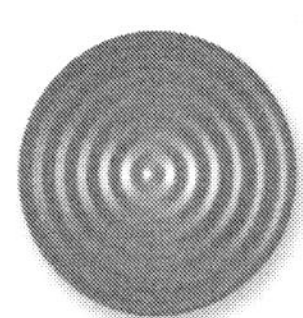

Practice Masters Level A

8.2 *Similar Polygons*

In Exercises 1–4, determine whether the polygons are similar. Explain your reasoning. If the polygons are similar, write a similarity statement.

1.

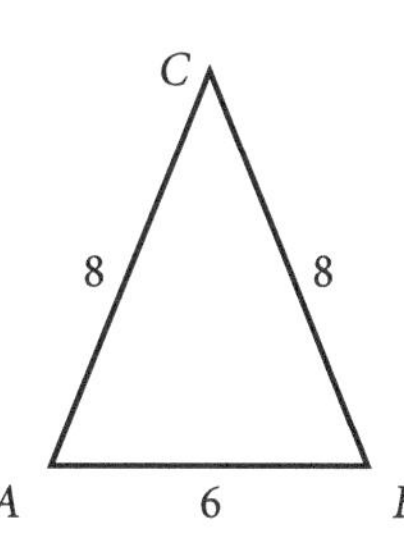

F
6 6
D 4 E

____

____

2.

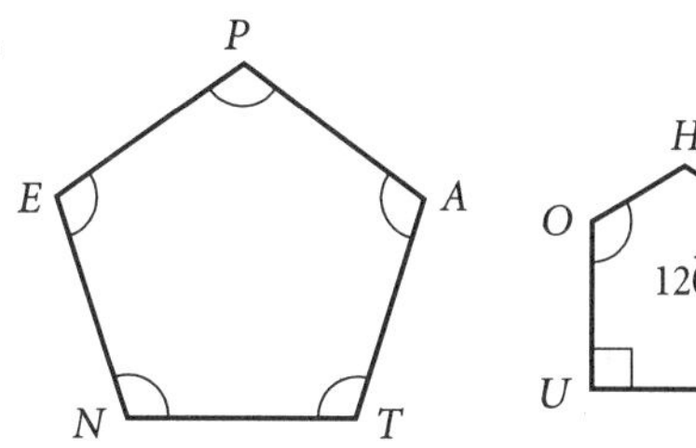

I
120°
U S

____

____

3.

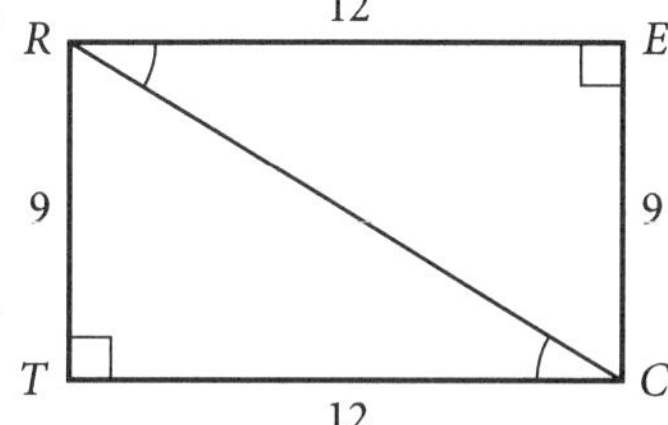

____

____

4.

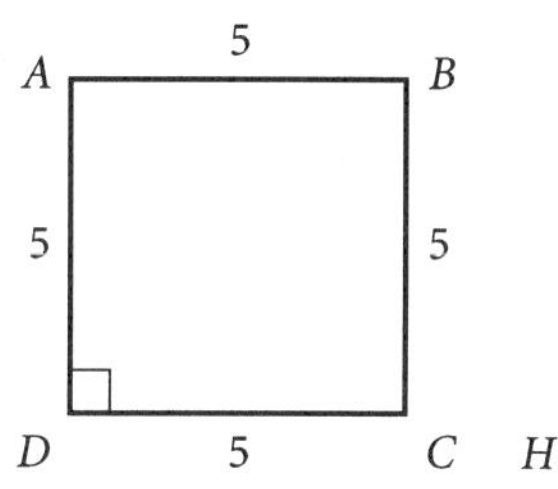

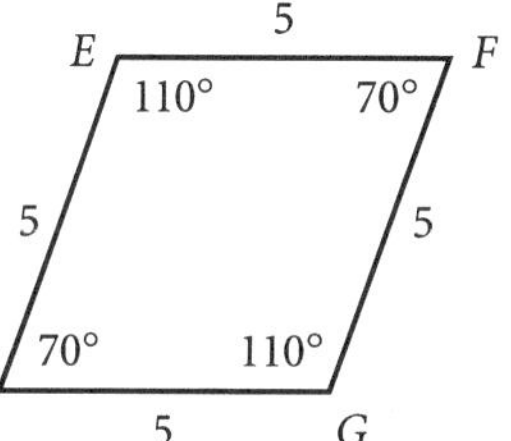

____

____

In Exercises 5 and 6, the polygons in each pair are similar. Find *x*.

5.

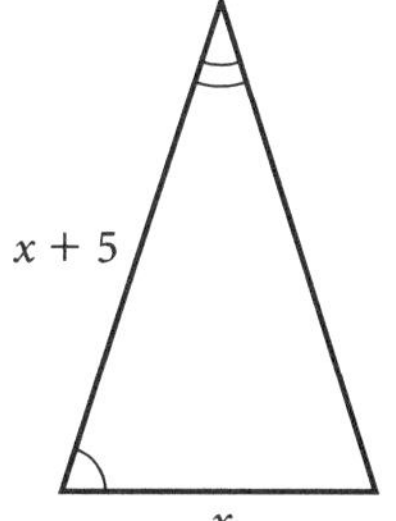

5
3

____

____

6.

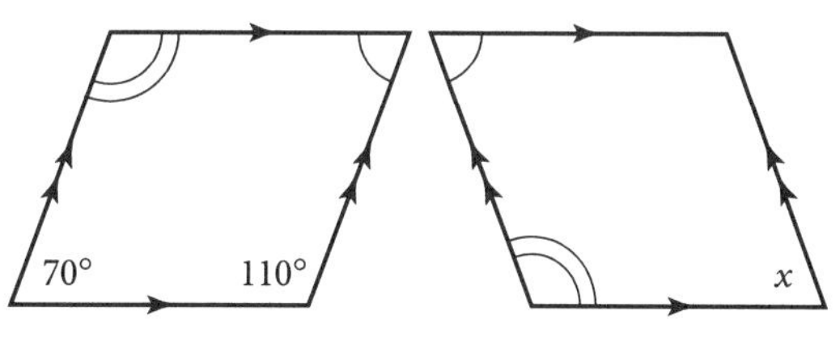

____

____

Solve each proportion for *y*.

7. $\frac{3y}{4.1} = \frac{6}{5}$

____

8. $\frac{36}{y + 2} = \frac{24}{y}$

____

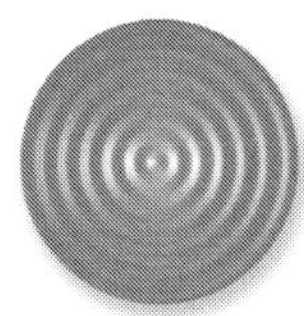

Practice Masters Level B

8.2 *Similar Polygons*

In Exercises 1 and 2, determine whether the polygons are similar. Explain your reasoning. If the polygons are similar, write a similarity statement.

1.

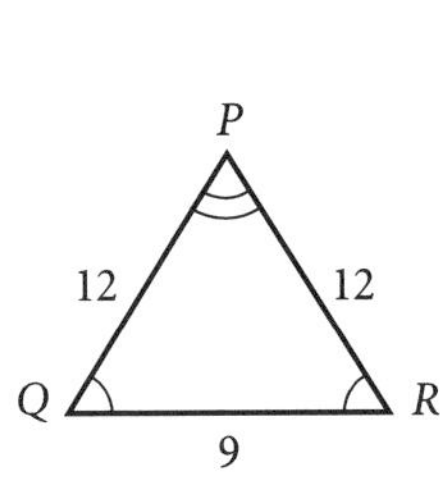

S
16
16
T
12
U

____

____

2.

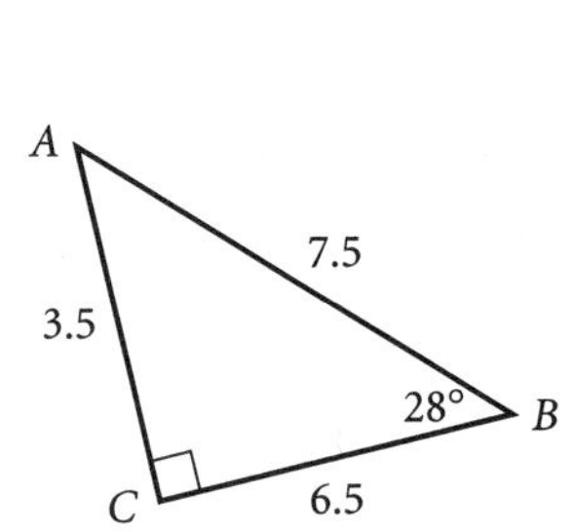

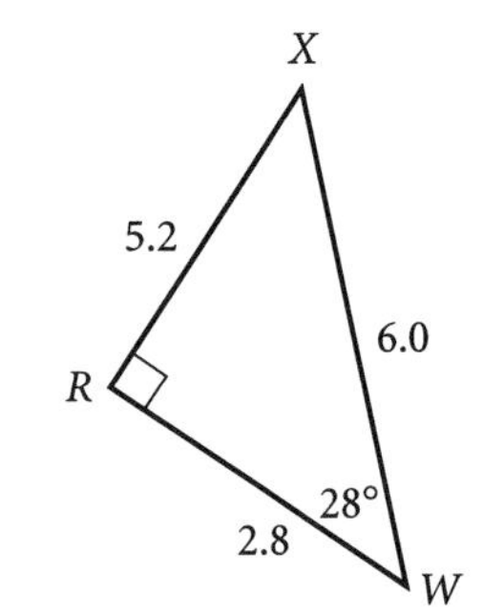

____

____

In Exercises 3 and 4, the polygons in each pair are similar. Find *x*.

3.

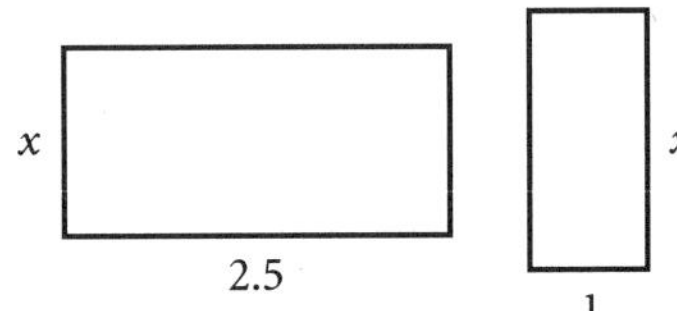

____

4.

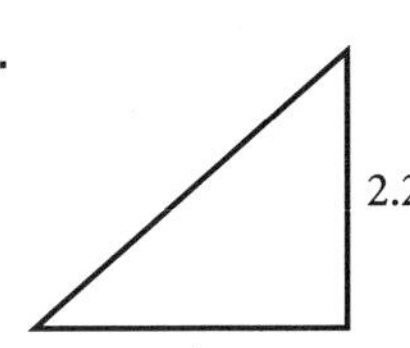

1.5
$x - 2$

____

Solve each proportion for *y*.

5. $\frac{2y + 1}{15} = \frac{3y - 2}{3}$ ______________________

6. $\frac{5}{2y} = \frac{18y}{5}$ ______________________

7. Carlos decides to make a scale drawing to help him plan how to arrange his furniture in his room in a new house. His new room will be 10 feet wide and 14 feet long, and he makes his scale drawing 5 inches wide and 7 inches long. His desktop measures 2 feet by 4.5 feet. What size rectangle should he use to represent his desk? ______________________

8. Verify the "Add-One" Property for the proportion $\frac{2}{5} = \frac{6}{15}$.

____

NAME ______________________ CLASS ____________ DATE ____________

Practice Masters Level C

8.2 Similar Polygons

In Exercises 1 and 2, determine whether a) the polygons are similar, b) the polygons are not similar, or c) not enough information is given. Explain your reasoning.

1.

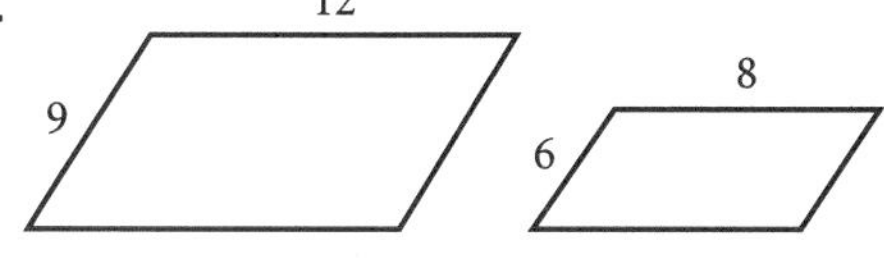

____

____

2.

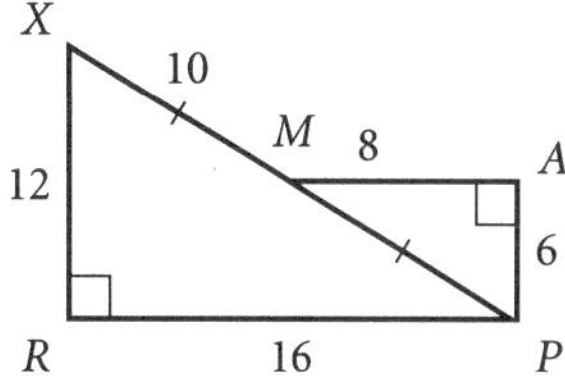

____

____

In Exercises 3 and 4, the polygons are similar. Find *x* and *y*.

3.

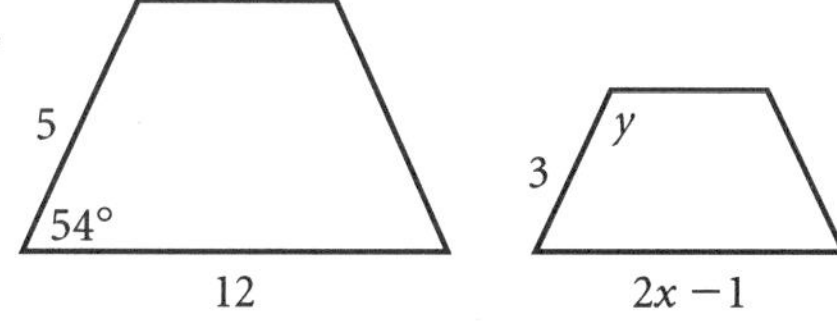

$x =$ ____________, $y =$ ____________

4.

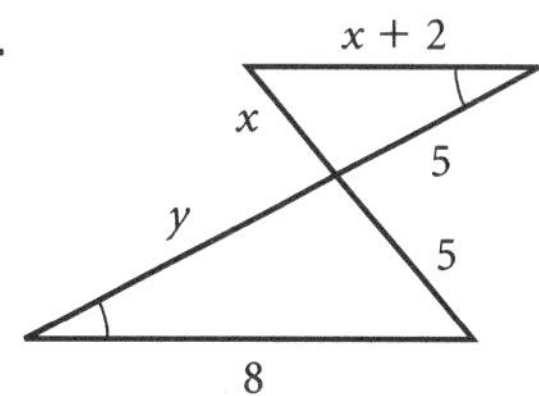

$x =$ ____________, $y =$ ____________

Use proportions to solve.

5. If a mix of hard candies sells at 2 pounds for $3.84, how much would 5 pounds cost? ____________

6. Suppose the exchange rate between Canadian and U.S. money is 1.4 Canadian dollars for each U.S. dollar.

a. How much Canadian money should a tourist receive for $235 in U.S. dollars? **a.** ____________

b. Find the price in U.S. dollars for an item that cost $53.55 in Canadian dollars. **b.** ____________

Complete this proof of the Cross Multiplication Property for proportions.

$\frac{a}{b} = \frac{c}{d}$, where a, b, c, and d are real numbers, and b and $d \neq 0$.

$\frac{a}{b} \cdot bd = \frac{c}{d} \cdot bd$ by **7.** ______________________. Therefore

$a \cdot d = c \cdot b$ by **8.** ______________________.

NAME ______________________ CLASS ____________ DATE ____________

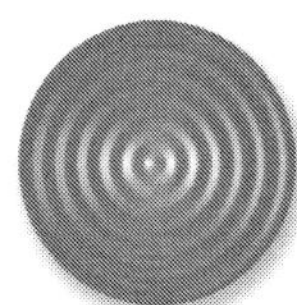

Practice Masters Level A

8.3 Triangle Similarity Postulates

Each pair of triangles can be proven similar by using AA, SAS, or SSS information. Write a similarity statement for each pair, and identify the postulate or theorem used.

1.

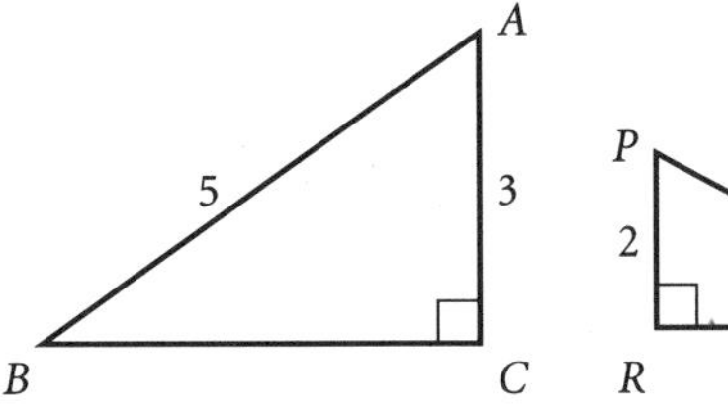

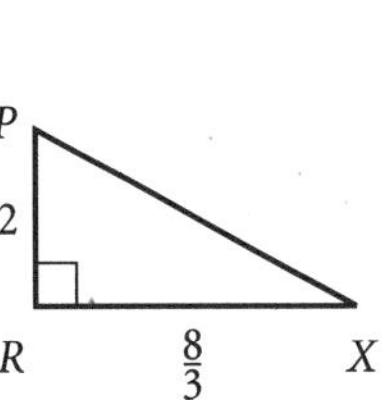

____

____

2.

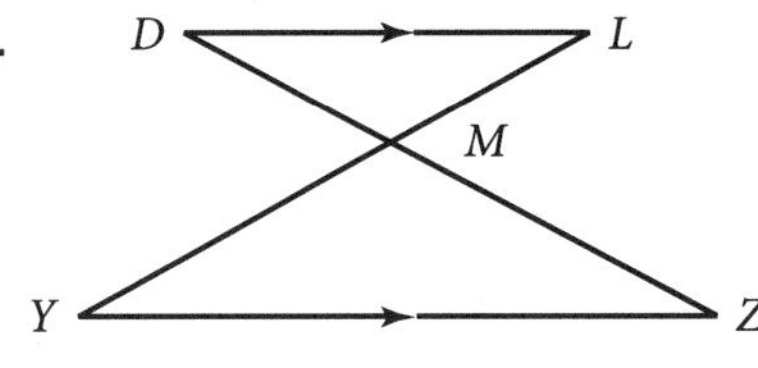

____

____

Determine whether each pair of triangles can be proven similar by using AA, SSS, or SAS. If so, write a similarity statement, and identify the postulate or theorem used.

3.

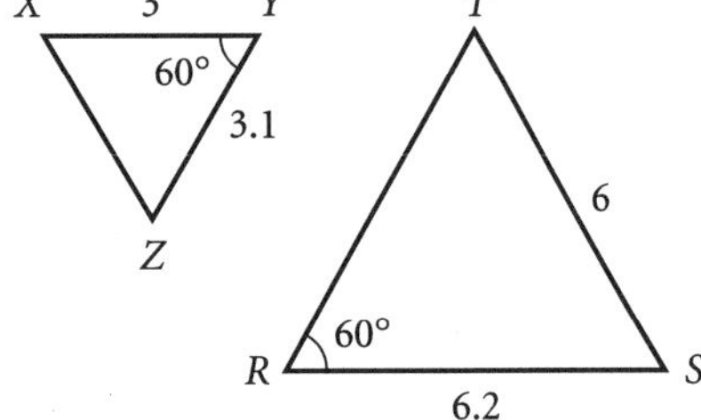

____

____

4.

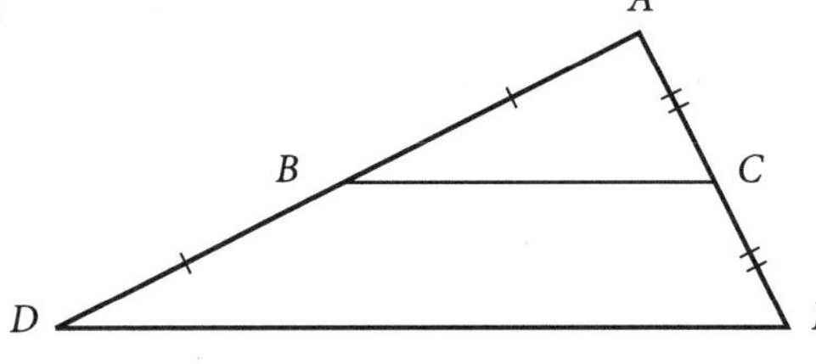

____

____

5.

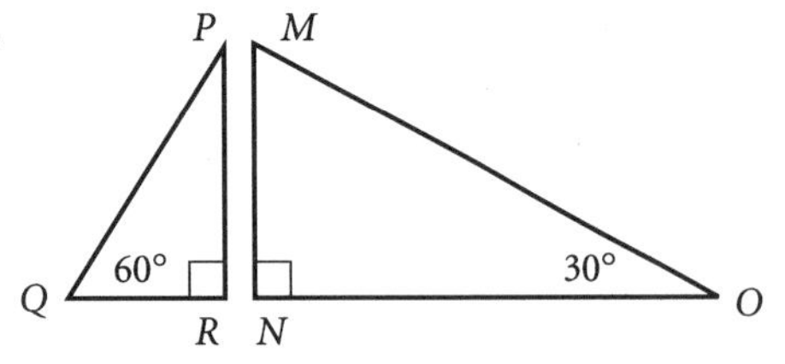

____

____

6.

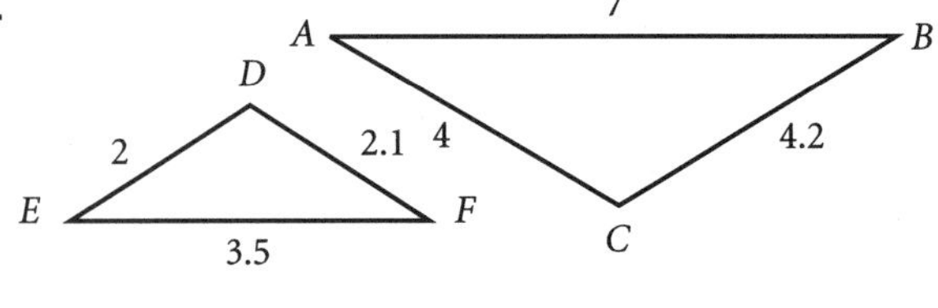

____

____

NAME ______________________ CLASS __________ DATE __________

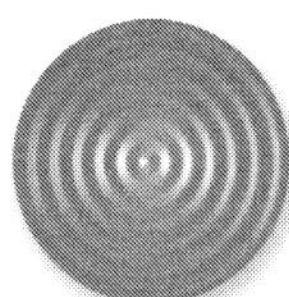

Practice Masters Level B

8.3 Triangle Similarity Postulates

Determine whether each pair of triangles can be proven similar by using AA, SSS, or SAS. If so, write a similarity statement, and identify the postulate or theorem used. If not, explain why not.

1.

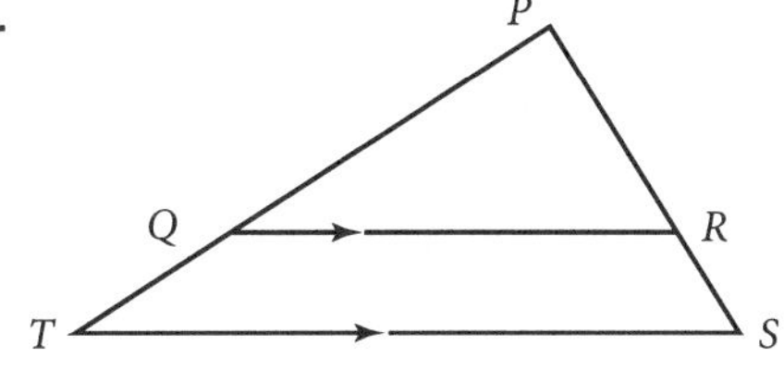

____

____

2.

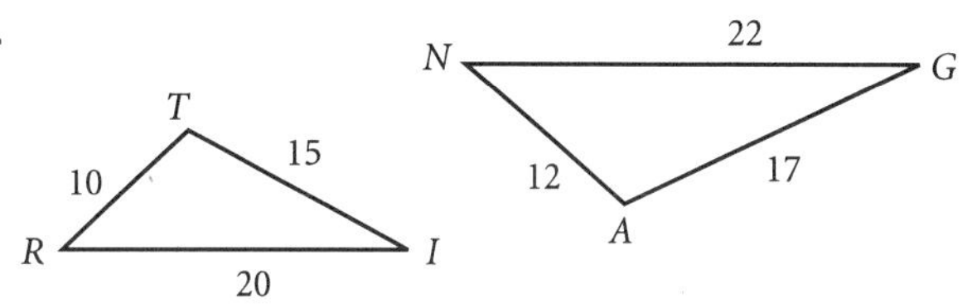

____

____

3.

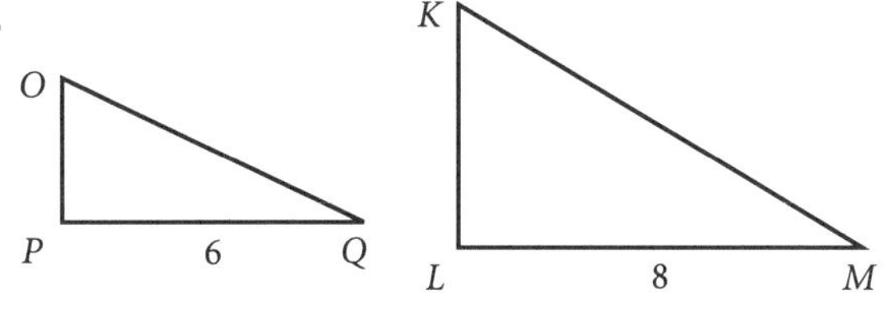

____

____

4.

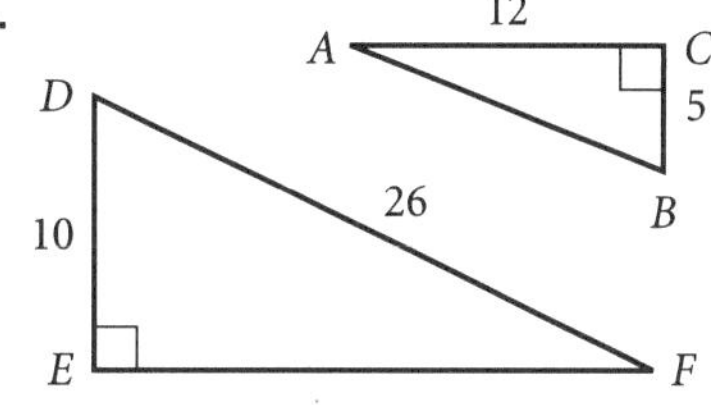

____

____

Complete the following proof.

Given: $\overline{AB} \parallel \overline{XC}$

Prove: $\triangle ABP \sim \triangle CXP$

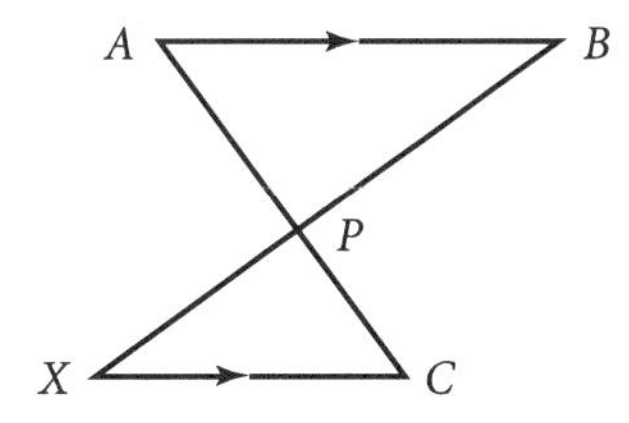

Statements	Reasons
$\angle APB \cong \angle CPX$	5. __________
6. __________	Given
$\angle B \cong \angle X$	7. __________
$\triangle ABP \sim \triangle CXP$	8. __________

9. It could be shown that $\triangle ABC \sim \triangle XYZ$ by SAS for the figures below. Complete each statement that also holds true.

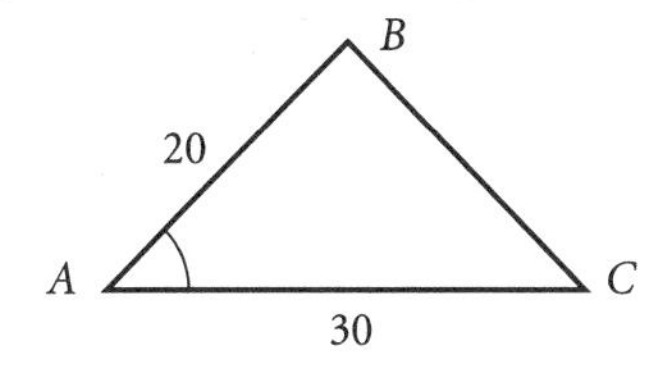

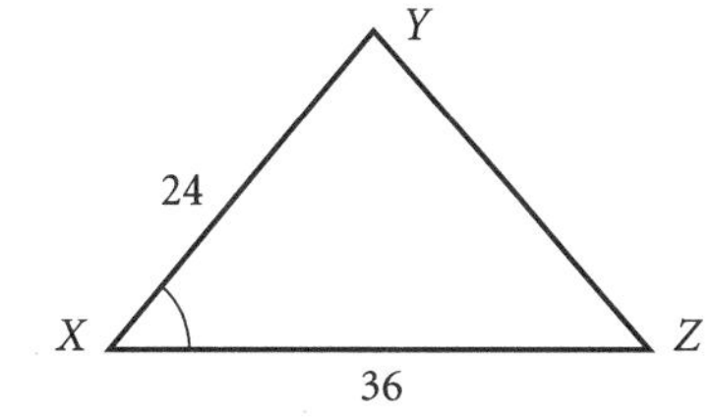

a. $\angle B \cong$ __________

b. $\angle C \cong$ __________

c. $\dfrac{BC}{YZ} =$ __________

NAME ______________________ CLASS ____________ DATE ____________

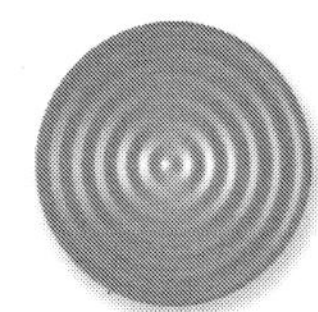

Practice Masters Level C

8.3 Triangle Similarity Postulates

Complete the following proof.

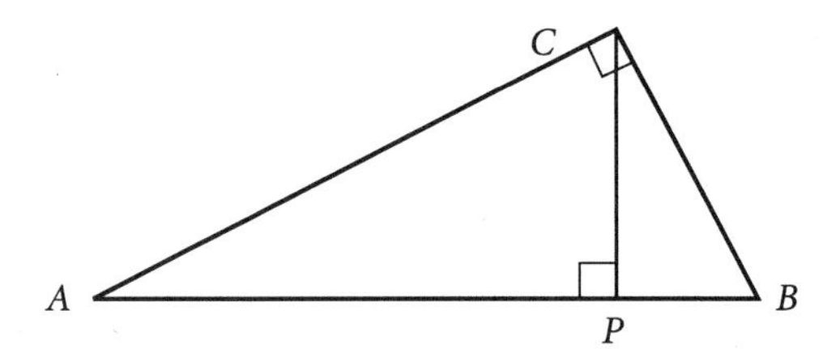

Given: $m\angle ACB = 90°$

$m\angle APC = 90°$

Prove: $\triangle APC \sim \triangle ACB$

Statements	Reasons
$m\angle ACB = 90°$, $m\angle APC = 90°$	1. ____________
$m\angle A = m\angle A$	2. ____________
$\triangle APC \sim \triangle ACB$	3. ____________

4. In a like manner, what other triangle could be proven similar to $\triangle ACB$? ____________

5. Must it be true that $\triangle APC$ is similar to $\triangle CPB$? Explain.

____

6. Write a statement, worded like a theorem, that describes what you have learned from Exercises 1–5.

____

____

A triangle *PQR* has vertices at *P*(−3, −1), *Q*(5, −1), and *R*(0, 3). Use this information for Exercises 7–10.

7. Use the distance formula to find the lengths of the sides of $\triangle PQR$.

$PQ =$ ________, $QR =$ ________, $RP =$ ________

8. Find the coordinates of the vertices of the image of $\triangle PQR$ under the dilation $D(x, y) = (2x, 2y)$.

P' ________, Q' ________, R' ________

9. Use the distance formula to find the lengths of the sides of the image $\triangle P'Q'R'$.

$P'Q' =$ ________, $Q'R' =$ ________, $R'P' =$ ________

10. Is it true that $\triangle PQR \sim \triangle P'Q'R'$? Explain why or why not.

____

Practice Masters Level A

8.4 The Side-Splitting Theorem

Use the Side-Splitting Theorem to find *x*.

1.

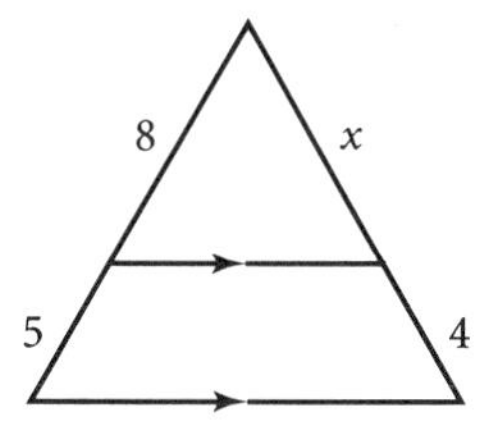

$x =$ ______________

2.

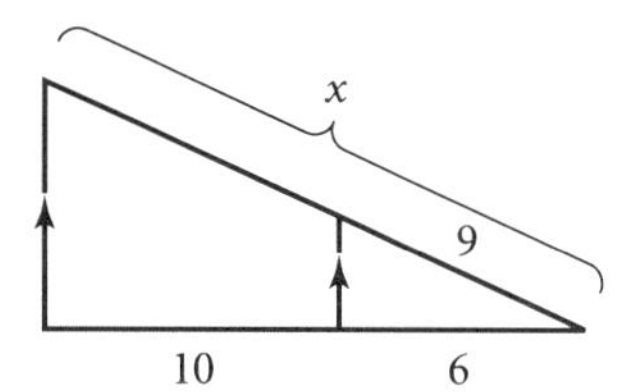

$x =$ ______________

3.

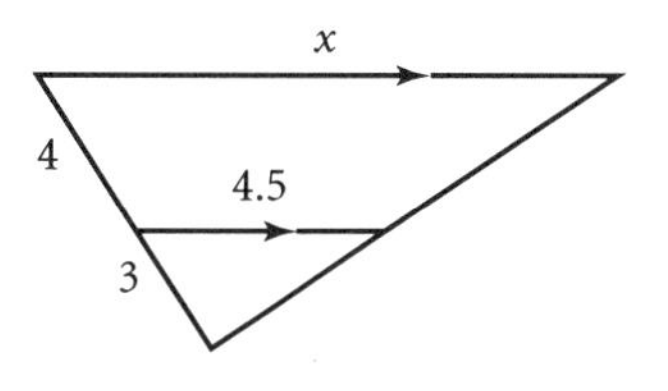

$x =$ ______________

4.

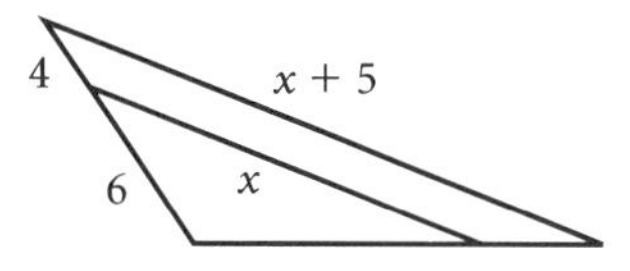

$x =$ ______________

Name all the similar triangles in each figure. State the postulate or theorem that justifies each similarity.

5.

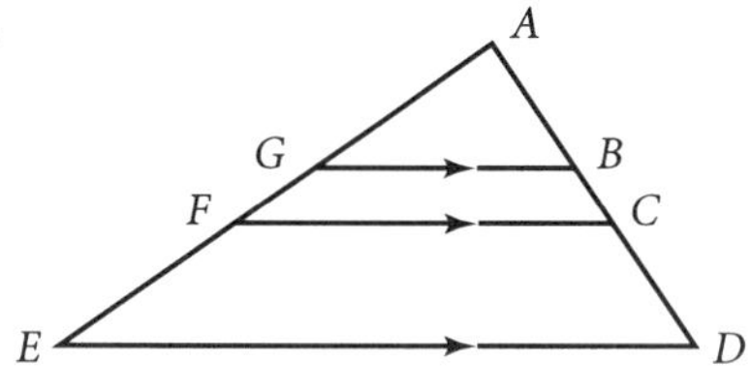

____

____

6.

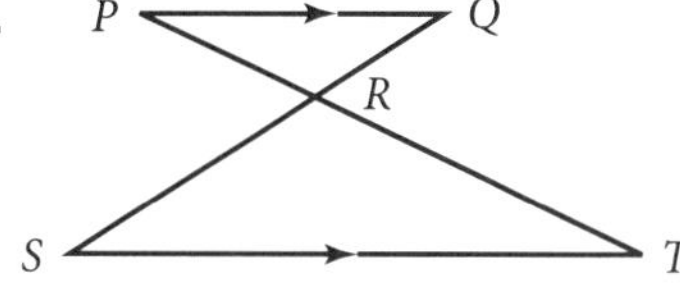

____

____

Use the Two-Transversal Proportionality Corollary to find *x*.

7.

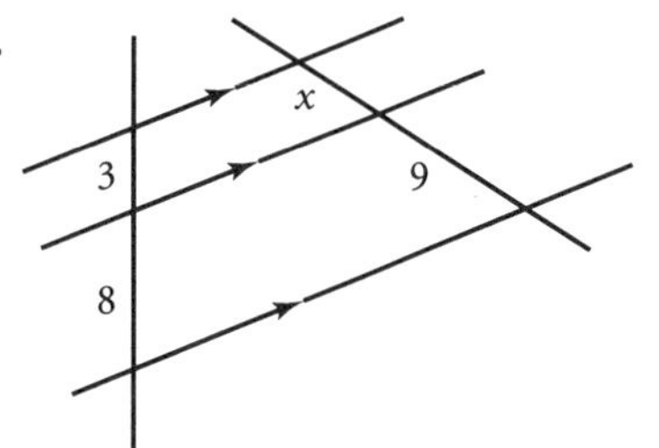

$x =$ ______________

8.

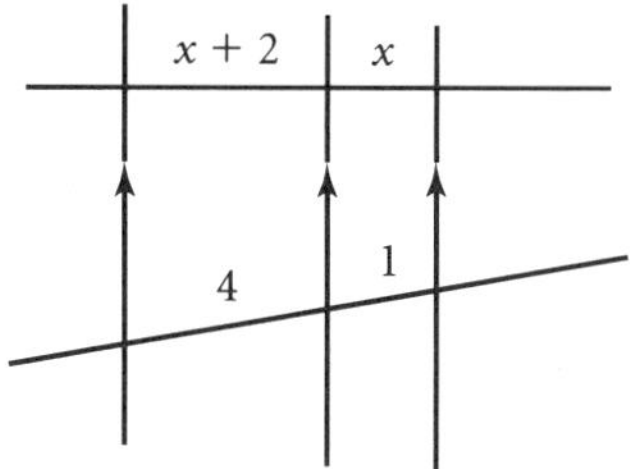

$x =$ ______________

NAME ______________________ CLASS ____________ DATE ____________

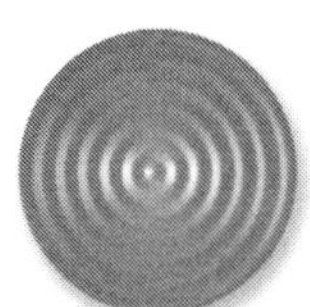

Practice Masters Level B

8.4 The Side-Splitting Theorem

Use the Side-Splitting Theorem to find *x*.

1.

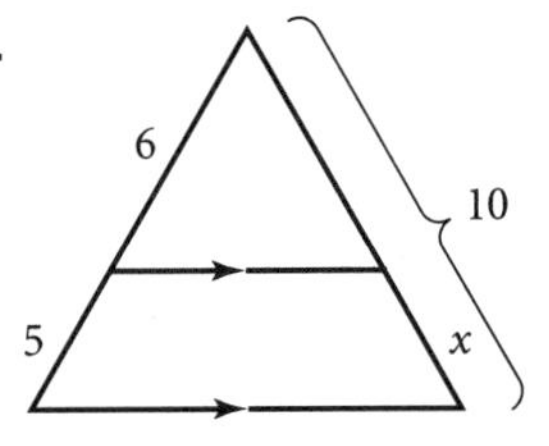

$x =$ ____________________

2. 

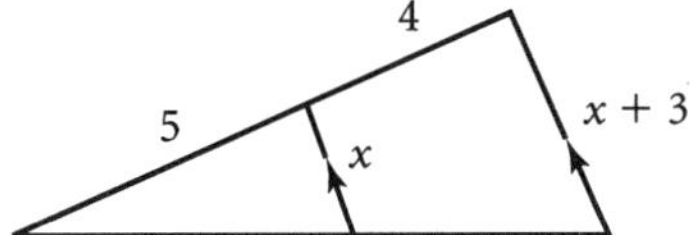

$x =$ ____________________

In Exercises 3–6, use what is given in each figure to find *x* and *y*.

3.

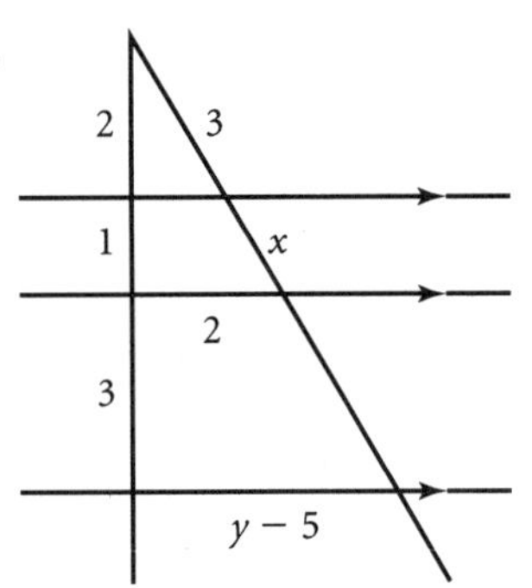

$x =$ ____________________

$y =$ ____________________

4.

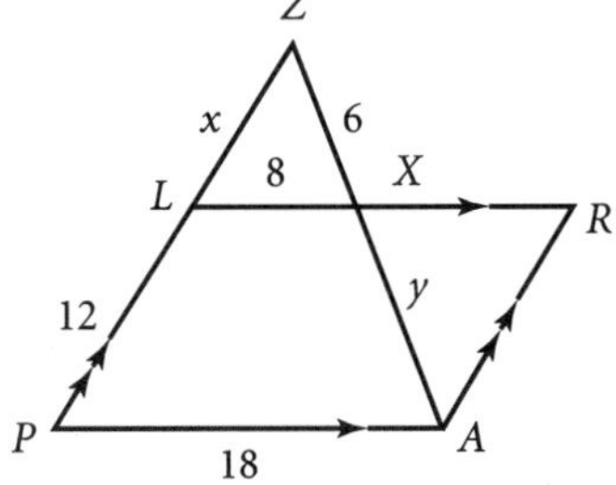

$x =$ ____________________

$y =$ ____________________

5.

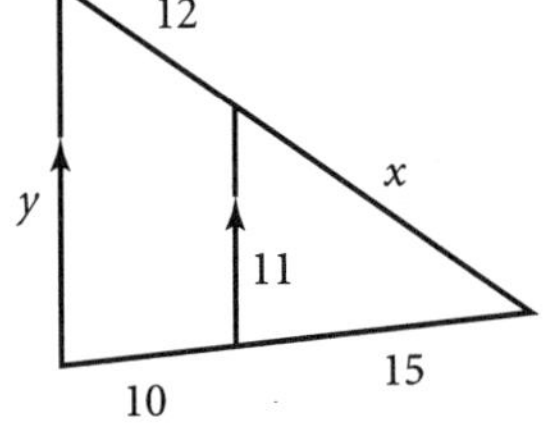

$x =$ ____________________

$y =$ ____________________

6.

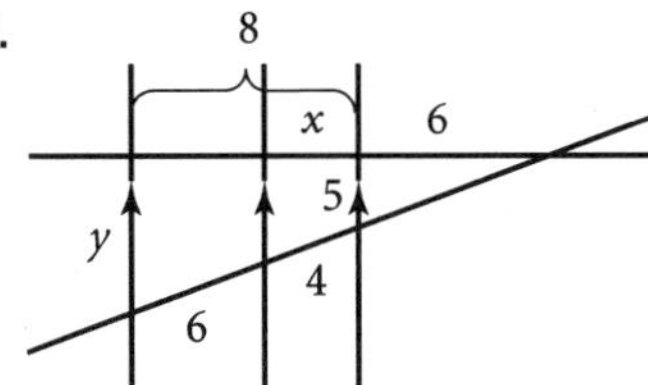

$x =$ ____________________

$y =$ ____________________

7. Use a compass and straightedge to construct lines that split $\overline{AB}$ into three segments whose lengths are in the ratio 1:3:2.

A ———————————— B

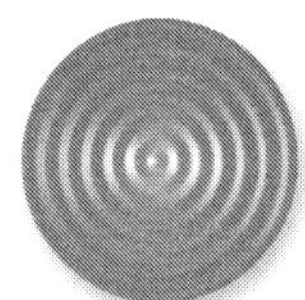

Practice Masters Level C

8.4 The Side-Splitting Theorem

In Exercises 1–4, find *x* and *y* for each figure.

1.

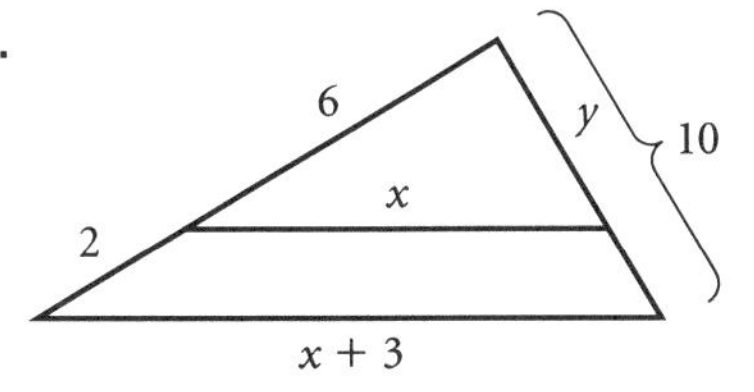

$x =$ ____________, $y =$ ____________

2.

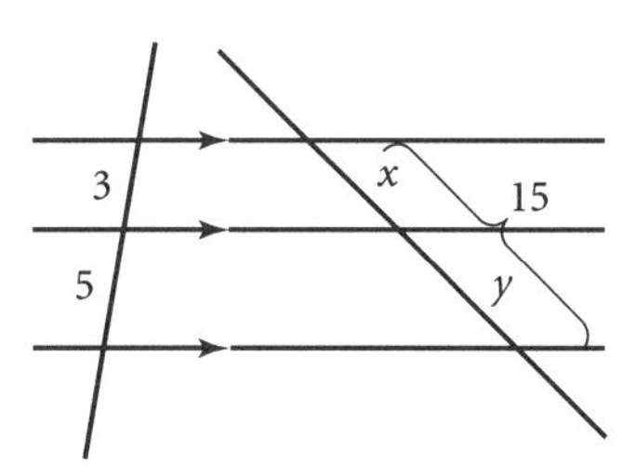

$x =$ ____________, $y =$ ____________

3.

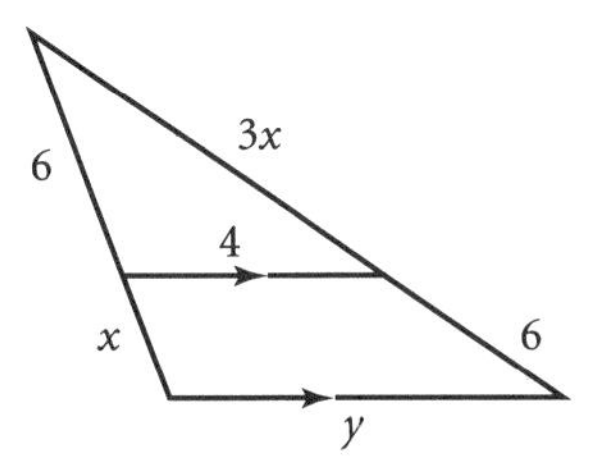

$x =$ ____________, $y =$ ____________

4.

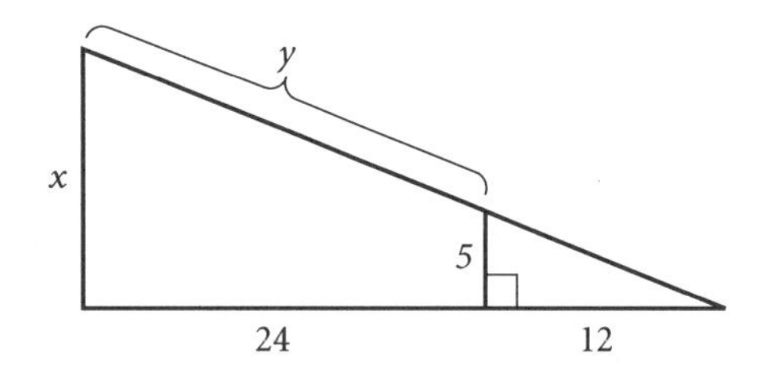

$x =$ ____________, $y =$ ____________

5. In the figure, $\overline{TR} \parallel \overline{PA} \parallel \overline{QS}$ and $\overleftrightarrow{BQ}$ and $\overleftrightarrow{BS}$ are two transversals that meet at point B.

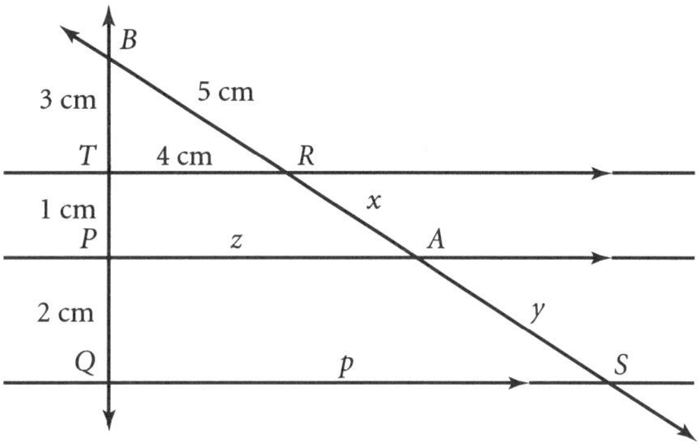

   a. Find the missing lengths x, y, z, and p.

   $x =$ ____________

   $y =$ ____________

   $z =$ ____________

   $p =$ ____________

   b. Determine whether trapezoid $TRAP$ is similar to trapezoid $PASQ$. Explain your answer.

____

____

____

____

____

____

NAME ______________________ CLASS ____________ DATE ____________

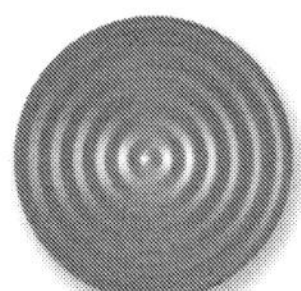

Practice Masters Level A

8.5 *Indirect Measurement and Additional Similarity Theorems*

In Exercises 1 and 2, complete the equation to make a true proportion.

1.

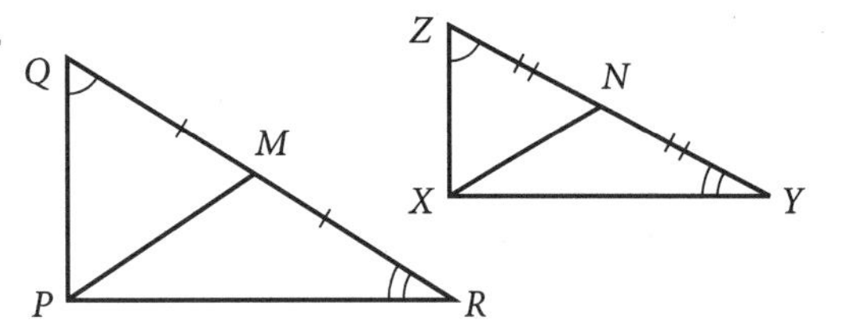

$\frac{PM}{XN} =$ ______________

2.

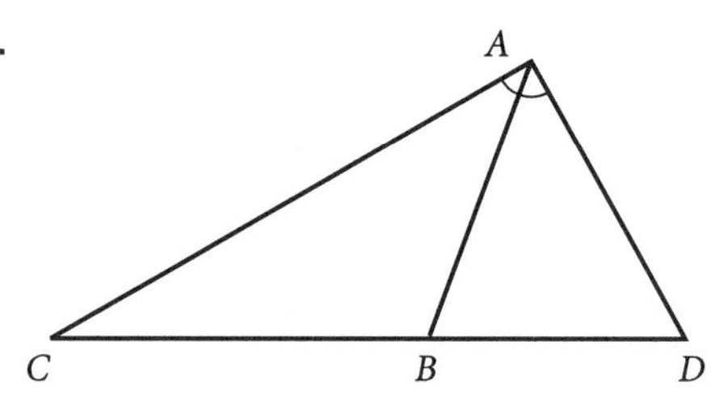

$\frac{BC}{BD} =$ ______________

In Exercises 3–6, the triangles are similar. Find *x*.

3.

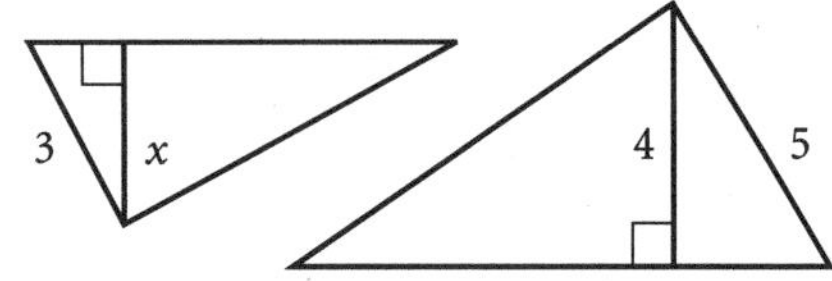

$x =$ ______________

4.

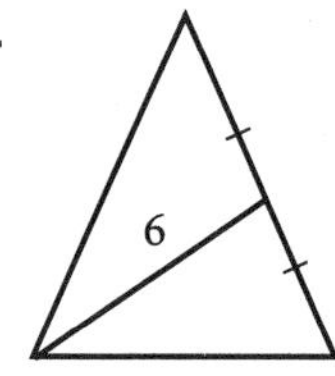

5

x

$x =$ ______________

5. Given: $\triangle APB \sim \triangle CPD$

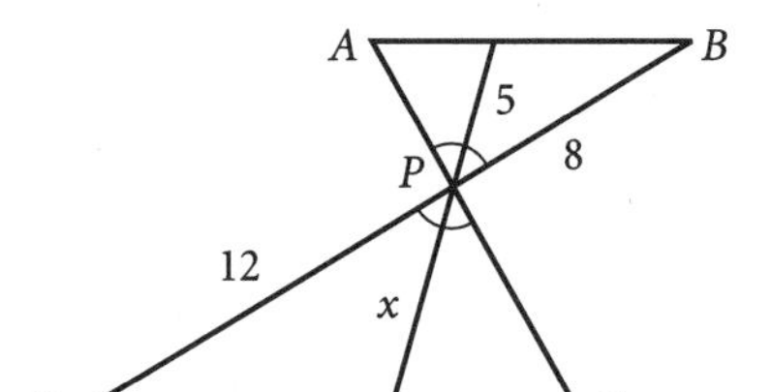

$x =$ ______________

6. Given: $\triangle WVS \sim \triangle YVZ$, $WS = 24$, $YZ = 30$

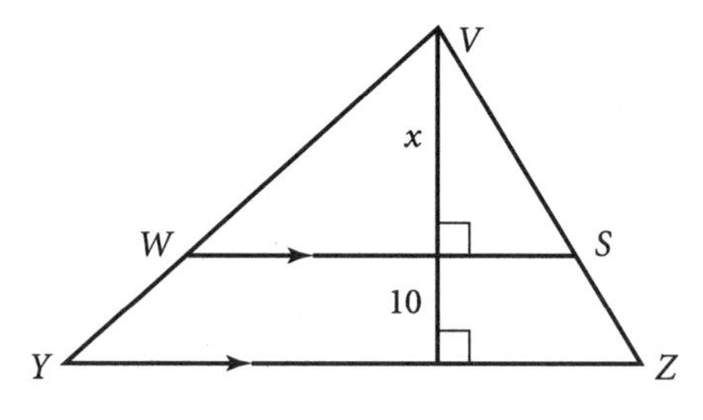

$x =$ ______________

7. When Susan stands 8 feet from the base of a street lamp, her shadow is 10 feet long. Susan is $5\frac{1}{2}$ feet tall. Find the height, h, of the lamp.

h

5.5 ft

8 ft

10 ft

____

NAME ______________________ CLASS ____________ DATE ____________

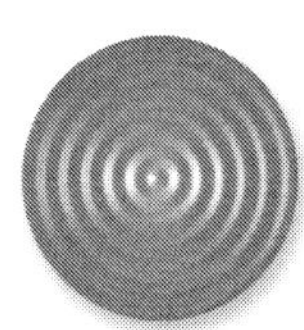

Practice Masters Level B

8.5 *Indirect Measurement and Additional Similarity Theorems*

In Exercises 1–4, apply a similarity theorem to find *x*.

1.

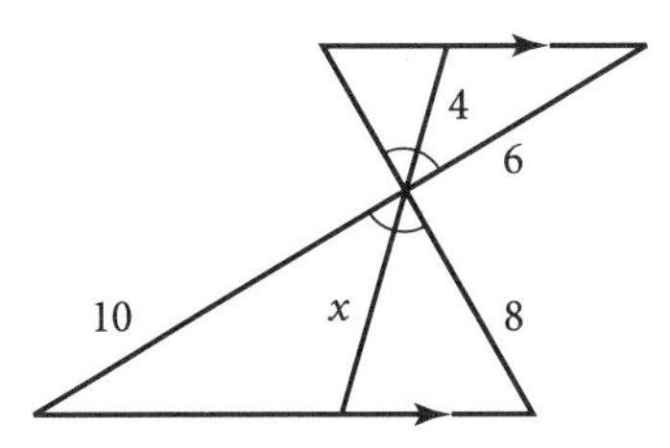

$x =$ ______________________

2.

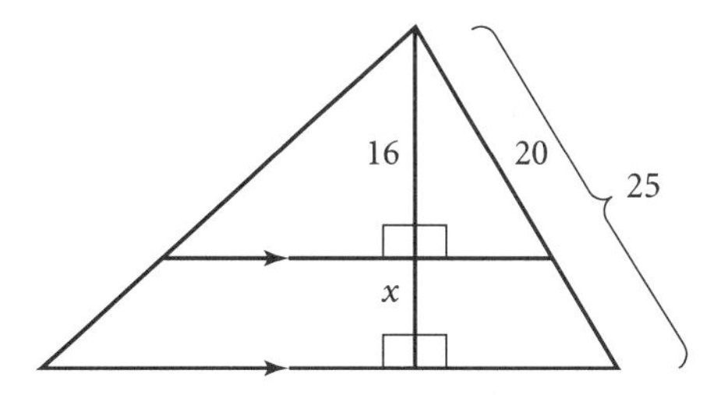

$x =$ ______________________

3.

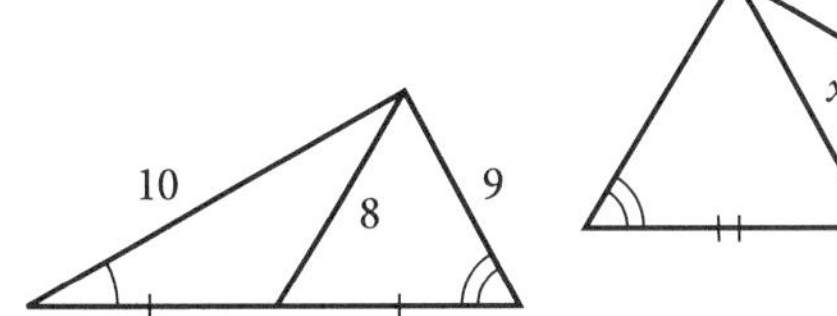

$x =$ ______________________

4.

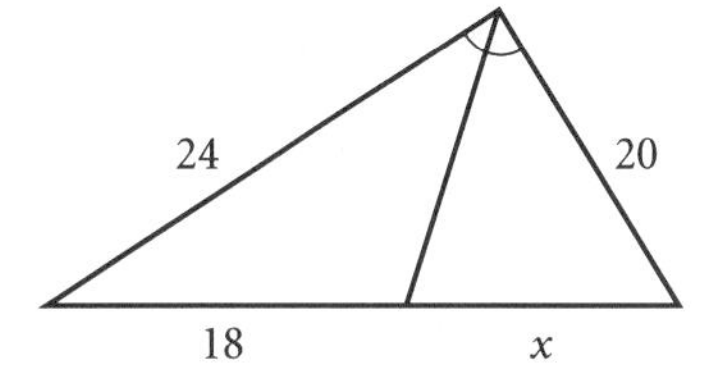

$x =$ ______________________

5. Use the diagram to find the width, *w*, of the river.

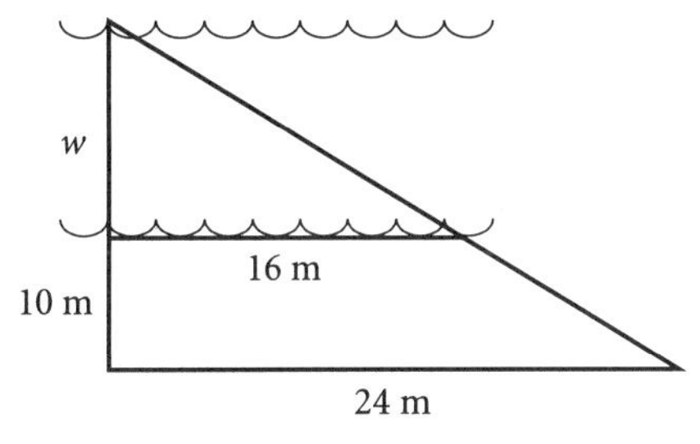

____

6. On a sunny day Maria, who is 5 feet tall, is standing near a tree. Her shadow is 12 feet long, while the shadow of the tree is 32 feet long. Use this information to find the height of the tree.

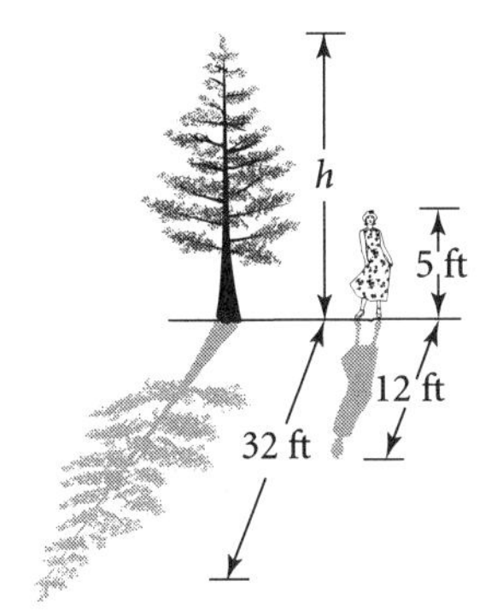

____

Practice Masters Level C

8.5 *Indirect Measurement and Additional Similarity Theorems*

In Exercises 1–3, find *x*.

1.

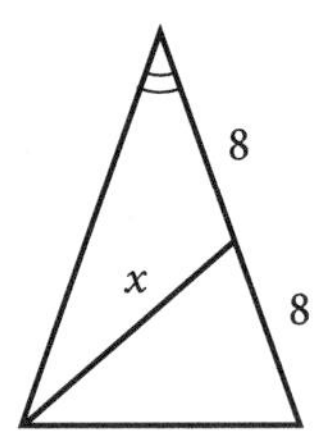

$x =$ ______________

2.

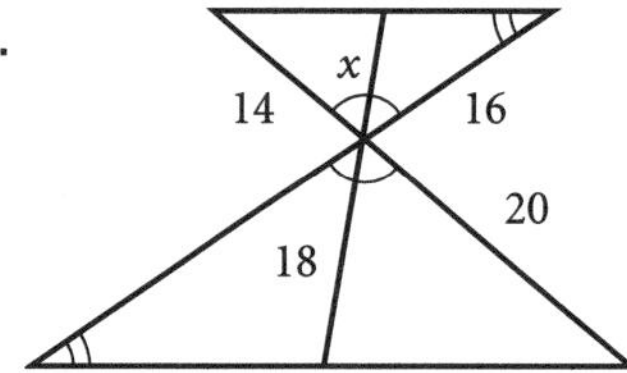

$x =$ ______________

3.

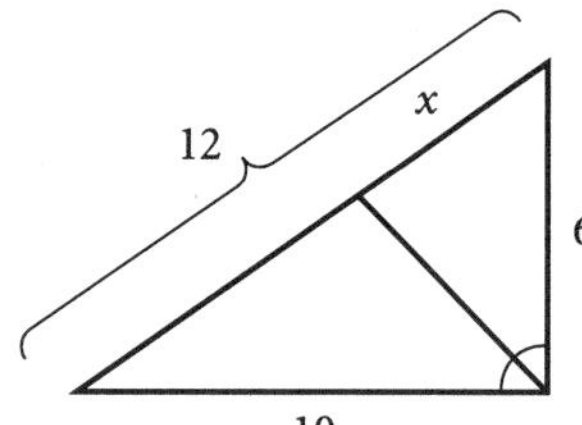

$x =$ ______________

A number, m, is called the geometric mean between two numbers a and b if m occupies the two middle positions, or means, between a and b in a true proportion. That is, $\frac{a}{m} = \frac{m}{b}$. Thus, 4 is the geometric mean between 2 and 8 because $\frac{2}{4} = \frac{4}{8}$.

Theorem: The length of the altitude to the hypotenuse of a right triangle is the geometric mean between the lengths of the two segments into which it divides the hypotenuse.

Complete the following proof.

Given: $\overline{AC} \perp \overline{BC}$, $\overline{CP} \perp \overline{AB}$

Prove: $\frac{PA}{PC} = \frac{PC}{PB}$

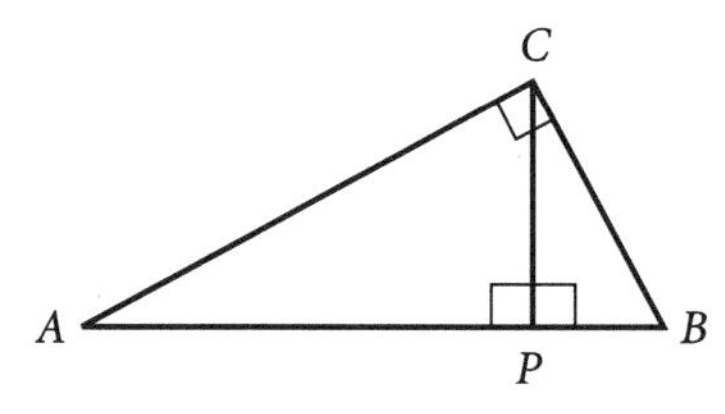

Statement	Reasons
$m\angle ACB = 90° = m\angle APC$	Definition of perpendicular lines
$m\angle A = m\angle A$	Reflexive property of equality
4. $\triangle ACB \sim \triangle$ ______	AA similarity postulate
$m\angle ACB = 90° = m\angle CPB$	5. ______________
$m\angle B = m\angle B$	6. ______________
7. $\triangle ACB \sim \triangle$ ______	8. ______________
$\triangle APC \sim \triangle CPB$	9. ______________
$\frac{PA}{PC} = \frac{PC}{PB}$	10. ______________

NAME ______________________ CLASS __________ DATE __________

Practice Masters Level A

8.6 Area and Volume Ratios

1. The ratio of the areas of two squares is $\frac{16}{25}$.

   a. Find the ratio of their sides. ______________

   b. The larger square has sides of length 10 centimeters. Find the side length of the smaller square. ______________

In Exercises 2–5, tell whether the quantity described varies with (a) the linear dimensions, or (b) the area, or (c) the volume of the object.

2. the weight of a statue ______________
3. the number of gallons of paint needed to paint a storage tank ______________
4. the cost to carpet a floor ______________
5. the amount of fencing needed to enclose a yard ______________

In Exercises 6–9, use area and/or volume ratios to solve each problem.

6. A trophy that is 8 inches tall weighs 4 pounds. A trophy of similar shape is 12 inches tall. How much does the larger trophy weigh? ______________

7. It costs $440 to carpet a room that measures 16 feet by 24 feet. How much would it cost to carpet a similar room that measures 12 feet by 18 feet? ______________

8. The radius of the Earth is about 3.7 times the radius of the moon. Since they both approximate spheres, consider that they are similar.

   a. Find the ratio of their surface areas. ______________

   b. Find the ratio of their volumes. ______________

9. Two similar triangles have areas in the ratio of $\frac{1}{3}$. The smaller triangle has an altitude of length 6 centimeters. Find the length of the corresponding altitude of the larger triangle. ______________

10. Consider the relationship between cross-sectional area, weight, and height. Explain why giant animals 10 times the size of normal animals and similar to them in shape and structure could not exist.

____

____

NAME ______________________ CLASS __________ DATE __________

Practice Masters Level B

8.6 *Area and Volume Ratios*

1. The ratio of the areas of two circles is $\frac{9}{16}$.

   a. Find the ratio of their radii. ______________

   b. The smaller circle has a radius of 6 centimeters. Find the radius of the larger circle. ______________

In Exercises 2–5, tell whether the quantity described varies with (a) the linear dimensions, or (b) the area, or (c) the volume of the object.

2. the amount of paper needed to wrap a box ______________
3. the amount of ribbon needed to tie around the box ______________
4. the amount of ingredients in a cake recipe ______________
5. the time needed to run a race ______________

In Exercises 6–9, use area and/or volume ratios to solve the problem.

6. Two rooms are similar in shape, with corresponding sides in the ratio of $\frac{2}{3}$. It takes 3 gallons of paint to cover the walls of the larger room. How much paint will be needed to paint the smaller room? ______________
7. A trophy that is 10 inches tall weighs 4 pounds. Estimate the height of a similar trophy that weighs 6 pounds. ______________
8. A mother has just sewn a cape for herself and plans to scale down the pattern to make a matching cape for her daughter. The mother is 5.5 feet tall and the daughter is 4 feet tall. The mother's cape required about 1.5 square yards of fabric. About how much fabric will be needed for the daughter's cape? ______________
9. Suppose that all pizzas have the same thickness and that cost and number of servings both depend only on the surface area. A pizza 10 inches in diameter costs $8.12 and serves 2 people.

   a. Find how much a 14-inch pizza should cost. ______________

   b. How many people would the 14-inch pizza serve? ______________

NAME ______________________ CLASS ______________ DATE ______________

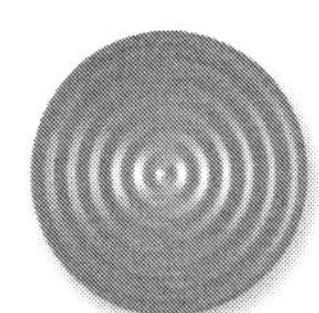

Practice Masters Level C

8.6 Area and Volume Ratios

1. Shea wants to adapt a favorite pie recipe for an 8-inch pan to her similar 10-inch pan.

   a. If the crust should be the same thickness for both pies, by what factor should she multiply all crust ingredients? ______________

   b. By what factor should she multiply all of the filling ingredients? ______________

Compare the quantities related to similar figures, and complete the chart.

Figure	Dimensions of preimage	Dimensions of image	Area of preimage	Area of image	Ratio of areas
Triangle	b, h	kb, kh	$A = \frac{1}{2}bh$	$A' = \frac{1}{2}(kb)(kh) = k^2\left(\frac{1}{2}bh\right) = k^2 A$	$\frac{A'}{A} = k^2$
Rectangle	l, w	2.	3.	4.	5.
Circle	r	6.	7.	8.	9.

10. What property of area formulas seems to assure that for two similar figures with scale factor of k, the ratio of areas will be k^2? ______________

Compare the quantities related to similar solids, and complete the chart.

Solid	Dimensions of preimage	Dimensions of image	Volume of preimage	Volume of image	Ratio of volumes
Box	l, w, h	kl, kw, kh	$V = lwh$	$V' = (kl)(kw)(kh) = k^3(lwh) = k^3 V$	$\frac{V'}{V} = k^3$
Square pyramid	s, s, h	11.	12.	13.	14.
Circle	r, h	15.	16.	17.	18.
Sphere	r	19.	20	21.	22.

23. What property of volume formulas seems to assure that for two similar solids with scale factor of k, the ratio of volumes will be k^3? ______________

NAME ______________________ CLASS __________ DATE __________

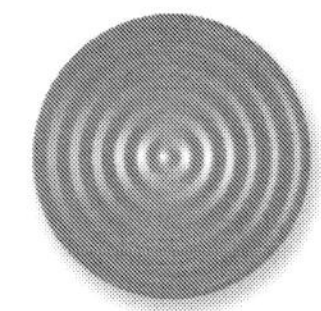

Practice Masters Level A

9.1 Chords and Arcs

Use the figure of $\odot M$ below for Exercises 1–3.

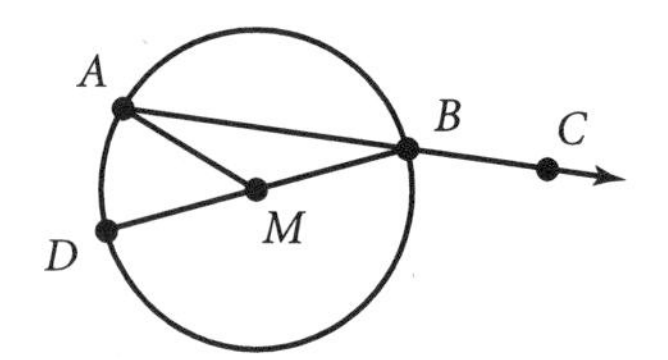

1. Name three radii of the circle. ____________

2. Name a diameter of the circle. ____________

3. Name a chord of the circle. ____________

Use the figure of $\odot P$ below for Exercises 4–7.

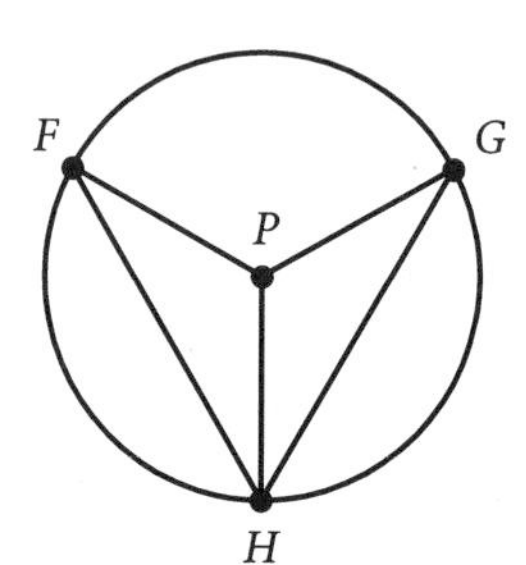

4. List three major arcs of the circle. ____________

5. List three minor arcs of the circle. ____________

6. If $m\angle FHG = 61°$, what is $m\angle FPG$? ____________

7. If $m\angle FPG = 96°$, what is $m\angle FHG$? ____________

Determine the length of the arc with the given central angle measure, $m\angle W$, in a circle with radius r. Round your answers to the nearest hundredth.

8. $m\angle W = 45°; r = 5$ ____________

9. $m\angle W = 90°; r = 10$ ____________

10. $m\angle W = 60°; r = 8$ ____________

11. $m\angle W = 120°; r = 20$ ____________

12. $m\angle W = 76°; r = 5.2$ ____________

13. $m\angle W = 196°; r = 12$ ____________

Determine the degree measure of an arc with the given length, L, in a circle with radius r. Round your answers to the nearest whole degree.

14. $L = 10; r = 7$ ________

15. $L = 14; r = 20$ ________

16. $L = 25; r = 12$ ________

17. $L = 36; r = 18$ ________

18. $L = 7; r = 13$ ________

19. $L = 4.2; r = 6$ ________

For Exercises 20–23, find the degree measures of each arc by using the central angle measures given in $\odot M$.

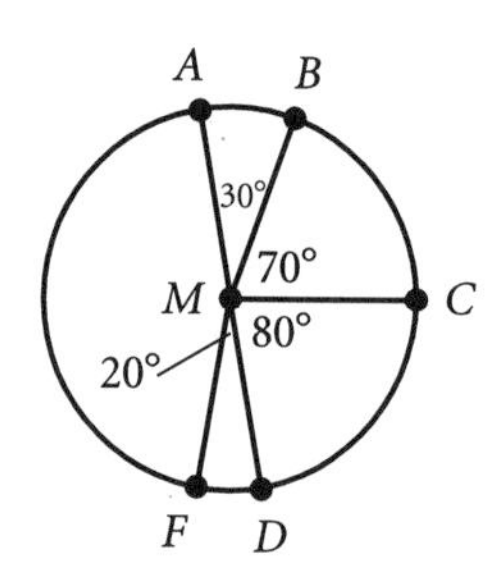

20. $m\overset{\frown}{AC}$ ____________

21. $m\overset{\frown}{DF}$ ____________

22. $m\overset{\frown}{ADC}$ ____________

23. $m\overset{\frown}{FBC}$ ____________

NAME ______________________ CLASS ____________ DATE ____________

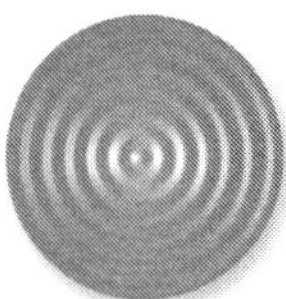

Practice Masters Level B

9.1 Chords and Arcs

For Exercises 1–6, find the degree measures of each arc by using the central angle measures given in ⊙*M*.

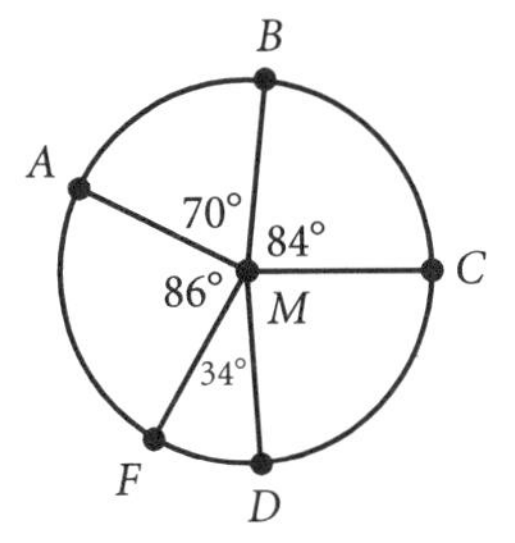

1. $m\overset{\frown}{AC}$ ____________
2. $m\overset{\frown}{FA}$ ____________
3. $m\overset{\frown}{CBF}$ ____________
4. $m\overset{\frown}{DB}$ ____________
5. $m\overset{\frown}{ADC}$ ____________
6. $m\overset{\frown}{DCA}$ ____________

7. Do arcs that are identified with the same letters necessarily have equal measures? Why or why not? ____________

____

Determine the length of the arc with the given central angle measure, m∠*W*, in a circle with radius *r*. Round your answers to the nearest hundredth.

8. $m\angle W = 240°;\ r = \frac{3}{4}$ ____________
9. $m\angle W = 360°;\ r = 4.7$ ____________
10. $m\angle W = 45°; r = x + 3$ ____________
11. $m\angle W = x°; r = 7$ ____________

Determine the degree measure of an arc with the given length, *L*, in a circle with radius *r*. Round your answers to the nearest hundredth.

12. $L = 15; r = 12$ ____________
13. $L = 11\frac{3}{4}; r = 8\frac{1}{2}$ ____________

Determine the length of the radius of a circle with the given central angle measure, m∠*W*, and the given arc length, *L*. Round your answers to the nearest hundredth.

14. $m\angle W = 61°; L = 12$ ____________
15. $m\angle W = 213°; L = 21$ ____________
16. $m\angle W = 114°; L = 16.5$ ____________
17. $m\angle W = 300°; L = 25$ ____________

NAME ______________________ CLASS __________ DATE __________

Practice Masters Level C

9.1 Chords and Arcs

Solve for x in each of the following where a circle has radius r, an arc of length L, and central angle $m\angle W$. Round your answers to the nearest hundredth.

1. $L = x + 2; r = 6; m\angle W = 31°$ ________
2. $L = 3x + 4; r = 2; m\angle W = 125°$ ________
3. $L = 2x - 3; r = 10; m\angle W = 75°$ ________
4. $L = x^2; r = 5; m\angle W = 210°$ ________
5. $L = 2x - 5; r = x + 1;$ $m\angle W = 150°$ ________
6. $L = 6x; r = 2x + 3;$ $m\angle W = 72°$ ________

Solve.

7. What central angle measure will give an arc length equal to the diameter of the circle? ________
8. What central angle measure will give an arc length three times the length of the radius of a circle? ________
9. What is the greatest integral ratio that can exist between a radius and an arc length that will produce a possible central angle; that is, one that is less than 360°? ________

In $\odot P$, the radius is 2; diameter $\overline{AD}$ is perpendicular to diameter $\overline{EC}$; $\overline{GD} \cong \overline{BD}$; and $m\overset{\frown}{BC} = 37°$. Find the following.

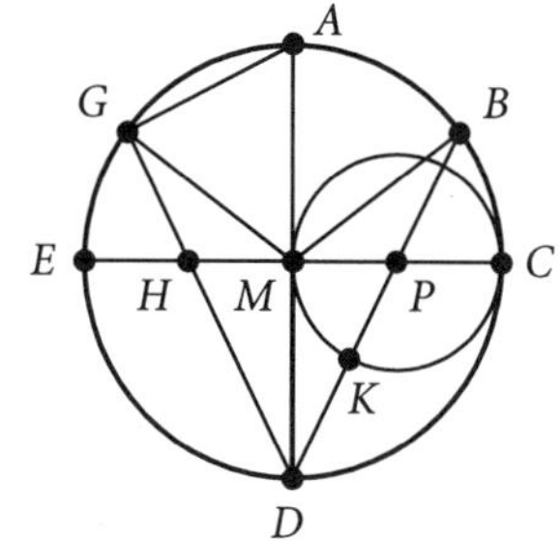

10. PD ________
11. GD ________
12. $m\angle ADB$ ________
13. $m\angle GDB$ ________
14. $m\angle BMD$ ________
15. $m\overset{\frown}{GD}$ ________

16. Write a paragraph proof of the following: In a circle, or in congruent circles, if two chords are congruent, then they are equidistant from the center of the circle.
Given: $AB = CD$; **Prove:** $RP = PS$ ________

____

____

____

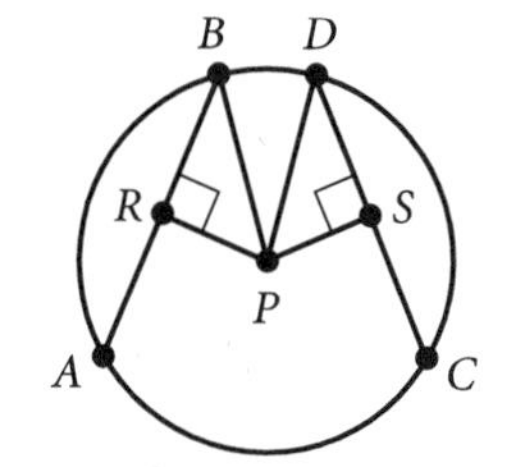

NAME ______________________ CLASS ____________ DATE ____________

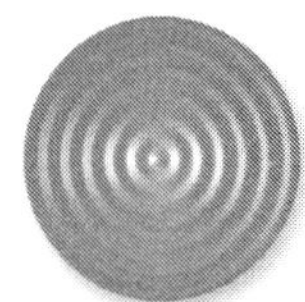

Practice Masters Level A

9.2 Tangents to Circles

Refer to ⊙K, in which $\overline{MS}$ is tangent to ⊙K, for Exercises 1–4.

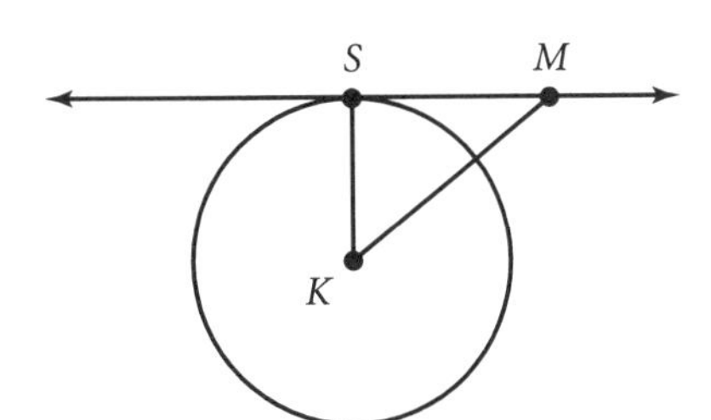

1. If $KS = 10$ and $MK = 26$, find SM. ______
2. If $KS = 3$ and $MK = 5$, find SM. ______
3. If $KS = 5$ and $MK = 13$, find SM. ______
4. If $KS = 6$ and $MK = 10$, find SM. ______

Refer to ⊙P, in which $\overline{PN} \perp \overline{QT}$ at M, for Exercises 5–9.

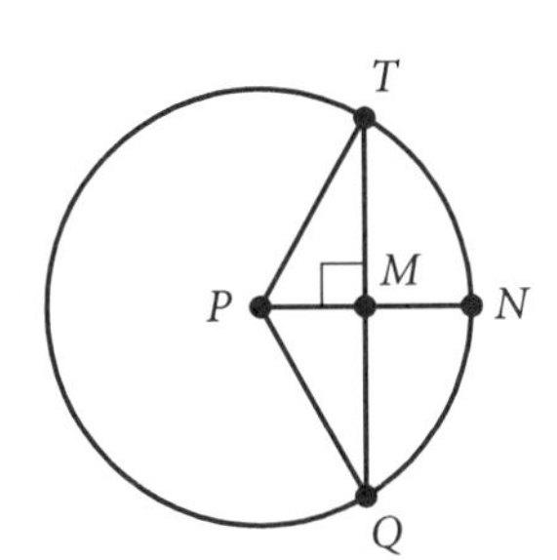

5. Name a segment congruent to $\overline{QM}$. ______
6. Name two segments congruent to $\overline{PN}$. ______
7. If $PT = 6$ and $PM = 3$, find QM. ______
8. If $PT = 4$ and $PM = 1$, find QM. ______
9. If $PQ = 13$ and $PM = 5$, find QM and MT. ______

Refer to ⊙M, in which $\overline{AG}$ is tangent to ⊙M, for Exercises 10–12.

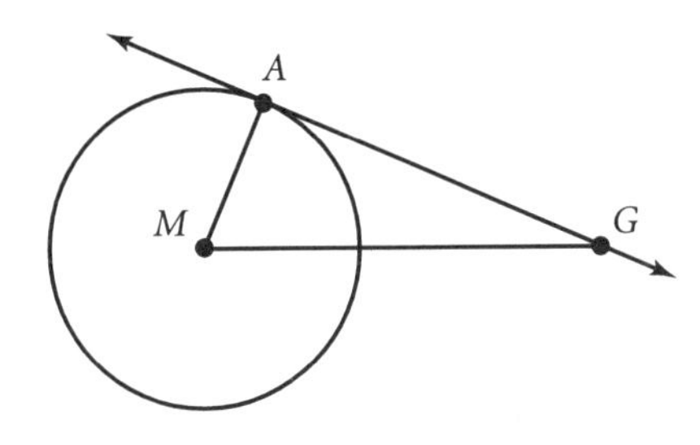

10. If $MA = 8$, and $MG = 10$, find AG. ______
11. If $MA = 22$ and $AG = 13$, find MG. ______
12. If $MG = 18$ and $AG = 16$, find the length of the diameter of the circle. ______

In ⊙M, $\overline{FM} \perp \overline{AB}$; $\overline{CD}$ is a diameter; $MD = 10$ and $FE = 2$.

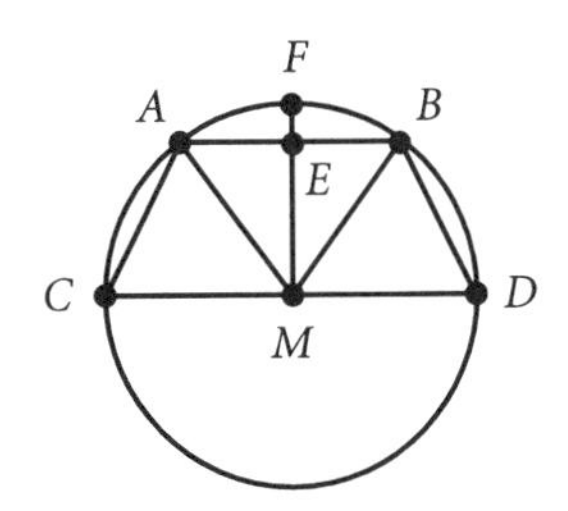

13. Find CD. ______
14. Find ME. ______
15. If $EB = 8$ and $BM = 17$, find ME. ______

Complete each statement. Assume all figures lie in the same plane.

16. A ____________ is a segment whose endpoints are on the circle.
17. A ____________ is a line that contains a chord.

NAME ______________________ CLASS ____________ DATE ____________

Practice Masters Level B

9.2 Tangents to Circles

Complete each statement. Assume all figures lie in the same plane.

1. A ____________ is a line that intersects the circle in exactly one point. ____________
2. A tangent is ____________ to a radius at its endpoints. ____________
3. A ____________ is the longest chord of a circle. ____________

In ⊙M, $\overline{FM} \perp \overline{AB}$ and $\overline{CD}$ is a diameter.

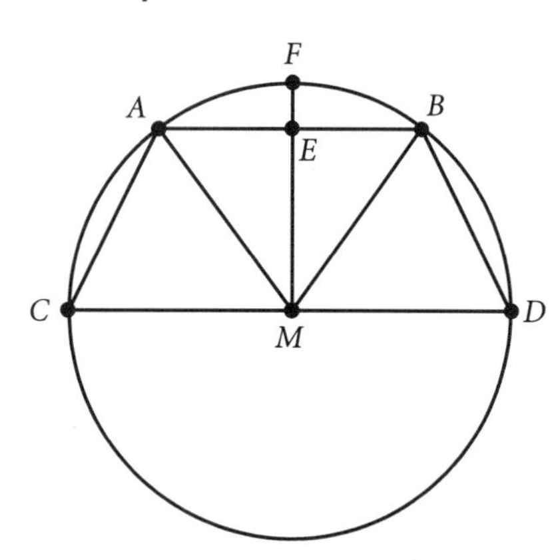

4. If $MD = 10$, find FE. ____________
5. If $AB = 33$ and $MD = 22$, find EM. ____________
6. If $AB = \sqrt{72}$ and $MD = \sqrt{54}$, find EM. ____________
7. If $MD = 10$ and $FE = 2$ find:
   a. the area of quadrilateral $ABCD$. ____________
   b. the area of $\triangle AMC$. ____________
   c. the area of $\triangle BMD$. ____________

In ⊙M, diameter $AD = 8$, $\overline{CM} \perp \overline{MB}$, m∠$BMA = 30°$.

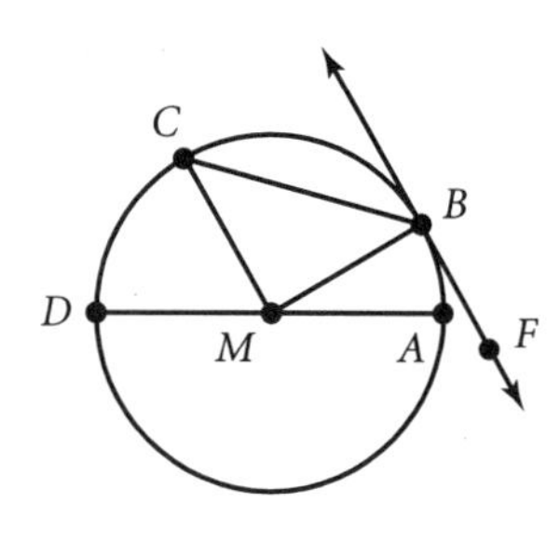

8. Find CD. ____________
9. Find CB. ____________
10. If $\overline{BF} \parallel \overline{CM}$, then what type of line is $\overline{BF}$? ____________

***AE* is the diameter of ⊙M and $\overline{CM} \perp \overline{BD}$.**

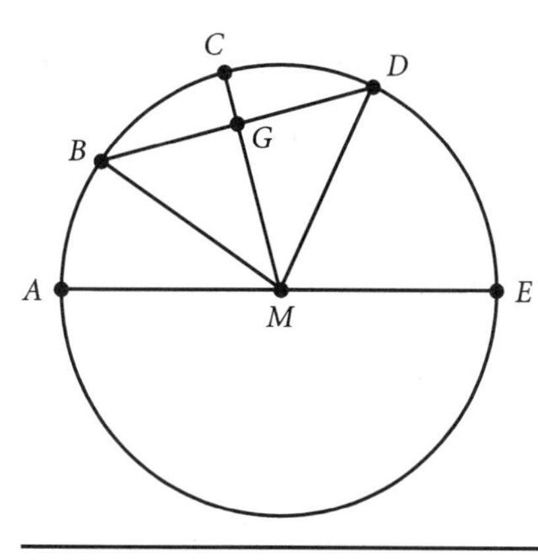

11. If $AE = 16$ and $CG = 2$, find GM. ____________
12. If $AE = 16$ and $CG = 2$, find BD. ____________
13. If $MB = \sqrt{x^2 + x}$, $BG = 2\sqrt{x}$, and $GM = \sqrt{10}$, find DM. ____________
14. Find the area of $\triangle BDM$. ____________
15. If $CG = 2$ and $BD = 8$, find AE. ____________
16. If $GC = 2$ and $GD = 6$, find CM. ____________

NAME ________________ CLASS ________ DATE ________

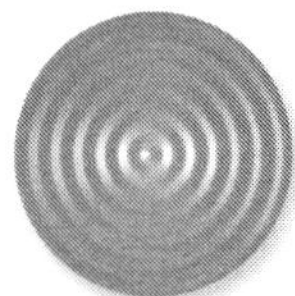

Practice Masters Level C

9.2 Tangents to Circles

In ⊙*M* the radius is 16, $\overline{ME} \perp \overline{GD}$.

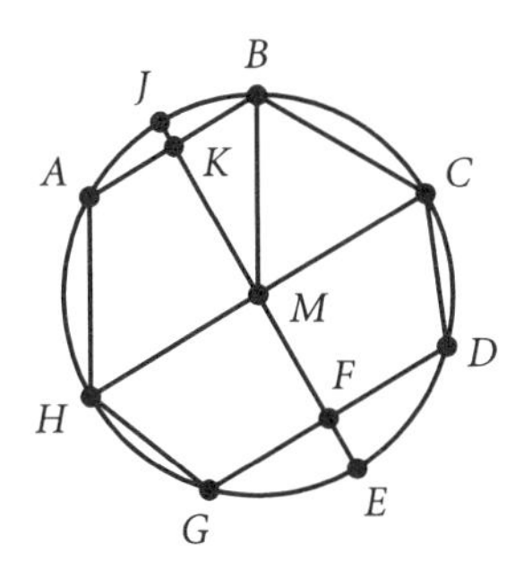

1. If $FE = 4$, find GD. ________________

2. IF $FE = 6$, find the area of $MCDF$. ________________

3. If the area of $ABCH$ is 250 and KM is $\frac{5}{9}$ of AB, find:

   KM ________________ AB ________________

4. If KB is $\frac{1}{2}KM$, find AB. ________________

In ⊙*M*, diameter $AF = 42$, $AC = 36$, $AB = 42$, $\overline{AC} \perp \overline{MB}$; $\overline{BA}$ and $\overline{BC}$ are tangent to the circle at points *A* and *C* respectively. Find the following:

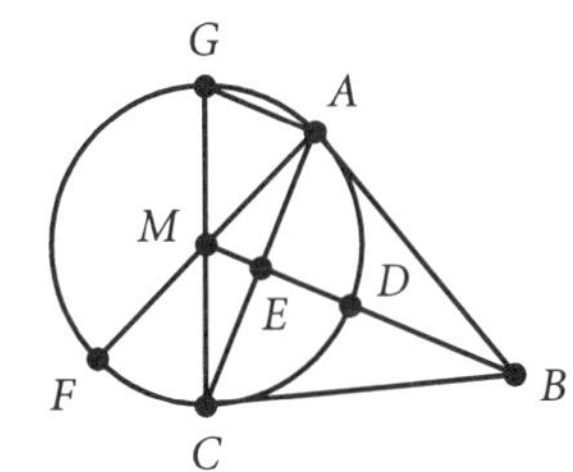

5. MB ________________

6. EM ________________

7. ED ________________

8. EB ________________

In ⊙*M*, $\overline{DC}$ is a diameter. $\overline{AB}$ is tangent to the circle at *A*.

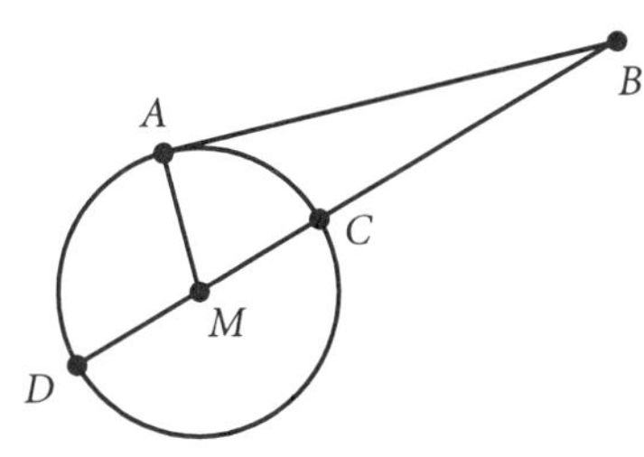

9. If $DC = 12$, $BD = 28$, and $AM = x + 2$, find AB. ________________

10. If $DB = 2x + 5$, $MC = x - 1.7$, and $AB = x + 4.9$, find MB. ________________

11. **Given:** ⊙M with $\overrightarrow{PA}$ and $\overline{PB}$ tangent to the circle at points A and B, respectively.

    Prove: $\overline{PA} \cong \overline{PB}$

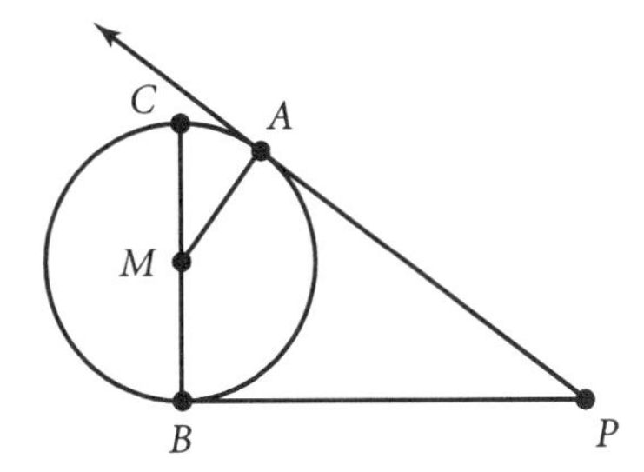

Statements	Reasons
1.	1.
2.	2.
3.	3.
4.	4.
5.	5.
6.	6.
7.	7.
8.	8.

NAME ______________________________ CLASS ______________ DATE ______________

Practice Masters Level A

9.3 *Inscribed Angles and Arcs*

1. Which of the following circles contains an inscribed angle?

a.

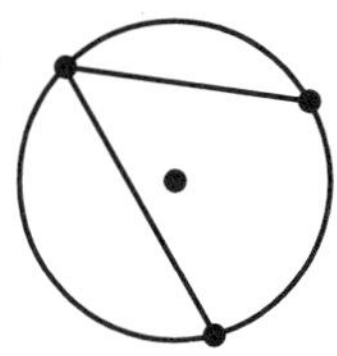

b.

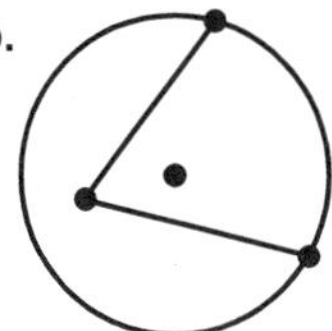

c.

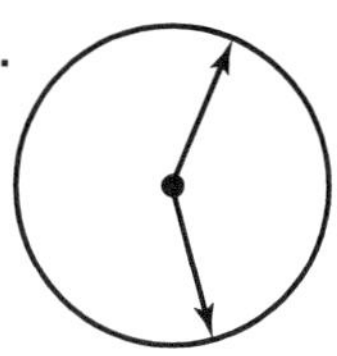

d.

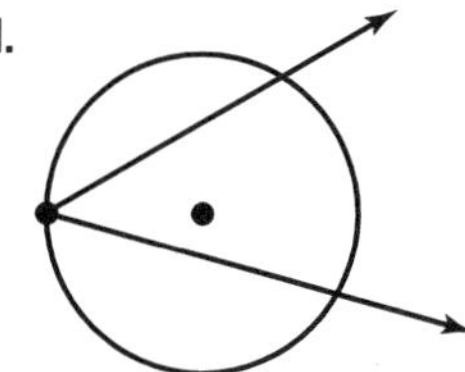

Refer to ⊙*P* for Exercises 2–5.

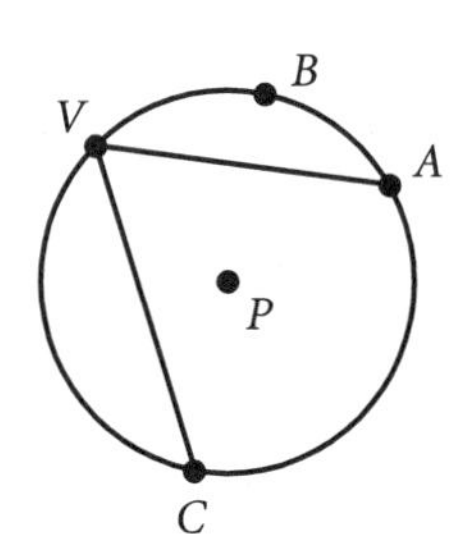

2. Identify the inscribed angle in ⊙*P*. ______________

3. Identify the major arc. ______________

4. If the intercepted arc of the inscribed angle is 130°, what is the measure of the inscribed angle? ______________

5. If $\overset{\frown}{BC}$ is a semicircle, then what is the $m\angle BAC$? ______________

In ⊙*M*, *AC* = 156°, $\overline{AB} \cong \overline{CB}$. Find the following:

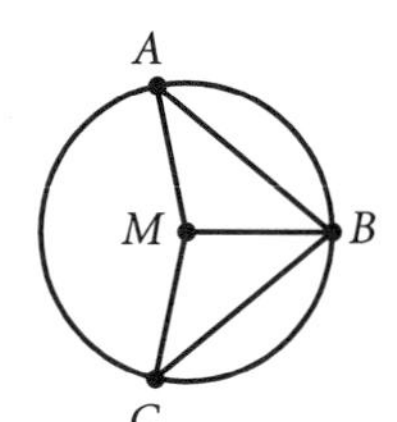

6. $m\angle ABC$ ______________

7. $m\angle AMC$ ______________

8. $m\angle BMC$ ______________

9. $m\angle BMA$ ______________

In ⊙*M*, $\overline{AC}$ is a diameter, $\overleftrightarrow{CF}$ is tangent to ⊙*M* at point *C*, and $m\angle BMA = 118°$. Find the following:

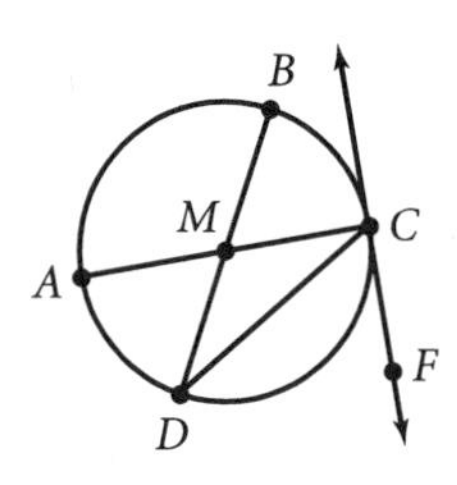

10. $m\overset{\frown}{AB}$ ______________

11. $m\overset{\frown}{BC}$ ______________

12. $m\angle DMC$ ______________

13. $m\angle CMB$ ______________

14. $m\angle MCD$ ______________

15. $m\angle FCD$ ______________

In ⊙*M*, $\overrightarrow{BA}$ and $\overrightarrow{BC}$ are tangents, $m\angle ADC = 64°$. Find the following:

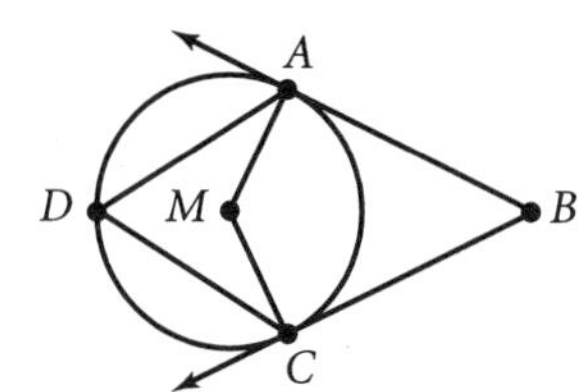

16. $m\angle AMC$ ______________

17. $m\angle MAB$ ______________

18. $m\angle ABC$ ______________

19. $m\overset{\frown}{ADC}$ ______________

NAME ______________________ CLASS ____________ DATE ____________

Practice Masters Level B

9.3 Inscribed Angles and Arcs

In ⊙*M*, chord $AD \cong AC$, $m\angle DMC = 154°$. Find the following:

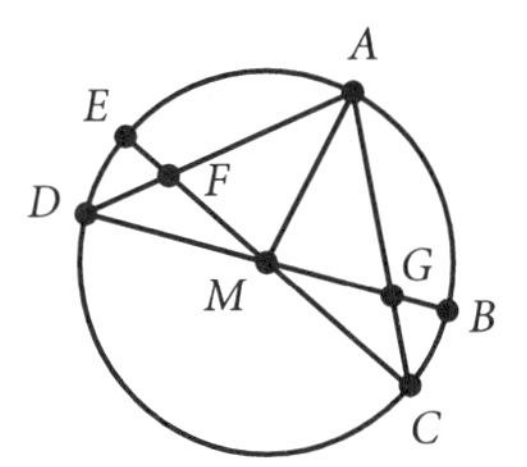

1. $m\angle BGC$ ____________
2. $m\angle DAC$ ____________
3. $m\overset{\frown}{DC}$ ____________
4. $m\overset{\frown}{AC}$ ____________
5. $m\overset{\frown}{BC}$ ____________
6. $m\angle ADM$ ____________

In ⊙*M*, $\overline{AC}$ is a diameter, $\overrightarrow{DC}$ is tangent to the circle at point C, and $m\overset{\frown}{BC} = 78°$. Find the following:

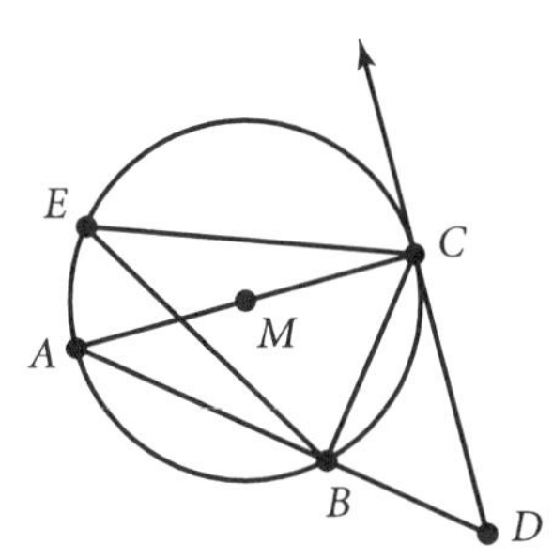

7. $m\angle BAC$ ____________
8. $m\angle BEC$ ____________
9. $m\overset{\frown}{AB}$ ____________
10. $m\angle ACB$ ____________
11. $m\angle ABC$ ____________
12. $m\angle ACD$ ____________

In ⊙*M*, if $BC = 12$, $CD = 16$, and $AC = 20$, find the following:

13. BD ____________
14. AD ____________

In the circle, $m\overset{\frown}{AB} = x + 1$, $m\angle DEB = 7x - 2$, $m\overset{\frown}{AE} = 7x - 9$, $m\overset{\frown}{ED} = 2x$, and $m\overset{\frown}{BC} = \frac{1}{2}\, m\overset{\frown}{CD}$. Find the following:

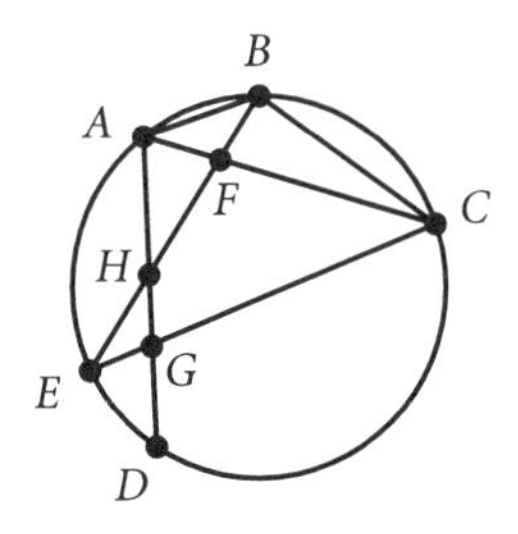

15. $m\angle BAD$ ____________
16. $m\angle BED$ ____________
17. $m\angle BEC$ ____________
18. $m\overset{\frown}{CD}$ ____________
19. $m\overset{\frown}{AB}$ ____________
20. $m\overset{\frown}{BAE}$ ____________
21. $m\angle ABC$ ____________
22. $m\angle AHB$ ____________

Decide whether the following statements are *always true, sometimes true,* or *never true.*

23. A secant is a chord. ____________
24. The measure of an inscribed angle is equal to the measure of a central angle. ____________
25. The measure of an inscribed angle is equal to one-half the measure of its intercepted arc. ____________

NAME ______________________ CLASS ____________ DATE ____________

Practice Masters Level C

9.3 Inscribed Angles and Arcs

In ⊙M, m$\angle AMC = 102°$, m$\angle ABC = 4x - 2$, m$\angle BCA = 64°$ and $\overline{AB} \cong \overline{BC}$. Find the following:

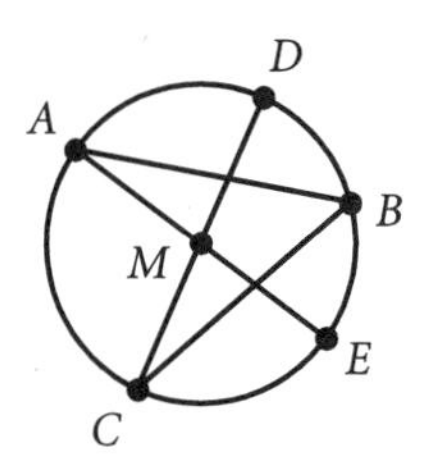

1. m$\angle ABC$ ____________
2. m$\angle BAE$ ____________
3. m$\overarc{AD}$ ____________
4. m$\overarc{BE}$ ____________

In ⊙M, $\overline{CE}$ is tangent to the inner circle at point E, $\overline{BD}$ and $\overline{CD}$ are secants. If m$\overarc{CB} = x^2 + 2x$, m$\overarc{BA} = 3x^2 - 4$, and m$\overarc{AD} = 4x + 4$, find:

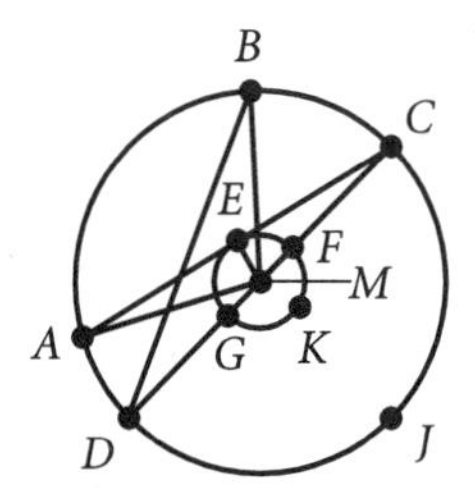

5. m$\overarc{AD}$ ____________
6. m$\overarc{AB}$ ____________
7. m$\overarc{GKF}$ ____________
8. m$\angle BDC$ ____________
9. m$\angle MEC$ ____________
10. m$\angle EMC$ ____________

The figure below contains concentric circles with a common center at M, $\overline{AB}$ and $\overline{DC}$ are tangent to the inner circle at points E and G, respectively, m$\overarc{EF} = 3x + 7$, m$\overarc{EGF} = 5x + 1$, and m$\angle ABD = x - 3$. Find the following:

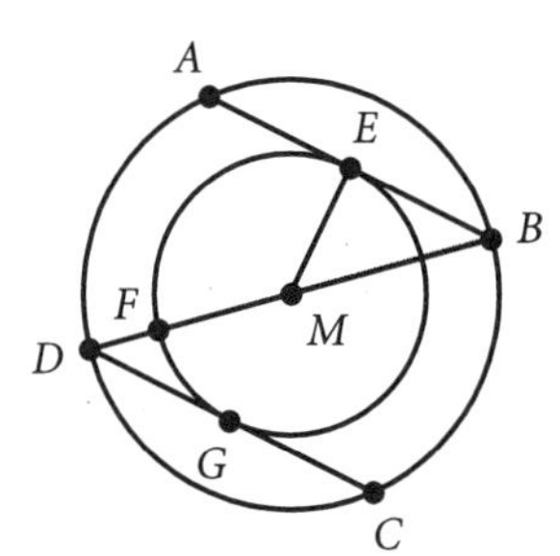

11. m$\angle ABD$ ____________
12. m$\overarc{AD}$ ____________
13. m$\angle BDC$ ____________
14. m$\angle EMB$ ____________
15. If m$\overarc{BC} = 3x + 7$, are the tangents parallel? Explain.

____

In the figures below, ⊙$M \cong$ ⊙N, $\angle BFC \cong \angle XVY$, $\angle ADB \cong \angle ZWY$, $BF = BD = WY = VY$, $VW = \frac{34x}{3}$, $VZ = 15x + 0.01$, and $ZY = 106x - 12$. Find the following:

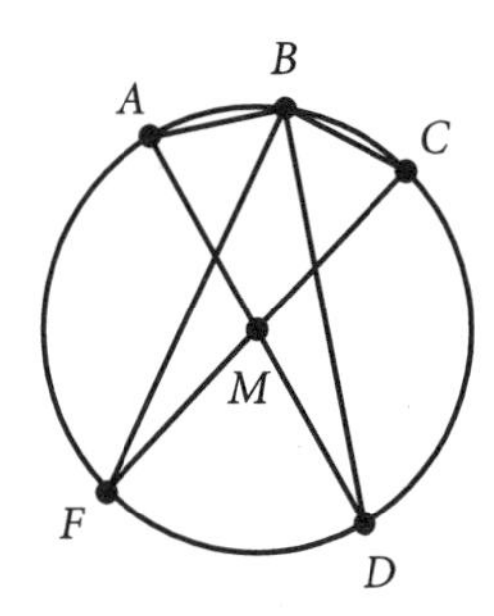

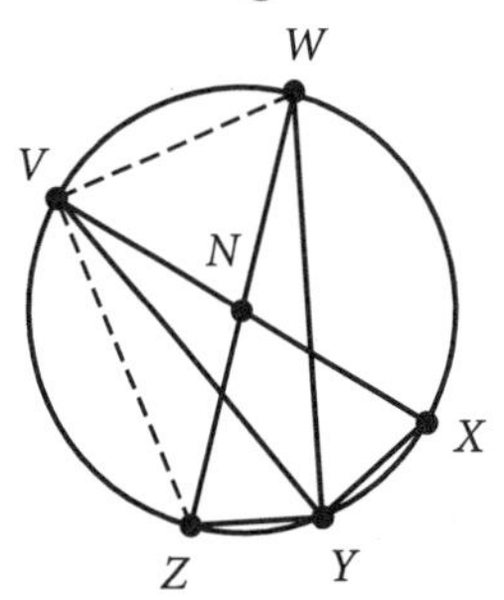

16. m$\overarc{FD}$ ____________
17. m$\angle ADB$ ____________
18. m$\angle FBD$ ____________
19. m$\angle WNX$ ____________
20. m$\angle CMD$ ____________

NAME ______________________ CLASS ____________ DATE ____________

Practice Masters Level A

9.4 Angles Formed by Secants and Tangents

In ⊙*S*, $\overrightarrow{QR}$ is tangent to ⊙*S* at *Q*.

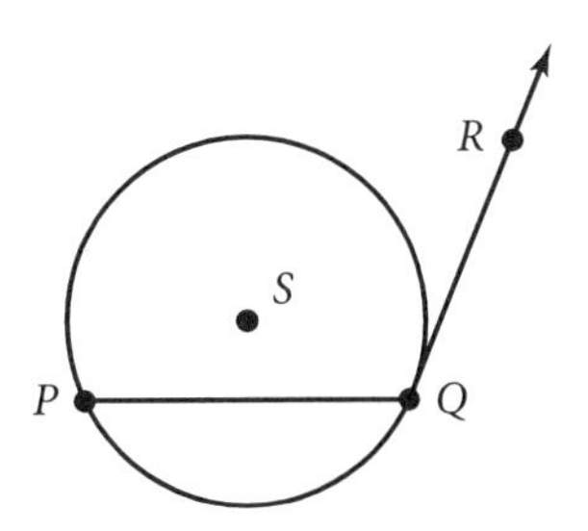

1. If $m\overset{\frown}{QP} = 115°$, find $m\angle RQP$. ____________
2. If $m\angle RQP = 70°$, find $m\overset{\frown}{QP}$. ____________
3. If $m\overset{\frown}{QP} = 65°$, find $m\angle RQP$. ____________
4. If $m\angle RQP = 145°$, find $m\overset{\frown}{QP}$. ____________
5. If $m\overset{\frown}{QP} = 111°$, find $m\angle RQP$. ____________

Find m∠*TRS* in each figure.

6.

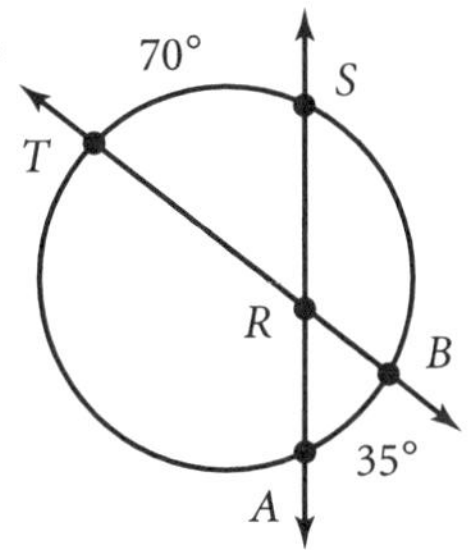

7.

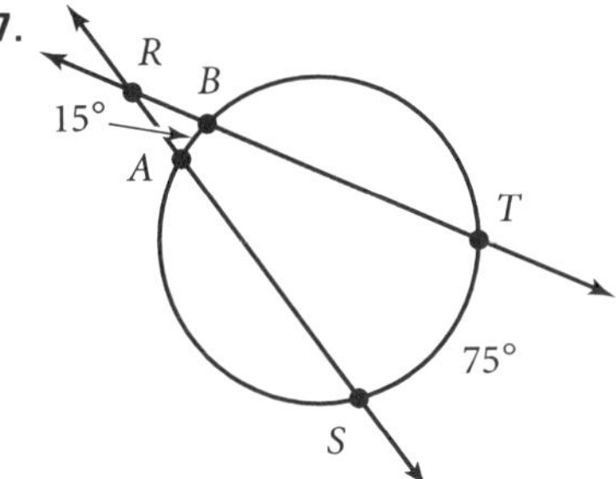

8.

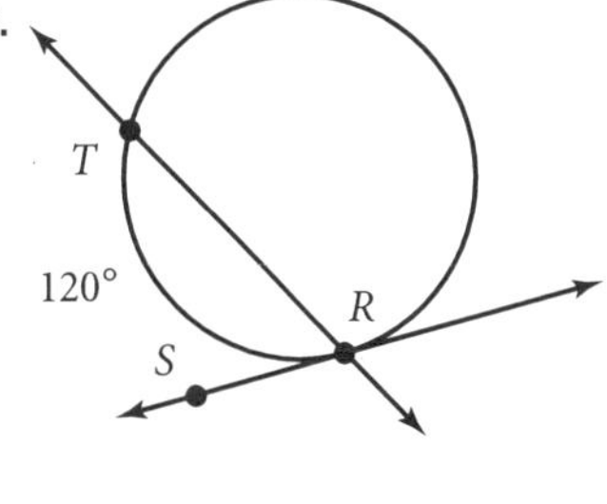

____________ ____________ ____________

In the figure, $\overrightarrow{CD}$ is tangent to ⊙*M* at point *D* and $m\overset{\frown}{AB} = 95°$. Find the following:

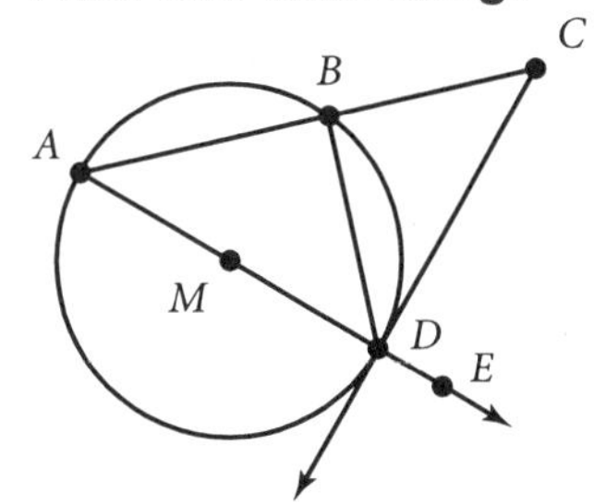

9. $m\angle ABD$ ____________
10. $m\angle ACD$ ____________
11. $m\angle BDC$ ____________
12. $m\angle ADB$ ____________

In the figure, *ABCD* is inscribed in ⊙*S*, m∠*C* = 75° and m∠D = 110°. Find the following:

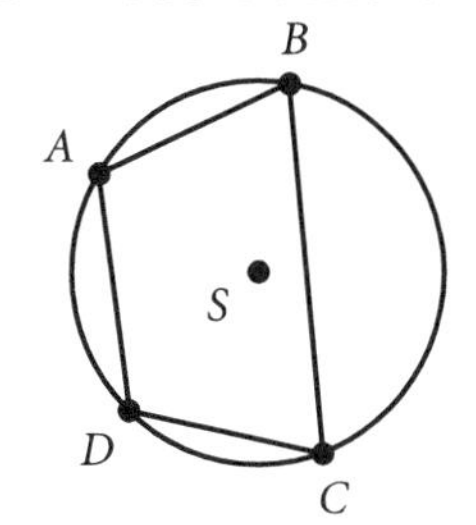

13. $m\overset{\frown}{ABC}$ ____________
14. $m\overset{\frown}{DAB}$ ____________
15. $m\overset{\frown}{BC}$ ____________

NAME ______________________ CLASS ____________ DATE ____________

Practice Masters Level B

9.4 Angles Formed by Secants and Tangents

In ⊙*M*, m$\overline{AC}$ and m$\overrightarrow{CB}$ are tangents to the circle at points *A* and *B*, respectively.

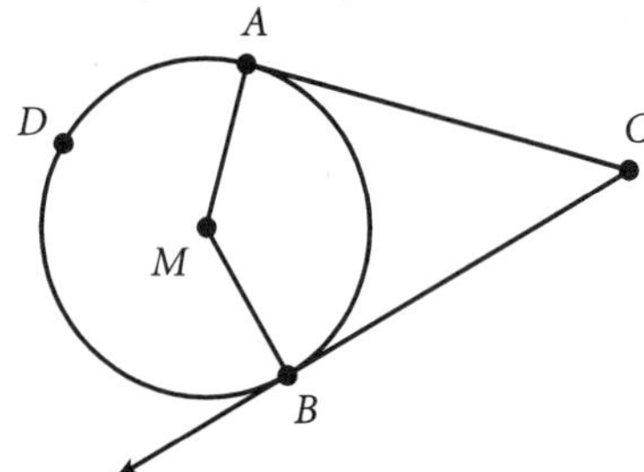

1. If $m\angle ACB = 58°$, find $m\angle AMB$. ______________
2. If $m\angle ACB = 71°$, find $m\overset{\frown}{ADB}$. ______________
3. If $m\overset{\frown}{AB} = 118°$, find $m\angle ACB$. ______________

In ⊙*M*, $m\overset{\frown}{AD} = m\overset{\frown}{CE}$, $m\overset{\frown}{DE} = 30°$, and $m\overset{\frown}{AEC} = 260°$. Find the following:

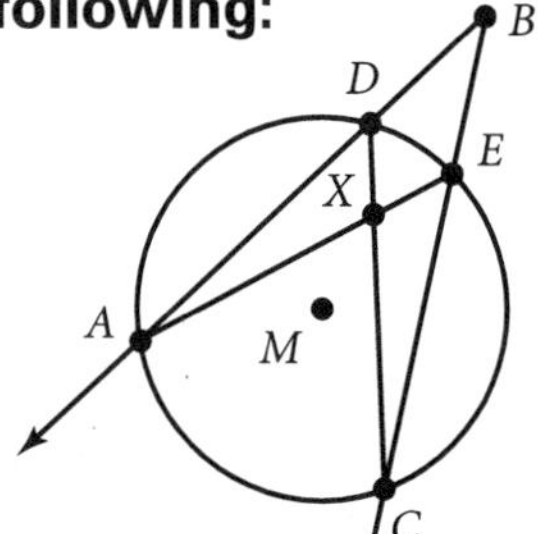

4. $m\overset{\frown}{AC}$ ______________
5. $m\angle AXD$ ______________
6. $m\angle ABC$ ______________
7. $m\angle DXE$ ______________

In ⊙*M*, $\overline{AB}$ and $\overline{BC}$ are tangent to the circle at points *A* and *C*, respectively. Find the following:

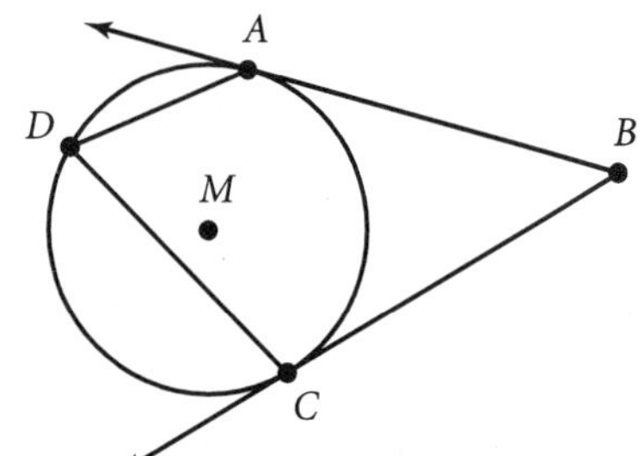

8. If $m\overset{\frown}{AC} = 166°$, find $m\angle ABC$. ______________
9. If $m\overset{\frown}{CDA} = 220°$, find $m\angle ABC$. ______________

In ⊙*M*, $m\overset{\frown}{AB} = 86°$, $m\overset{\frown}{CED} = 25°$, $\overline{AD} \cong \overline{BC}$, $\overrightarrow{FG}$ and $\overrightarrow{FH}$ are tangent at points *G* and *H* respectively, and $m\overset{\frown}{HEG} = 112°$. Find the following:

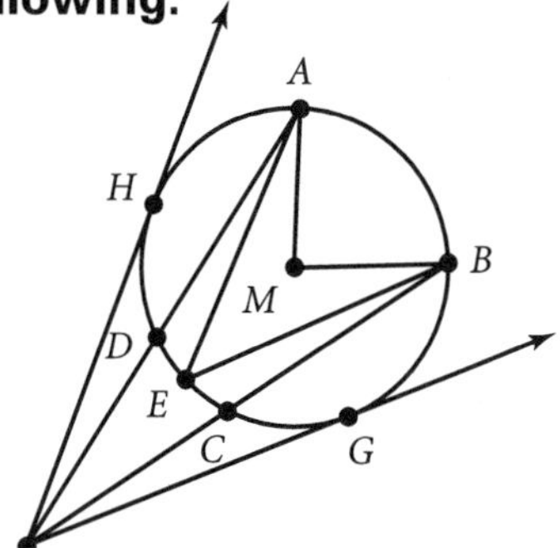

10. $m\angle AMB$ ______________
11. $m\angle AEB$ ______________
12. $m\angle AFB$ ______________
13. $m\overset{\frown}{BC}$ ______________
14. What type of angle is formed by tangent *FH* and *FG*?

____

NAME ______________________ CLASS ____________ DATE ____________

Practice Masters Level C

9.4 Angles Formed by Secants and Tangents

In ⊙S, $\overline{AB}$ is tangent to the circle at point A, $m\angle B = x$, $m\overset{\frown}{AC} = y$, $m\overset{\frown}{CD} = 3y$, $m\overset{\frown}{DEA} = x + 3y$. Find the following:

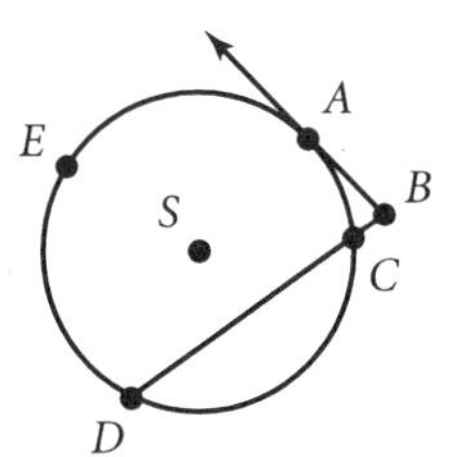

1. $m\overset{\frown}{CD}$ ____________
2. x ____________
3. y ____________
4. $m\overset{\frown}{AED}$ ____________

In the figure, $ABCD$ is an inscribed quadrilateral. The measure of $\overset{\frown}{AB} = 4x - 25$, $m\overset{\frown}{BC} = x$, $m\overset{\frown}{CD} = 2x + 20$, and $m\overset{\frown}{DA} = 3x - 35$. Find the following:

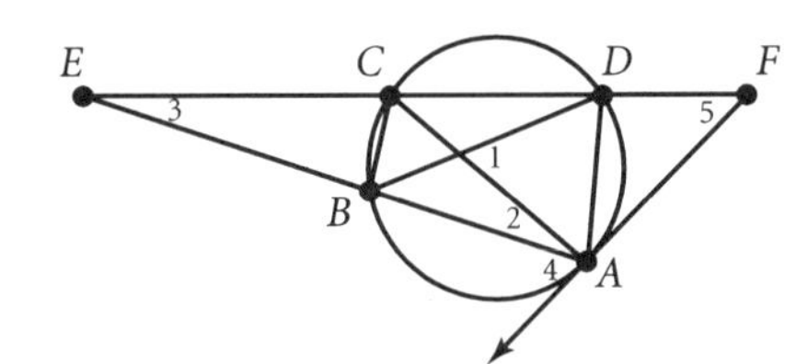

5. $m\angle 1$ ____________
6. $m\angle 2$ ____________
7. $m\angle 3$ ____________
8. $m\angle 4$ ____________
9. $m\angle 5$ ____________
10. $m\angle CAF$ ____________
11. $m\angle CDA$ ____________
12. $m\angle DCB$ ____________

In ⊙M, $\overline{AB}$ is a diameter, rays PC, PQ and BY are tangents to the circle at C, A, and B, respectively, $m\overset{\frown}{AC} = 45°$, $m\overset{\frown}{CF} = 40°$, $m\overset{\frown}{FG} = 35°$, and $m\angle ABD = 46°$. Find the following:

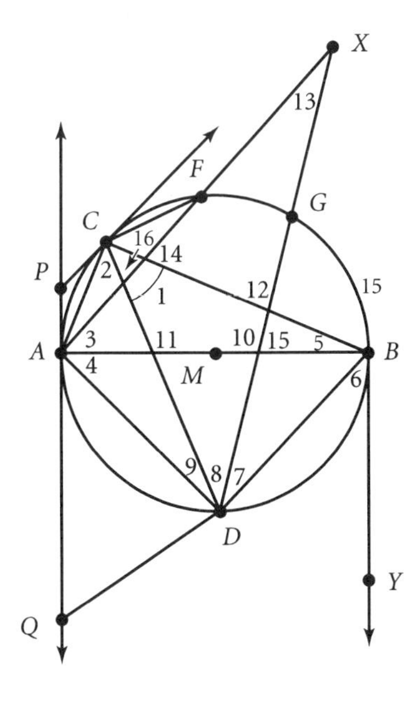

13. $m\angle CAD$ ____________
14. $m\angle CBD$ ____________
15. $m\angle ACF$ ____________
16. $m\angle ADB$ ____________
17. $m\angle 1$ ____________
18. $m\angle 2$ ____________
19. $m\angle 3$ ____________
20. $m\angle 4$ ____________
21. $m\angle 5$ ____________
22. $m\angle 6$ ____________
23. $m\angle 7$ ____________
24. $m\angle 8$ ____________
25. $m\angle 9$ ____________
26. $m\angle 10$ ____________
27. $m\angle 11$ ____________
28. $m\angle 12$ ____________
29. $m\angle 13$ ____________
30. $m\angle 14$ ____________
31. $m\angle 15$ ____________
32. $m\angle 16$ ____________

NAME ______________________ CLASS ______________ DATE ______________

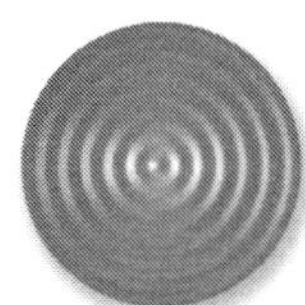

Practice Masters Level A

9.5 Segments of Tangents, Secants, and Chords

In $\odot M$, $\overrightarrow{CD}$ is tangent to the circle at point *D*, and $\overline{AH} \cong \overline{HB}$.

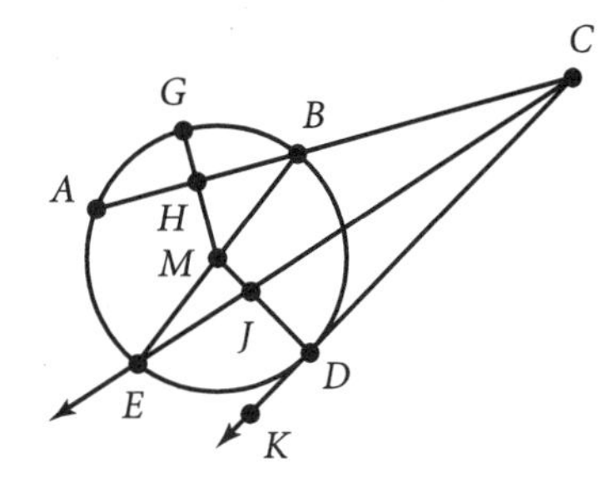

1. Name two chords. ______________
2. Name two secants. ______________
3. Name the right angles. ______________
4. Name a tangent. ______________
5. Name all external secant segments. ______________

Refer to $\odot P$ for Exercises 6–8.

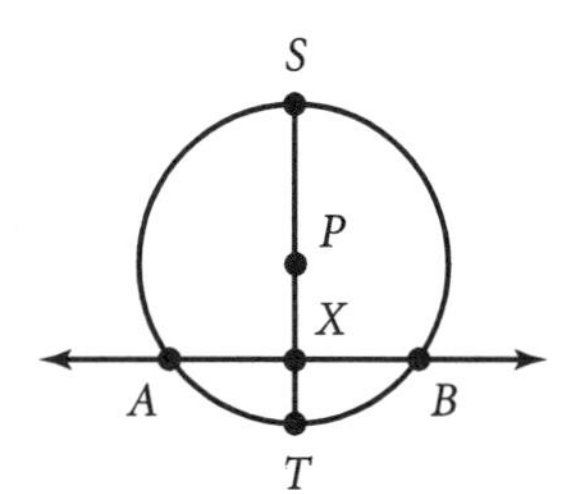

6. If $AX = BX = 20$ and $TX = 6$, find TS. ______________
7. If $AX = BX = 15$ and $TX = 5$, find TS. ______________
8. If $AX = BX = 14$ and $TX = 7$, find TS. ______________

Refer to $\odot S$ for Exercises 9–12. $\overline{WX}$ is a tangent and $\overline{WR}$ is a secant to $\odot S$.

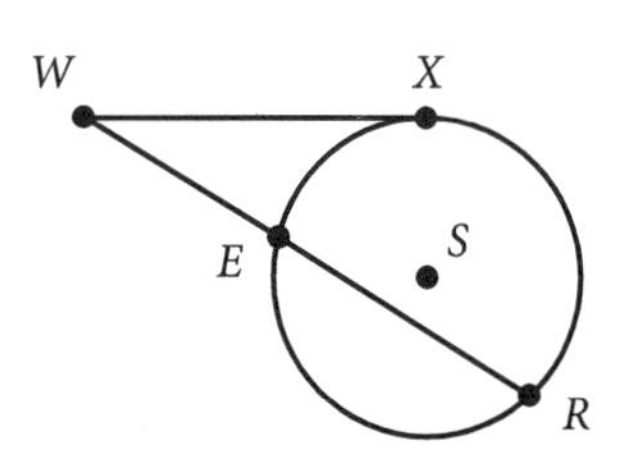

9. If $ER = 16$ and $WX = 8$, find WE. ______________
10. If $WE = 20$ and $ER = 5$, find WX. ______________
11. If $ER = 15$ and $WE = 12$, find WX. ______________
12. If $WX = WE = 5$, find ER. ______________

In $\odot M$, $\overline{CE}$ is tangent to the circle at point *E*.

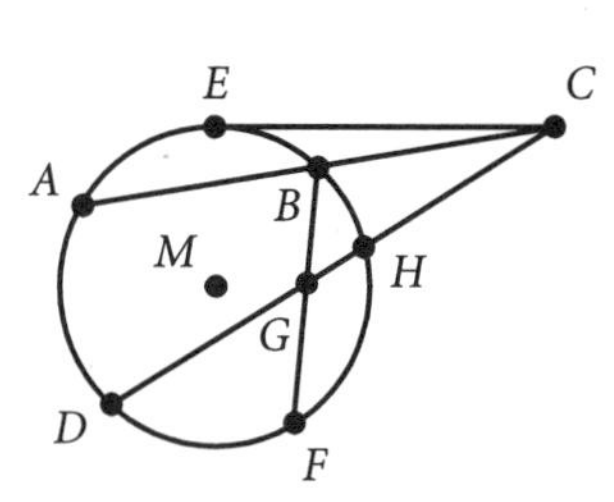

13. If $AB = 5$, $BC = 4$, find EC. ______________
14. If $HC = 3$, $AB = 4$, $BC = 2$, find DC. ______________

15. A tangent and secant are drawn to a circle from the same external point. The exterior tangent segment equals 4 while the internal segment of the secant segment is 6. Find the length of the external secant segment. ______________

Practice Masters Level B

9.5 Segments of Tangents, Secants, and Chords

1. If the location of the vertex of two secant segments is located outside of ⊙Q, then what can be said about the product of the lengths of one secant segment and its exterior segment? ______________

2. If two tangent segments are congruent and they meet at a common vertex, S, then where is the vertex located in reference to ⊙Q? ______________

$\overline{CE}$ is tangent to ⊙M at point E, $\overline{CE} \parallel \overline{AB}$, $m\overset{\frown}{DB} = 60°$, $m\overset{\frown}{AE} = 120°$, $MF = 2$, $DC = 1.96$, $DB = 2.02$, and $AG = 1.73$. Find the following:

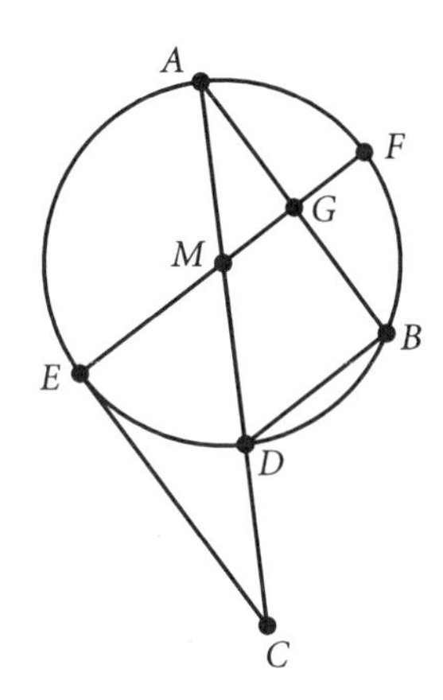

3. $m\angle EMA$ ______________
4. $m\overset{\frown}{BA}$ ______________
5. $m\angle DAB$ ______________
6. $m\angle DBA$ ______________
7. $m\angle ECA$ ______________
8. $m\angle AMF$ ______________
9. EF ______________
10. EC ______________
11. AB ______________
12. GF ______________
13. GB ______________
14. EG ______________
15. A pair of congruent angles. ______________

In ⊙M, $\overline{CD}$ is tangent to the circle at D; $BC = 1.5$; CD is six more than BC.

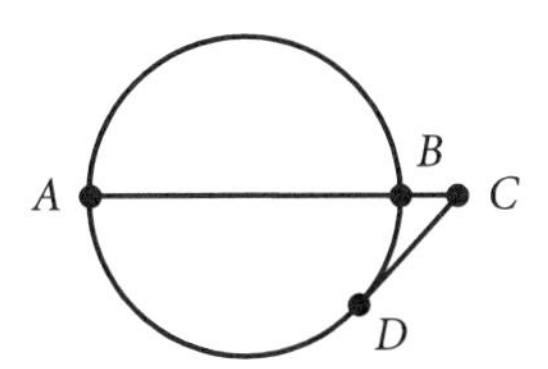

16. What type of segment is $\overline{AB}$? ______________
17. Find CD. ______________
18. Find AC. ______________

Solve.

19. Point A is 15 inches from the center of a circle with a radius of 9 inches. Find the length of the tangent from point A to the circle. ______________

20. Chords $\overline{CD}$ and $\overline{EF}$ intersect at P inside circle M. If $CP = 4$, $EP = 2$, and $PD = 9$, find EF. ______________

NAME ______________________________ CLASS ______________ DATE ______________

Practice Masters Level C

9.5 *Segments of Tangents, Secants, and Chords*

Ray FD is tangent to $\odot M$ at point D, $\overline{GC} \parallel \overrightarrow{FD}$, $m\overset{\frown}{BC} = 76°$, $GF = 4$, $FD = 6$, $ED = 2$, $m\overset{\frown}{GD} = 52°$, and $BM = 5$. Find the following:

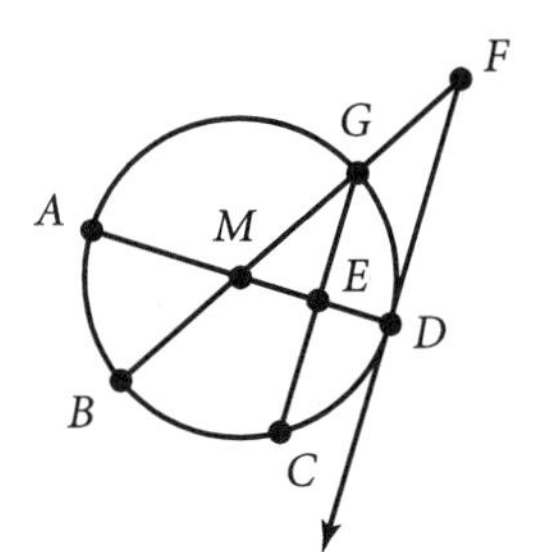

1. $m\angle BFD$ ______________
2. $m\angle DEC$ ______________
3. $m\overset{\frown}{AG}$ ______________
4. $m\overset{\frown}{AB}$ ______________
5. GM ______________
6. GC ______________

In $\odot M$, $\overline{AB}$ and $\overline{BC}$ are tangent to the circle at points A and C, respectively. If $m\overset{\frown}{AC} = 3x$, $m\angle ABC = x^2$, and $m\angle GMA = 12x$, find the following:

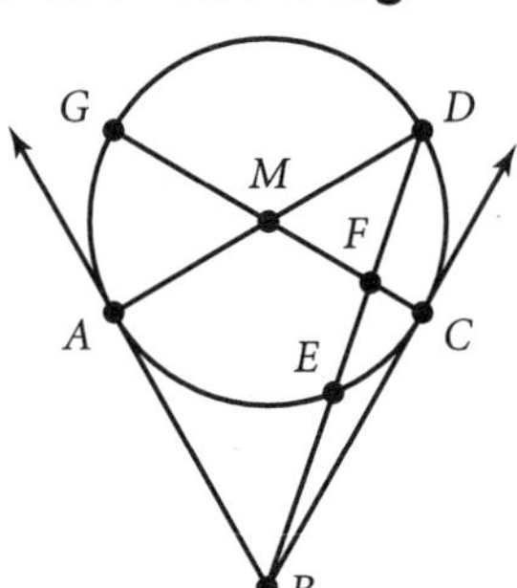

7. $m\overset{\frown}{AC}$ ______________
8. $m\angle ABC$ ______________
9. $m\overset{\frown}{CD}$ ______________
10. $m\overset{\frown}{GD}$ ______________

In $\odot M$, if $AB = 2x$, $BE = x$, $ED = 12$, $CF = x$, $GC = 8x$, $AM = 9$, $FC = 2$, $ED = 12$, $BE = 3$, and $\overline{DF}$ is twice the length of $\overline{EF}$, find the following.

11. GF ______________
12. BD ______________
13. DF ______________
14. AD ______________

In the circle, $\overline{AB}$ is tangent to the circle at A, $AB = 3x$, $CB = x + 5$, and $DC = 8x - 23$. Find the following:

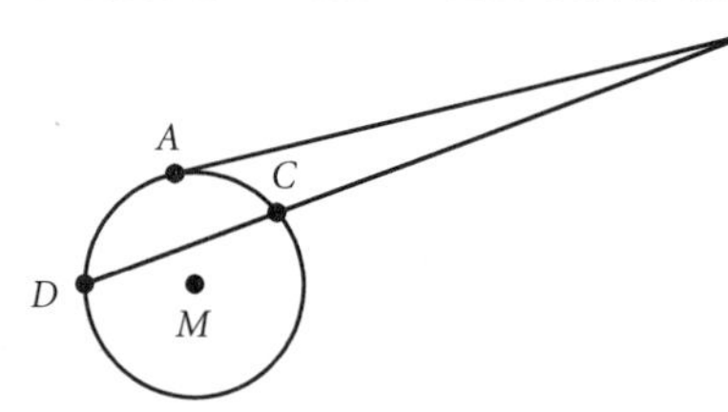

15. AB ______________
16. CB ______________
17. DC ______________

Solve.

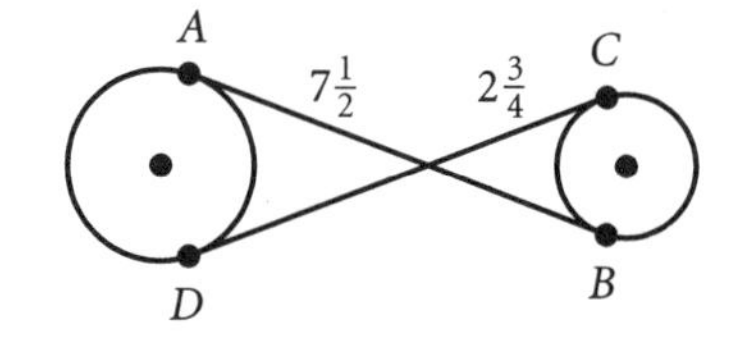

18. In the figure at the right, $\overline{AB}$ and $\overline{CD}$ are internally tangent to the circles at A, B, C, and D. Find AB. ______________

19. $\overline{PA}$ is tangent to circle M at A. $\overline{PC}$ is a secant to circle M that intersects the circle at points B and C. If $PB = 10$ and $PC = 40$, find PA. ______________

NAME ______________________ CLASS ____________ DATE ____________

Practice Masters Level A

9.6 Circles in the Coordinate Plane

Find the center and radius of each circle.

1. $x^2 + y^2 = 169$

____

____

2. $x^2 + y^2 = 72$

____

____

3. $(x + 2)^2 + y^2 = 36$

____

____

4. $x^2 + (y - 4)^2 = 1$

____

____

5. $(x - 6)^2 + (y - 2)^2 = 25$

____

____

6. $(x + 2)^2 + (y + 7)^2 = 24$

____

____

Write an equation for the circle with the given center and radius.

7. center $(-3, 4)$, radius 3 ______________________

8. center $(2, 13)$, radius 8 ______________________

9. center $(4, 6)$, radius 8 ______________________

10. center $(-5, -4)$ radius 6 ______________________

Find the *x*- and *y*-intercepts for the graph of each equation.

11. $x^2 + y^2 = 169$

____

____

12. $x^2 + y^2 = 72$

____

____

13. $(x + 2)^2 + y^2 = 36$

____

____

14. $x^2 + (y - 4)^2 = 1$

____

____

15. $(x - 6)^2 + (y - 2)^2 = 2$

____

____

16. $x^2 + y^2 = 81$

____

____

NAME ______________________ CLASS ____________ DATE ____________

Practice Masters Level B

9.6 Circles in the Coordinate Plane

Find the center and radius of each circle.

1. $x^2 + y^2 = 27$

2. $\dfrac{(x-1)^2}{3} + \dfrac{(y+2)^2}{3} = 1$

3. $(x + \sqrt{5})^2 + (y - \sqrt{2})^2 = 3$

4. $(x - m)^2 + (y + n)^2 = w$

Write the equation of the circle with the given characteristics.

5. center $\left(-2, \dfrac{1}{2}\right)$, radius $\sqrt{2}$ ____________

6. center (2, 11), (2, 8) as one endpoint of the diameter ____________

7. center $(-6, 3)$, tangent to the y-axis ____________

8. center $\left(\dfrac{3}{4}, 1\dfrac{1}{2}\right)$, tangent to the x-axis ____________

9. center (3, 4), contains the origin as a point ____________

10. List four equations of the circles with a radius of 4, tangent to both axes.

Write the equation for each circle.

11.

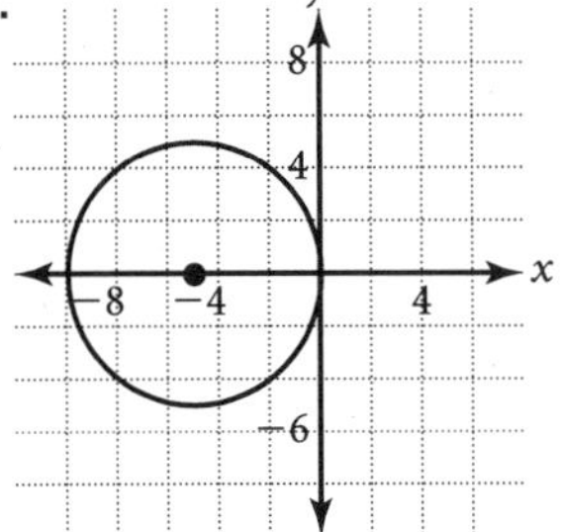

12.

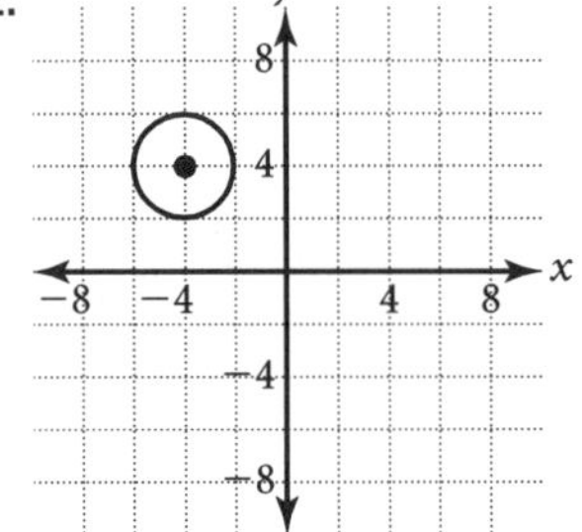

13.

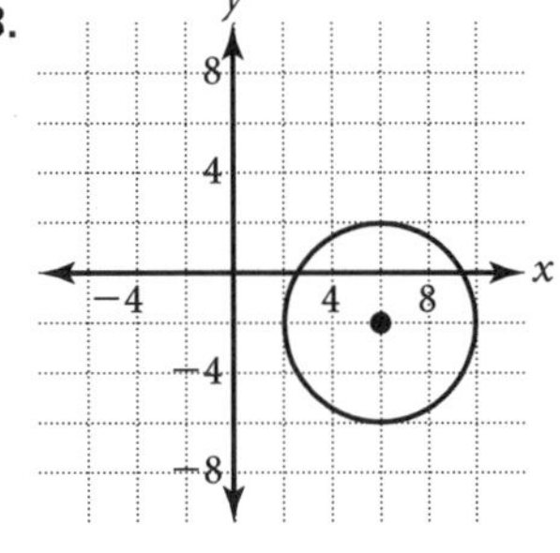

NAME ______________________ CLASS ____________ DATE ____________

Practice Masters Level C

9.6 Circles in the Coordinate Plane

Find the center and radius of each circle.

1. $x^2 + y^2 + 4x - 10y = -20$

2. $-2x^2 + 24x + 4y = 2y^2 + 72$

3. $3x^2 + 3y^2 + 6x + 12y = 33$

4. $4x^2 + 4y^2 - 16x + 24y = 12$

Write the equation for the circle with the given characteristics.

5. center $(2m, 4)$, radius 7 ______________________

6. center $(-3, 0)$, $(9, 2\sqrt{10})$ as one endpoint of the diameter ______________________

7. endpoints of the diameter are $(4, 2)$ and $(6, 2)$ ______________________

8. center $(7, 3)$, tangent to the x-axis ______________________

9. center $(7, 3)$, tangent to the y-axis ______________________

10. center on the line $y = x$, tangent to the x-axis at 5 ______________________

11. center on $y = 2x$, tangent to the y-axis at 2 ______________________

12. center $(7, -2)$ and contains the point $(3, 3)$ ______________________

13. When both squared terms are put on the same side of the equation of a circle, what do you notice about the coefficients? ______________________

14. What do you notice about the signs of the coefficients from Exercise 13? ______________________

15. Points $(-3, 4)$,$(3, -4)$ and $(0, 5)$ are all contained by the circle $x^2 + y^2 = 25$. Describe the transformation and find the corresponding points contained by the circle $(x - 1)^2 + (y - 3)^2 = 25$. ______________________

NAME ______________________ CLASS __________ DATE __________

Practice Masters Level A

10.1 Tangent Ratios

In Exercises 1 and 2, measure the sides of the triangle to find tan *A*.

1.

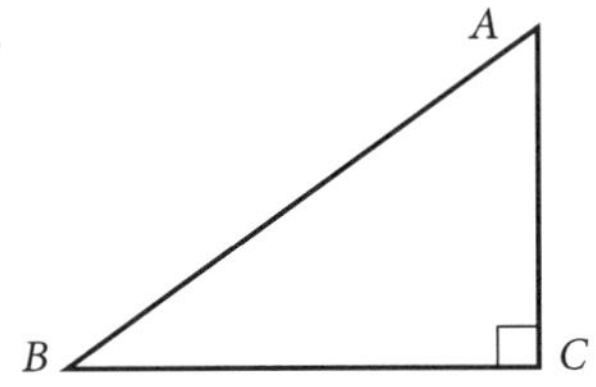

$\tan A \approx$ ______________

2.

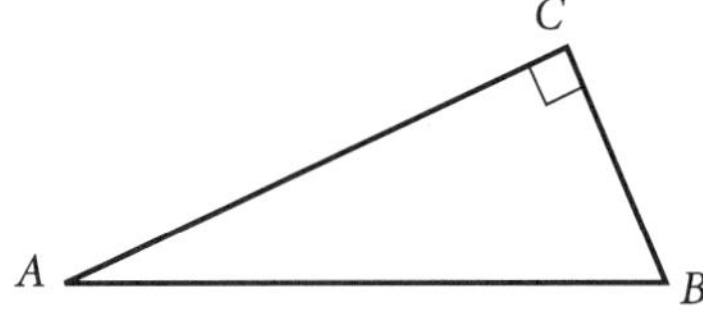

$\tan A \approx$ ______________

3. Use the triangle in Exercise 1.

a. Use a protractor to find $m\angle A$. ________

b. Use the $\tan^{-1}$ key on your calculator, with the value of tan *A* found in Exercise 1, to find $m\angle A$. ________

4. Use the triangle in Exercise 2.

a. Use a protractor to find $m\angle A$. ________

b. Use the $\tan^{-1}$ key on your calculator, with the value of tan *A* found in Exercise 2, to find $m\angle A$. ________

In Exercises 5 and 6, find tan *B* for each triangle.

5.

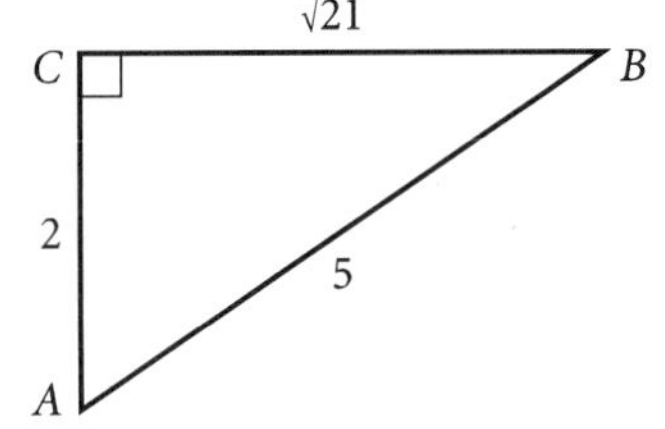

$\tan B =$ ______________

6.

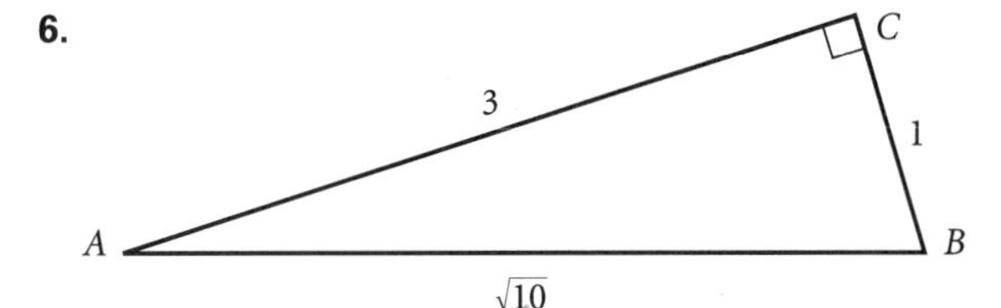

$\tan B =$ ______________

7. Use a scientific or graphics calculator to find the tangent of each angle. Round to the nearest hundredth.

a. $\tan 56° =$ ______________

b. $\tan 12° =$ ______________

c. $\tan 85° =$ ______________

d. $\tan 60° =$ ______________

8. Use a scientific or graphics calculator to find the inverse tangent of each ratio. Round to the nearest degree.

a. $\tan^{-1}\left(\frac{2}{3}\right) =$ ______________

b. $\tan^{-1}(0.4) =$ ______________

c. $\tan^{-1}(1.426) =$ ______________

d. $\tan^{-1}\left(\frac{13}{3}\right) =$ ______________

NAME ______________________ CLASS __________ DATE __________

Practice Masters Level B

10.1 Tangent Ratios

In Exercises 1 and 2, find tan *B* for each triangle.

1.

$2\sqrt{10}$

C B

3

7

A

$\tan B =$ ______________

2.

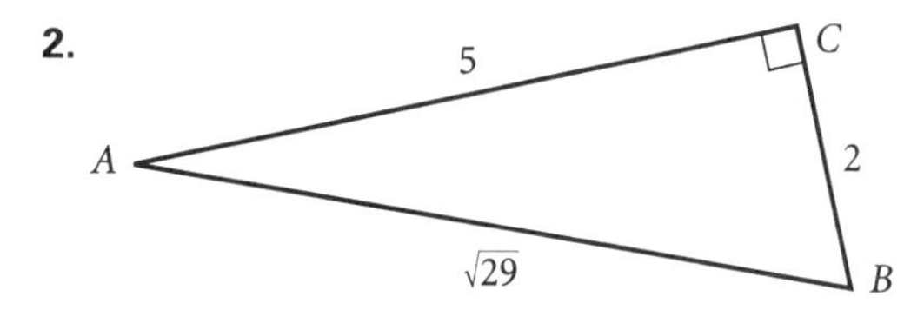

$\tan B =$ ______________

3. Use a scientific or graphics calculator to find the tangent of each angle. Round to the nearest hundredth.

a. $\tan 73° =$ ______________

b. $\tan 18° =$ ______________

c. $\tan 89° =$ ______________

d. $\tan 11° =$ ______________

4. Use a scientific or graphics calculator to find the inverse tangent of each ratio. Round to the nearest degree.

a. $\tan^{-1}\left(\frac{4}{5}\right) =$ ______________

b. $\tan^{-1}(0.7) =$ ______________

c. $\tan^{-1}(2.75) =$ ______________

d. $\tan^{-1}\left(\frac{15}{4}\right) =$ ______________

For Exercises 5 and 6, use the definition of the tangent ratio to write an equation involving *x*. Find the tangent of the given angle with a calculator, and solve the equation to find the unknown side of the triangle. Round your answers to the nearest hundredth.

5.

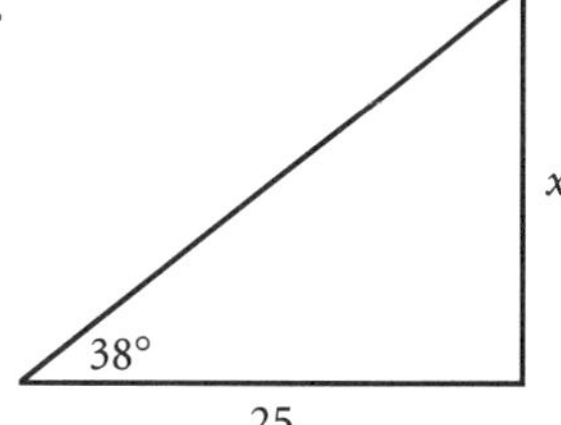

____

____

6.

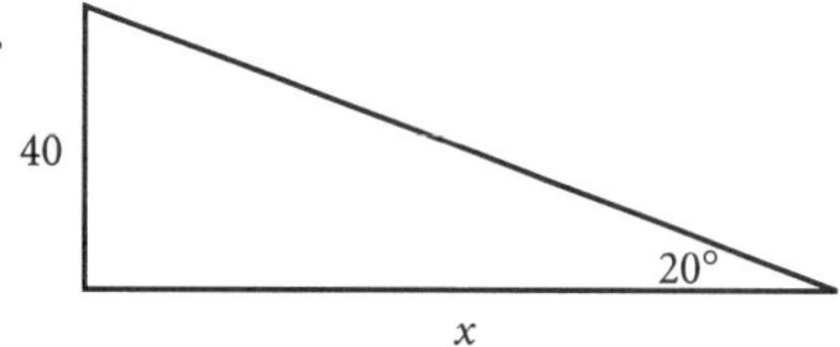

____

____

NAME ______________________ CLASS ____________ DATE ____________

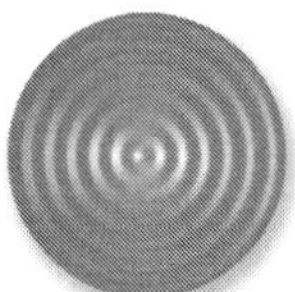

Practice Masters Level C

10.1 Tangent Ratios

For Exercises 1 and 2, use the definition of the tangent ratio to write an equation involving *x*. Find the tangent of the given angle with a calculator, and solve the equation to find the unknown side of the triangle. Round your answers to the nearest hundredth.

1.

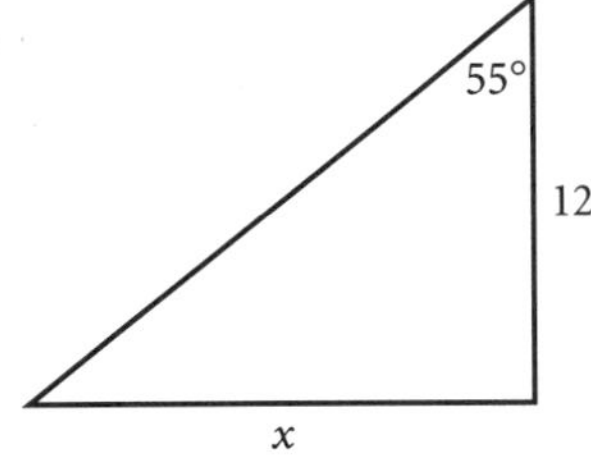

____

____

2.

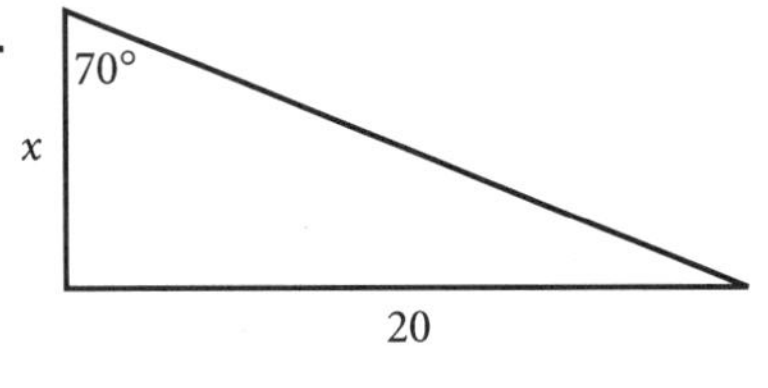

____

____

In Exercises 3 and 4, use the definition of the tangent ratio to write an equation involving the unknown angle *θ*. Use a calculator to solve that equation for *θ*. Round your answers to the nearest degree.

3.

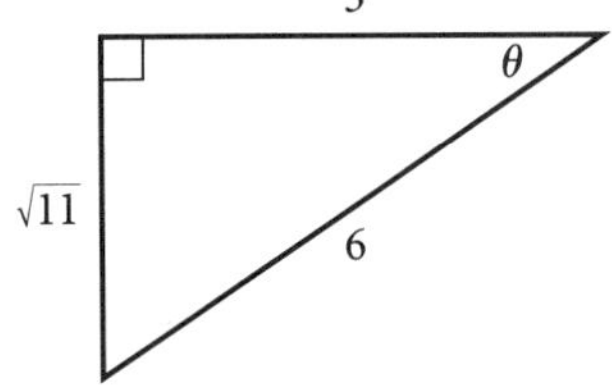

____

____

4.

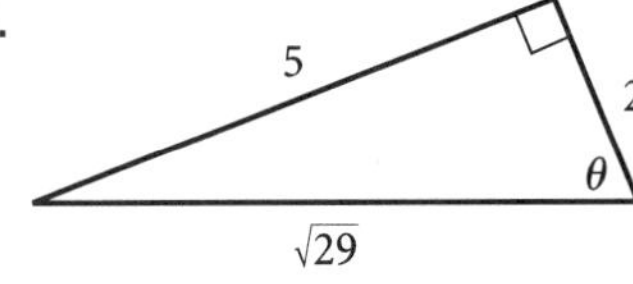

____

____

5. Use the tangent ratio and the Pythagorean Theorem to find x and y in the triangle at the right. Round answers to the nearest tenth.

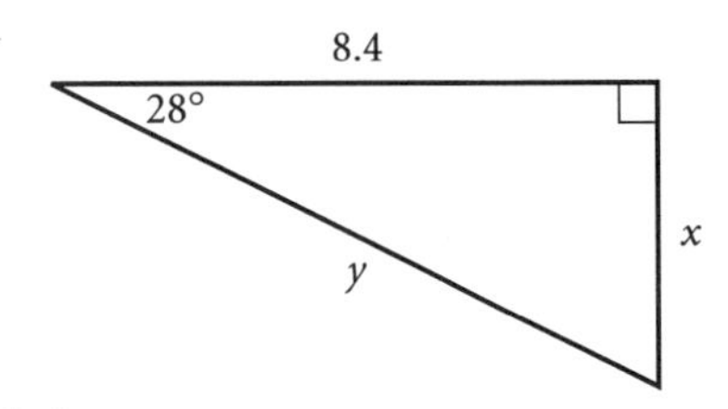

$x =$ ____________ $y =$ ____________

6. On a standard staircase, the depth of each step is 10 inches and the height of each riser is 8 inches. A plank is to be laid on the staircase to form a ramp. Find the angle that the ramp will make with the ground.

____

NAME ______________________ CLASS ______________ DATE ______________

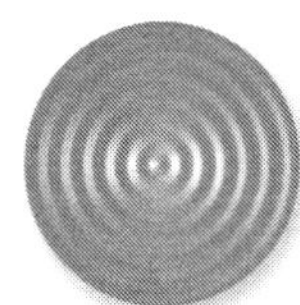

Practice Masters Level A

10.2 Sines and Cosines

1. Refer to $\triangle ABC$.

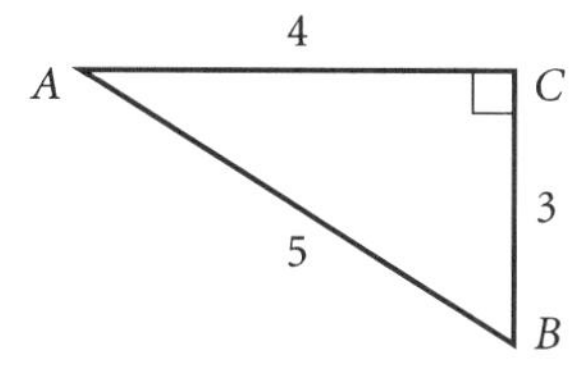

Find each of the following.

a. $\sin A$ ______________________

b. $\cos B$ ______________________

c. $\tan B$ ______________________

2. Refer to $\triangle DEF$.

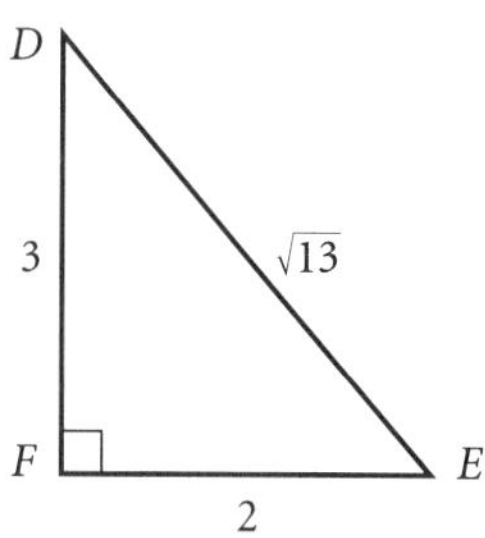

Find each of the following.

a. sin ________ $= \dfrac{3}{\sqrt{13}}$

b. cos ________ $= \dfrac{2}{\sqrt{13}}$

c. tan ________ $= \dfrac{2}{3}$

In Exercises 3 and 4, use a scientific or graphics calculator.

3. Round answers to four decimal places.

a. $\sin 74°$ ______________________

b. $\cos 22°$ ______________________

c. $\tan 48°$ ______________________

4. Round answers to the nearest degree.

a. $\sin^{-1} \dfrac{1}{2}$ ______________________

b. $\cos^{-1} \dfrac{1}{2}$ ______________________

c. $\tan^{-1} 3$ ______________________

In Exercises 5 and 6, use a trigonometric ratio to find the height of the triangle.

5.

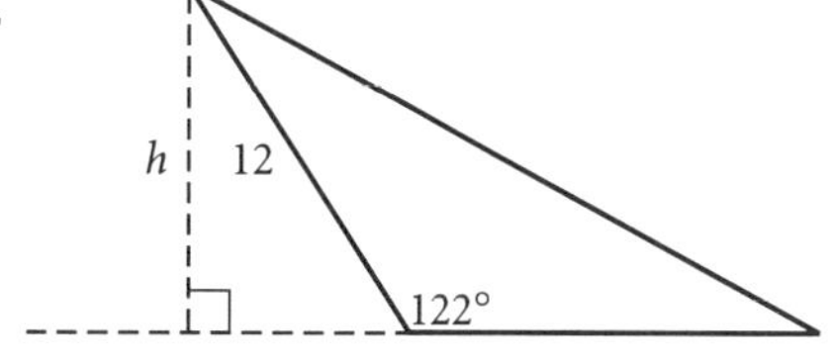

____

6.

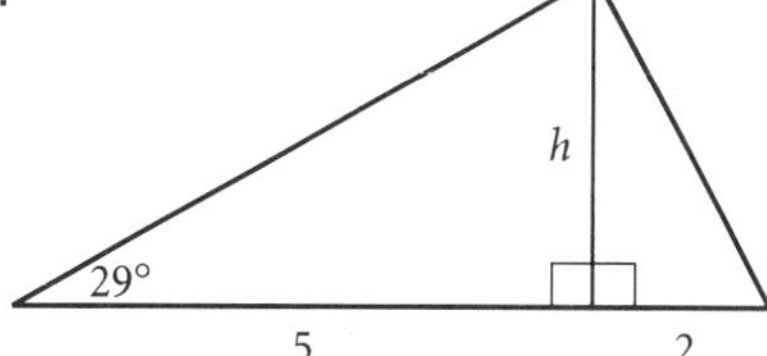

____

7. Robby is flying his kite. The hand holding the string is 4 feet above ground level, and the string makes an angle of 55° above horizontal. When he has 100 feet of string out, how high is the kite above the ground?

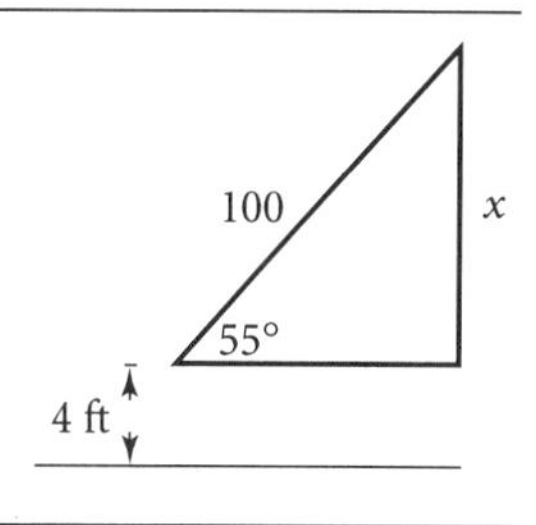

____

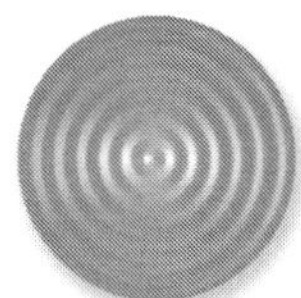

Practice Masters Level B

10.2 Sines and Cosines

1. Refer to $\triangle ABC$.

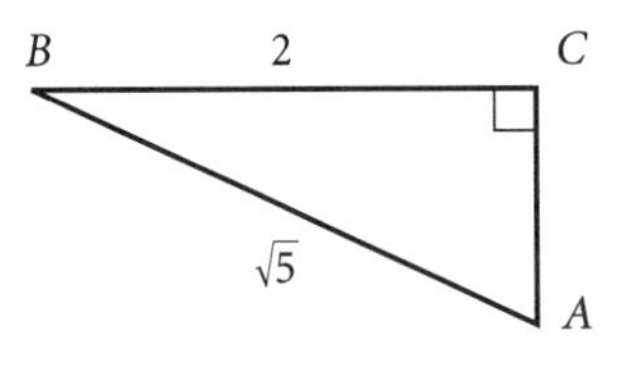

Find each of the following.

a. $\sin A$ ______________________

b. $\cos B$ ______________________

c. $\tan B$ ______________________

2. Refer to $\triangle DEF$.

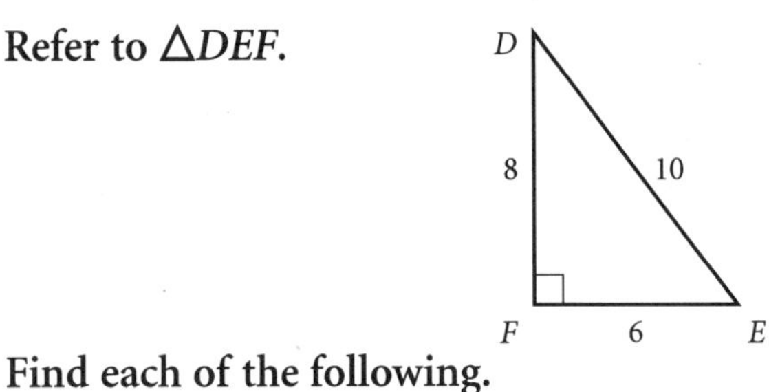

Find each of the following.

a. sin ________ $= 0.6$

b. cos ________ $= 0.6$

c. tan ________ $= \frac{3}{4}$

In Exercises 3 and 4, use a scientific or graphics calculator.

3. Round answers to four decimal places.

a. $\sin 56°$ ______________________

b. $\cos 34°$ ______________________

c. $\tan 18°$ ______________________

4. Round answers to the nearest degree.

a. $\sin^{-1} \frac{2}{3}$ ______________________

b. $\cos^{-1} \frac{1}{\sqrt{2}}$ ______________________

c. $\tan^{-1} 2.345$ ______________________

In Exercises 5 and 6, use a trigonometric ratio to find the height of the triangle.

5.

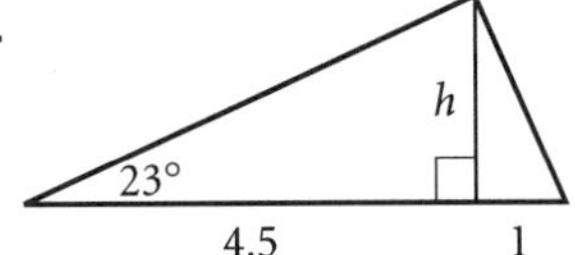

____

6.

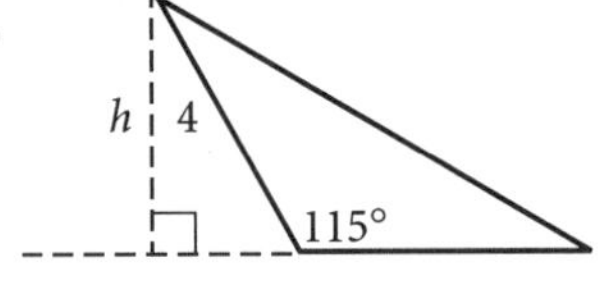

____

In Exercises 7 and 8, use a trigonometric ratio to find the angle measure θ. Round your answers to the nearest degree.

7.

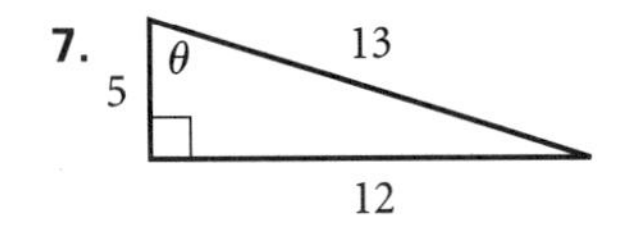

____

8.

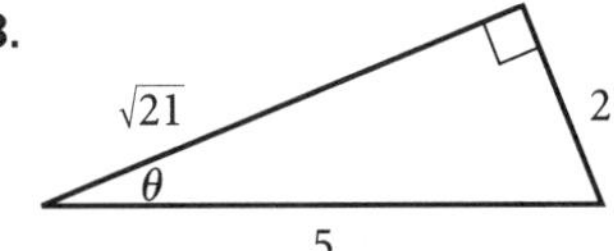

____

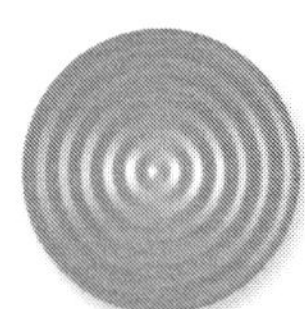

Practice Masters Level C

10.2 Sines and Cosines

In Exercises 1–4, use a trigonometric ratio to find *x* for each triangle.

1.
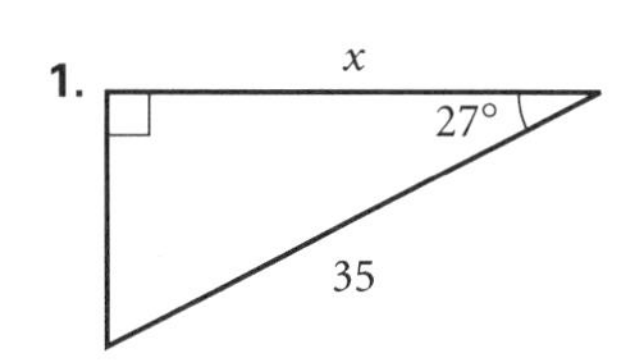

2.
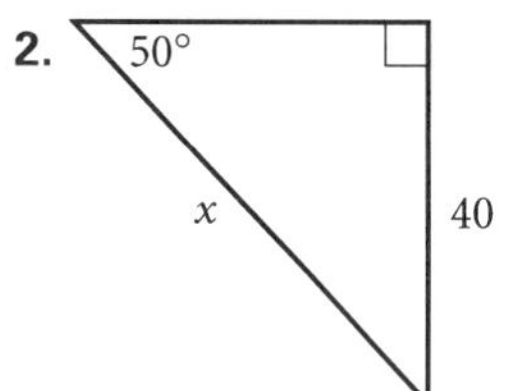

3.
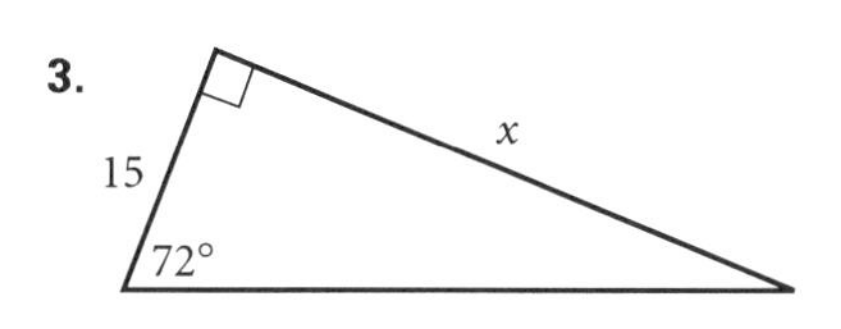

4.
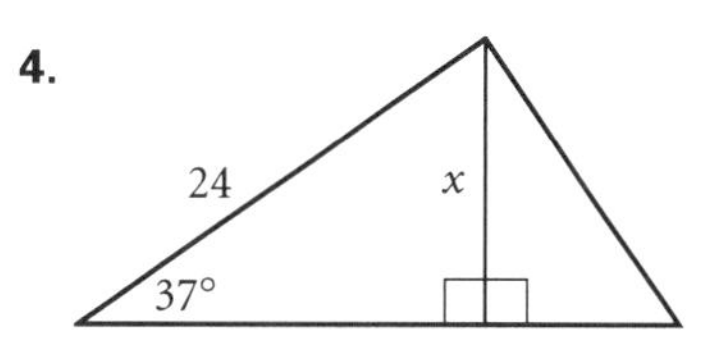

In Exercises 5 and 6, use a trigonometric ratio to find angle θ. Round to the nearest degree.

5.
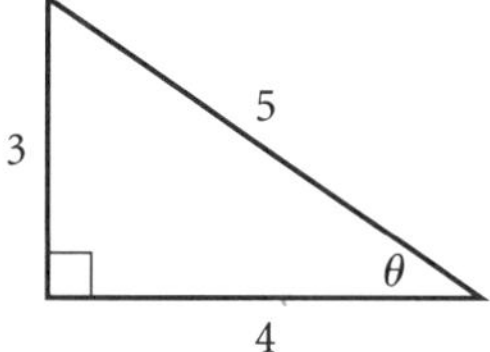

6.
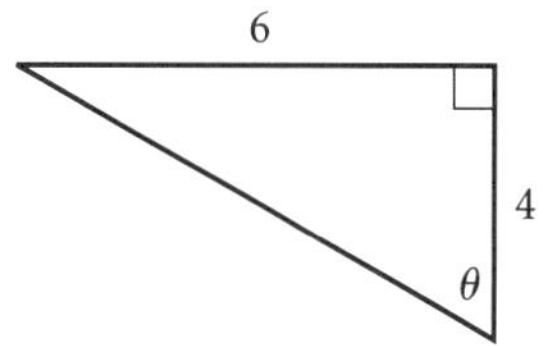

7. Joey is walking home from the library. He can either walk for 1 mile along the highway, then turn right and walk another 1.5 miles on his street, or he can cut across a large field straight to his house. At what angle, θ, should he head off the highway, and how far, d, will he walk if he cuts across the field?

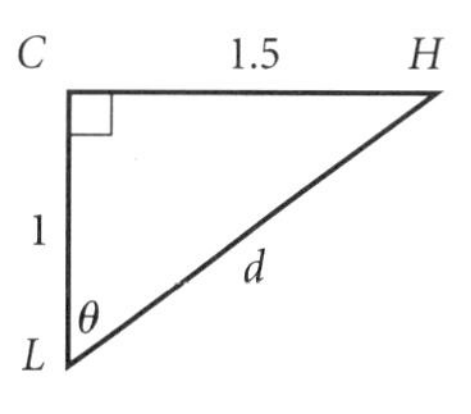

$\theta =$ ____________ $d =$ ____________

In Exercises 8–9, determine which of the following statements are identities. Use the identities given in this lesson to prove which are identities and a counterexample to prove which are *not* identities.

8. $\tan \theta \sin \theta = \cos \theta$

9. $\tan(90 - \theta) = \dfrac{\cos \theta}{\sin \theta}$

NAME ______________________ CLASS __________ DATE __________

Practice Masters Level A

10.3 Extending the Trigonometric Ratios

In Exercises 1–6, sketch a ray with the given angle θ with the positive *x*-axis. Label the coordinates of the point on the ray at a distance of 1 from the origin. Use these values and the unit circle definitions of sine and cosine to give the sine and cosine of each angle. Leave your answers in simplified radical form.

1. $60°$

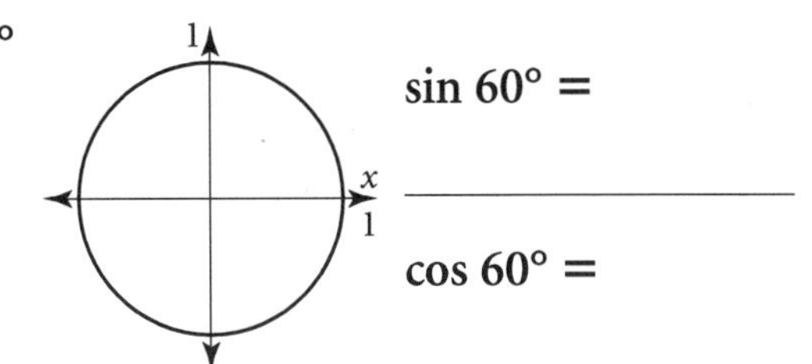

$\sin 60° =$ __________

$\cos 60° =$ __________

2. $90°$

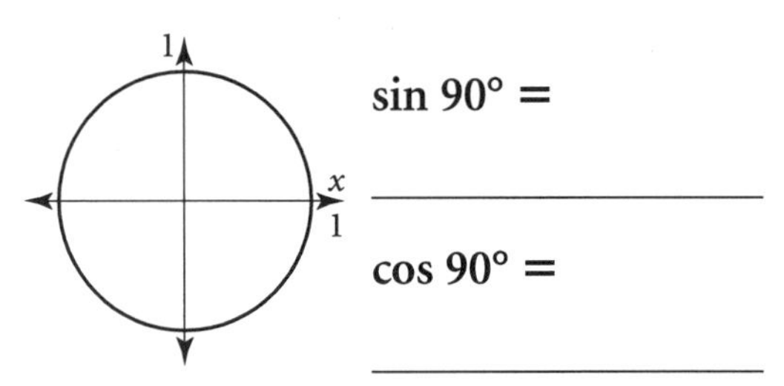

$\sin 90° =$ __________

$\cos 90° =$ __________

3. $120°$

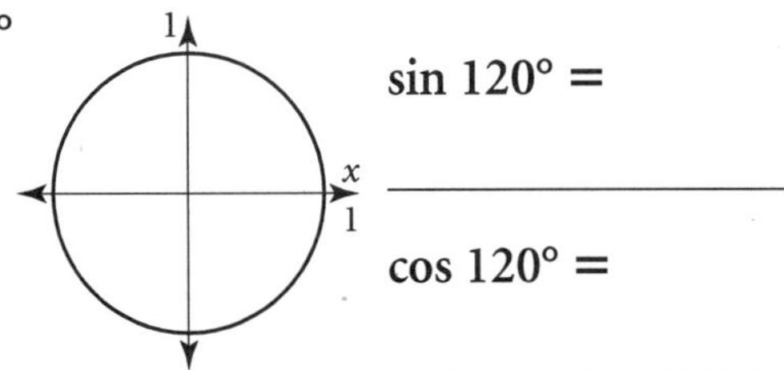

$\sin 120° =$ __________

$\cos 120° =$ __________

4. $-45°$

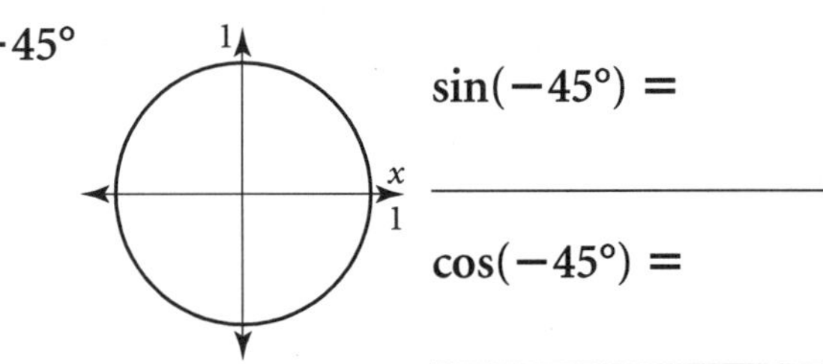

$\sin(-45°) =$ __________

$\cos(-45°) =$ __________

5. $210°$

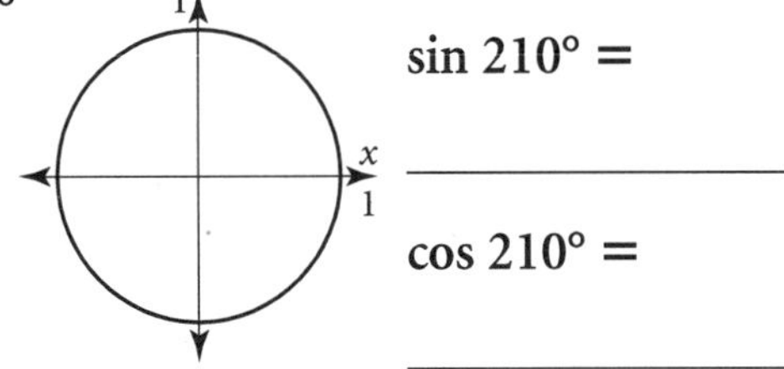

$\sin 210° =$ __________

$\cos 210° =$ __________

6. $180°$

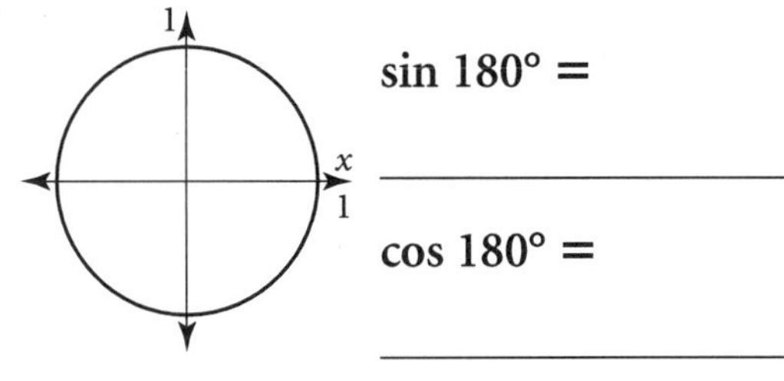

$\sin 180° =$ __________

$\cos 180° =$ __________

In Exercises 7–12, use a calculator to find the sine and cosine of each angle to four decimal places. Compare with the values you found in Exercises 1–6.

7. $\sin 60°$ __________ $\cos 60°$ __________

8. $\sin 90°$ __________ $\cos 90°$ __________

9. $\sin 120°$ __________ $\cos 120°$ __________

10. $\sin(-45°)$ __________ $\cos(-45°)$ __________

11. $\sin 210°$ __________ $\cos 210°$ __________

12. $\sin 180°$ __________ $\cos 180°$ __________

13. a. If $\sin \theta$ is positive, in what quadrant(s) will the terminal ray of angle θ be located? __________

    b. Find all possible values of θ between $0°$ and $360°$ such that $\sin \theta = 0.7623$.

    ____

14. a. If $\cos \theta$ is negative, in what quadrant(s) will the terminal ray of angle θ be located? __________

    b. Find all possible values of θ between $0°$ and $360°$ such that $\cos \theta = -0.2468$.

    ____

Practice Masters Level B

10.3 Extending the Trigonometric Ratios

In Exercises 1–4, sketch a ray with given angle θ with the positive x-axis. Label the coordinates of the point on the ray at a distance of 1 from the origin. Use these values and the unit circle definitions of sine and cosine to give the sine and cosine of each angle. Leave your answers in simplified radical form.

1. 150°

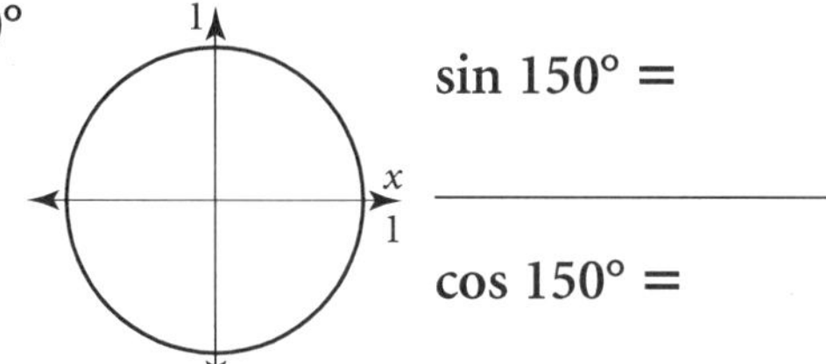

$\sin 150° =$ ______________

$\cos 150° =$ ______________

2. 300°

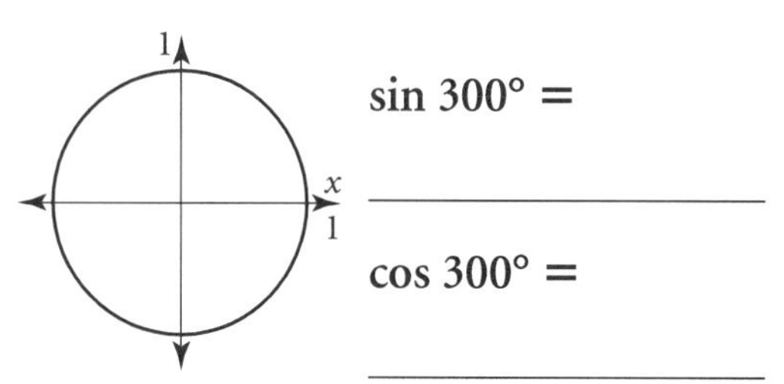

$\sin 300° =$ ______________

$\cos 300° =$ ______________

3. 270°

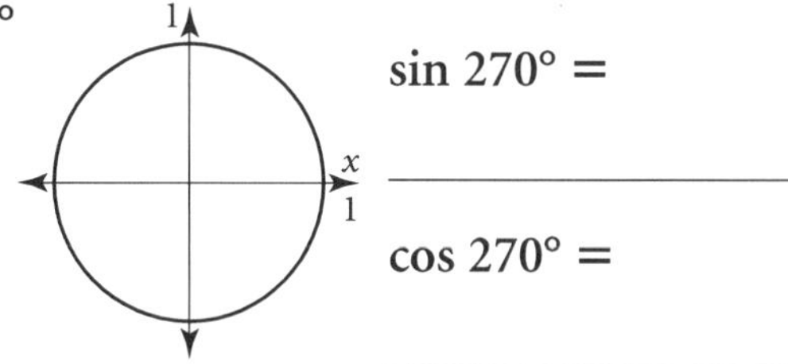

$\sin 270° =$ ______________

$\cos 270° =$ ______________

4. 225°

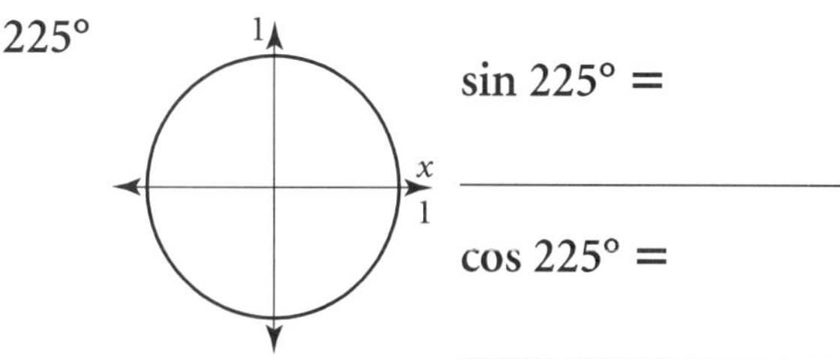

$\sin 225° =$ ______________

$\cos 225° =$ ______________

In Exercises 5–8, use a calculator to find the sine and cosine of each angle to four decimal places. Compare with the values you found in Exercises 1–4.

5. $\sin 150°$ ________ $\cos 150°$ ________
6. $\sin 300°$ ________ $\cos 300°$ ________
7. $\sin 270°$ ________ $\cos 270°$ ________
8. $\sin 225°$ ________ $\cos 225°$ ________

9. a. If $\sin \theta$ is positive, in what quadrant(s) will the terminal ray of angle θ be located?

   ____

   b. Find all possible values of θ between 0° and 360° such that $\sin \theta = 0.2195$.

   ____

10. a. If $\cos \theta$ is negative, in what quadrant(s) will the terminal ray of angle θ be located?

    ____

    b. Find all possible values of θ between 0° and 360° such that $\cos \theta = -0.8530$.

    ____

11. The second hand on a clock turns at the rate of 6° per second. Assume that the length of the hand is 1 unit. Write an equation for the vertical position of point x if it starts from the horizontal position at $t = 0$ seconds. Recall that the clockwise direction is considered to be negative. ______________________________

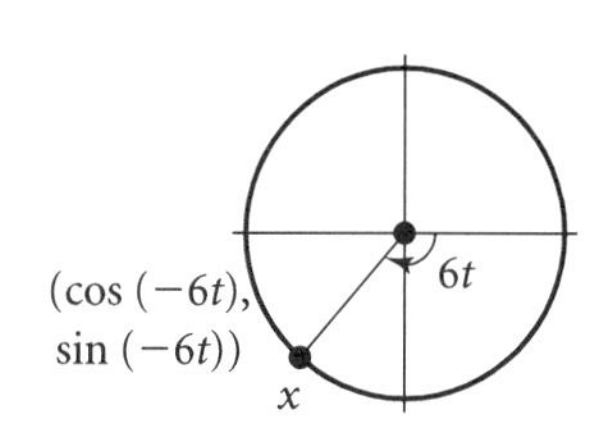

Practice Masters Level C

10.3 Extending the Trigonometric Ratios

In Exercises 1–4, sketch a ray with the given angle θ with the positive x-axis. Label the coordinates of the point on the ray at a distance of 1 from the origin. Use these values and the unit circle definitions of sine and cosine to give the sine and cosine of each angle. Leave your answers in simplified radical form.

1. 330°

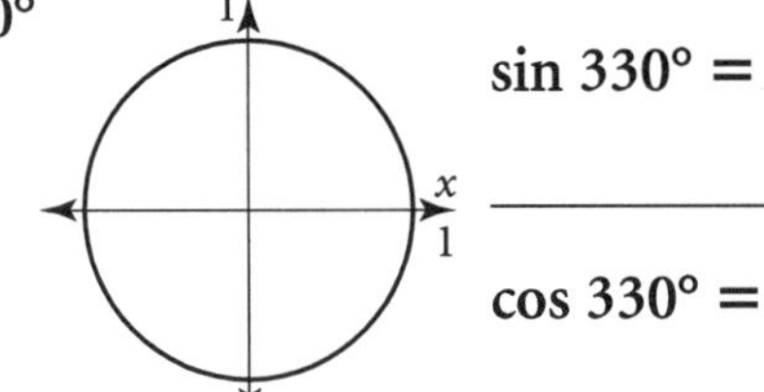

$\sin 330° =$ ____________

$\cos 330° =$ ____________

2. −120°

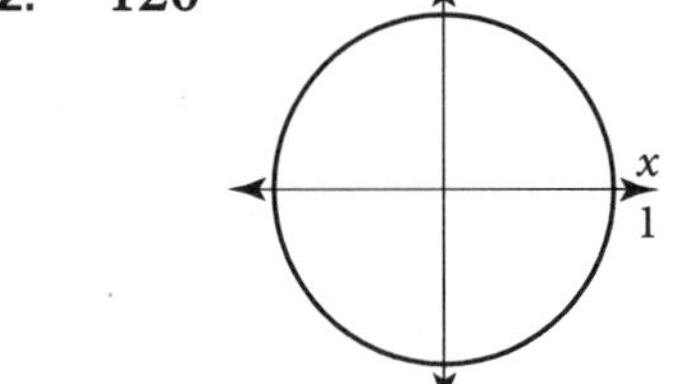

$\sin(-120°) =$ ____________

$\cos(-120°) =$ ____________

3. 135°

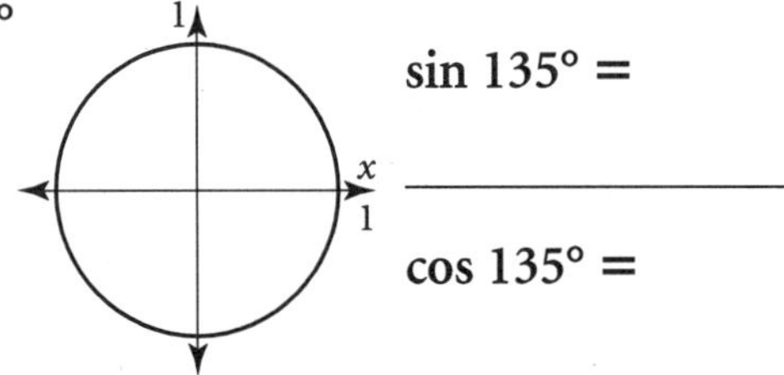

$\sin 135° =$ ____________

$\cos 135° =$ ____________

4. 90°

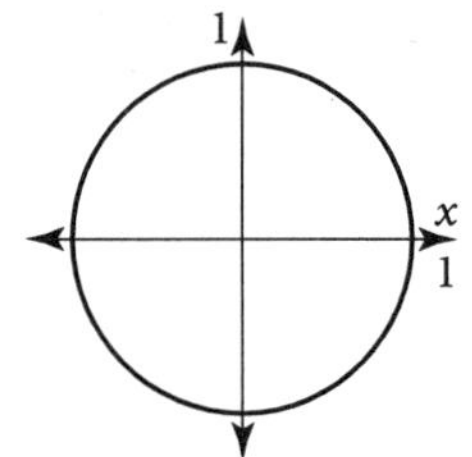

$\sin 90° =$ ____________

$\cos 90° =$ ____________

In Exercises 5–8, use a calculator to find the sine and cosine of each angle to four decimal places. Compare with the values you found in Exercises 1–4.

5. $\sin 330°$ ________ $\cos 330°$ ________
6. $\sin(-120°)$ ________ $\cos(-120°)$ ________
7. $\sin 135°$ ________ $\cos 135°$ ________
8. $\sin 90°$ ________ $\cos 90°$ ________
9. What would be a natural way to extend the definition of the tangent ratio so it would apply to angles of any size? ____________
10. If $\tan \theta$ is positive, in what quadrant(s) will the terminal ray of angle θ be located? ____________
11. If $\tan \theta$ is negative, in what quadrant(s) will the terminal ray of angle θ be located? ____________
12. Consider the identity $(\cos \theta)^2 + (\sin \theta)^2 = 1$, which was proven using the Pythagorean Theorem, for acute angles of a right triangle. Give a proof that shows it to be true for angles of any size. (Hint: Use the distance formula.) ____________

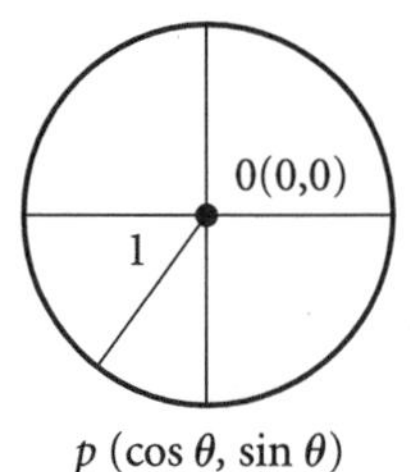

$p\ (\cos \theta, \sin \theta)$

NAME ______________________ CLASS ____________ DATE ____________

Practice Masters Level A

10.4 The Law of Sines

In Exercises 1–5, find the indicated measures. Round your answers to the nearest tenth.

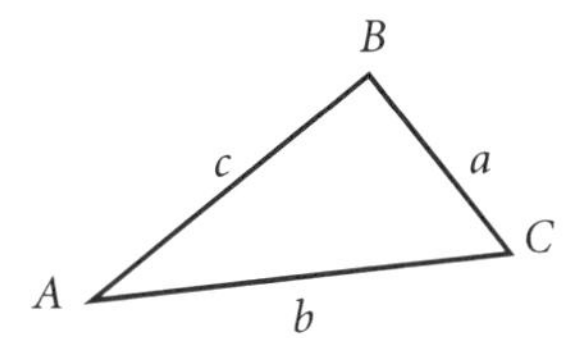

1. $m\angle A = 28°$, $m\angle C = 52°$, $a = 15.3$, $c = ?$ ______________
2. $m\angle A = 31°$, $m\angle C = 70°$, $a = 10.8$, $b = ?$ ______________
3. $m\angle B = 85°$, $m\angle C = 67°$, $a = 6.2$, $c = ?$ ______________
4. $m\angle B = 98°$, $b = 14.2$, $c = 12.5$, $m\angle C = ?$ ______________
5. $m\angle C = 63°$, $a = 4.5$, $c = 8.8$, $m\angle A = ?$ ______________

In Exercises 6 and 7, the measures of △*ABC* given are two side lengths and the angle measure opposite one side. Find the two possible values for m∠*B*.

6. $m\angle A = 30°$, $a = 8.4$, $b = 12.2$ $m\angle B =$ __________ or __________
7. $m\angle C = 58°$, $b = 6.8$, $c = 6.2$ $m\angle B =$ __________ or __________

In Exercises 8–11, two sides of a triangle, *a* and *b*, and an angle opposite one side, ∠*A*, are given. Explain whether the given measurements determine one triangle, two possible triangles, or no triangles. It may be helpful to sketch the triangle roughly to scale.

8. $m\angle A = 105°$, $a = 18$, $b = 14$ ______________________________
9. $m\angle A = 92°$, $a = 10.5$, $b = 16$ ______________________________
10. $m\angle A = 48°$, $a = 8.6$, $b = 7.2$ ______________________________
11. $m\angle A = 65°$, $a = 4.3$, $b = 6.7$ ______________________________

In Exercises 12 and 13, solve each triangle. If the triangle is ambiguous, give both possible angles and all unknown parts of the two triangles possible. It may be helpful to sketch each triangle roughly to scale.

12. $m\angle A = 56°$, $m\angle B = 73°$, $c = 6.0$ Find: $m\angle C$ __________ b __________ a __________

13. $m\angle A = 25°$, $a = 12.6$, $b = 8.3$ Find: $m\angle B$ __________ $m\angle C$ __________ c __________

NAME ______________________ CLASS ____________ DATE ____________

Practice Masters Level B

10.4 The Law of Sines

In Exercises 1–3, find the indicated measures. Round your answers to the nearest tenth.

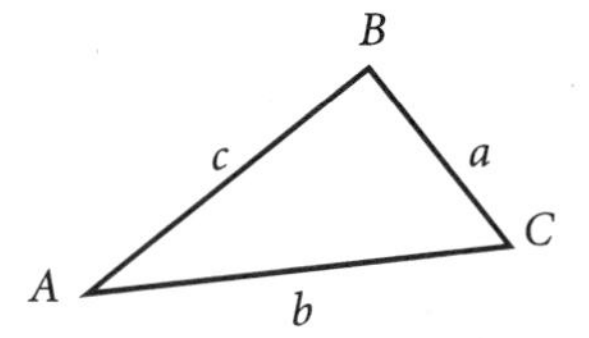

1. $m\angle A = 32°$, $m\angle C = 61°$, $a = 12.3$, $b = ?$ ____________

2. $m\angle B = 92°$, $m\angle C = 58°$, $a = 10.5$, $b = ?$ ____________

3. $m\angle B = 102°$, $b = 14.6°$, $c = 10.5$, $m\angle C = ?$ ____________

In Exercises 4 and 5, the measures given are two sides and the angle opposite one side. Find the two possible values for m∠B.

4. $m\angle A = 34°$, $a = 6.5$, $b = 10.2$ $m\angle B =$ ________ or ________

5. $m\angle C = 53°$, $b = 4.2$, $c = 3.9$ $m\angle B =$ ________ or ________

In Exercises 6–9, two sides of a triangle, *a* and *b*, and an angle opposite one side, ∠*A*, are given. Explain whether the given measurements determine one triangle, two possible triangles, or no triangles. It may be helpful to sketch the triangle roughly to scale.

6. $m\angle A = 110°$, $a = 12$, $b = 14$ ____________

7. $m\angle A = 96°$, $a = 14.5$, $b = 11$ ____________

8. $m\angle A = 62°$, $a = 6.0$, $b = 6.4$ ____________

9. $m\angle A = 25°$, $a = 2.4$, $b = 8.8$ ____________

In Exercises 10 and 11, solve each triangle. If the triangle is ambiguous, give both possible angles and all unknown parts of the two triangles possible. It may be helpful to sketch each triangle roughly to scale.

10. $m\angle A = 48°$ Find: $m\angle C$ ________
$m\angle B = 65°$ b ________
$c = 8.7$ a ________

11. $m\angle A = 56°$ Find: $m\angle B$ ________
$a = 21.8$ $m\angle C$ ________
$b = 24.0$ c ________

12. Two observers on the ground view a hot-air balloon between them at angles of 52° and 67°, respectively. The observers are $\frac{1}{2}$ mile (2640 feet) apart. Find the distance between the balloon and the closest observer.

____

NAME ______________________ CLASS ____________ DATE ____________

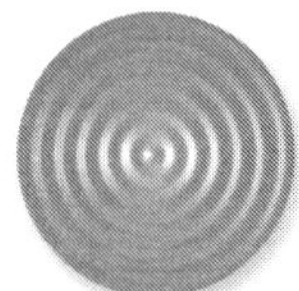

Practice Masters Level C

10.4 The Law of Sines

In Exercises 1–4 the measures of three parts of $\triangle ABC$ are given. Explain whether the given measurements determine one triangle, two possible triangles, or no triangles. It may be helpful to sketch the triangle roughly to scale.

1. $m\angle A = 65°$, $m\angle B = 21°$, $c = 8.0$ ____________

2. $m\angle A = 46°$, $a = 12.5$, $b = 10.2$ ____________

3. $m\angle A = 96°$, $a = 8.0$, $b = 8.4$ ____________

4. $m\angle A = 32°$, $a = 3.6$, $b = 4.8$ ____________

In Exercises 5 and 6, the measures given are two sides and the angle opposite one side. Find the two possible values for $m\angle B$.

5. $m\angle A = 54°$, $a = 7.3$, $b = 8.2$ $m\angle B =$ ________ or ________

6. $m\angle C = 22°$, $b = 5.8$, $c = 3.9$ $m\angle B =$ ________ or ________

In Exercises 7 and 8, find all unknown sides and angles of each triangle. If the triangle is ambiguous, give both possible angles and all unknown parts of the two triangles possible. It may be helpful to sketch each triangle roughly to scale.

7. $m\angle A = 52°$ Find: $m\angle B$ ________
   $m\angle C = 45°$ a ________
   $b = 6.4$ c ________

8. $m\angle A = 44°$ Find: $m\angle B$ ________
   $a = 9.5$ $m\angle C$ ________
   $b = 12.0$ c ________

9. Show that in the case of a right triangle, where $m\angle C = 90°$, the law of sines reverts to the definitions of sin A and sin B. ____________

10. Two observers on the ground view a hot-air balloon between them at angles of 52° and 67°, respectively. The observers are $\frac{1}{2}$ mile (2640 feet) apart. Find the height of the balloon above the ground. ____________

11. From his seat 10 feet above water level, a lifeguard sees two swimmers at angles of 12° and 20°, respectively, below a horizontal line. Find the distance between the swimmers. (Hint: You may need to solve several triangles.)

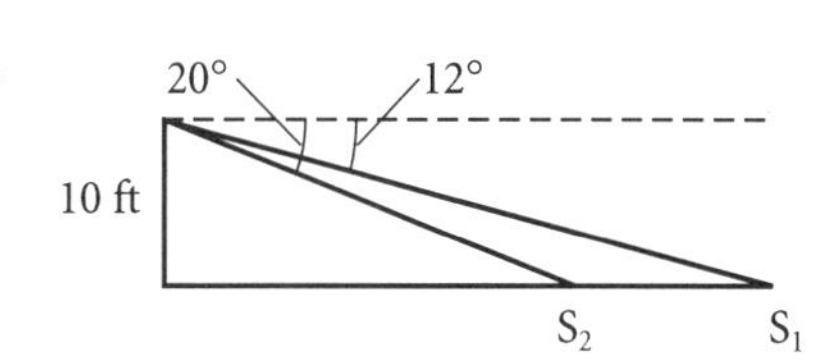

____

NAME ______________________ CLASS ____________ DATE ____________

Practice Masters Level A

10.5 The Law of Cosines

In Exercises 1 and 2, which rule should you use, the law of sines or the law of cosines, to find each indicated measurement? Explain your reasoning.

1.

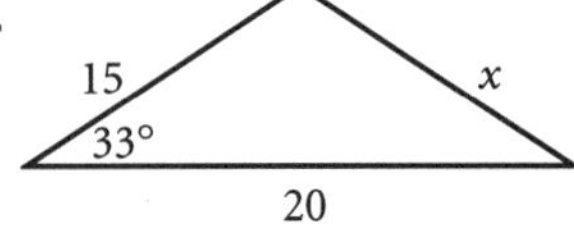

2.

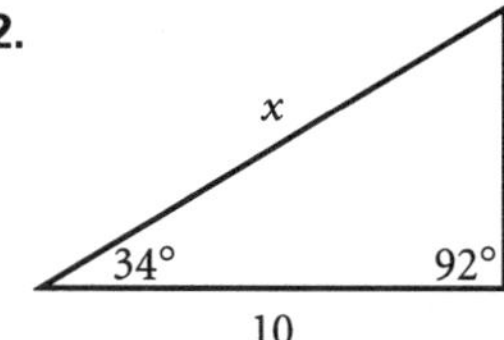

______________________ ______________________

______________________ ______________________

In Exercises 3–6, find the indicated measures. Round your answers to the nearest tenth.

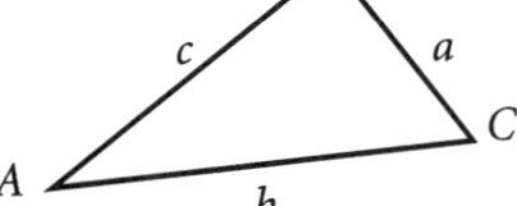

3. $m\angle A = 35°, \quad b = 15.5, \quad c = 12.4, \quad a = ?$ ______________

4. $m\angle B = 94°, \quad m\angle A = 28°, \quad b = 8.5, \quad a = ?$ ______________

5. $a = 4.1, \quad b = 8.3, \quad c = 7.2, \quad m\angle B = ?$ ______________

6. $m\angle C = 65°, \quad m\angle A = 32°, \quad b = 10.8, \quad c = ?$ ______________

In Exercises 7–10, use the law of cosines and/or the law of sines to solve each triangle. Round answers to the nearest tenth.

7.

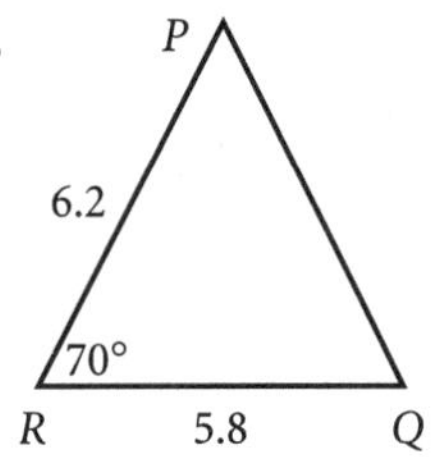

____

____

____

8.

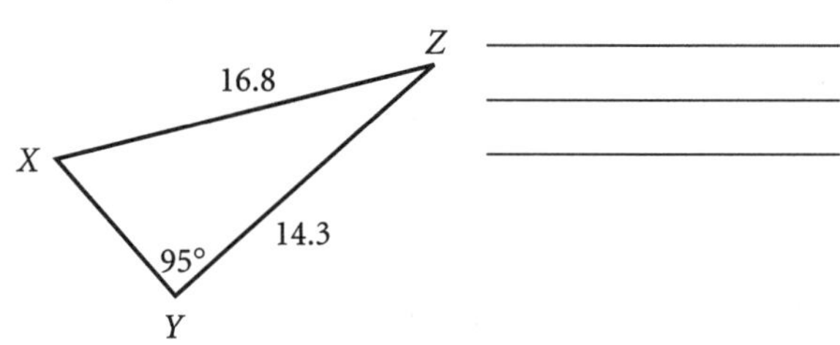

____

____

____

9.

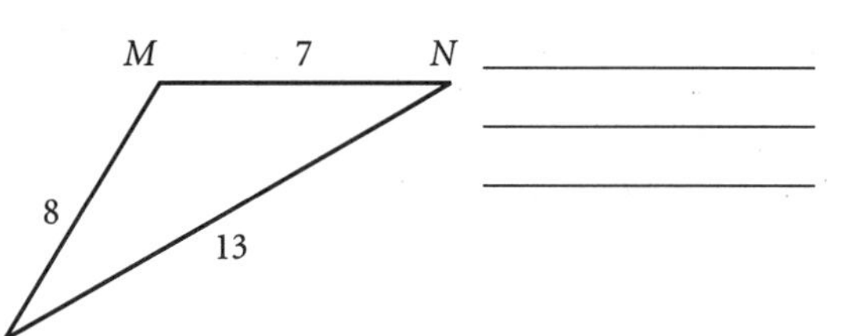

____

____

____

10.

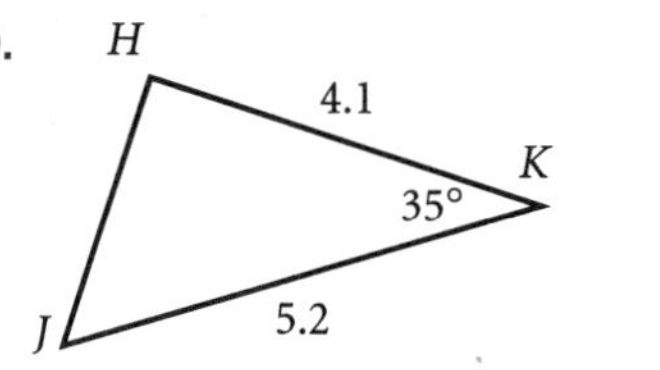

____

____

____

NAME ________________ CLASS ________ DATE ________

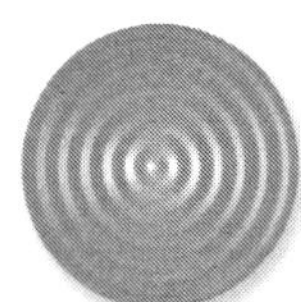

Practice Masters Level B

10.5 The Law of Cosines

In Exercises 1 and 2, which rule should you use, the law of sines or the law of cosines, to find each indicated measurement? Explain your reasoning.

1.

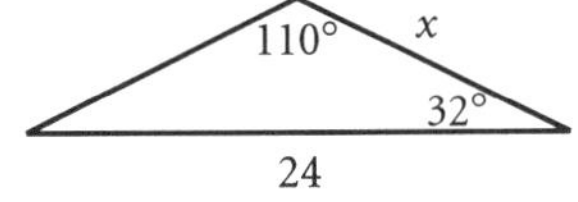

____

____

2.

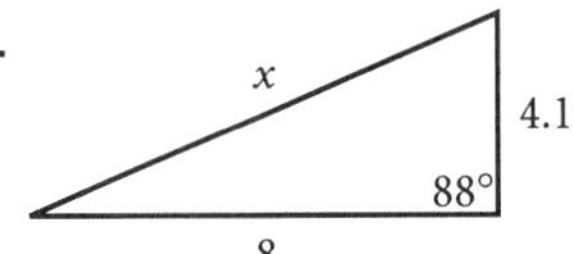

____

____

In Exercises 3–5, find the indicated measures. Round your answers to the nearest tenth.

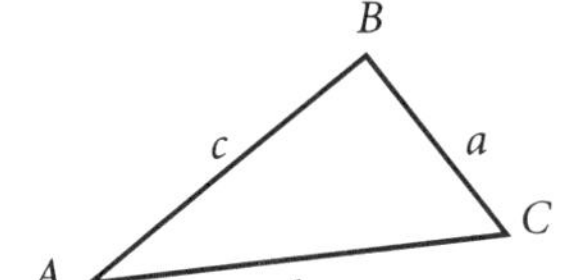

3. $m\angle C = 52°, \quad b = 10.3, \quad a = 6.1, \quad c = ?$ ________

4. $m\angle C = 68°, \quad m\angle A = 28°, \quad b = 24, \quad c = ?$ ________

5. $a = 3.2, \quad b = 6.5 \quad c = 5.0, \quad m\angle C = ?$ ________

In Exercises 6–9, use the law of cosines and/or the law of sines to solve each triangle. Round answers to the nearest tenth.

6.

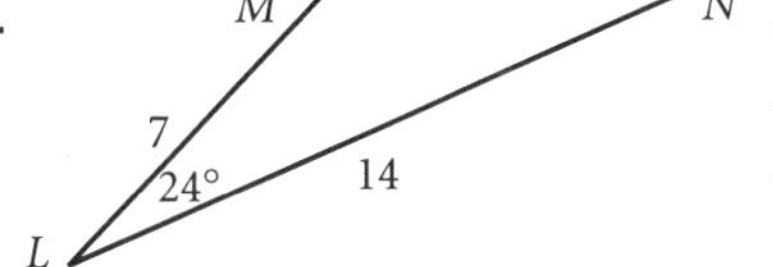

____

____

____

7.

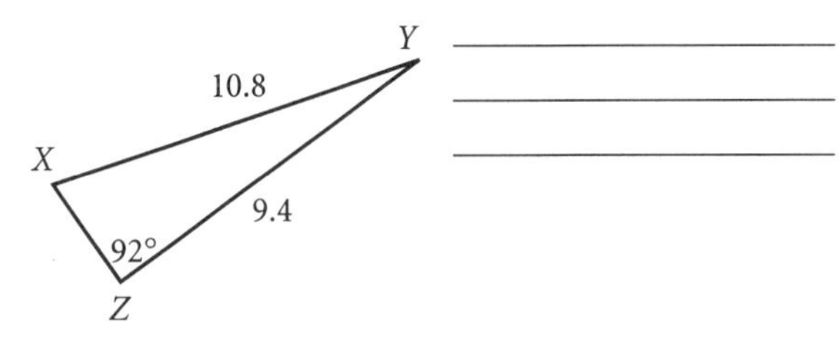

____

____

____

8.

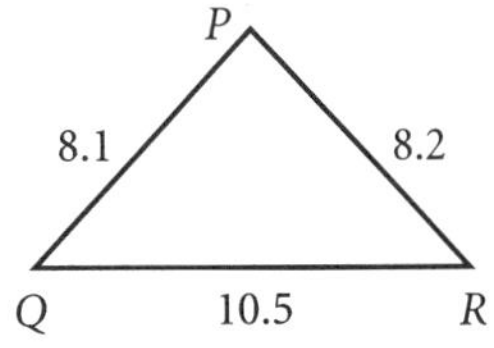

____

____

____

9.

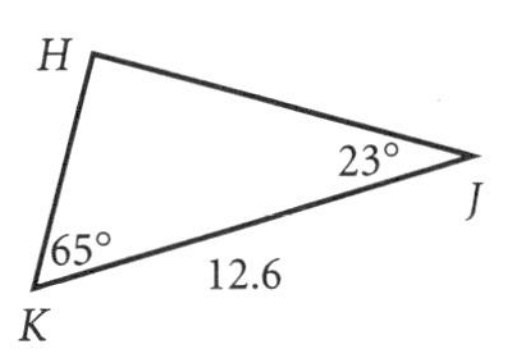

____

____

____

10. Two trains depart from the same station on tracks that form a 65° angle. Train *A* leaves at noon and travels at an average speed of 52 miles per hour. Train *B* leaves at 1 P.M. and travels at an average speed of 60 miles per hour. How far apart are the trains at 3 P.M.? ________

Practice Masters Level C

10.5 The Law of Cosines

In Exercises 1 and 2, which rule should you use, the law of sines or the law of cosines, to find each indicated measurement? Explain your reasoning.

1.

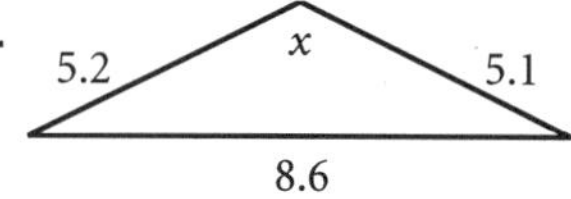

____

____

2.

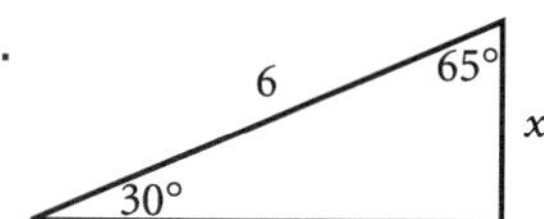

____

____

In Exercises 3–5, find the indicated measures. Round your answers to the nearest tenth.

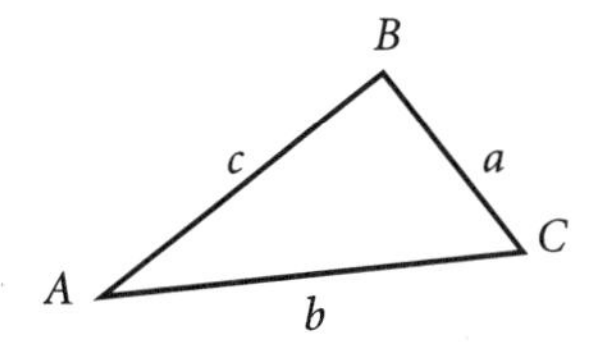

3. $m\angle B = 86°$, $b = 10.3$, $c = 8.4$, $m\angle C = ?$ ______________

4. $m\angle C = 64°$, $a = 8.7$, $b = 16.5$, $c = ?$ ______________

5. $a = 3.0$, $b = 6.2$, $c = 5.0$, $m\angle B = ?$ ______________

In Exercises 6–9, use the law of cosines and/or the law of sines to solve each triangle. Round answers to the nearest tenth.

6.

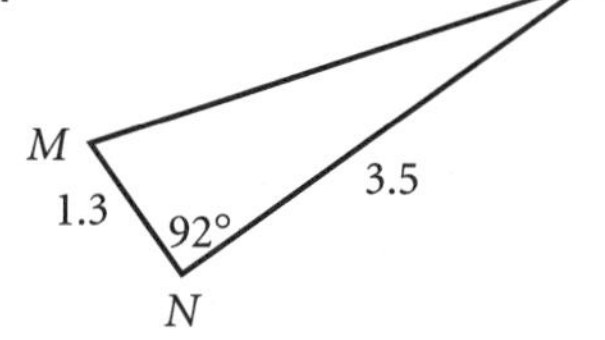

____

____

____

7.

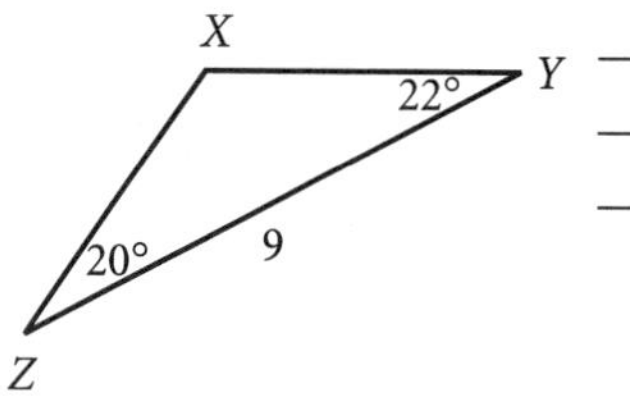

____

____

____

8.

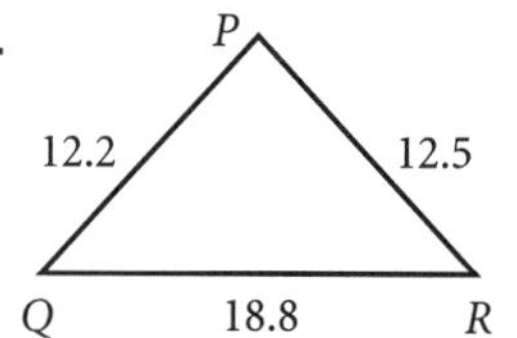

____

____

____

9.

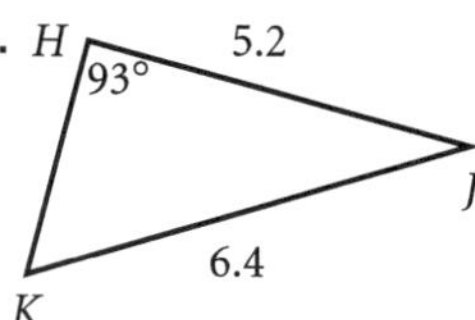

____

____

____

10. The vertices of $\triangle ABC$ are located on a coordinate plane at A(2, −3), B(6, 0) and C(−1, 4).

a. Use the distance formula to find the lengths of the sides.

AB ______________, *BC* ______________, *CA* ______________

b. Use the law of cosines to find the measures of the angles.

$m\angle A$ ______________, $m\angle B$ ______________, $m\angle C$ ______________

NAME ____________ CLASS ________ DATE ________

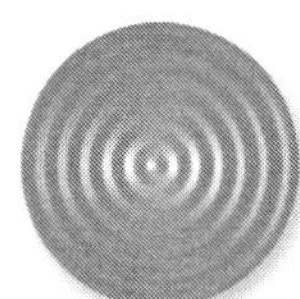

Practice Masters Level A

10.6 Vectors in Geometry

In Exercises 1–3, draw both sum vectors, $\overrightarrow{a} + \overrightarrow{b}$ and $\overrightarrow{b} + \overrightarrow{a}$, by using the head-to-tail method.

1. $\overrightarrow{a} + \overrightarrow{b}$ $\overrightarrow{b} + \overrightarrow{a}$

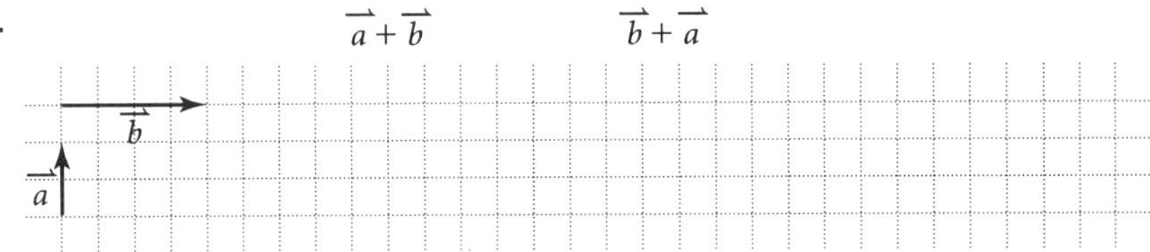

2.

3.

4. a. From your results in Exercises 1–3, what seems to be the relationship between $\overrightarrow{a} + \overrightarrow{b}$ and $\overrightarrow{b} + \overrightarrow{a}$? ____________

   b. By what name (from algebra) might you call this property of vector addition? ____________

In Exercises 5 and 6, draw the vector sum $\overrightarrow{a} + \overrightarrow{b}$ by using the parallelogram method. You may need to translate the vectors.

5.

6.

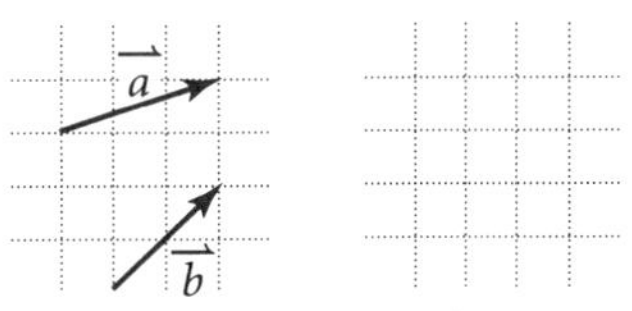

7. Vectors $\overrightarrow{p}$ and $\overrightarrow{q}$ are given at the right.

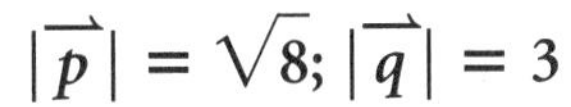

$|\overrightarrow{p}| = \sqrt{8}$; $|\overrightarrow{q}| = 3$

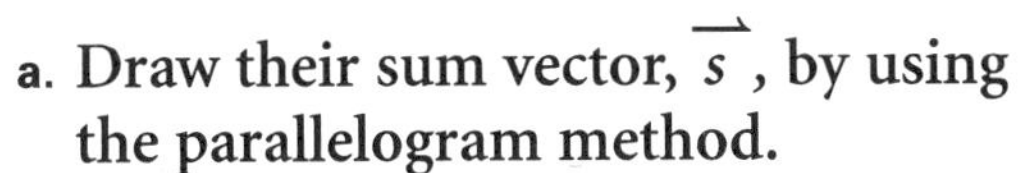

   a. Draw their sum vector, $\overrightarrow{s}$, by using the parallelogram method.

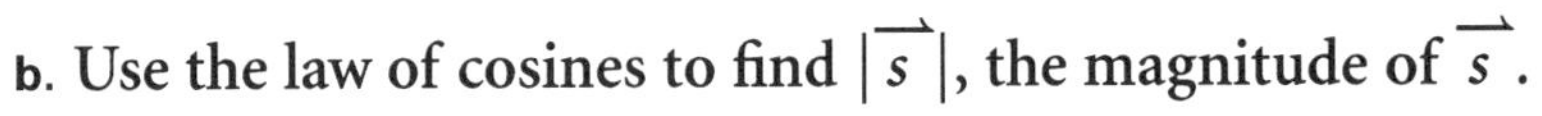

   b. Use the law of cosines to find $|\overrightarrow{s}|$, the magnitude of $\overrightarrow{s}$. ____________

   c. Use the law of sines to find the angle that $\overrightarrow{s}$ makes with $\overrightarrow{q}$. ____________

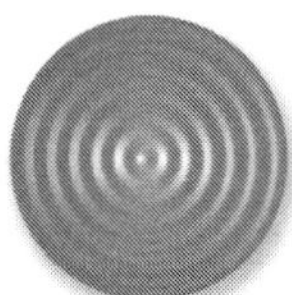

Practice Masters Level B

10.6 Vectors in Geometry

The opposite of vector $\vec{b}$, denoted by $-\vec{b}$, is the vector with the same magnitude as $\vec{b}$, but the opposite direction. Vector subtraction is defined in terms of addition as $\vec{a} - \vec{b} = \vec{a} + (-\vec{b})$.

In Exercises 1–3, draw difference vectors, $\vec{a} - \vec{b}$ and $\vec{b} - \vec{a}$, by using the head-to-tail method.

1. 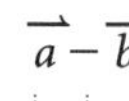 $\vec{b} - \vec{a}$

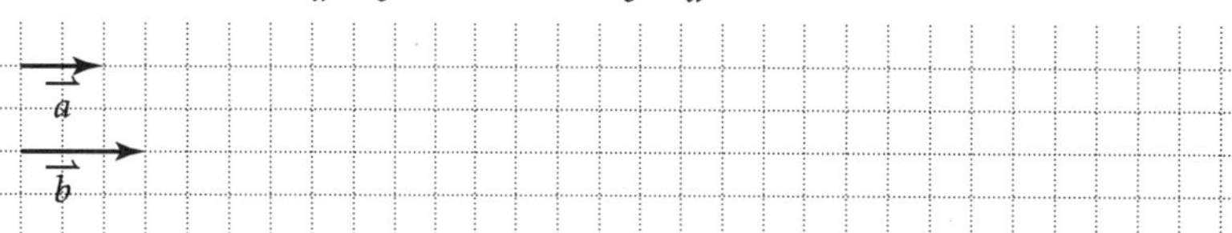

2.

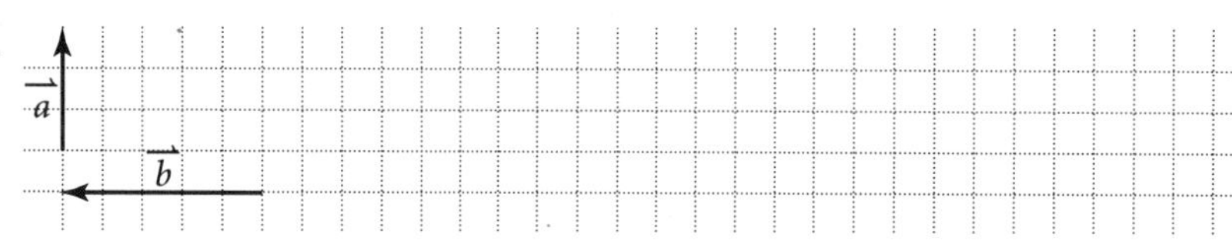

3.

4. From your results in Exercises 1–3, what seems to be the relationship between $\vec{a} - \vec{b}$ and $\vec{b} - \vec{a}$? ____________

In Exercises 5 and 6, draw the vector sum $\vec{a} + \vec{b}$ by using the parallelogram method. You may need to translate the vectors.

5.

6.

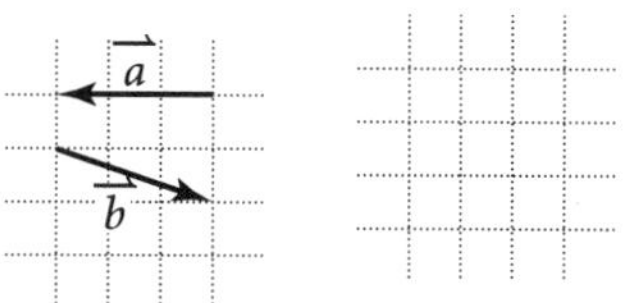

7. Vectors $\vec{p}$ and $\vec{q}$ are given at the right.

$|\vec{p}| = \sqrt{8}$; $|\vec{q}| = 5$

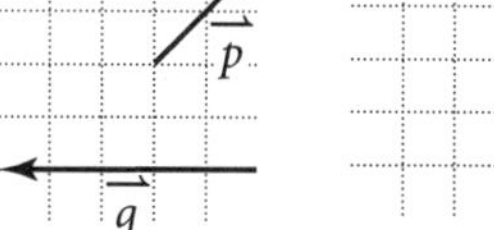

a. Draw their sum vector, $\vec{s}$, using the parallelogram method.

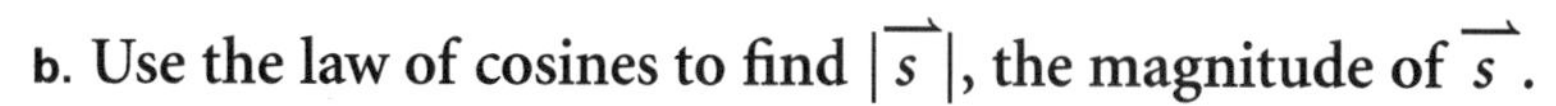

b. Use the law of cosines to find $|\vec{s}|$, the magnitude of $\vec{s}$. ____________

c. Use the law of sines to find the angle that $\vec{s}$ makes with $\vec{q}$. ____________

NAME ______________________ CLASS ____________ DATE ____________

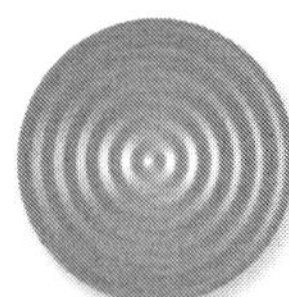

Practice Masters Level C

10.6 Vectors in Geometry

Multiplication of a vector by a scalar (number) can be thought of in terms of repeated addition. Thus, for this vector *a*, $3\vec{a} = \vec{a} + \vec{a} + \vec{a}$.

In Exercises 1–3, use the head-to-tail method to draw all indicated vectors.

1. $2\vec{a} + 2\vec{b}$ $\vec{a} + \vec{b}$ $2(\vec{a} + \vec{b})$

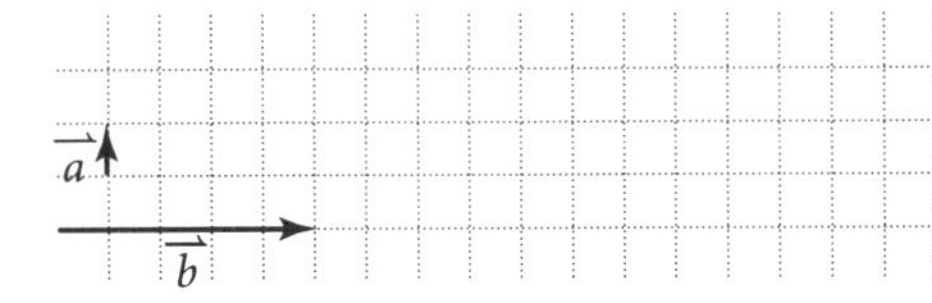

2.

3.

4. a. From your results in Exercises 1–3, what seems to be the relationship between $2\vec{a} + 2\vec{b}$ and $2(\vec{a} + \vec{b})$? ____________

   b. By what name (from algebra) might you call this property of scalar multiplication over vector addition? ____________

5. Vectors $\vec{p}$ and $\vec{q}$ are given at the right.

   $|\vec{p}| = \sqrt{8}$; $|\vec{q}| = \sqrt{18}$.

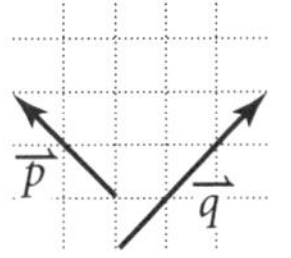

   a. Draw their sum vector, $\vec{s}$, using the parallelogram method.

   b. Use the law of cosines to find $|\vec{s}|$, the magnitude of $\vec{s}$. ____________

   c. Use the law of sines to find the angle that $\vec{s}$ makes with $\vec{q}$. ____________

6. A swimmer heads in a downstream direction at an angle of 20° with the direction of a 1.5 miles per hour current. The speed of the swimmer in still water is 2.2 miles per hour. Find the following:

   a. the swimmer's actual speed ____________

   b. the swimmer's direction angle, θ, with respect to the direction of the current ____________

NAME ______________________ CLASS ____________ DATE ____________

Practice Masters Level A

10.7 Rotations in the Coordinate Plane

In Exercises 1–4, a point, *P*, and an angle of rotation are given. Determine the coordinates of the image point *P′* in the following two ways. Round your answers to the nearest tenth. Show that you get equivalent answers by either method.

a. by drawing and using your knowledge of 30–60–90 or 45–45–90 triangles

b. by using the transformation equations

1. $P(4, 0); \theta = 60°$

   a. ______________________

   b. ______________________

2. $P(6, 6); \theta = 45°$

   a. ______________________

   b. ______________________

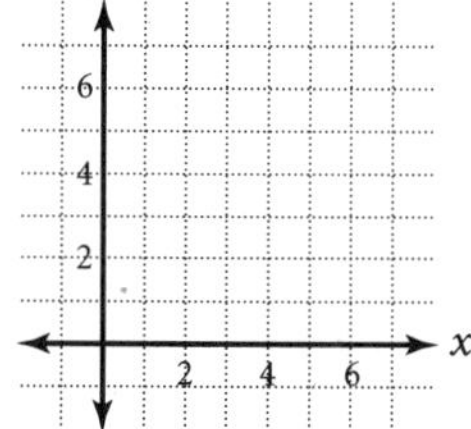

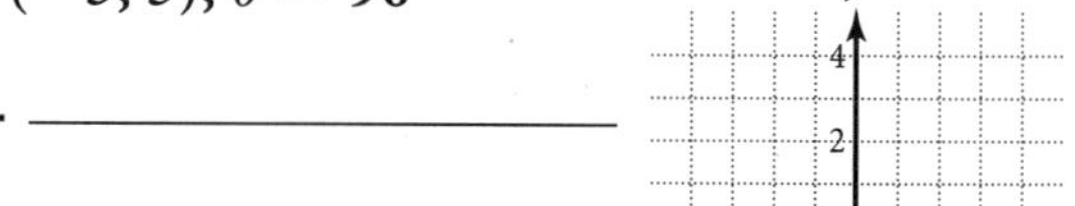

3. $P(0, -2); \theta = 210°$

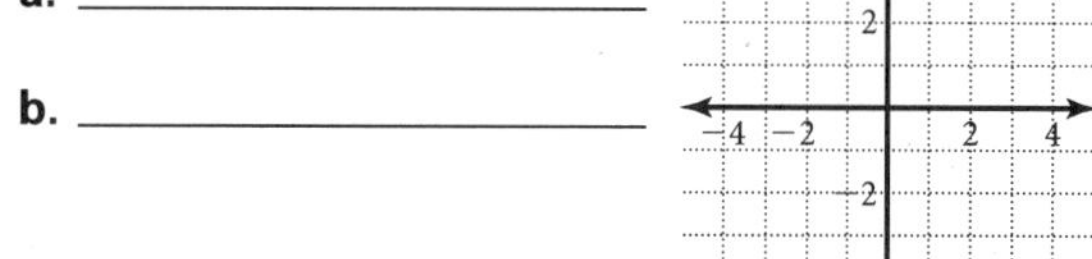

   a. ______________________

   b. ______________________

4. $P(-3, 3); \theta = 90°$

   a. ______________________

   b. ______________________

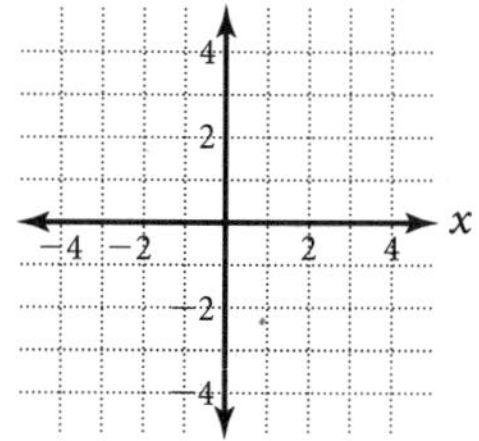

For Exercises 5 and 6, use the transformation equations to find the coordinates of the image point under rotation through the given angle. Round your answer to the nearest tenth.

5. $P(3, -4); \theta = 106°$ ______________________

6. $P(-1, -4); \theta = -15°$ ______________________

For Exercises 7 and 8:

a. Find the rotation matrix for each angle of rotation, and

b. apply the rotation matrix to the vertices of $\triangle ABC$ to find its image under the rotation through angle $\boldsymbol{\theta}$.

Use these vertices for $\triangle ABC$: A(5, 0), B(0, 5), C(–1, 2). Round your answers to the nearest tenth.

7. $\theta = 30°$

   a. matrix ______________________

   b. A' ________, B' ________, C' ________

8. $\theta = 125°$

   a. matrix ______________________

   b. A' ________, B' ________, C' ________

NAME ______________________ CLASS ____________ DATE ____________

Practice Masters Level B

10.7 Rotations in the Coordinate Plane

In Exercises 1–4, a point, *P*, and an angle of rotation are given. Determine the coordinates of the image point *P′* in the following two ways. Round your answers to the nearest tenth. Show that you get equivalent answers by either method.

a. by drawing and using your knowledge of 30–60–90 or 45–45–90 triangles

b. by using the transformation equations

1. $P(-2, 0); \theta = 150°$

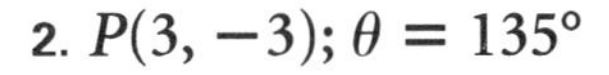

a. ______________

b. ______________

2. $P(3, -3); \theta = 135°$

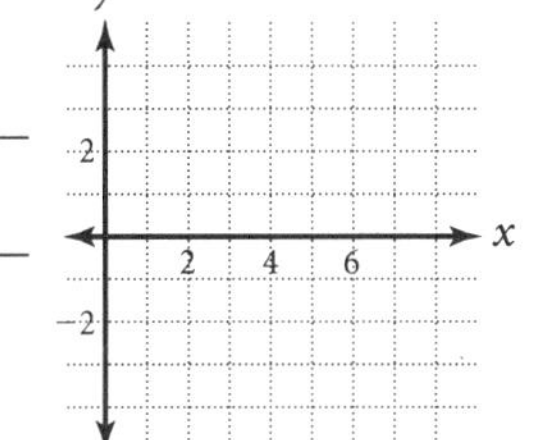

a. ______________

b. ______________

3. $P(0, 4); \theta = -120°$

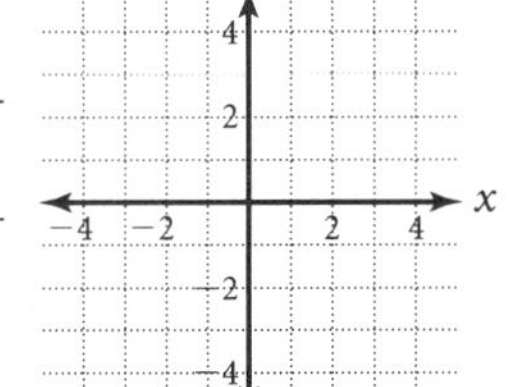

a. ______________

b. ______________

4. $P(5, 5); \theta = 90°$

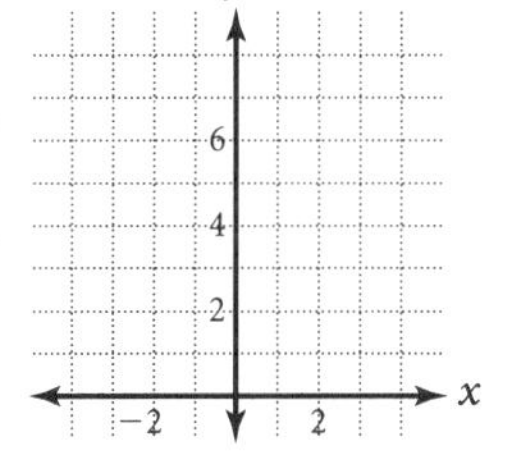

a. ______________

b. ______________

For Exercises 5 and 6, use the transformation equations to find the coordinates of the image point under rotation through the given angle. Round your answer to the nearest tenth.

5. $P(-4, 3); \theta = 74°$

____

6. $P(-2, -5); \theta = -98°$

____

For Exercises 7 and 8:

a. Find the rotation matrix for each angle of rotation, and

b. apply the rotation matrix to the vertices of $\triangle ABC$ to find its image under the rotation through angle θ.

Use these vertices for $\triangle ABC$: A(–2, 0), B(3, 5), C(4, 2). Round your answer to the nearest tenth.

7. $\theta = 120°$

a. matrix ______________

b. A' ______, B' ______, C' ______

8. $\theta = -50°$

a. matrix ______________

b. A' ______, B' ______, C' ______

NAME ______________________ CLASS ______________ DATE ______________

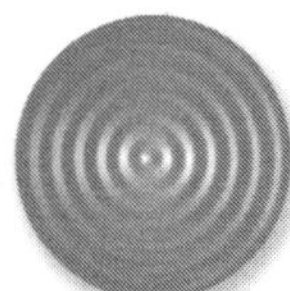

Practice Masters Level C

10.7 Rotations in the Coordinate Plane

In Exercises 1–4, a point, P, and an angle of rotation are given. Determine the coordinates of the image point P' in the following two ways. Round your answers to the nearest tenth. Show that you get equivalent answers by either method.

a. by drawing and using your knowledge of 30–60–90 or 45–45–90 triangles
b. by using the transformation equations

1. $P(-7, -7)$; $\theta = -90°$

   a. ______________________

   b. ______________________

2. $P(0, 6)$; $\theta = 210°$

   a. ______________________

   b. ______________________

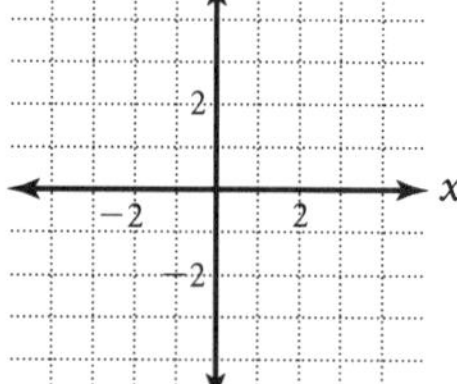

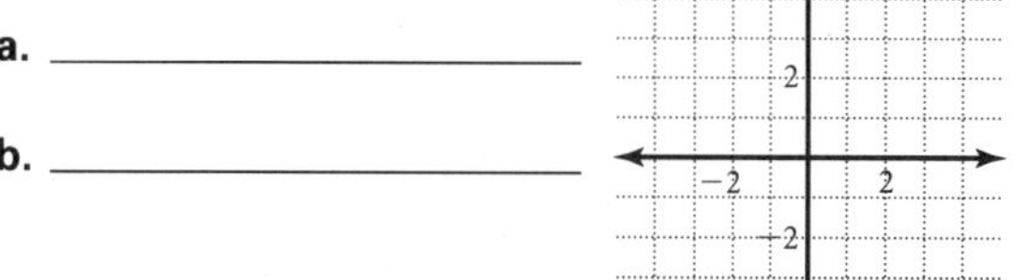

3. $P(5, -5)$; $\theta = -225°$

   a. ______________________

   b. ______________________

4. $P(-4, 0)$; $\theta = 240°$

   a. ______________________

   b. ______________________

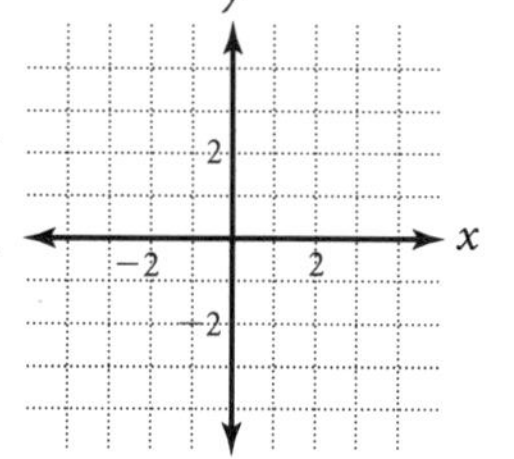

For Exercises 5 and 6, use the transformation equations to find the coordinates of the image point under rotation through the given angle. Round your answer to the nearest tenth.

5. $P(-3, 2)$; $\theta = 100°$ ______________________

6. $P(5, -1)$; $\theta = 380°$ ______________________

7. An animator wants to simulate the waving of a triangular pennant on a computer screen. In its initial position, the pennant will appear attached to a pole along the line $y = 2x$ at the points $(2, 4)$ and $(3, 6)$. The third vertex is at $(0.5, 6)$.

   a. Find the coordinates of the vertices of the triangle under a 30° rotation.

   ____

   b. Draw the pennant in both its initial and final positions and verify by measurement that the pole is rotated through 30°.

   ____

   ____

NAME ______________________ CLASS ____________ DATE ____________

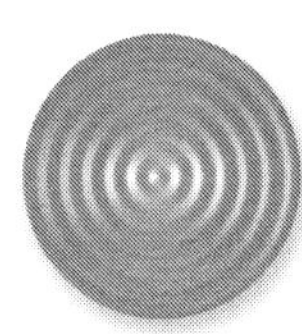

Practice Masters Level A

11.1 Golden Connections

Determine the indicated side length(s) of each golden rectangle. Round your answers to the nearest hundredth.

1.

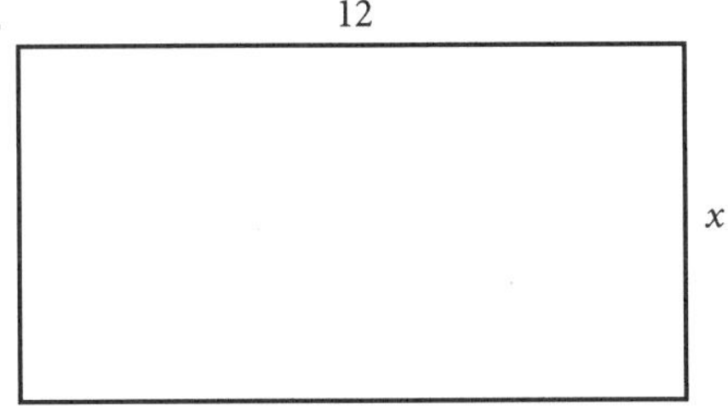

____

2.

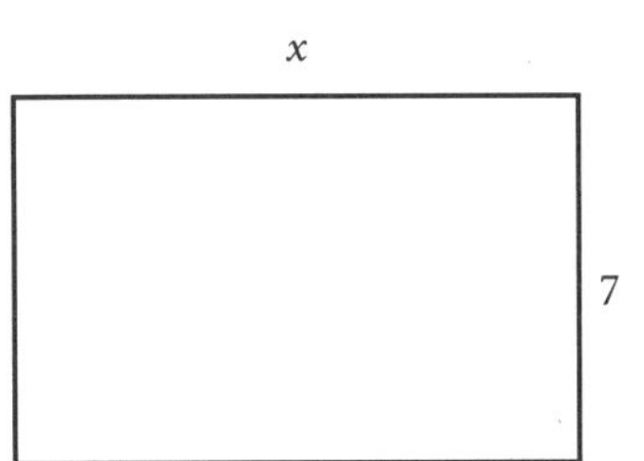

____

3.

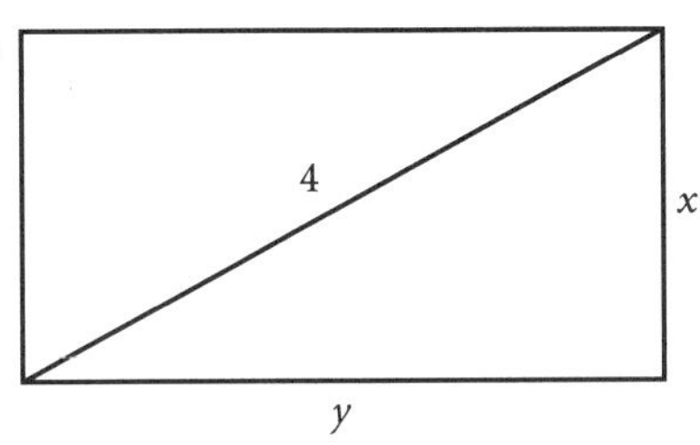

____

4.

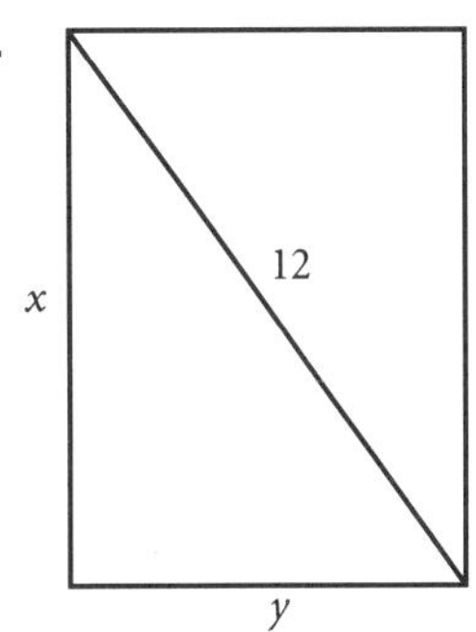

____

5.

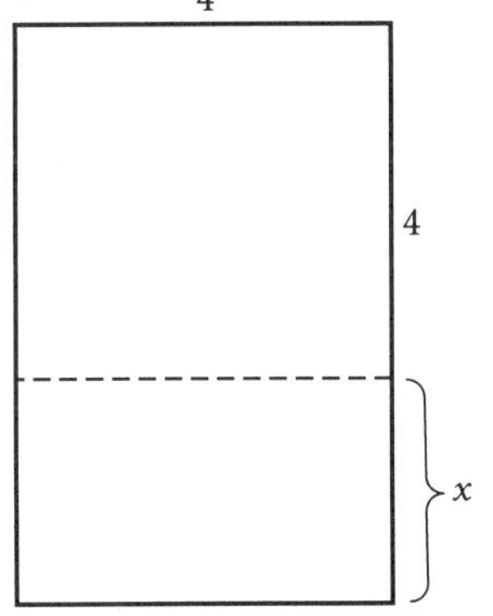

____

6.

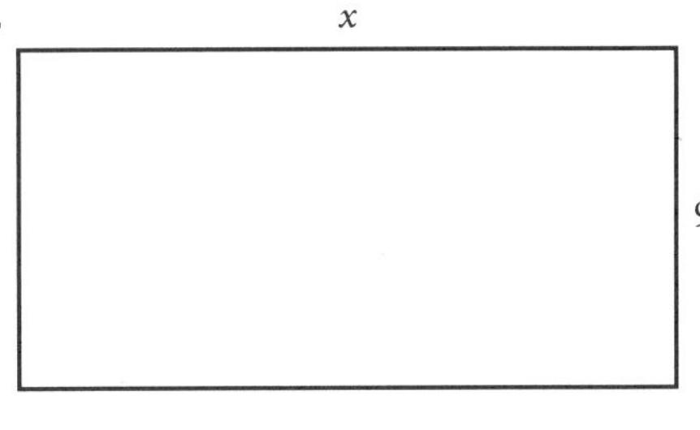

____

Solve. Round your answers to the nearest hundredth.

7. One side of a golden rectangle is 1. Determine the two possible lengths for the other side. ______________________

8. One side of a golden rectangle is $\sqrt{5}$. Determine the two possible lengths for the other side. ______________________

NAME ______________________ CLASS ____________ DATE ____________

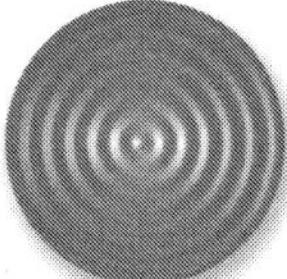

Practice Masters Level B

11.1 Golden Connections

Determine the indicated side length(s) of each golden rectangle. Round your answers to the nearest hundredth.

1.

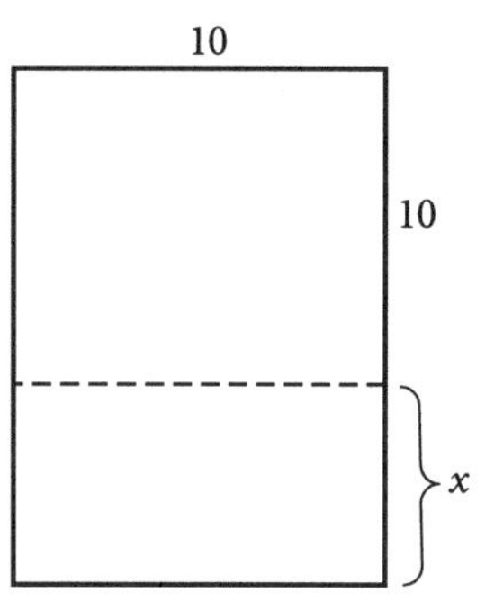

____

2.

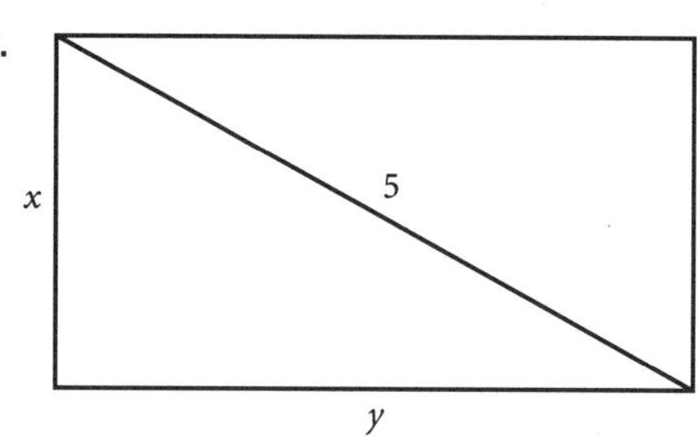

____

3.

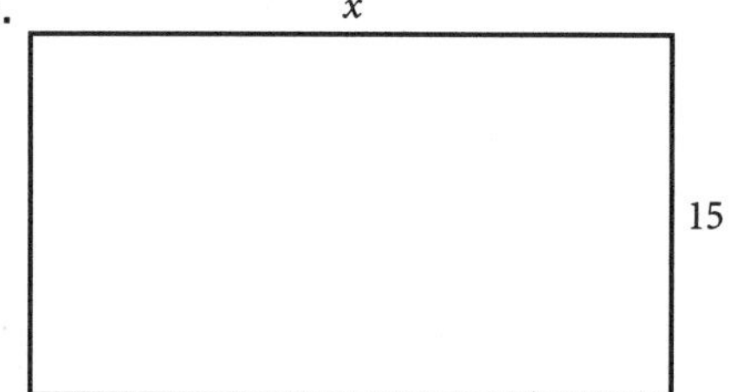

____

4.

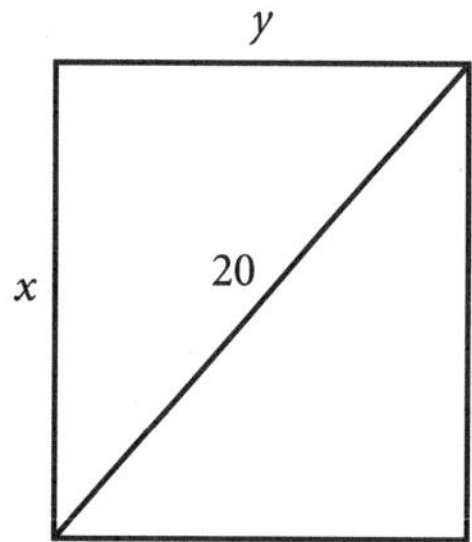

____

5.

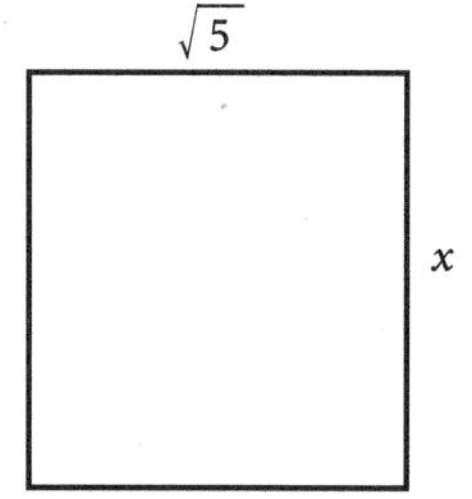

____

6.

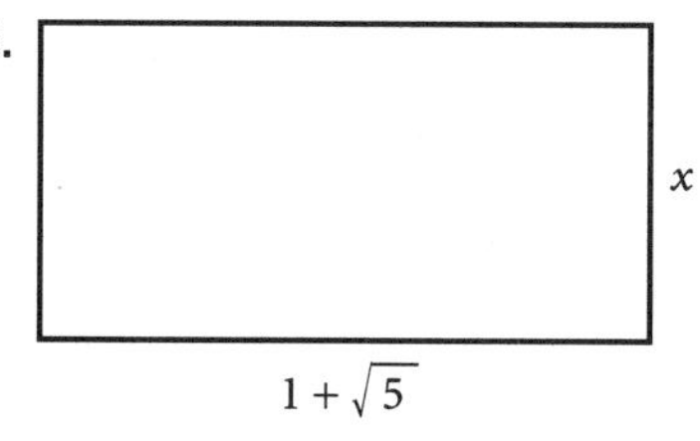

____

Solve. Round your answers to the nearest hundredth.

7. The shorter side of a golden rectangle is 1. Determine the length of the diagonal. ______________________

8. The longer side of a golden rectangle is 1. Determine the length of the diagonal. ______________________

9. The diagonal of a golden rectangle is 1. Determine the lengths of the rectangle's sides. ______________________

Practice Masters Level C

11.1 Golden Connections

Determine the indicated side length(s) of each golden rectangle. Round your answers to the nearest hundredth.

1\.

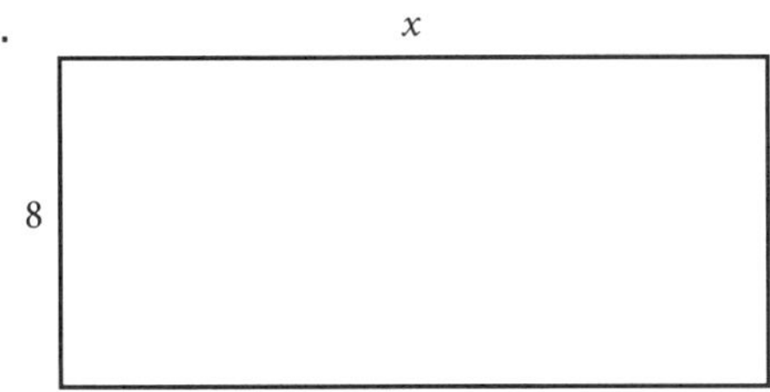

2\.

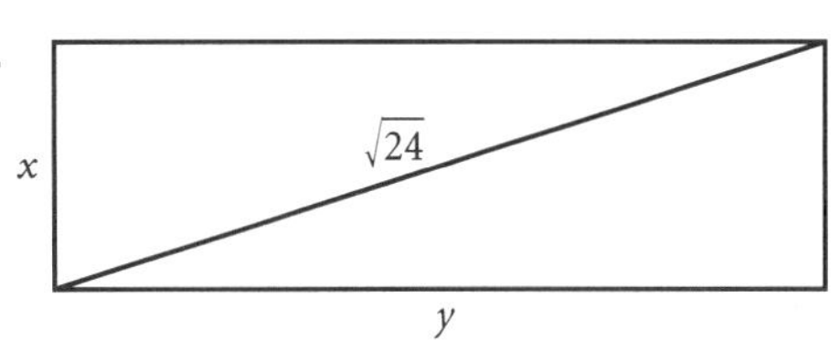

For each rectangle below, decide how much more must be added to the length to make it a golden rectangle.

3\.

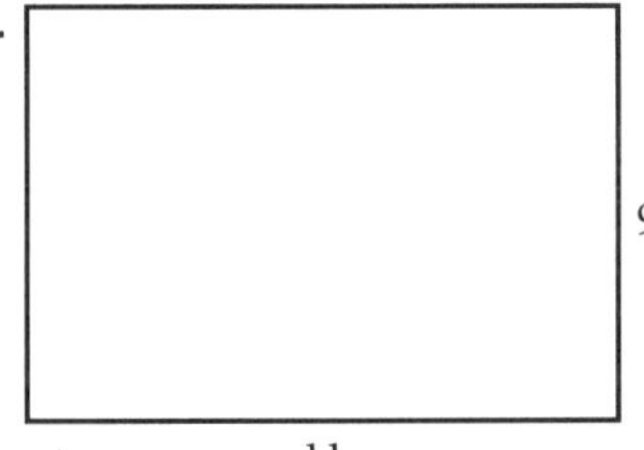

4\.

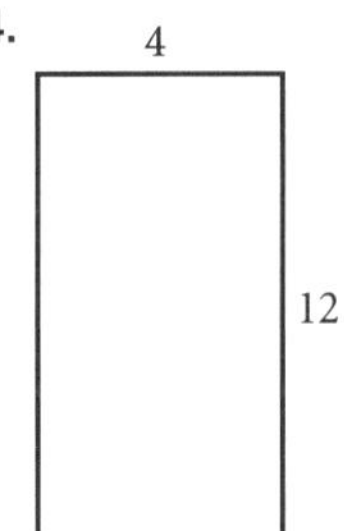

Use the golden rectangle shown for Exercises 5–10. Let e represent the golden ratio $\frac{1+\sqrt{5}}{2}$. Write numerical answers in terms of e, when necessary.

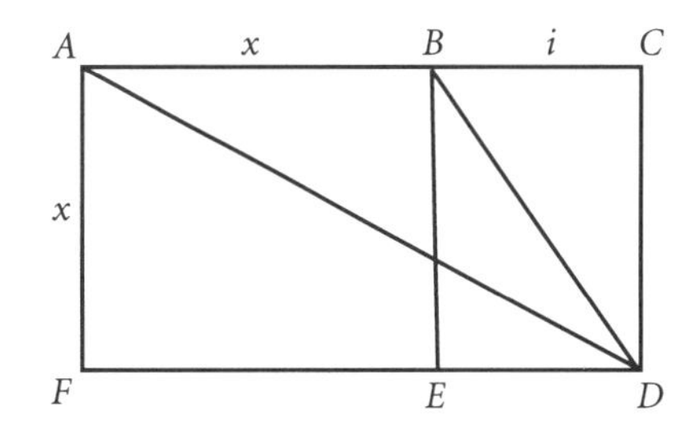

5\. Find the ratio of AC to AF. ______________

6\. Find the ratio of AD to BD. ______________

7\. The length of a segment to AB is $e : 1$. Find the segment length. ______________

8\. Find the area of $ABEF$. Write your answer as a sum. ______________

9\. Find the ratio of the area of $ABEF$ to $BCDE$. ______________

10\. The area of a triangle to that of $\triangle BCD$ is $e : 1$. Find the area of the triangle. ______________

NAME ______________________ CLASS ____________ DATE ____________

Practice Masters Level A

11.2 Taxicab Geometry

Find the taxidistance between each pair of points.

1. $(4, 11)$ and $(6, 8)$ ______________________

2. $(-2, 12)$ and $(8, -14)$ ______________________

3. $(7, -5)$ and $(-1, 1)$ ______________________

4. $(-12, 8)$ and $(8, -12)$ ______________________

5. $(2, 0)$ and $(-6, -6)$ ______________________

6. $(10, 8)$ and $(4, 2)$ ______________________

For Exercises 7–10, use the points (2, 3) and (4, 6).

7. Using different colored pencils, show the different ways to move the minimum distance from one point to the other.

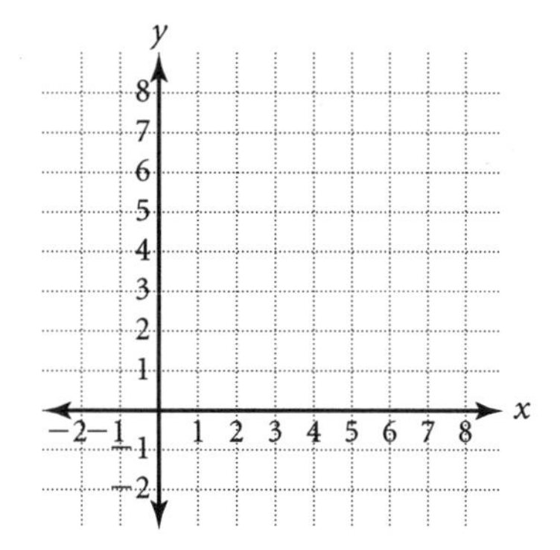

8. What is the taxidistance? ______________________

9. How many different pathways are possible? ______________________

10. Are there any pathways that are longer or shorter than the taxidistance? ______________________

In Exercises 11 and 12, label the grid, plot the taxicab circle described onto the grid, and find its circumference.

11. center C at $(3, 1)$; radius of 3 units

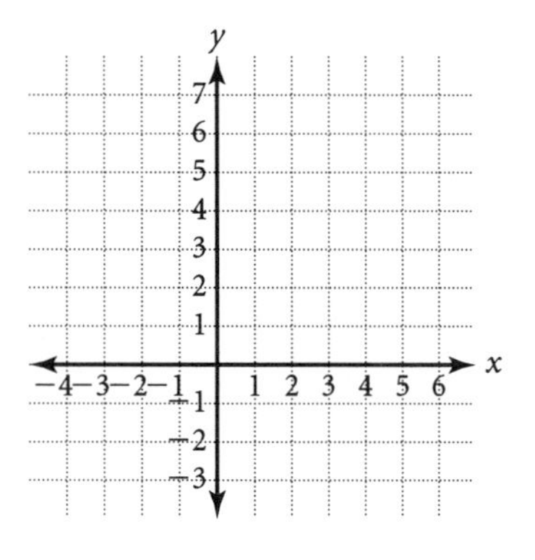

____

12. center O at $(-4, -2)$; radius of 8 units

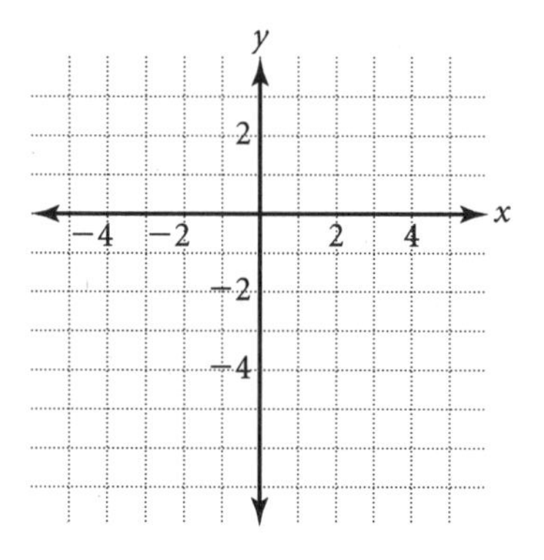

____

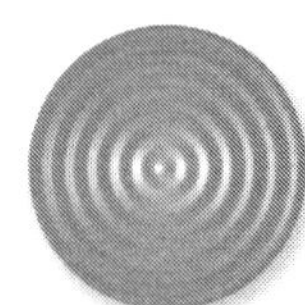

Practice Masters Level B

11.2 Taxicab Geometry

Find the taxidistance between each pair of points.

1. $(-3, 5)$ and $(2, -2)$ ______________
2. $(4, -1)$ and $(3, 0)$ ______________
3. $(-5, 2)$ and $(-2, 1)$ ______________
4. $(0, 3)$ and $(1, 6)$ ______________

In Exercises 5 and 6, label the grid and plot the taxicab circle described onto the grid and find its circumference.

5. center at C $(4, 5)$; radius of 4 units

6. center O at $(-3, 1)$; radius of 6 units

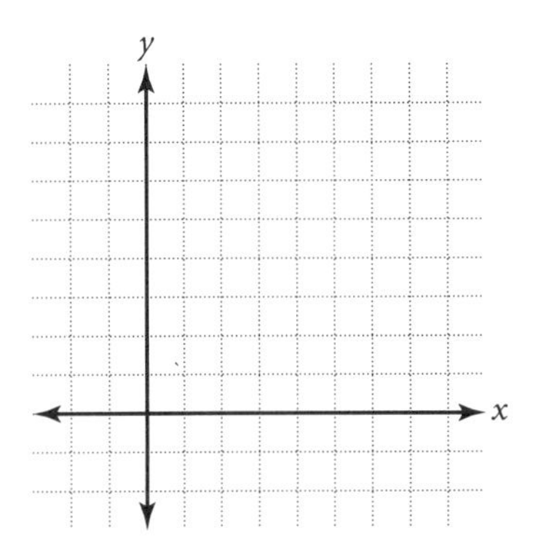

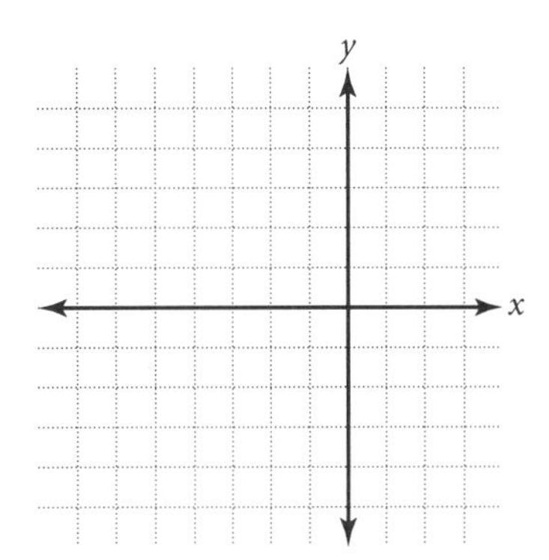

______________________ ______________________

Find all possible values for *a* when *d* is the taxidistance between the pair of points.

7. $(0, 0)$ and $(a, 10)$; $d = 11$ ______________
8. $(a, 9)$ and $(4, -2)$; $d = 12$ ______________
9. $(-6, 15)$ and $(12, a)$; $d = 29$ ______________
10. $(a, -5)$ and $(6, -2)$; $d = 5$ ______________
11. $(-20, a)$ and $(3, -9)$; $d = 32$ ______________
12. $(-1, 38)$ and $(a, 25)$; $d = 13$ ______________

Find the circumference on the taxicab circle with the given radius.

13. $r = 3$ ________
14. $r = 5$ ________
15. $r = 6$ ________
16. $r = 7$ ________

Find the number of points on the taxicab circle with the given radius.

17. $r = 6$ ______________
18. $r = 7$ ______________

NAME ______________________ CLASS ____________ DATE ____________

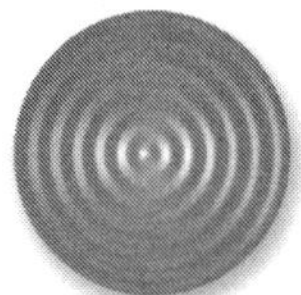

Practice Masters Level C

11.2 Taxicab Geometry

For Exercises 1–10, count the number of shortest pathways from (0, 0) to the indicated point. To keep from having to recount, write the number of shortest pathways at each intersection in the grid at the right.

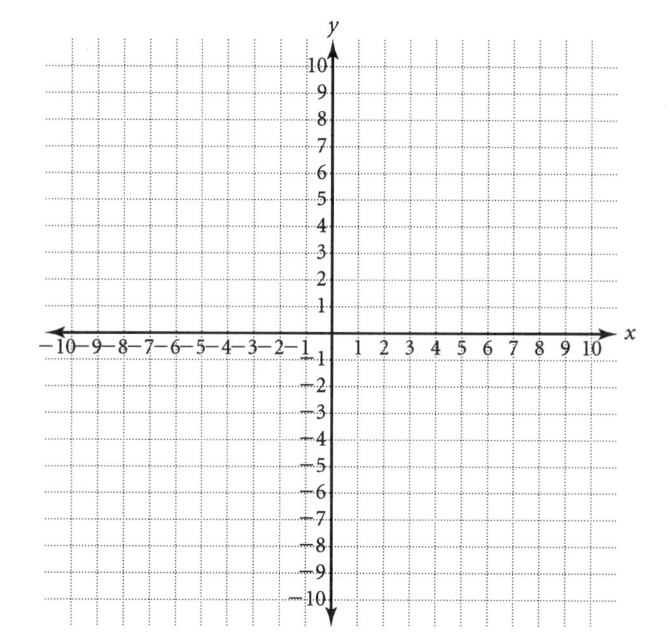

1. $(1, 0)$ ____________
2. $(2, 0)$ ____________
3. $(3, 0)$ ____________
4. $(4, 0)$ ____________
5. $(1, 1)$ ____________
6. $(2, 1)$ ____________
7. $(3, 1)$ ____________
8. $(4, 1)$ ____________
9. $(2, 2)$ ____________
10. $(3, 2)$ ____________
11. $(4, 2)$ ____________
12. $(5, 2)$ ____________
13. $(6, 2)$ ____________
14. $(7, 2)$ ____________
15. $(8, 2)$ ____________
16. $(9, 2)$ ____________

Find all possible values for *a* when *d* is the taxidistance between the pair of points.

17. $(0, -12)$ and $(4, -a)$; $d = 9$ ____________
18. $(10, -a)$ and $(-15, -3)$; $d = 41$ ____________

Find the circumference of each taxicab circle with the given radius.

19. $r = 6$ ____________
20. $r = 7$ ____________

Find the number of points on the taxicab circle with the given radius.

21. $r = 5$ ____________
22. $r = 6$ ____________

NAME ______________________ CLASS __________ DATE __________

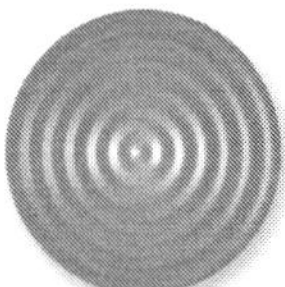

Practice Masters Level A

11.3 Graph Theory

Determine whether the graphs below contain an Euler path, an Euler circuit, or neither. Where would you need to start in order to trace the figure without lifting your finger?

1.

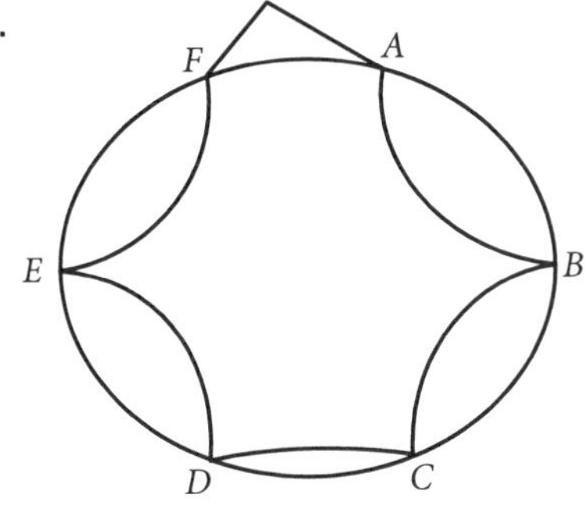

____

____

2.

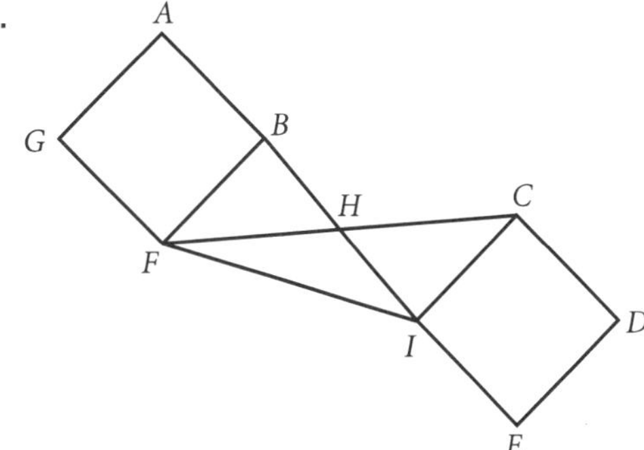

____

____

3.

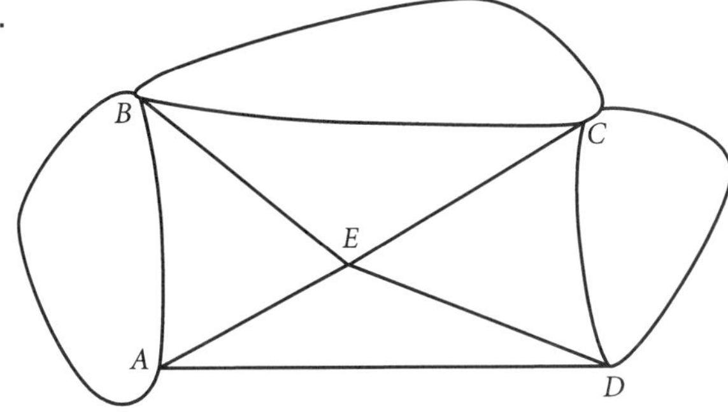

____

____

4.

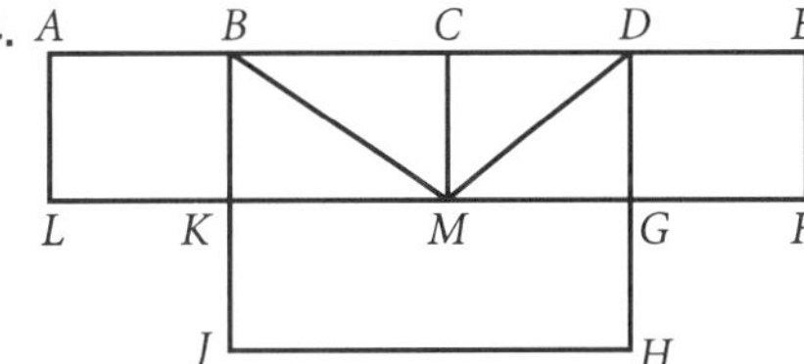

____

____

5.

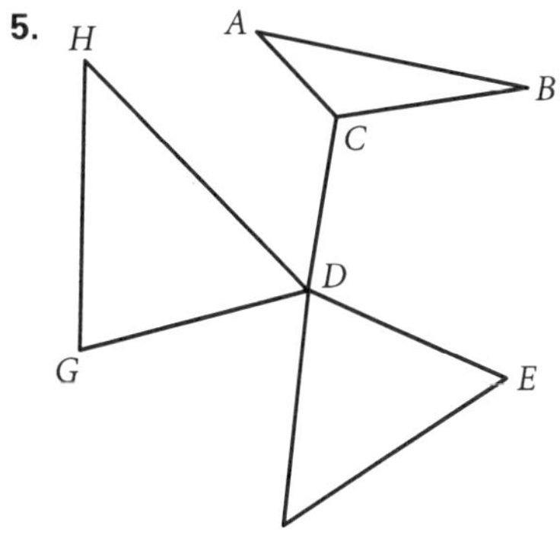

____

____

NAME ______________________ CLASS __________ DATE __________

Practice Masters Level B

11.3 Graph Theory

Determine whether the graphs below contain an Euler path, an Euler circuit, or neither. Decide if you can trace the entire figure without lifting your pencil. If so, tell where you would have to start for it to work.

1.

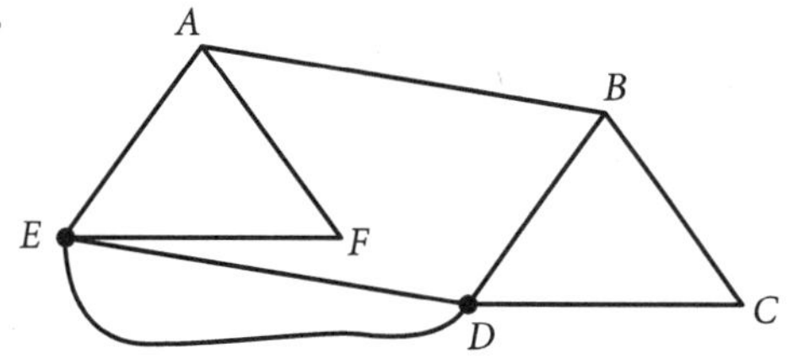

____

____

2.

____

____

3.

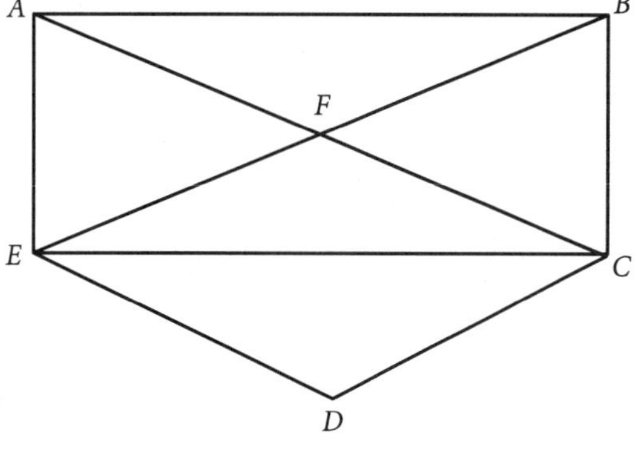

____

____

4.

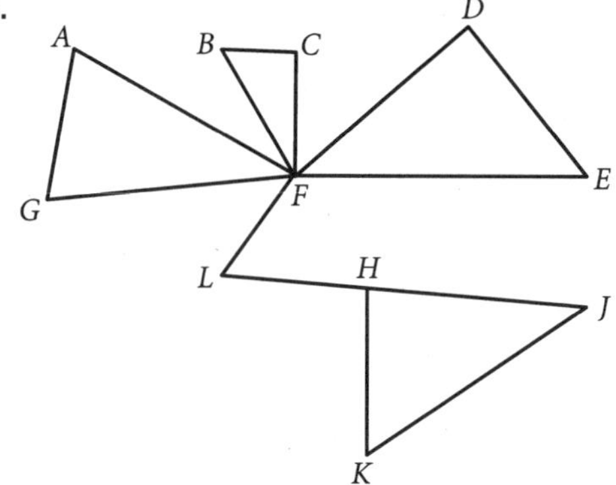

____

____

5. Is it possible to walk through all the doors in the house by going through each door exactly once? Make a model to help you answer the question. Then mark the path you would take. You may find it helpful to identify each room by a different letter when you make your model.

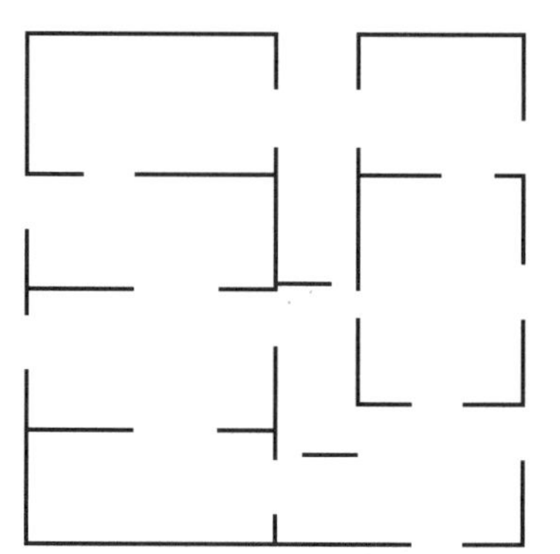

____

NAME ______________________ CLASS ____________ DATE ____________

Practice Masters Level C

11.3 Graph Theory

Determine whether the graphs below contain an Euler path, an Euler circuit, or neither. Decide if you can trace the entire figure without lifting your pencil. If so, tell where you would have to start for it to work.

1.

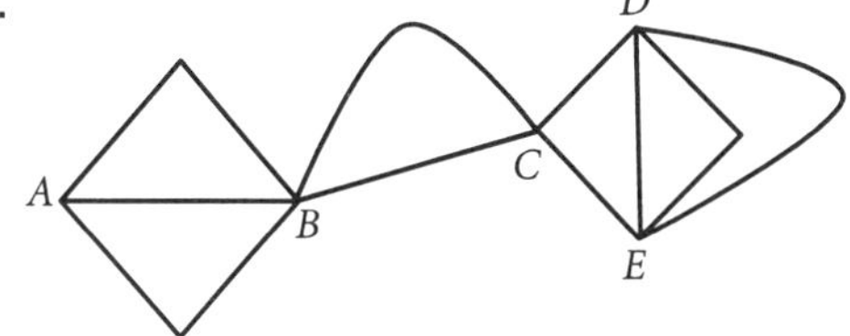

2.

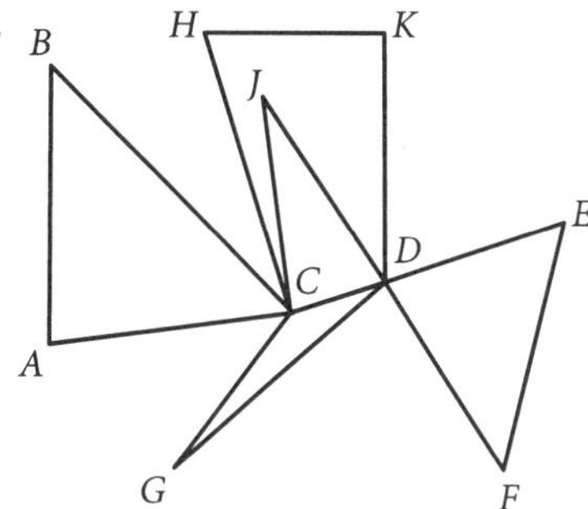

____

____

3. Create an Euler path that has 12 edges and is *not* an Euler circuit.

____

This figure shows distances in kilometers between pairs of 5 cities. A route between cities *A, B,* and *C* can be described as *ABC*. Use this information and the figure to answer the questions.

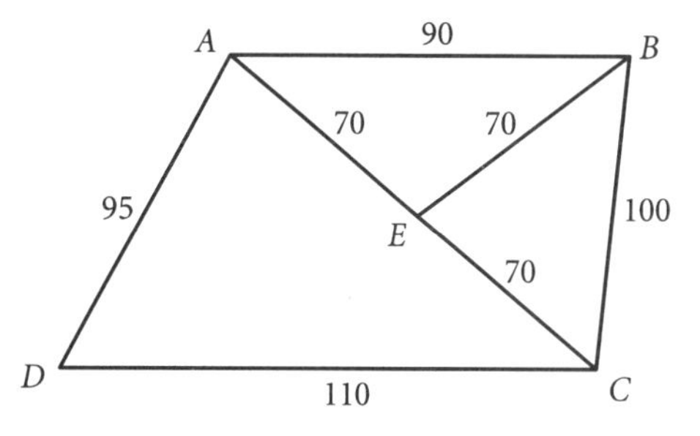

4. The shortest direct route distance between any two cities is ______________.

5. The longest direct route between any two cities is ______________.

6. What is the shortest segment that you can remove to cause the figure to have an Euler path? ______________

7. What is the longest segment that you can remove to cause the figure to have an Euler path? ______________

8. What is the least number of segments you can remove to cause the figure to have an Euler circuit? Consider $\overline{AC}$ as the two segments $\overline{AE}$ and $\overline{EC}$. ______________

9. Describe the shortest route from *A* to *C*. ______________

NAME ______________________ CLASS __________ DATE __________

Practice Masters Level A

11.4 Topology

Determine the number of regions into which the plane is divided by the curve.

1.

____

2.

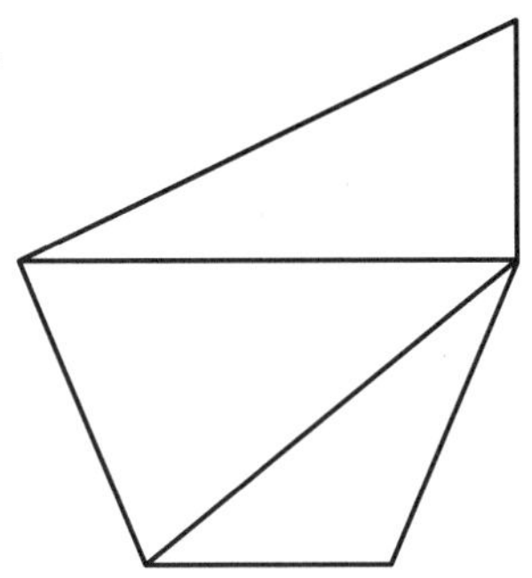

____

Use the figures below for Exercises 3–5.

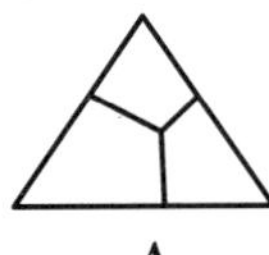

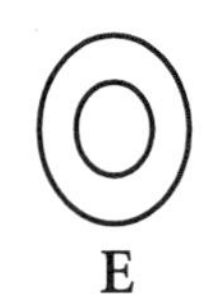

A B C D E F

3. List any of the shapes that are topologically equivalent to F. ______________________

4. List any of the shapes that are topologically equivalent to D. ______________________

5. List any of the shapes that are topologically equivalent to A. ______________________

6. How to you determine whether shapes are topologically equivalent? ______________________

____

Refer to the simple closed curve for Exercises 7–9.

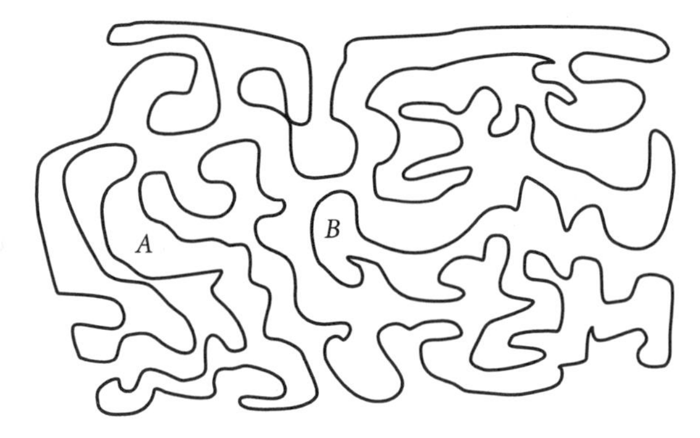

7. Is point A on the inside or the outside of the curve?

____

8. Is point B on the inside or the outside of the curve?

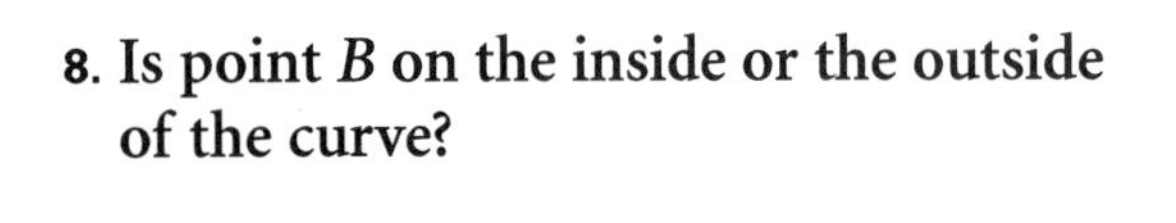

____

9. How do you determine whether a point is inside or outside of a simple closed curve?

____

NAME ______________________ CLASS __________ DATE __________

Practice Masters Level B

11.4 Topology

Determine the number of regions into which the plane is divided.

1. 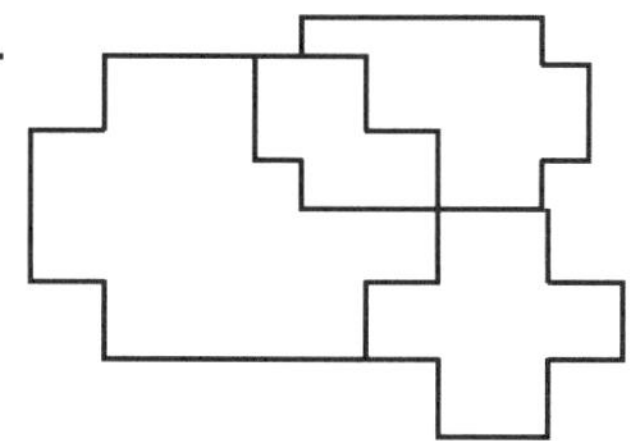

____

2. 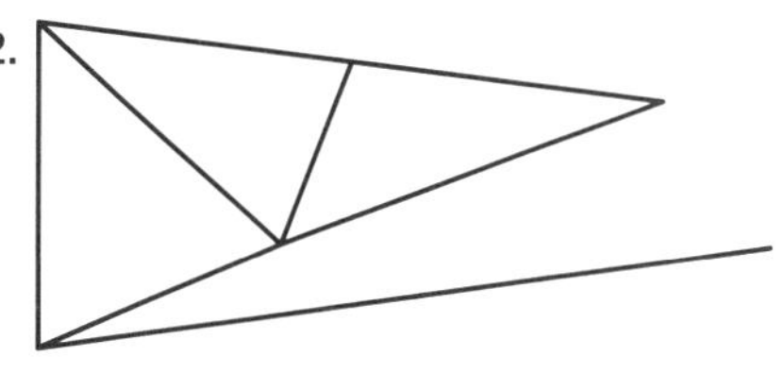

____

Use the figures below for Exercises 3–5.

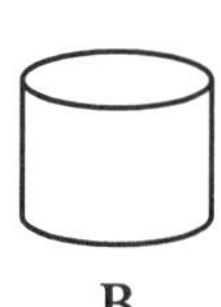

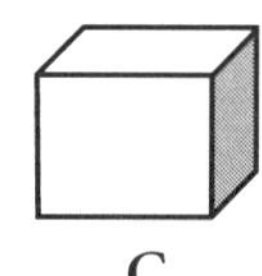

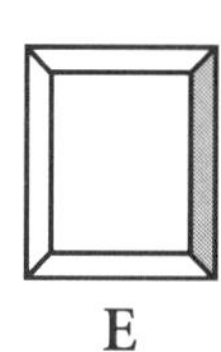

3. List any of the shapes that are topologically equivalent to A. _______________

4. List any of the shapes that are topologically equivalent to B. _______________

5. List any of the shapes that are topologically equivalent to C. _______________

Refer to the simple closed curve for Exercises 6 and 7.

6. Is point A on the inside or outside of the curve?

____

7. Is point B on the inside or outside of the curve?

____

Verify Euler's formula for each polyhron below.

8. 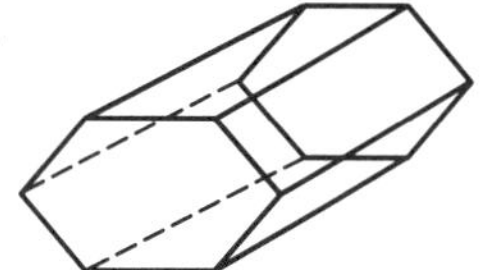

____

9. 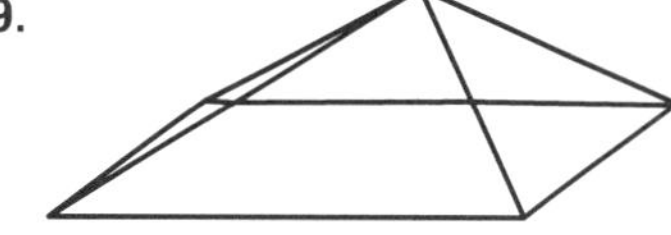

____

NAME ______________________ CLASS ____________ DATE ____________

Practice Masters Level C

11.4 Topology

Determine the number of regions into which the plane is divided.

1.

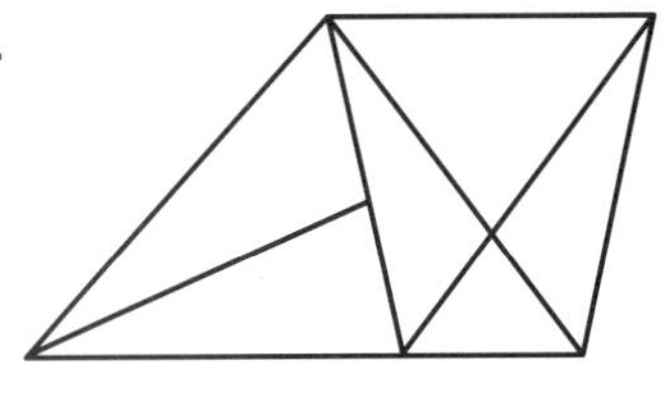

____

2.

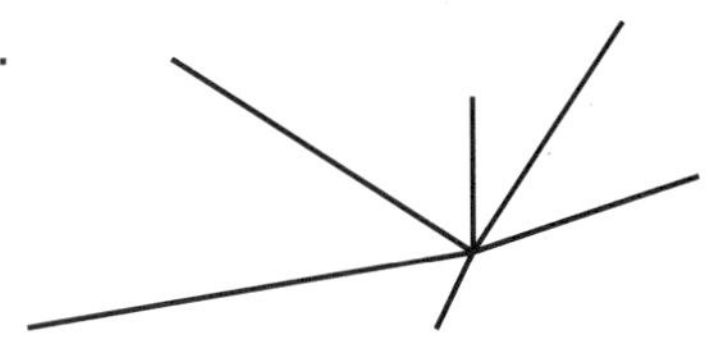

____

Use the figures below for Exercises 3 and 4.

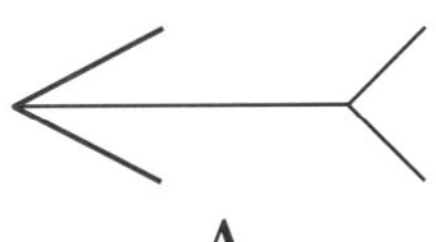

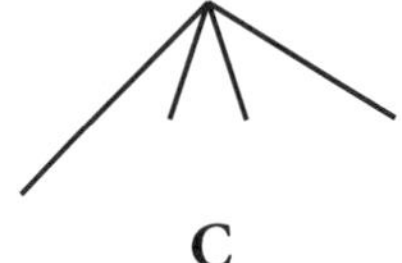

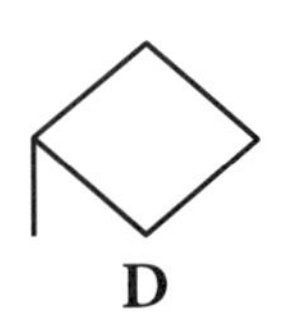

A B C D E

3. List the shapes that are topologically equivalent to A. ______________

4. List the shapes that are topologically equivalent to E. ______________

For Exercises 5–9, consider the alphabet in capital block letter form:

A B C D E F G H I J K L M N O P Q R S T U V W X Y Z

5. Which of the letters are topologically equivalent to **A**? ______________

6. Which of the letters are topologically equivalent to **I**? ______________

7. Which one letter is topologically equivalent to **O**? ______________

8. Letters **F** and **G** are topologically equivalent. Which other three letters are topologically equivalent to **F** and **G**? ______________

9. Look at the letters that are left. Determine which are topologically equivalent to each other. ______________

10. Decide whether points *A*, *B*, and *C* are inside or outside of the simple closed curve.

A ______________

B ______________

C ______________

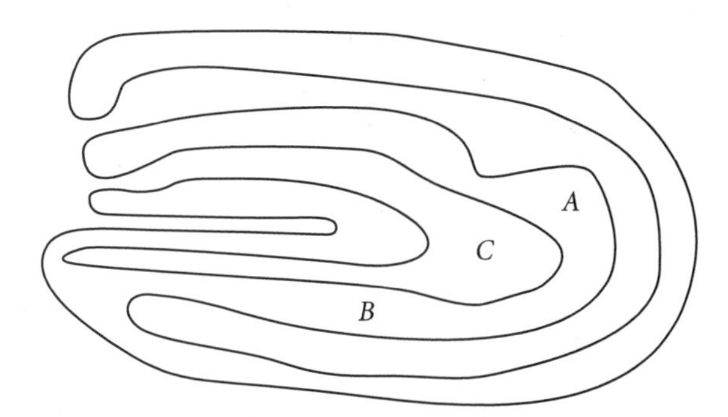

NAME ______________________ CLASS __________ DATE __________

Practice Masters Level A

11.5 Euclid Unparalleled

For Exercises 1–6, refer to the figure that shows $\overleftrightarrow{AB}$ and point *C* on the surface of Poincare's model of hyperbolic geometry.

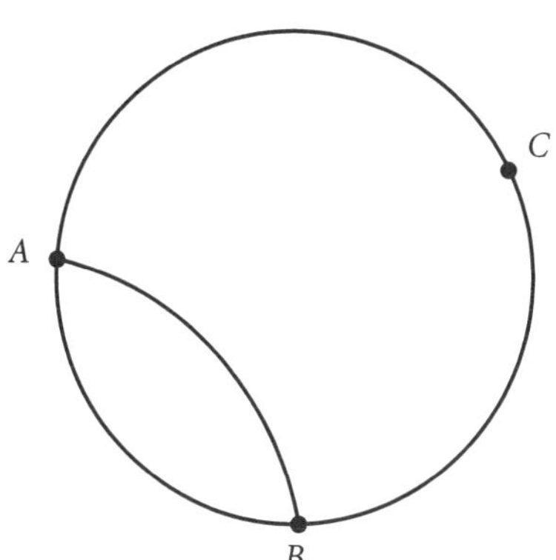

1. Describe a line in Poincare's model. Is it infinite or measurable? ______________________

2. How can you draw another Poincare segment, $\overline{CD}$, through point C that is congruent to $\overline{AB}$?

____

3. What is the least number of lines needed to form a polygon?

____

4. What is the least number of lines needed to form a quadrilateral? ______________________

5. Explain how you could create a parallelogram whose opposite sides are both parallel and congruent. ______________________

____

6. Does a parallelogram whose opposite sides are parallel necessarily have equal opposite sides? ______________________

For Exercises 7–11, refer to the figure that shows line *l* on the surface of a sphere, Riemann's model of spherical geometry.

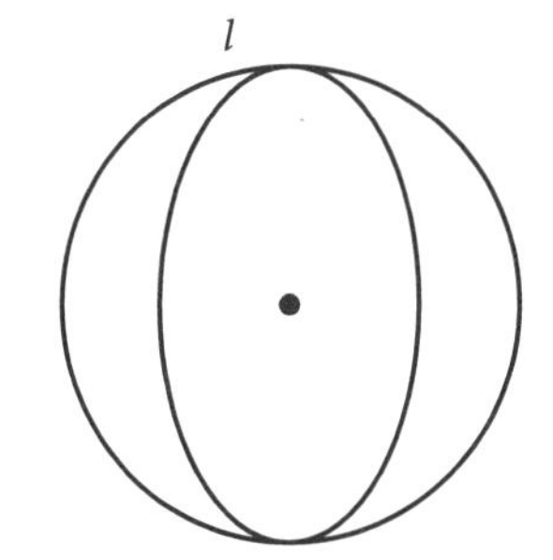

7. Describe a line in Riemann's model. Is it infinite or measurable? ______________________

8. What is the least number of sides of a polygon in the Riemann's model? ______________________

9. What is the greatest sum of the angles of a triangle in the Riemann model? ______________________

10. Do all Riemann quadrilaterals have the same interior angle sum? ______________________

11. The diameter of a great circle of a sphere is 18 centimeters. What is the radius of the sphere? ______________________

NAME ______________________ CLASS ____________ DATE ____________

Practice Masters Level B

11.5 Euclid Unparalleled

For Exercises 1–6, refer to the figure that shows $\overleftrightarrow{AB}$ and $\overleftrightarrow{CD}$ on the surface of Poincare's model of hyperbolic geometry.

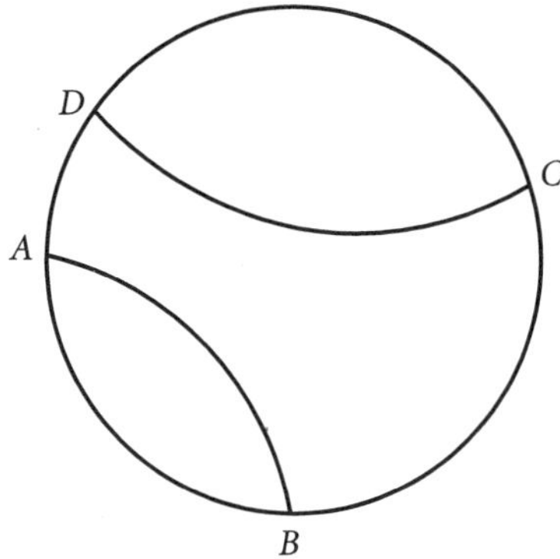

1. Compare $\overleftrightarrow{AB}$ and $\overleftrightarrow{CD}$. How can you verify that this is true?

   ____

   ____

2. The closer the vertices of a polygon are to ____________ of Poincare's model, the more it resembles a Euclidean polygon.

3. The closer the vertices of a polygon are to ____________ of Poincare's model, the less it resembles a Euclidean polygon.

4. Is a straight line possible in Poincare's model? ______________________

5. What would be the first step in drawing $\overleftrightarrow{BC}$? ______________________

6. What is the maximum whole-number sum of the angles of a triangle in Poincare's model? ______________________

For Exercises 7–8, refer to the figure that shows line *l* on the surface of a sphere, Riemann's model of spherical geometry.

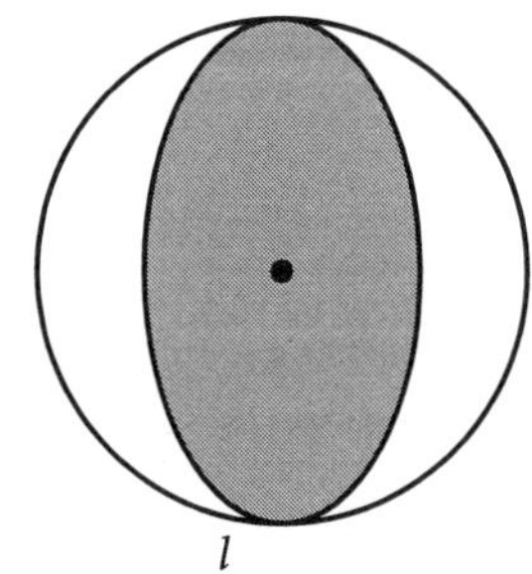

7. In Riemann's model, the closer the vertices of a polygon are to ____________________, the more it resembles a Euclidean polygon.

8. In Riemann's model, the farther away the vertices of a polygon are from ____________________, the less it resembles a Euclidean polygon.

9. What is the least whole-number sum of the angles of a triangle in the Riemann model? ______________________

10. What is the least whole-number sum of the angles of a quadrilateral in the Riemann model? ______________________

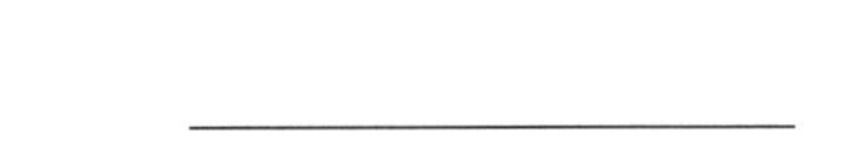

11. There can be no parallelograms in a Riemann model because ____________________ do not exist in the Riemann model.

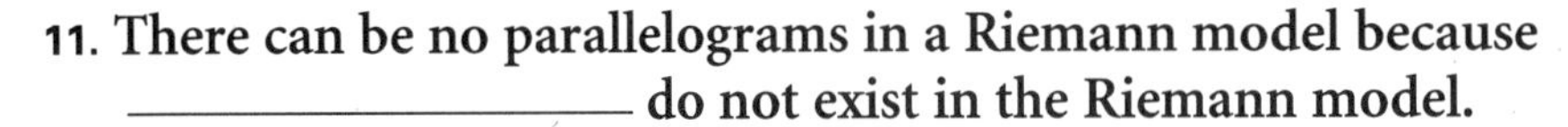

NAME ______________________ CLASS __________ DATE __________

Practice Masters Level C

11.5 Euclid Unparalleled

For Exercises 1–5, refer to the figure that shows $\overleftrightarrow{AB}$ on the surface of Poincare's model of hyperbolic geometry. Write answers to the nearest hundredth, if necessary.

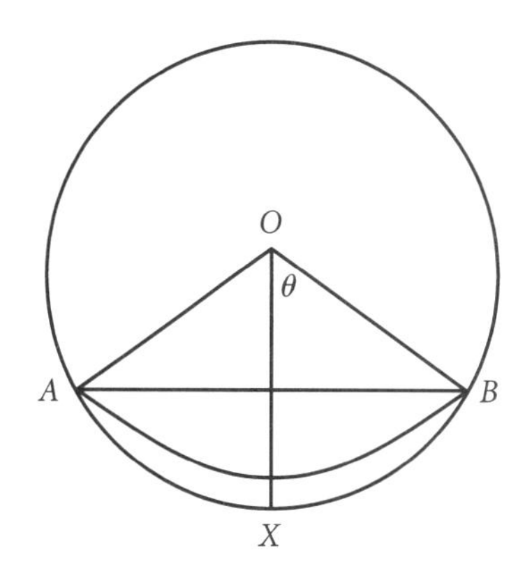

1. Can the center of any arc that is a Poincare line lie inside the Poincare plane?

   ____

2. The radius of a circle is 10 centimeters. Points A and B are infinitely close to the edge of Poincare's plane. Euclidean $AB = 12$ centimeters. Find Poincare AB.

   ____

3. The radius of a circle is 12 inches. Points A and B are infinitely close to the edge of Poincare's plane. Euclidean $AB = 20$ inches. Find Poincare AB. __________

4. The radius of a circle is 20 centimeters and $\theta = 75°$. Find Euclidean AB. Find Poincare AB. __________

5. Suppose you want Poincarean line CD to intersect Poincarean line AB, with points C and D infinitely close to the edge of the plane. Can both points C and D be on the major arc of AB? __________

For Exercises 6–9, refer to the figure showing Riemann's model. One great circle exists at its "Equator." Another great circle is perpendicular to it. Point *N* is at the "North Pole." Point *S* is at the "South Pole."

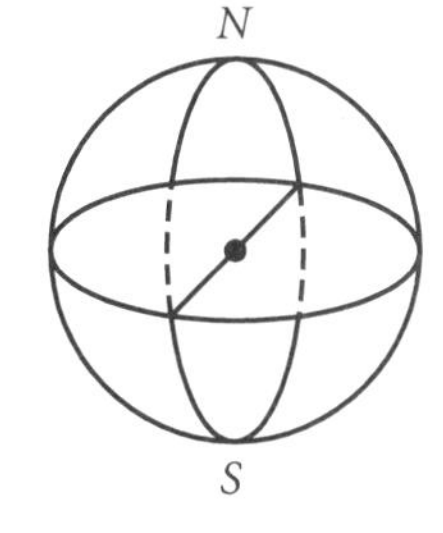

6. Points S and N are as far from the Equator as possible. How many Riemannian lines can pass through both S and N?

   ____

7. Describe the location of a point that has exactly one line passing through it perpendicular to the equator. __________

8. Is it true in Riemann's model that two lines perpendicular to the same line is parallel to that line? __________

9. Does the exterior angle theorem hold for a triangle in Riemann's model?

   ____

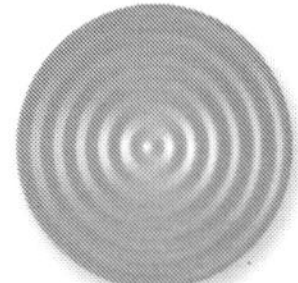

Practice Masters Level A

11.6 Fractal Geometry

Here are the first two levels in the construction of a fractal.

Level 0:

Level 1: Divide the segment into fourths. Bend the middle 2 fourths up to create the sides of a box. Add a top that is the same length as the other fourth.

1. In the space provided, complete two more levels of the fractal.

Level 2: ______________________

Level 3: ______________________

2. What do you notice is happening? ______________________

____

3. Write a description of the steps to follow to get from Level 1 to Level 2.

Level 1 Level 2

____

____

____

____

NAME ______________________________ CLASS ______________ DATE ______________

Practice Masters Level B

11.6 Fractal Geometry

Refer to Pascal's triangle for Exercises 1–8.

1
1 1
1 2 1
1 3 3 1
1 4 6 4 1
1 5 10 10 5 1
1 6 15 20 15 6 1
1 7 21 35 35 21 7 1
1 8 28 56 70 56 28 8 1
1 9 36 84 126 126 84 36 9 1
1 10 45 120 210 252 210 120 45 10 1

Use a pencil with an eraser to help answer Exercises 1–3.

1. Circle all even numbers in Pascal's triangle. Describe the results.

____

2. Now, place an *X* on all numbers divisible by 3 in Pascal's triangle. Is the pattern of numbers divisible by 3 similar to the pattern of numbers divisible by 2? ______________

3. What is the greatest number that divides evenly into all the numbers having both a circle and an *X*? ______________

4. What pattern emerges for the numbers divisible by both 3 and 2? ______________

5. Name the first divisible number greater than 3 that would result in a similar pattern. ______________

6. In Row 8, a triangle of numbers divisible by 2 begins. Each side of the triangle is 7 "units" or numerical entries long. It overlaps a triangle of numbers divisible by 3. In what row does that triangle start? How many units is the length of each of its sides? ______________

7. In what row does the first triangle of numbers divisible by 6 begin? How many units is the length of each side? ______________

8. What would happen if you erased all numbers divisible by 1? ______________

NAME ______________________ CLASS ____________ DATE ____________

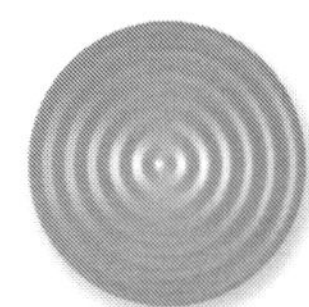

Practice Masters Level C

11.6 Fractal Geometry

Use the cube at the right for Exercises 1–3.

1. Find the surface area and the volume of the cube.

   Surface area: ____________ Volume: ____________

2. Multiply the sides of the cube by 3. Find the surface area and volume of the new cube.

   Surface area: ____________ Volume: ____________

3. As the surface area of a Euclidean cube increases, the

   volume ____________.

10 cm

This sponge cube has a hole in the shape of a rectangular prism. Use this cube for Exercises 4–10.

10 cm, 2 cm, 2 cm, 10 cm

4. Is the surface area of this figure less than or greater than the surface area of the cube above? Is the volume less than or greater than the volume of the cube above?

   Surface area: ____________ Volume: ____________

5. Find the surface area and the volume of the cube with the hole.

   Surface area: ____________ Volume: ____________

6. Suppose you create a second hole at the center of the top face that passes through to the bottom face of the cube. Describe what you remove and include the dimensions. ____________

7. What is the surface area and volume of the cube now that it has two holes?

   Surface area: ____________ Volume: ____________

8. A third hole is created at the center of the side face and passes through to the opposite face. Find the new surface area and volume.

   Surface area: ____________ Volume: ____________

9. Consider the three holes as one iteration. As the surface area of a sponge cube increases, the volume ____________.

10. Compare your answers to Exercises 3 and 9. Are they the same? ____________

Practice Masters Level A

11.7 Other Transformations and Projective Geometry

For Exercises 1 and 2, sketch the preimage and image for each affine transformation on the given coordinate plane.

1. triangle: $W(3, 2)$; $X(1, 5)$; $Y(0, 1)$

   $T(x, y) = (2x, -y)$

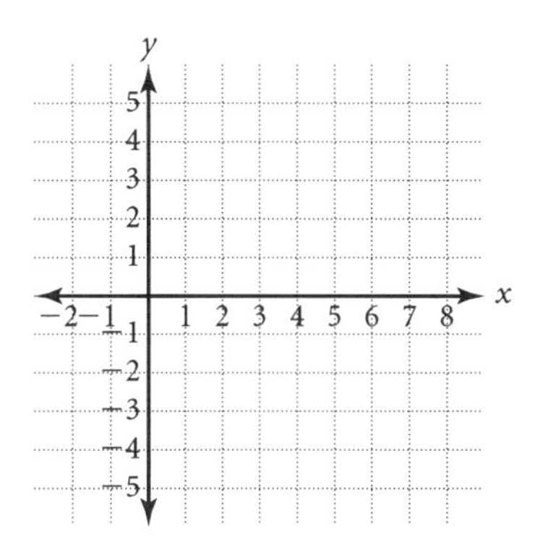

2. square: $J(1, 4)$; $K(3, 4)$; $L(3, 6)$; $M(1, 6)$

   $T(x, y) = (-2x, \frac{y}{2})$

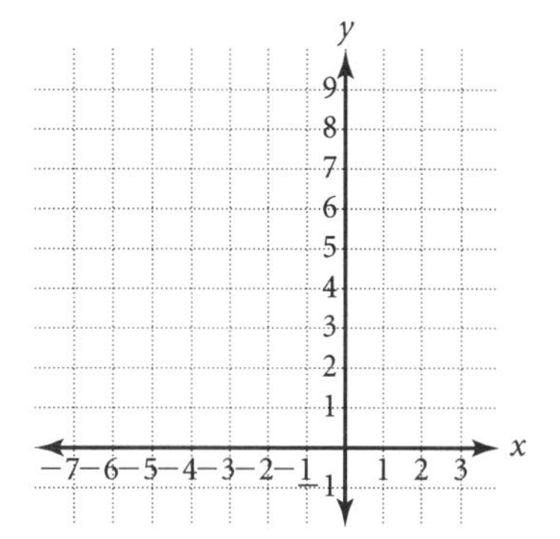

For Exercises 3–8, use the figure at the right.

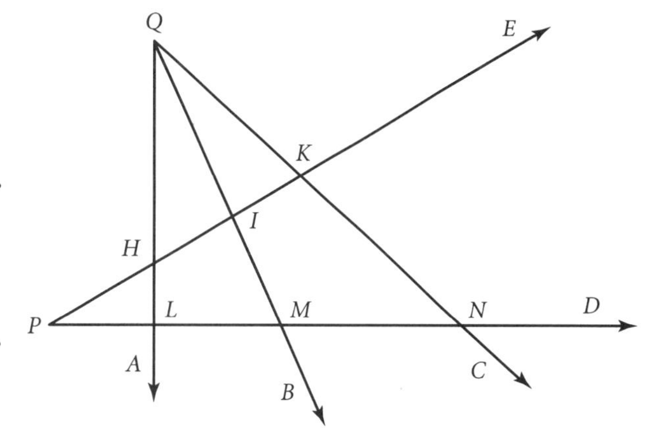

If point P is the center of projection, then:

3. the projective rays are ______________.

4. the projection of L onto $\overrightarrow{QB}$ is ______________.

5. the projection of I onto $\overrightarrow{QC}$ is ______________.

If point Q is the center of projection, then:

6. the projective rays are ______________.

7. the projection of I onto $\overrightarrow{PD}$ is ______________.

8. the projection of H onto $\overrightarrow{PD}$ is ______________.

Practice Masters Level B

11.7 Other Transformations and Projective Geometry

Use the graph of *ABCD* for Exercises 1–4.

1. Write the coordinates for the image of *ABCD* after the transformation $T(x, y)=(-\frac{x}{2}, -y)$, Then find the ratio of the areas of *ABCD* and its image.

   Image ______________________

   Area ratio ______________________

2. Write the coordinates for the image of *ABCD* after the transformation $T(x, y) = (2x, 3y)$. Then find the ratio of the areas of *ABCD* and its image.

   Image ______________________

   Area ratio ______________________

3. Write the coordinates for the image of *ABCD* after the transformation $T(x, y) = (-x, 2y)$. Then find the ratio of the areas of *ABCD* and its image.

   Image ______________________ Area ratio ______________

4. Write the coordinates for the image of *ABCD* after the transformation $T(x, y) = (ax, by)$. Then find the ratio of the areas of *ABCD* and its image.

   Image ______________________ Area ratio ______________

Use the graph of $\triangle ABC$ for Exercises 5–7.

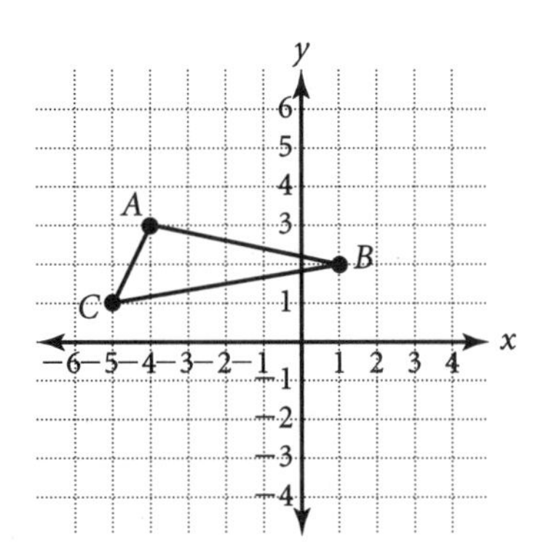

5. If $\triangle ABC$ is projected onto the graph of $y = 6$ from the point (0, 0), what will be the coordinates of its image?

   ____

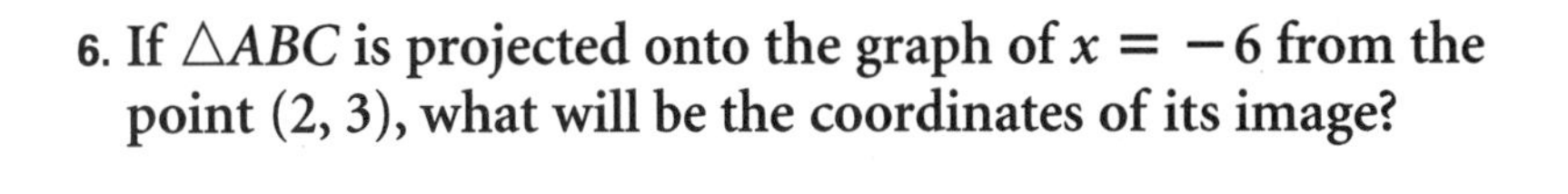

6. If $\triangle ABC$ is projected onto the graph of $x = -6$ from the point (2, 3), what will be the coordinates of its image?

   ____

7. Suppose $\triangle ABC$ is projected onto the graph of $y = 2x - 4$, and its image is $A'(5, 6)$, $B'(3, 2)$, and $C'(1, -2)$. Name the coordinates of the center of projection. ______________________

NAME ______________________ CLASS ____________ DATE ____________

Practice Masters Level C

11.7 Other Transformations and Projective Geometry

The endpoints of the base of a triangle are (1, 1) and (1, 5). The area of the triangle is 6 square units. Use this information for Exercises 1–4.

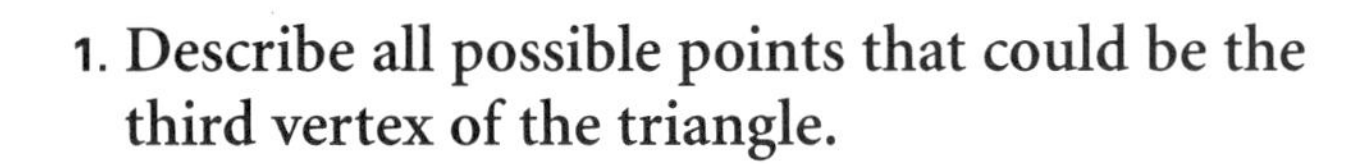

1. Describe all possible points that could be the third vertex of the triangle.

   ____

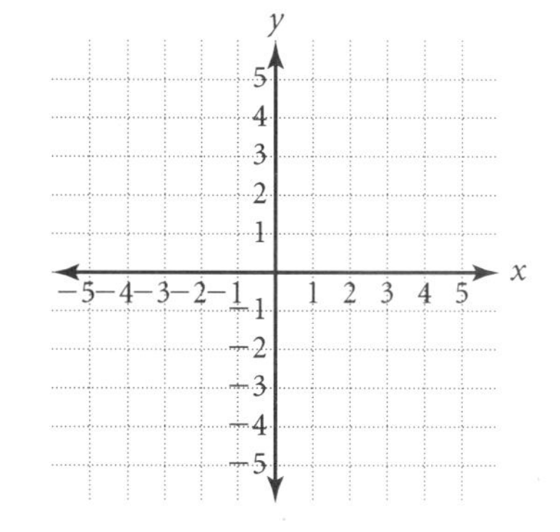

2. What are the endpoints of the image triangle's base after the transformation $T(x, y) = (3x, -2y)$?

   ____

3. Describe all possible points that could be the third vertex of the triangle's image after the transformation in Exercise 2.

   ____

4. What is the area of the new triangle? ____________

Points *A*, *B*, and *C* were projected from a point *O* onto the graph of the equation $y = -x + 7$. The image points were then projected onto $x = 10$ from $P(-2, 3)$, resulting in $A''(10, 9)$, $B''(10, 3)$ and $C''(10, 0)$. Use this information for Exercises 5–10.

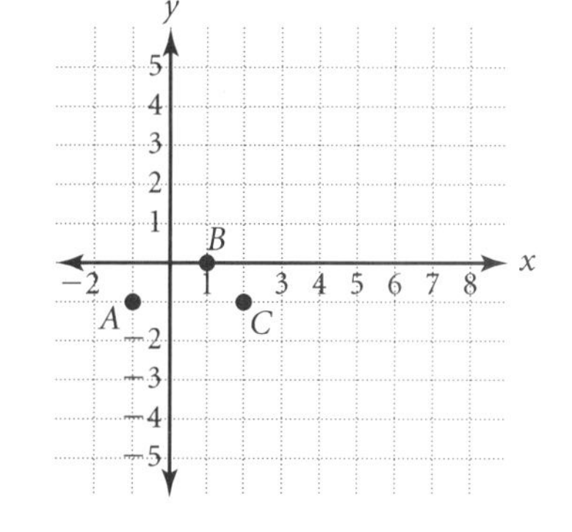

5. Name the coordinate pairs of the first image points *A*, *B*, and *C*.

   ____

6. Name the coordinate pair for *O*. ____________

7. Name the coordinate pair for the intersection of AA'' and BB''. ____________

8. Name the coordinate pair for the intersection of AA'' and CC''. ____________

9. Name the coordinate pair for the intersection of BB'' and CC''. ____________

10. The points A', B', and C' cannot be the first image points of *A*, *B*, and *C* because their projective rays do not ____________.

NAME ______________________ CLASS __________ DATE __________

Practice Masters Level A

12.1 Truth and Validity in Logical Arguments

In Exercises 1–3, write a valid conclusion from the given premises. Identify the form of the argument.

1. If today is Friday, then today is the last day of John's work week. Today is Friday.

   ____

2. If a quadrilateral is a square, then the quadrilateral has four right angles. *ABCD* does not have four right angles.

   ____

3. If you get three strikes in baseball, then you're out. Jose did not get out.

   ____

In Exercises 4–7, you are given the following premises:

If a person lives in Illinois, then that person lives in the U.S.A.
Abe lives in Illinois.
Barbara lives in the U.S.A.
Carol does not live in Illinois
David does not live in the U.S.A.

Which of the following conclusions are valid? Give the traditional name of the form of the argument.

4. Abe lives in the U.S.A. ______________________
5. Barbara lives in Illinois. ______________________
6. Carol does not live in the U.S.A. ______________________
7. David does not live in Illinois. ______________________

In Exercises 8–11, you are given the following premises:

If a number is 2, then the square of the number is 4.
$w = 2 \quad x \neq 2 \quad y^2 = 4 \quad z^2 \neq 4$

Which of the following conclusions are valid? Give an example to show that each conclusion that is *not* valid may in fact be false, even when the premises are true.

8. $w^2 = 4$ ______________________
9. $x^2 \neq 4$ ______________________
10. $y = 2$ ______________________
11. $z \neq 2$ ______________________

NAME ______________________ CLASS __________ DATE __________

Practice Masters Level B

12.1 Truth and Validity in Logical Arguments

In Exercises 1–3, write a valid conclusion from the given premises. Identify the form of the argument.

1. If Sally studies for the test, then she will do well. Sally studies for the test.

____

2. If a triangle is equilateral, then it is isosceles. Triangle *ABC* is not isosceles.

____

3. If I sleep through my alarm, then I will be running late. If I'm running late, then I'll miss the bus. If I miss the bus, then I'll be late for school. I sleep through my alarm.

____

In Exercises 4 and 5, use two of the given premises to write a valid conclusion. Identify the form of the argument you used.

4. If a quadrilateral is a square, then the quadrilateral has four congruent sides. *ABCD* has four congruent sides. *DEFG* is a square.

____

5. If I go to the grocery superstore today, then I will buy milk. I bought milk. I did not buy milk.

____

In Exercises 6–9, you are given the following premises:

If a number is divisible by 4, then the number is divisible by 2.
a is divisible by 4.
b is divisible by 2
c is not divisible by 4.
d is not divisible by 2.

Which of the following conclusions are valid? Give an example to show that each conclusion that is *not* valid may in fact be false, even when the premises are true.

6. a is divisible by 2. ______________________________

7. *b* is divisible by 4. ______________________________

8. *c* is not divisible by 2. ______________________________

9. *d* is not divisible by 4. ______________________________

NAME ______________________ CLASS ____________ DATE ____________

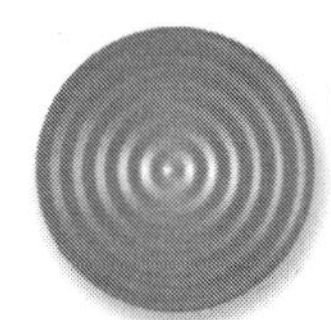

Practice Masters Level C

12.1 Truth and Validity in Logical Arguments

In Exercises 1 and 2, write a valid conclusion from the given premises. If no valid conclusion can be drawn, write *no valid conclusion.*

1. If two triangles are congruent, then they are similar. $\triangle ABC \sim \triangle DEF$

____

2. If the measure of an angle is 30°, then the sine value for that angle is 0.5. $\sin \theta = 0.5$.

____

Use two of the given premises to write a valid conclusion. Identify the form of the argument you used.

3. If today is Wednesday, then the cafeteria is serving pizza. The cafeteria is serving pizza. The cafeteria is not serving pizza.

____

In Exercises 4–7, you are given the following premises:

If a person wins the district 100 meter race, then that person is a good runner.
Abby did not win the district 100 meter race.
Bob is a good runner.
Cathy won the district 100 meter race.
David is not a good runner.

Which of the following conclusions are valid? Give the traditional name of the form of the argument.

4. Abby is not a good runner. ______________________
5. Bob won the district 100 meter race. ______________________
6. Cathy is a good runner. ______________________
7. David did not win the district 100 meter race. ______________________

In Exercises 8–11, you are given the following premises:

If a triangle is equilateral, then it is isosceles.
$\triangle ABC$ is equilateral.
$\triangle DEF$ is isosceles.
$\triangle GHI$ is not equilateral.
$\triangle JKL$ is not isosceles.

Which of the following conclusions are valid? Give an example to show that each of the conclusions that is not valid may in fact be false, even when the premises are true.

8. $\triangle ABC$ is isosceles. ______________
9. $\triangle DEF$ is equilateral. ______________
10. $\triangle GHI$ is not isosceles. ______________
11. $\triangle JKL$ is not equilateral. ______________

NAME ______________________ CLASS ____________ DATE ____________

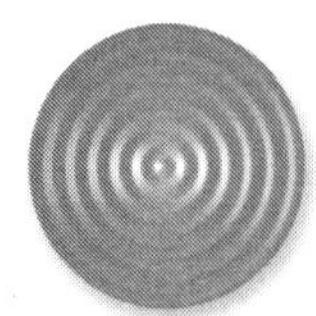

Practice Masters Level A

12.2 And, Or, and Not in Logical Arguments

In Exercises 1–3, indicate whether each compound statement is true or false. Explain your reasoning.

1. All squares are rectangles and all triangles are isosceles.

____

2. All squares are rectangles or all triangles are isosceles.

____

3. Fish can swim or birds can sing.

____

In Exercises 4 and 5, write a conjunction for each pair of statements. Determine whether the conjunction is true or false.

4. All cats were once kittens. Blue is a color. ______________________

5. Six is a prime number. Five divides into 16 evenly. ______________________

In Exercises 6 and 7, write a disjunction for each pair of statements. Determine whether each disjunction is true or false.

6. Elephants can fly. Dogs can bite. ______________________

7. Corn is a fruit. Apples grow on vines. ______________________

In Exercises 8–10, write the statement expressed by the symbols, where *p, q, r,* and *s* represent the statements shown below.

p: Dudley is a muggle. $\quad$ *q:* $1 + 1 = 3$

r: $x + 3 > 4$ $\quad$ *s:* $x < 0$

8. $\sim s$ ______________________

9. p OR $\sim q$ ______________________

10. $\sim (r \text{ AND } s)$ ______________________

11. Complete the following truth table:

p	q	$\sim p$	$\sim p$ OR q
T	T		
T	F		
F	T		
F	F		

NAME ______________________ CLASS __________ DATE __________

Practice Masters Level B

12.2 And, Or, and Not in Logical Arguments

In Exercises 1–3, indicate whether each compound statement is true or false. Explain your reasoning.

1. Three is a prime number and four is a perfect square.

____

2. Three is a prime number or four is a perfect square.

____

3. Fish are mammals or bats are reptiles.

____

In Exercises 4 and 5, write a conjunction for each pair of statements. Determine whether the conjunction is true or false.

4. All paper is white. All pen ink is blue. ______________________________

5. Two is greater than three. $5 + 6 = 11$ ______________________________

In Exercises 6 and 7, write a disjunction for each pair of statements. Determine whether each disjunction is true or false.

6. Two is greater than three. $5 + 6 = 11$ ______________________________

7. Six is an even integer. Six divides into 18. ______________________________

In Exercises 8–10, write the statement expressed by the symbols, where *p, q, r,* and *s* represent the statements shown below.

p: Dudley is a muggle. q: $1 + 1 = 2$

r: $x + 5 < 4$ s: $x > 0$

8. $\sim s$ ______________________________

9. p OR q ______________________________

10. $\sim$ (r AND s) ______________________________

11. Complete the following truth table:

p	q	$\sim q$	p OR $\sim q$
T	T		
T	F		
F	T		
F	F		

NAME ______________________ CLASS ____________ DATE ____________

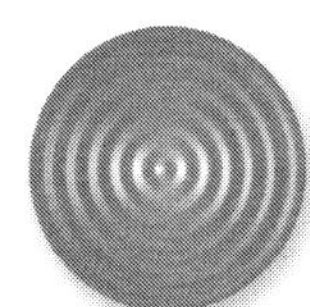

Practice Masters Level C

12.2 And, Or, and Not in Logical Arguments

In Exercises 1–3, indicate whether each compound statement is true or false. Explain your reasoning.

1. Friday follows Thursday and all weeks have 8 days.

____

2. Friday follows Thursday or all weeks have 8 days.

____

3. Five is a prime number or two is an even integer.

____

In Exercises 4 and 5, write a conjunction for each pair of statements. Determine whether the conjunction is true or false.

4. Every square is a rhombus. All triangles have three sides. ____________

5. For all numbers x, $x^2 > 0$. $2 \cdot 5 = 10$ ____________

In Exercises 6 and 7, write a disjunction for each pair of statements. Determine whether each disjunction is true or false.

6. Three is greater than two. $5 + 4 = 9$ ____________

7. Bears hibernate in winter. Snakes are mammals. ____________

In Exercises 8–10, write the statement expressed by the symbols, where *p, q, r,* and *s* represent the statements shown below.

p: Steve likes to play baseball.

q: $x > 4$

r: $1 + 6 = 5$

s: $x + 3 < 5$

8. $\sim q$ ____________

9. p OR $(\sim r)$ ____________

10. $\sim (q \text{ AND } s)$ ____________

11. Complete the following truth table and the statement below:

p	q	p OR q	$\sim(p$ OR $q)$	$(\sim p)$ AND $(\sim q)$
T	T			
T	F			
F	T			
F	F			

$\sim (p$ OR $q)$ is truth functionally equivalent to ____________.

NAME ______________________ CLASS ______________ DATE ______________

Practice Masters Level A

12.3 A Closer Look at If-Then Statements

For each conditional in Exercises 1-3, explain why it is true or false. Then write the converse, inverse, and contrapositive, and explain why each is true or false.

1. Conditional: If two angles are vertical angles, then the two angles are congruent. ______________________

   Converse: ______________________

   Inverse: ______________________

   Contrapositive: ______________________

2. Conditional: For numbers a, b, and c, if $a + c < b + c$, then $a < b$. ______________________

   Converse: ______________________

   Inverse: ______________________

   Contrapositive: ______________________

3. Conditional: If $1 + 1 = 2$, then elephants can fly. ______________________

   Converse: ______________________

   Inverse: ______________________

   Contrapositive: ______________________

For Exercises 4–6, write each statement in if-then form.

4. When the car runs out of gas, it will stop. ______________________
5. All puppies are cute. ______________________
6. We'll go to the park if the weather is nice. ______________________

NAME ______________________ CLASS ____________ DATE ____________

Practice Masters Level B

12.3 A Closer Look at If-Then Statements

For each conditional in Exercises 1-3, explain why it is true or false. Then write the converse, inverse, and contrapositive, and explain why each is true or false.

1. Conditional: For numbers a, b, and c, if $a > b$, then $ac > bc$. ______________________

   Converse: ______________________

   Inverse: ______________________

   Contrapositive: ______________________

2. Conditional: If $\angle A \cong \angle B$, then $\sin A = \sin B$. ______________________

   Converse: ______________________

   Inverse: ______________________

   Contrapositive: ______________________

3. Conditional: If $2 = 3$, then $5 = 6$. ______________________

   Converse: ______________________

   Inverse: ______________________

   Contrapositive: ______________________

For Exercises 4–6, write each statement in if-then form.

4. I'll pay you when you finish the job. ______________________
5. All lizards are reptiles. ______________________
6. Following this diet will help me lose weight. ______________________

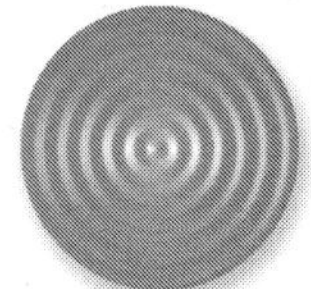

Practice Masters Level C

12.3 A Closer Look at If-Then Statements

For each conditional in Exercises 1–3, explain why it is true or false. Then write the converse, inverse, and contrapositive, and explain why each is true or false.

1. Conditional: If two angles are complementary angles, then they are both acute. ______________________

   Converse: ______________________

   Inverse: ______________________

   Contrapositive: ______________________

2. Conditional: If $AX = XB$, then X is the midpoint of segment AB. ______________________

   Converse: ______________________

   Inverse: ______________________

   Contrapositive: ______________________

3. Conditional: If a tail is a leg, then horses have 5 legs. ______________________

   Converse: ______________________

   Inverse: ______________________

   Contrapositive: ______________________

For Exercises 4–6, write each statement in if-then form.

4. All desserts are sweet. ______________________
5. We'll play baseball if it doesn't rain. ______________________
6. Numbers divisible by five end in either 5 or 0. ______________________

NAME ______________________ CLASS ____________ DATE ____________

Practice Masters Level A

12.4 Indirect Proof

In Exercises 1–4, form a contradiction by using each statement and its negation.

1. $\triangle ABC$ is an obtuse triangle. ______________________

2. All integers are even. ______________________

3. $x > 3$ ______________________

4. Two lines meet in exactly one point. ______________________

In Exercises 5–7, determine whether the given argument is an example of indirect reasoning. Explain why or why not.

5. If Shannon had been in Florida, then she'd have a good suntan. But Shannon does not have a good tan. Therefore, she has not been in Florida.

   ____

6. If Joey misses his bus, then he is late to school. Joey is late to school, so Joey must have missed his bus.

   ____

7. If Chris falls asleep in class, he will surely fail the next test. Chris does not fall asleep in class. Therefore, Chris will not fail the next test.

   ____

Complete the indirect proof below.

Prove: $2(5 + 6x) \neq 4(3x + 2)$ for any real number x.

Indirect Proof: Suppose that $2(5 + 6x) = 4(3x + 2)$ for some real number x.

8. Then $10 + 12x = 12x + 8$ by the ____________ property of algebra.

9. And then $10 = 8$ by the ______________________.

   But $10 = 8$ is a false statement, a contradiction.

10. Therefore, the opposite of the assumption is true, that is ______________________.

NAME ______________________ CLASS ____________ DATE ____________

Practice Masters Level B

12.4 Indirect Proof

In Exercises 1–4, form a contradiction by using each statement and its negation.

1. $x < 2$ ______________________

2. $a = 5$ ______________________

3. Some parties are fun. ______________________

4. All people are honest. ______________________

In Exercises 5 and 6, determine whether the given argument is an example of indirect reasoning. Explain why or why not.

5. If Charlie works hard all day, he'll complete the painting of the fence. Charlie did not work hard all day, so he must not have finished painting the fence.

____

6. If the neighbors were planning a cookout today, they would have their grill out by now. Their grill is not out, so they must not be having a cookout.

____

Complete the indirect proof below.

Theorem: The hypotenuse of a right triangle is its longest side.

Given: Right triangle *ABC*, with side lengths $a > 0$, $b > 0$, and $c > 0$. The length of the hypotenuse is *c*.

Prove: $c > b$. (A similar argument could be used to prove $c > a$.)

Indirect Proof: Suppose that $c \not> b$, that is, that $c \leq b$.

7. From algebra, for positive numbers, squaring preserves inequality, so ____________.

8. Now $a^2 + b^2 = c^2$ by the ______________________,

9. so $a^2 + b^2 \leq b^2$, using ______________________.

10. Subtracting b^2 from both sides gives $a^2 \leq 0$, which is a contradiction of: ____________

So, the assumption that $c \leq b$ is false; it leads to a contradiction. That is, $c > b$ is true.

NAME ______________________ CLASS ____________ DATE ____________

Practice Masters Level C

12.4 Indirect Proof

In Exercises 1–4, form a contradiction by using each statement and its negation.

1. $2x + 1 = 3$ ______________________
2. $y^2 > 0$ ______________________
3. Some people have blue eyes. ______________________
4. All squares are rectangles. ______________________

In Exercises 5 and 6, determine whether the given argument is an example of indirect reasoning. Explain why or why not.

5. If those dark clouds meant a severe thunderstorm, the warning siren would have sounded. The warning siren has not sounded, so there will not be a severe thunderstorm.

____

6. If Keisha plays soft music while she studies, she can concentrate better. Keisha was not able to play soft music while studying at a friend's house, so she couldn't concentrate.

____

Complete the indirect proof below.

Uniqueness of Perpendiculars Theorem: From a point outside a line, there is only one line perpendicular to the given line.

Given: Line l with point P not on l. $\overline{PX} \perp$ at point X on l.

Prove: There is no other point Y on l such that $\overline{PY} \perp l$.

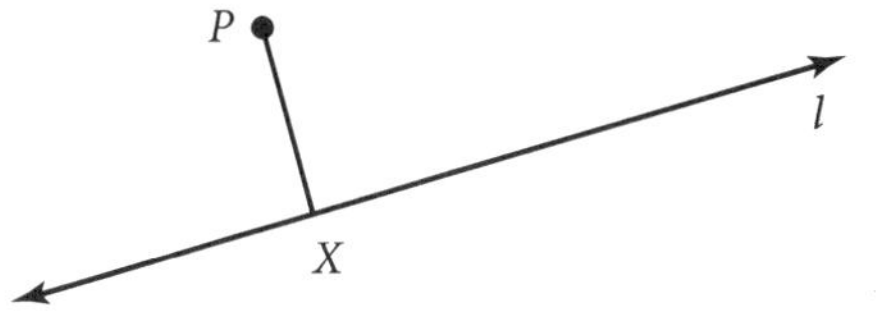

7. **Indirect Proof:** Suppose

______________________.

(Hint: Start by assuming the negation of what is to be proven. Then continue drawing logical conclusions until you arrive at a contradiction. Be sure you add your assumed line segment to the figure above.)

____

____

____

NAME ______________________ CLASS ______________ DATE ______________

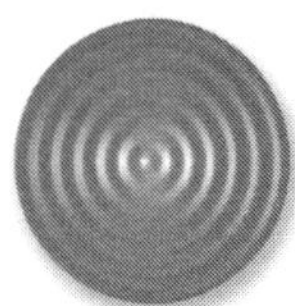

Practice Masters Level A

12.5 Computer Logic

In Exercises 1 and 2, create a logical expression that corresponds to each network.

1\.

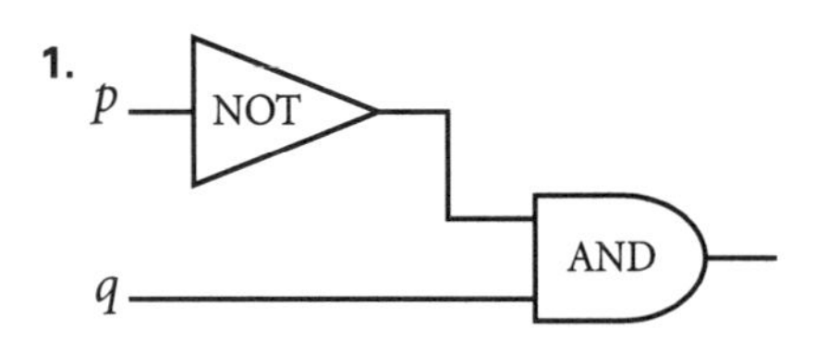

____

2\. p q OR NOT

____

In Exercises 3 and 4, construct a network of logic gates for each expression.

3\. p OR (NOT q)

____

4\. NOT (p AND q)

____

In Exercises 5–16, complete the input-output table for each network of logical gates.

	p	q	NOT p	NOT p AND q
5.	1	1		
6.	1	0		
7.	0	1		
8.	0	0		

p NOT q AND

	p	q	r	p OR q	(p OR q) OR r
9.	1	1	1		
10.	1	1	0		
11.	1	0	1		
12.	1	0	0		
13.	0	1	1		
14.	0	1	0		
15.	0	0	1		
16.	0	0	0		

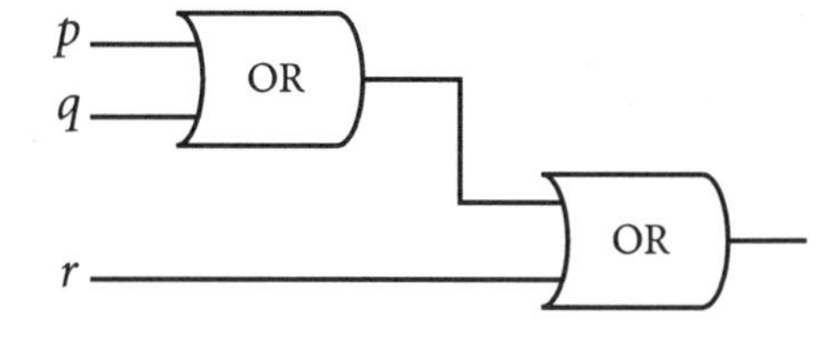

NAME ______________________ CLASS ____________ DATE ____________

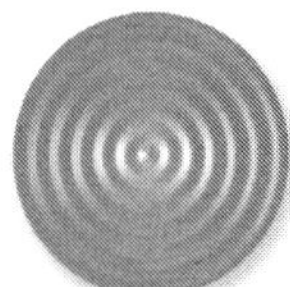

Practice Masters Level B

12.5 Computer Logic

In Exercises 1 and 2, create a logical expression that corresponds to each network.

1. 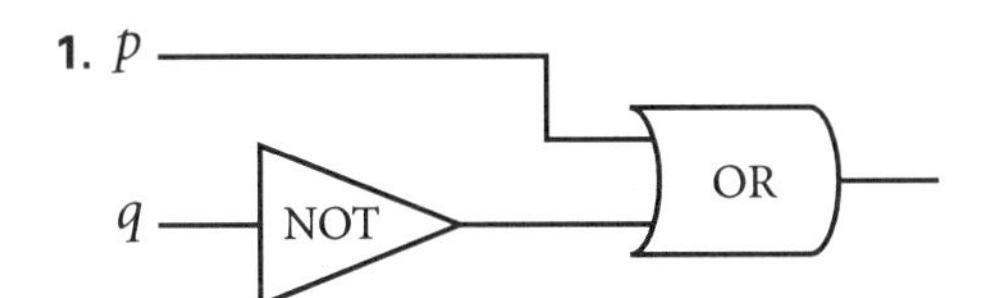

____

2. 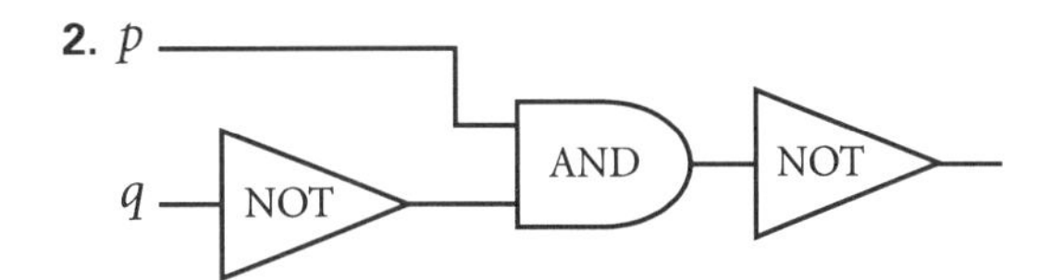

____

In Exercises 3 and 4, construct a network of logic gates for each expression.

3. (NOT p) AND q

____

4. NOT (NOT p AND q)

____

In Exercises 5–16, complete the input-output table for each network of logical gates.

	p	q	NOT p	NOT q	(NOT p) AND (NOT q)
5.	1	1			
6.	1	0			
7.	0	1			
8.	0	0			

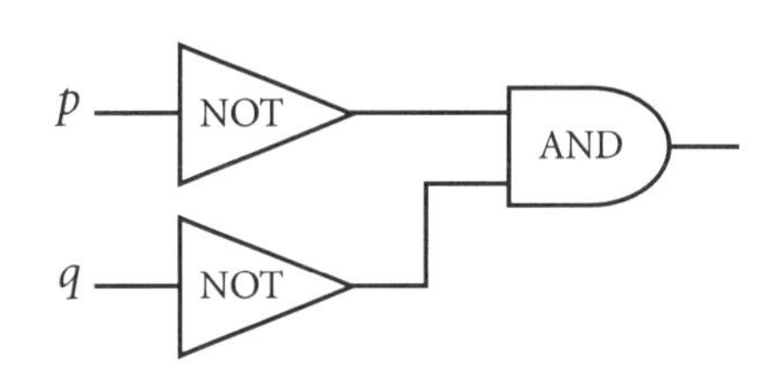

	p	q	r	p OR q	NOT r	(p OR q) AND NOT r
9.	1	1	1			
10.	1	1	0			
11.	1	0	1			
12.	1	0	0			
13.	0	1	1			
14.	0	1	0			
15.	0	0	1			
16.	0	0	0			

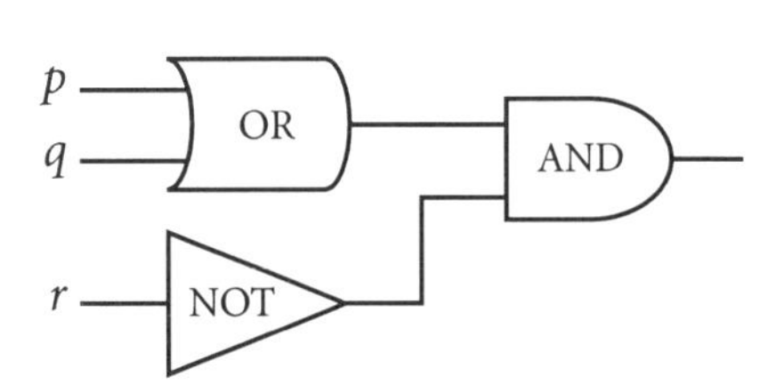

Practice Masters Level C

12.5 Computer Logic

1. How many sequences of 1s and 0s are possible as outputs of a four-line input-output table? ____________

The input-output tables for the three basic logic gates are as follows: Use these tables for Exercises 2–15.

p	NOT p
1	0
0	1

p	q	p OR q
1	1	1
1	0	1
0	1	1
0	0	0

p	q	p AND q
1	1	1
1	0	0
0	1	0
0	0	0

In Exercises 2–5, create a network of logic gates that corresponds to each input-output table.

2.

p	q	???
1	1	0
1	0	0
0	1	0
0	0	1

____

3.

p	q	???
1	1	1
1	0	0
0	1	1
0	0	1

____

4.

p	q	???
1	1	0
1	0	1
0	1	0
0	0	0

____

5.

p	q	???
1	1	1
1	0	1
0	1	0
0	0	0

____

For Exercises 6–15, list the other ten possible four-line input-output tables.

6.

p	q	
1	1	
1	0	
0	1	
0	0	

7.

p	q	
1	1	
1	0	
0	1	
0	0	

8.

p	q	
1	1	
1	0	
0	1	
0	0	

9.

p	q	
1	1	
1	0	
0	1	
0	0	

10.

p	q	
1	1	
1	0	
0	1	
0	0	

11.

p	q	
1	1	
1	0	
0	1	
0	0	

12.

p	q	
1	1	
1	0	
0	1	
0	0	

13.

p	q	
1	1	
1	0	
0	1	
0	0	

14.

p	q	
1	1	
1	0	
0	1	
0	0	

15.

p	q	
1	1	
1	0	
0	1	
0	0	

Answers

Lesson 1.1
Level A

1. $\overline{AB}$, $\overline{BC}$, $\overline{CD}$, $\overline{AD}$
2. $\overrightarrow{DC}$, $\overrightarrow{DA}$
3. $\angle 1$, $\angle BAD$, $\angle DAB$; $\angle 2$, $\angle ABC$, $\angle CBA$; $\angle 3$, $\angle BCD$, $\angle DCB$; $\angle 4$, $\angle ADC$, $\angle CDA$
4. drawing of a ray with endpoint at Q
5. drawing of an angle with vertex at A
6. segment with endpoints R and S
7. ray ("line" is acceptable)
8. plane
9. point
10. line
11. A, C, B
12. E
13. False; Postulates are statements that are accepted as true without proof.
14. False; Two lines intersect only at one point.

Lesson 1.1
Level B

1. L
2. $\angle JIG$, $\angle GIK$, $\angle HIK$, $\angle JIH$
3. G, I, and H, or J, I, and K
4. one
5. False; A line contains an infinite number of points.
6. True
7. True
8. answers may vary: sample answer—the tip of your pencil
9. answers may vary: sample answer—the intersection of the wall and the ceiling in your classroom
10. Check student's drawings.
11. Check student's drawings.
12. Check student's drawings.
13. $\overline{PO}$, $\overline{PI}$, $\overline{PN}$, $\overline{PT}$, $\overline{OI}$, $\overline{ON}$, $\overline{OT}$, $\overline{IN}$, $\overline{IT}$, $\overline{NT}$
14. False
15. 8, 2 from each of O, I, N, and 1 from each of P and T.

Lesson 1.1
Level C

1. 15
2. A, C, B
3. vertex
4. B and V, S and B
5. two
6. true
7. S
8. Check student's drawings.
9. Check student's drawings.
10. False, a line is considered undefined.
11. False, $\overleftrightarrow{AB}$ is the same as $\overleftrightarrow{AC}$.
12. False, three planes can intersect at 0, 1, 2, or 3 lines.
13. False, three points are needed to name a plane.
14. True, if two points are in a plane, then the segment that contains those points must also be in the plane.

Answers

Lesson 1.2
Level A

1. 3
2. 7
3. 3
4. Sample answer:
   A plotted at -3 and B plotted at 1
5. 17;

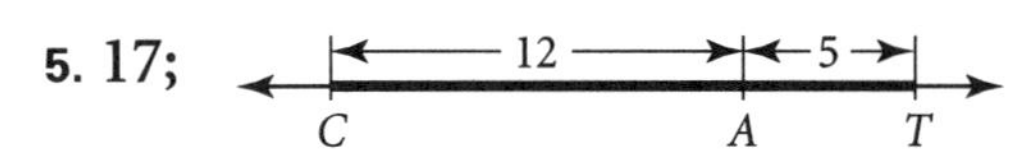

6. 2.5;

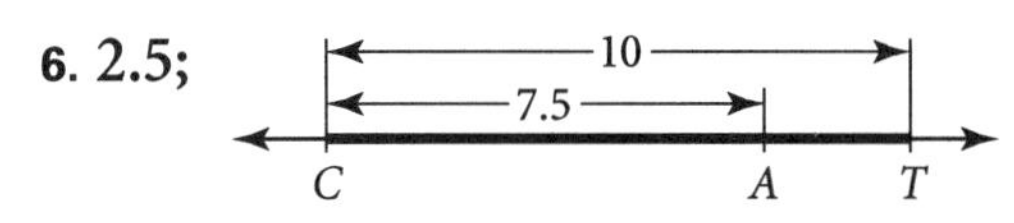

7. 0.7;

9.4
8.7
C A T

8. $\overline{FA} \cong \overline{FB} \cong \overline{FC} \cong \overline{FD} \cong \overline{FE}$
   $\overline{AB} \cong \overline{BC} \cong \overline{CD} \cong \overline{DE} \cong \overline{AE}$
9. 90 miles
10. 2.7
11. 5.4

Lesson 1.2
Lewvel B

1. 12
2. 7
3. $R = -15$ or $R = -7$; Check student's plot.
4. $\overline{DA} \cong \overline{DC}, \overline{EA} \cong \overline{EC}, \overline{AB} \cong \overline{BC}$
5. 21
6. 8.8;

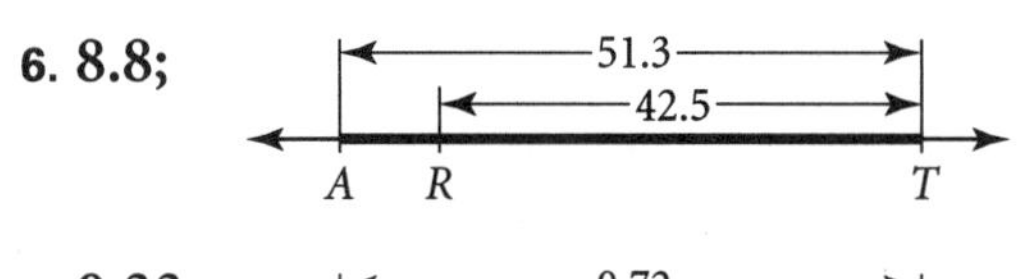

7. 0.33;

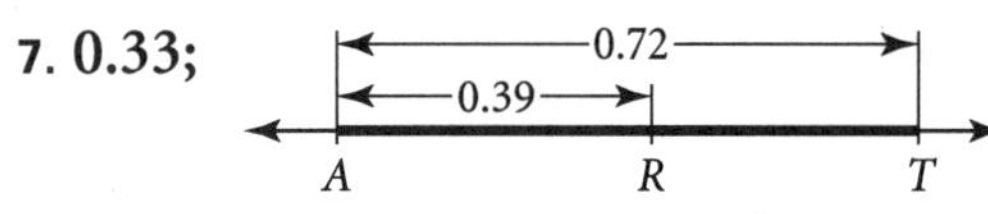

8. 214 miles
9. $x = 30$
10. $DC = 26$
11. $CE = 65$
12. $x = 7$
13. $CI = 28$
14. $IJ = 32$

Lesson 1.2
Level C

1. 57
2. 55
3. R could be located at 4 or -82; Check student's plot.
4. $\overline{CD} \cong \overline{EF} \cong \overline{GH}$
5. $\overline{EA} \cong \overline{EB} \cong \overline{EC} \cong \overline{ED}$,
   $\overline{AB} \cong \overline{BC} \cong \overline{CD} \cong \overline{AD}$
6. $15\frac{1}{3}$
7. 213.4 miles
8. $x = 12$
9. $CE = 21$
10. $CD = 53$
11. $x = 23$
12. $RS = 55$
13. $ST = 77$
14. $RT = 132$

Lesson 1.3
Level A

1. $\approx 30°$
2. $\approx 140°$
3. $\approx 90°$

Answers

4. Check student's drawings.

5. Check student's drawings.

6. False, if two angles are *supplementary*, they (sometimes) form a linear pair.

7. False, supplementary angles are only congruent if they are both right angles.

8. true

9. 62°

10. 146°

11. $\angle ZAX$ and $\angle XAR$, $\angle ZAP$ and $\angle PAR$, $\angle ZAQ$ and $\angle QAR$

12. $x = 43$

13. 58°

14. 32°

15. complementary

Lesson 1.3
Level B

1. ≈130°

2. ≈20°

3. ≈30°

4. ≈160°

5. They form a linear pair and are supplementary.

6. Sample drawing:

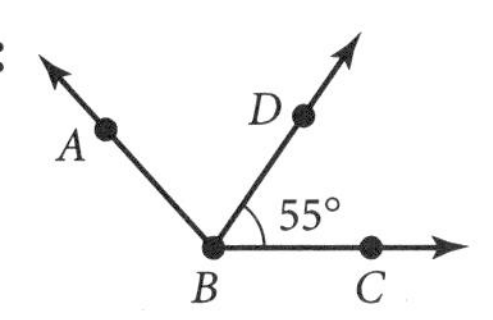

7. 64°

8. 116°

9. 45°

10. 45°

11. ≈110° (answers may vary)

12. 67.5°

13. $x = 6$

14. 105°

15. 27°

16. 132°

17. $x = 7$

18. 90°

19. 61°

20. 29°

Lesson 1.3
Level C

1. The drawing should be a quadrilateral.

2. 106°

3. 40°

4. 34°

5. 146°

6. 74°

7. 140°

8. 180°, 360°; Sample pattern: The sum of the exterior angles is double that of the sum of the interior angles.

9. $x = 11$

10. 49°

11. 131°

12. 180°

13. They form a linear pair and they are supplementary.

14. 37.5°

15. 78.75°

16. $x = 6$

17. 39°

18. 66°

19. 105°

20. 144°

Lesson 1.4
Level A

1. angle bisector
2. 74°
3. congruent
4. $x = 5$
5. $AC = 22$
6. $CB = 22$
7. $AB = 44$
8. Fold the paper once through A and B.
9. Fold the paper through A so that line l matches up with itself. This is line m. Fold the paper through B so that line l matches up with itself. This is line n.
10. Fold the paper so that A matches up with B. This is point M.
11. Fold along M so that l matches up with itself.
12. Line m is parallel to line t which is parallel to line n.

Lesson 1.4
Level B

1. $\overline{EG}$
2. False
3. $\overleftrightarrow{GE}$ is the perpendicular bisector of $\overline{AB}$.
4. $\overleftrightarrow{GE}$ is the angle bisector of $\angle AGB$.
5. $GA = GB$ since G lies on the perpendicular bisector of $\overline{AB}$.
6. $x = 8$
7. 39°
8. 78°
9. 33
10. 33
11. 66
12. folded segments RS, ST, and RT
13. The perpendicular bisectors intersect at the same point.
14. measurements will vary
15. measurements will vary, but should be about the same as Exercise 14
16. measurements will vary, but should be about the same as Exercise 14 and Exercise 15
17. The three measurements are equal.

Lesson 1.4
Level C

1. $\overline{DC} \cong \overline{DA}$, $\overline{BC} \cong \overline{BA}$, $\overline{EC} \cong \overline{EA}$
2. $\overline{DC} \cong \overline{DA}$ and $\overline{BC} \cong \overline{BA}$ because B and D lie on the perpendicular bisector of $\overline{AB}$; $\overline{EC} \cong \overline{EA}$ since $\overline{BD}$ is the perpendicular bisector of $\overline{AB}$.
3. $\angle CDB \cong \angle ADB$; $\angle CBD \cong \angle ABD$; all angles at E are congruent. Sample reason: The perpendicular bisector forms four right congruent angles, $\overline{DB}$ is the angle bisector for $\angle ADB$ and $\angle CDB$.
4. Construct $\overline{RS}$, then construct the perpendicular bisector of $\overline{RS}$. Mark point T on the perpendicular bisector, then fold over $\overline{RS}$ to find U.

Answers

5. $\overline{RU} \cong \overline{US} \cong \overline{ST} \cong \overline{TR}$

6. Since $x = 6$, $AC \neq BC$.

7. $x = 7.5$

8. 34°

9. 68°

10. The distances from a point on the angle bisector to the sides or the angle are equal.

11. $\overline{AT}$ is the perpendicular bisector of $\overline{MN}$.

Lesson 1.5
Level A

1.

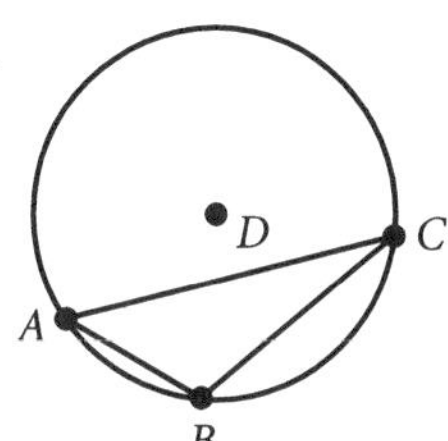

2.

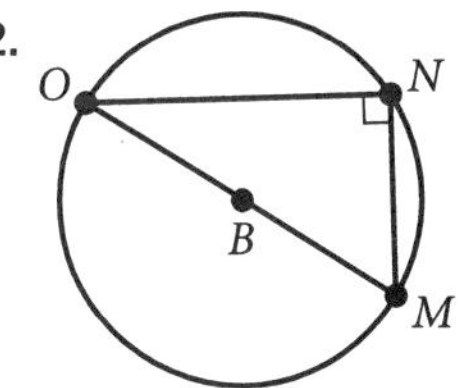

3.

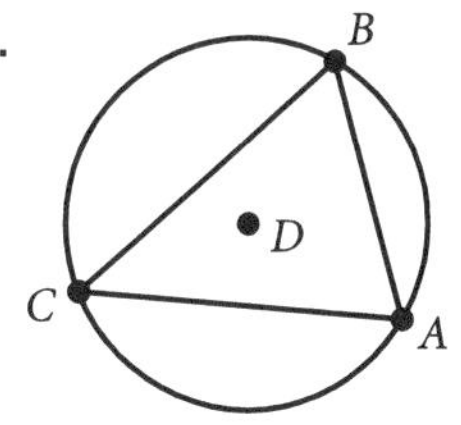

4. obtuse: outside the triangle
right: midpoint of the hypotenuse
acute: inside the triangle

5.

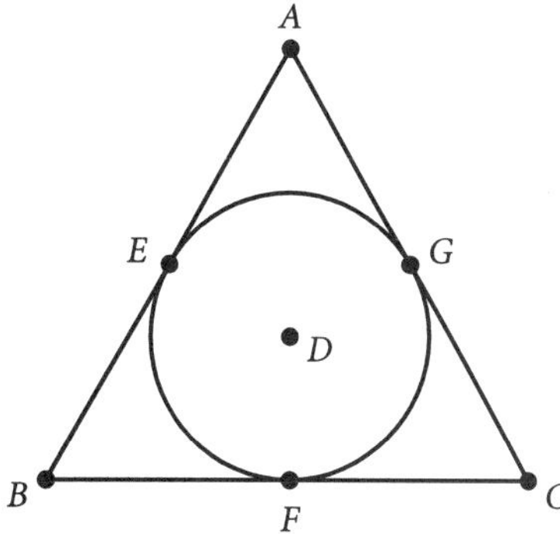

6.

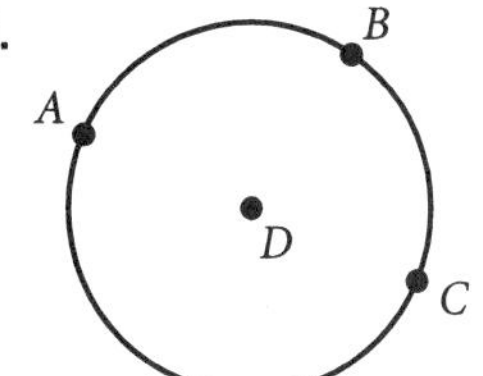

7. False; $\overline{AE}$ is a median, not an angle bisector.

8. True

9. False; it is the centroid of the triangle.

10. False; they are medians of the triangle.

11. True

Lesson 1.5
Level B

1. The drawing should be either an isosceles or equilateral triangle.

2. Check student's drawings.

3. The drawing should be an obtuse triangle.

4. The drawing should be either a right or an acute triangle.

5.

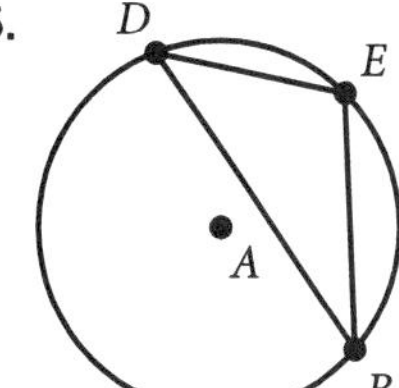

6. 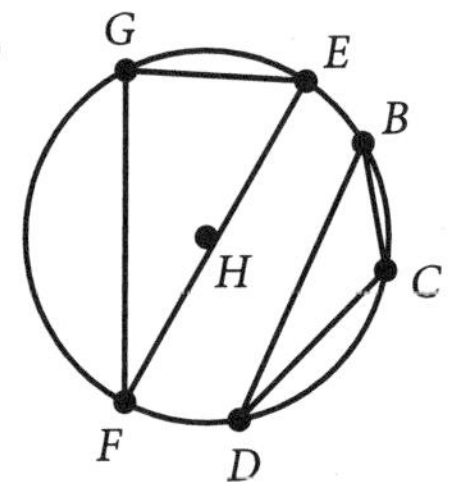

7. 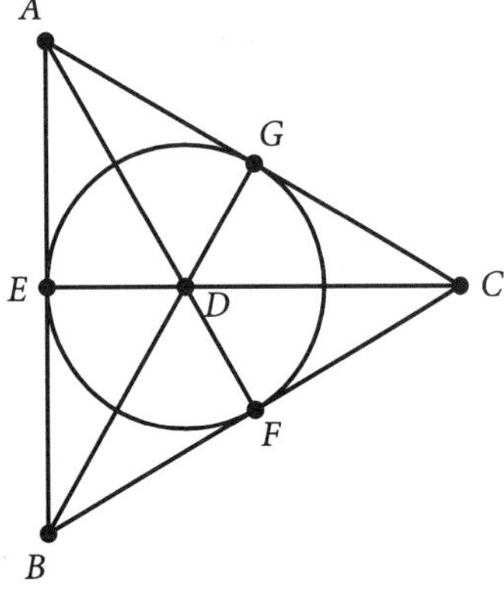

Lesson 1.5
Level C

1. 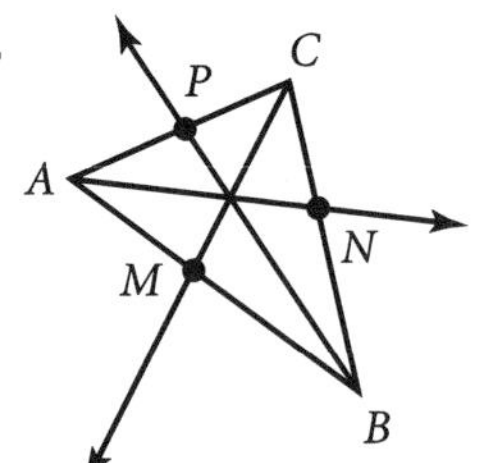

2. 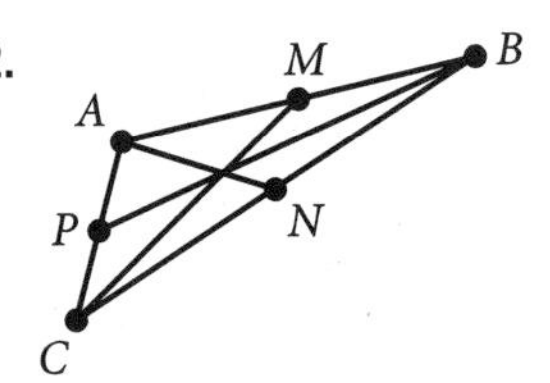

3. 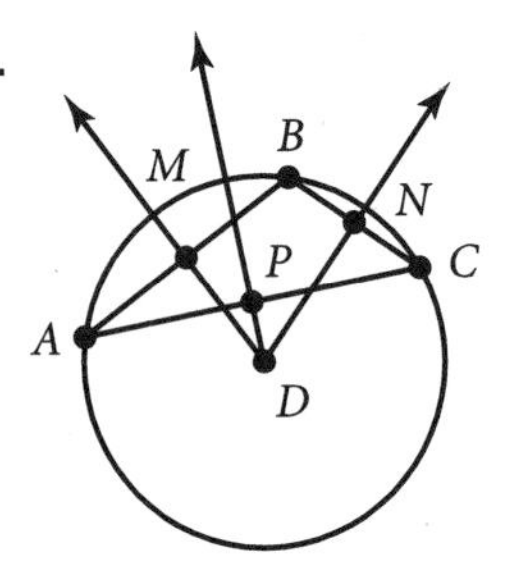

4. *Approximate* measurements:

	AM	*MB*	*BN*	*NC*	*CP*	*PA*
Fig. 1	1.4 cm	1.9 cm	1.8 cm	1.2 cm	1.0 cm	1.2 cm
Fig. 2	1.5 cm	1.5 cm	1.9 cm	1.9 cm	0.8 cm	0.8 cm
Fig. 3	1 cm	1 cm	0.7 cm	0.7 cm	1.4 cm	1.4 cm

5–7. *Approximate* measurements:

		$\frac{AM}{MB}$	$\frac{BN}{NC}$	$\frac{CP}{PA}$	$\frac{AM}{MB} \bullet \frac{BN}{NC} \bullet \frac{CP}{PA}$
5.	Figure 1	0.75	1.5	0.83	≈ 1
6.	Figure 2	1	1	1	1
7.	Figure 3	1	1	1	1

8. product of the ratios is close or equal to 1

9. always

10. always

Lesson 1.6
Level A

1. rotation
2. translation
3. reflection
4. translation
5. reflection
6. rotation
7. 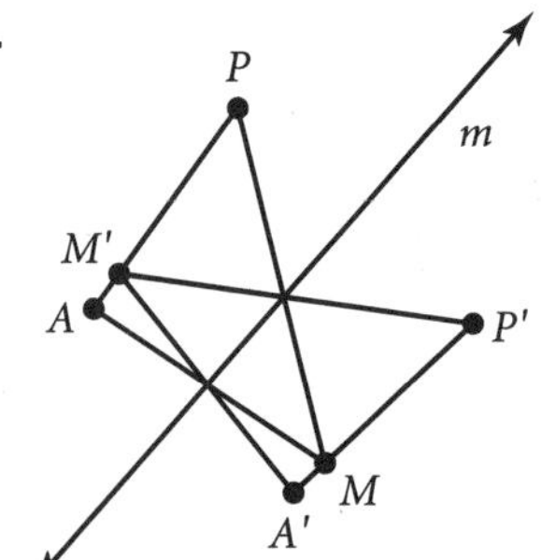

8. Drawing depends on angle of rotation used.

Answers

9.

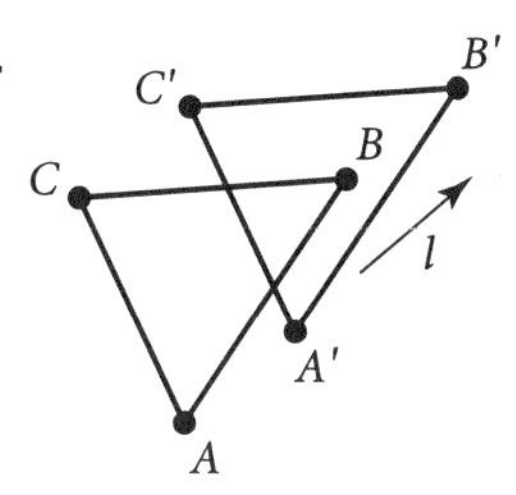

10. True

11. False; This is true for a reflection by a translation.

12. True

Lesson 1.6
Level B

1. Sample answer: the image you see in a mirror

2. Sample answer: the motion of a ferris wheel ride

3. Sample answer: a car as it moves down the highway

4. Sample answer: your foot prints in the sand as you walk down the beach

5.

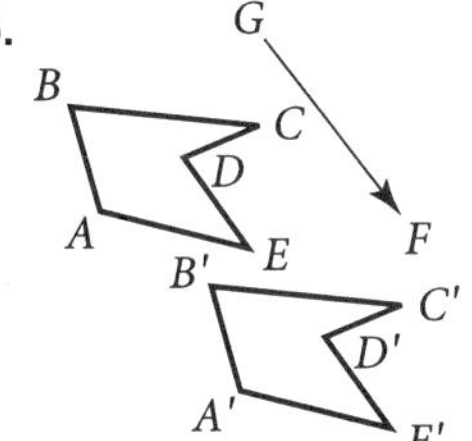

6.

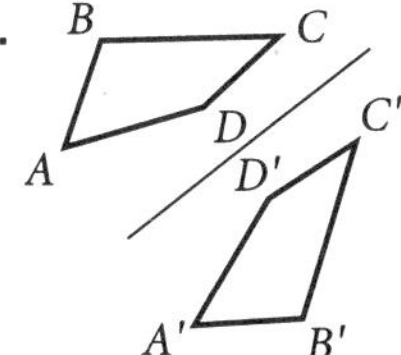

7. $\overline{AD} \cong \overline{A'D'}$

8. The line of reflection is the perpendicular bisector of $\overline{BB'}$.

9.

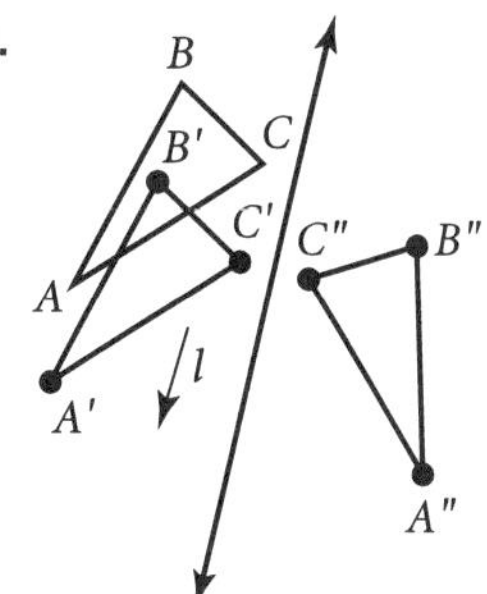

10. Drawings will very depending on angle of rotation the student used.

11. glide reflection.

Lesson 1.6
Level C

1–2.

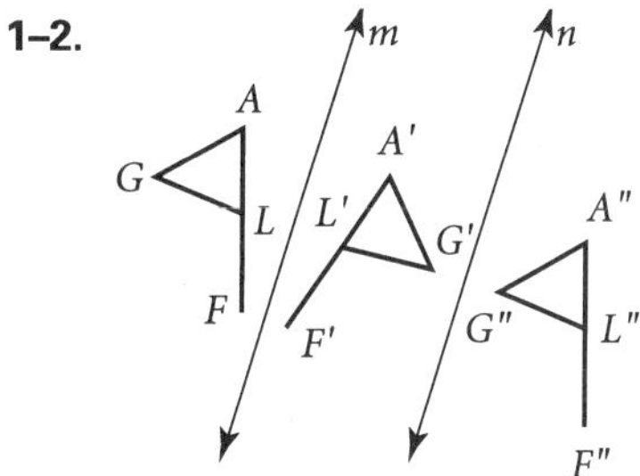

3. Translation

4. approximately 2.3cm

5. approximately 4.6cm

6. The distance between the image and the preimage should be twice the distance between the lines.

7.

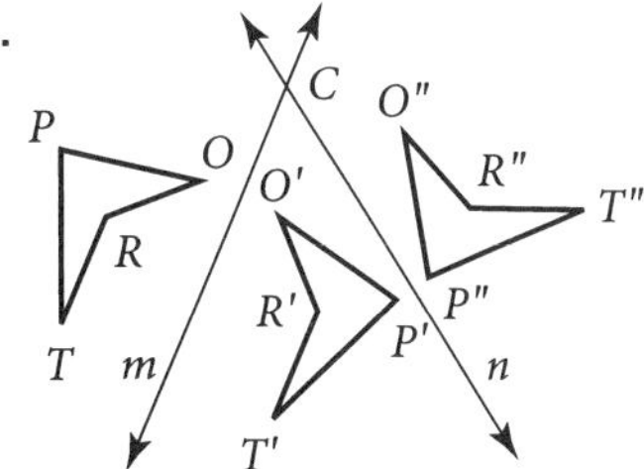

8. See answer for Exercise 7.

9. Rotation

10. approximately 110°

11. approximately 55°

12. The measure of the angle between corresponding points of the image and preimage should be double the measure of the acute angle formed by intersecting lines.

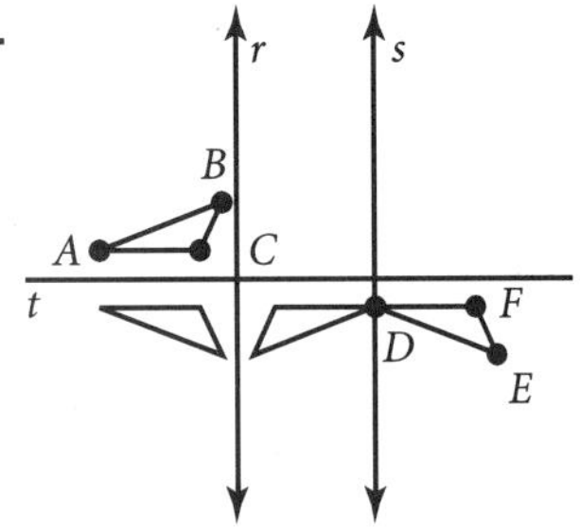

16. glide reflection

Lesson 1.7
Level A

1. down 3
2. left 6 and down 2
3. right 4 and down 12
4. left 7
5. -2
6. -2
7.

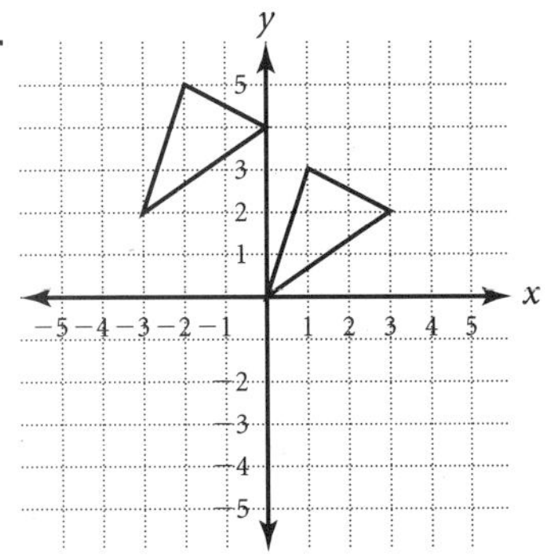

8.

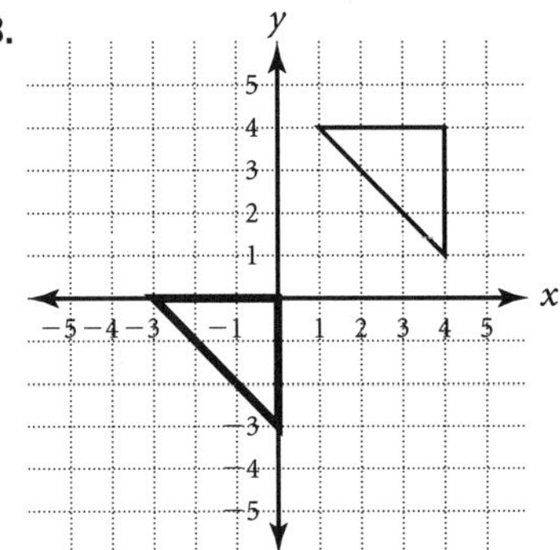

9. $T(x, y) = (x - 4, y - 2)$
10. $T(x, y) = (-y, -x)$
11. reflection (over the line $y = -x$)

Lesson 1.7
Level B

1.

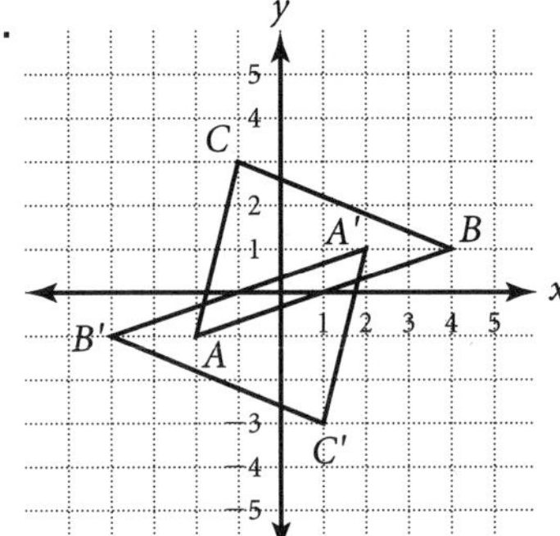

2.

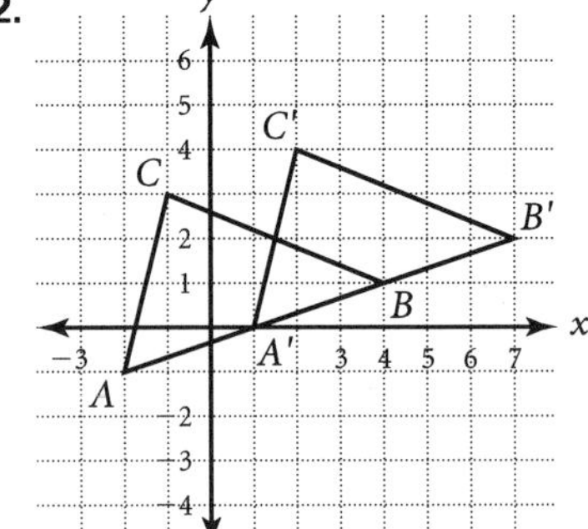

3. 180°
4. $m = \frac{1}{3}$
5. $P(x, y) = (-x, y)$
6. $P(x, y) = (x - 4, y + 3)$

Answers

7. $T(x, y) = (x + 3, y + 5)$

8. $(-3, 1), (-1, 1), (-1, 4)$

9. opposite y-coordinates

10.

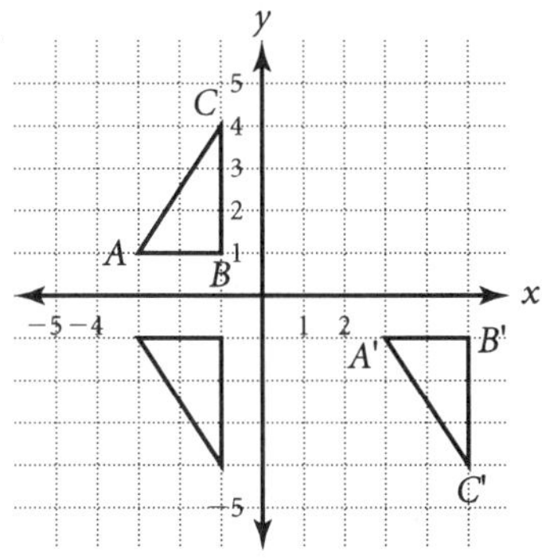

11. $T(x, y) = (x + 6, -y)$

Lesson 1.7
Level C

1.

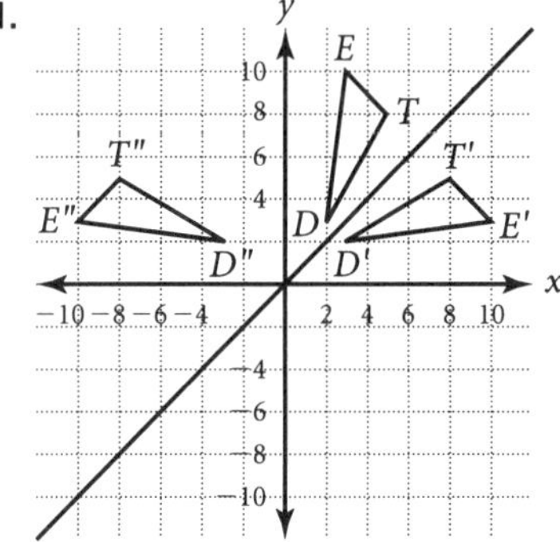

2. $R(x, y) = (y, x)$

3. See answer for Exercise 1.

4. They all measure 90°.

5. 45°

6. rotation of 90°

7. $T(x, y) = (-y, x)$

8. $A'(-1, 3), B'(-3, 1), C'(-5, 5)$

9.

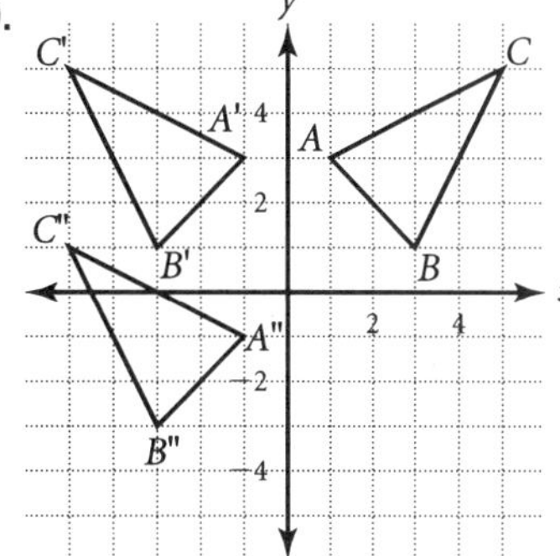

10. See graph in Exercise 9; $T(x, y) = (-x, y - 4)$

11. glide reflection

12. a glide reflection that reflects a figure over the x-axis and translates the image two units to the right

13. reflecting a figure over the line $y = -x$

14. rotating a figure 180°

15. reflecting a figure over the line $y = x$, then translating the image down 8 units

16. the identify transformation, no movement

Answers

Lesson 2.1
Level A

1–6.

Number of Sides	Number of Triangles
3	1
4	2
5	3
6	4
7	5
8	6
9	7

7. The number of triangles is two fewer than the number of sides.

8. 8

9. number of triangles $= n - 2$

10. Measurements will vary, but they should be equal. $AC \approx 2.4$ cm, $AC' \approx 2.4$ cm

11. The student's conjecture is correct. Explanations will vary. Sample answer: Because AC is congruent to AC'.

12. Conjectures will vary. Sample answer: The base angles of an isosceles triangle are congruent. To prove this conjecture, show base angles are congruent.

Lesson 2.1
Level B

1–4. Check student's drawings.

Number of Sides	Number of Diagonals
3	0
4	2
5	5
6	9

5. Descriptions may vary. Sample pattern: The difference in the number of diagonals from one figure to the next keeps increasing by one.

6. 35

7. Number of diagonals $= \frac{n(n-3)}{2}$

8. AX, BX, CX, DX, AC, BD, and all angles whose vertex is X. $AX = BX = CX = DX = 2.1$ cm; $AC = 4.2$ cm; $BD = 4.2$ cm; $\angle AXD = \angle BXC = 110°$; $\angle AXB = \angle DXC = 70°$

9. The student's conjecture is incorrect. The diagonals *are* bisectors of each other, but not perpendicular.

10. Conjectures and proofs will vary. Sample answer: The diagonals are congruent. Show that $\overline{AX}$ is congruent to $\overline{XC}$.

Lesson 2.1
Level C

1.

A	B	C	D
2	4	8	16
32	64	128	256
512	1024	2048	4096

2. The units digit in Column A is 2, column B is 4, column C is 8, and column D is 6.

3. Column C

4. To show that the statement will be true at all times.

5. Descriptions may vary. Sample answer: Construct the diagonals of the parallelogram and the point where they intersect. Measure the distances from the intersection point to each of the vertices of the parallelogram. If this statement is true, the distance on each diagonal should be equal.

6. 26

7. 27; Sample answer: Add one red cube to the 26 green cubes to get the total number of small cubes in this large cube.

Answers

8.

Layer #	Number of additional cubes needed	Total number of small cubes
1 (red)	—	1
2 (green)	26	27
3 (yellow)	98	125

9. 6 layers

10. 343 blue cubes

Lesson 2.2
Level A

1. If an animal is a dog, then it is a mammal.

2. Hypothesis: An animal is a dog. Conclusion: An animal is a mammal.

3. 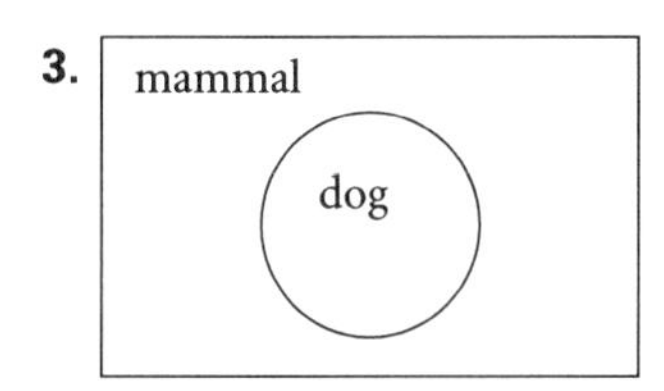

4. If an animal is a mammal, then it is a dog. False. Counterexample: cat

5. If m is the perpendicular bisector of $\overline{AB}$, then $PA = PB$.

6. If $PA = PB$, then m is the perpendicular bisector of $\overline{AB}$. True statement (as drawn); false if P lies on $\overline{AB}$.

7. If it rains on Saturday, then I won't get wet.

8. conditional statements

9. If p then q. $p \Rightarrow q$

Lesson 2.2
Level B

1. If a figure is a rectangle, then it is a parallelogram.

2. Hypothesis: A figure is a rectangular. Conclusion: It is a parallelogram.

3. 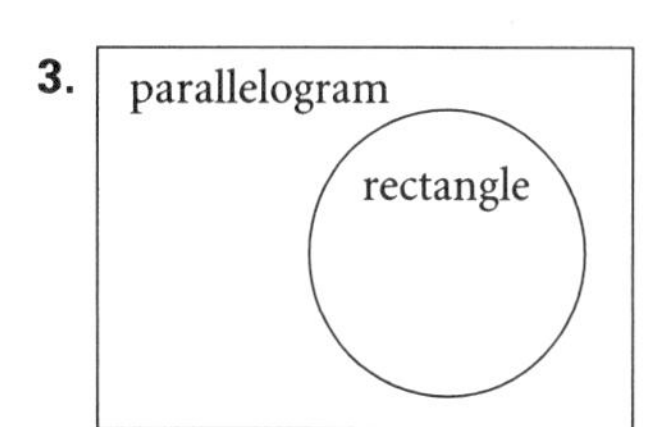

4. If a figure is a parallelogram, then it is a rectangle. False statement. Counterexample: any non-rectangular parallelogram.

5. If a and b are even integers, then $a + b$ is an even integer. True statement: since a and b are divisible by 2, their sum will be divisible by 2.

6. If $a + b$ is an even integer, then a and b are even integers. False statement; counterexample: a and b are any two odd integers.

7. You may use the car tonight.

8. Examples may vary. Explanation should include an "if" and a "then." Sample answer: If I pass this test, then my parents will be happy. Sample explanation: the conditional statements include an "if" and a "then".

Answers

Lesson 2.2
Level C

1. If $x^2 = 9$, then $x = 3$. It has an "if" and a "then." False statement; counterexample: $x = -3$.
2. If $x = 3$, then $x^2 = 9$. True statement.
3. If $MA = MB$, then M, A, and B are collinear. False statement; counterexample: $\angle AMB$
4. If M, A, and B are collinear, then $MA = MB$. False statement; counterexample: Any segment containing M, A, and B, in which M is not the midpoint.
5. No conclusion can be drawn from these statements.

Lesson 2.3
Level A

1. $\angle KXG$ and $\angle GXH$; $\angle GXH$ and $\angle HXI$; $\angle HXI$ and $\angle IXJ$.
2. 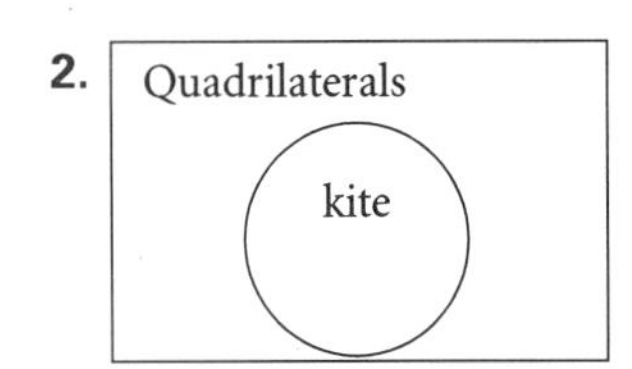

3. A and D are kites. (D is also a rhombus). Sample definition: A quadrilateral with two sets of adjacent sides congruent.
4. If a figure is a triangle, then it is formed by three segments.
5. If a figure is formed by three segments, then it is a triangle.
6. A figure is a triangle iff it is formed by three segments.
7. This statement is *not* a definition. Sample explanation: You can draw a figure with three segments that all overlap and this will not be a triangle.

Lesson 2.3
Level B

1. $\angle ABC$ and $\angle CBE$; $\angle DCB$ and $\angle BCE$
2. A, C, and D are equilateral.
3. A figure is equilateral if all sides are congruent.
4. If a figure is a square, it has four congruent sides.
5. If a figure has four congruent sides, then it is a square.
6. A figure is a square iff it has four congruent sides.
7. This statement is *not* a definition. Sample explanation: A rhombus has four congruent sides but it is not a square—definition should include all right angles.

Lesson 2.3
Level C

1. $\angle CDB$ and $\angle BDA$; $\angle DCA$ and $\angle ACB$; $\angle CBD$ and $\angle DBA$; $\angle DAC$ and $\angle CAB$.
2. B and C are trapezoids.
3. A trapezoid is a quadrilateral with only one pair of parallel sides.
4. If angles are a linear pair, then they are adjacent and supplementary.
5. If angles are adjacent and supplementary, then they form a linear pair.

6. Angles are a linear pair iff they are adjacent and supplementary.
7. This is a definition. Sample explanation: The conditional statement and the converse are both true and this makes the biconditional true at all times.

Lesson 2.4
Level A

1. f
2. g
3. d
4. a
5. e
6. c
7. b
8. h
9. $m\angle EAL$
10. $12°$
11. $60°$
12. Reflexive Property
13. Transitive Property
14. Subtraction Property
15. Substitution Property
16. Subtraction Property

Lesson 2.4
Level B

1. Subtraction Property
2. Division Property
3. $x = 11$
4. $CE = 23$
5. $EF = 54$
6. $CD = 100$
7. $90°$
8. $x = 8.5$
9. $m\angle GXL = 59°$
10. $m\angle GXN = 31°$
11. AD, AR
12. CB, BT
13. Transitive Property (or Substitution)
14. Substitution
15. Reflexive Property
16. Subtraction Property

Lesson 2.4
Level C

1. Multiplication Property
2. Subtraction Property
3. Division Property
4. Substitution
5. $6 = 0$
6. Division Property: $a - 6 = 0$, and you cannot divide by zero.
7. $74°$
8. $16°$
9. $53°$
10. $37°$
11. $127°$
12. $37°$

13. Proofs will vary. Statements and reasons given.

$\frac{1}{4}x - 7 = 2$	*Given*
$\frac{1}{4}x + 0 = 9$	*Addition Property*
$\frac{1}{4}x = 9$	*Additive Identity*
$4\left(\frac{1}{4}\right)x = 4(9)$	*Multiplication Property*
$1x = 36$	*Distributive Property*
$x = 36$	*Multiplicative Identity*

Lesson 2.5
Level A

1. $\angle 3$ and $\angle 4$ are vertical angles.
2. Linear Pair Properties
3. $\angle 1$ and $\angle 4$ are supplementary.
4. $\angle 3 \cong \angle 4$; Angles supplementary to the same angle are congruent.
5. They are vertical angles and are congruent.
6. 7
7. 53°
8. 53°
9. 37°
10. 22.5°
11. 157.5°
12. 157.5°
13. Congruent Supplements Theorem
14. induction: You are *assuming* that it was snowing and that is why James wore his coat.
15. deduction: The conclusion follows logically from the given statements.

Lesson 2.5
Level B

1. $m\angle 3$ and $m\angle 4$ are vertical angles; Given
2. $m\angle 1 + m\angle 4 = 180°$
   $m\angle 1 + m\angle 3 = 180°$;
   Linear Pair Property
3. $\angle 1$ and $\angle 4$ are supplementary
   $\angle 3$ and $\angle 1$ are supplementary;
   definition of supplementary angles
4. $\angle 3 \cong \angle 4$; Congruent Supplements Theorem
5. $\angle 1$ and $\angle 3$, $\angle 2$ and $\angle 4$, $\angle 6$ and $\angle 8$, $\angle 5$ and $\angle 7$
6. Congruent Supplements Theorem
7. $\angle 1 \cong \angle 3 \cong \angle 6 \cong \angle 8$
8. $\angle 2 \cong \angle 4 \cong \angle 5 \cong \angle 7$
9. $x = 5.6$
10. 47.2°
11. 132.8°
12. 47.2°
13. You used inductive reasoning.
14. No, this is not a proof. Explanations will vary. Sample answer: A specific example was shown. In order to be a proof, it needs to be shown true for all rectangles.

Answers

Lesson 2.5
Level C

1. Proofs may vary but should include the following:
   m$\angle$3 and m$\angle$4 are vertical angles; Given
   m$\angle 1 + m\angle 4 = 180°$
   m$\angle 2 + m\angle 3 = 180°$;
   Linear Pair Property
   $\angle 1$, $\angle 4$ are supplementary
   $\angle 2$, $\angle 3$ are supplementary;
   definition of supplementary angles
   $\angle 3 \cong \angle 4$; Congruent Supplements Theorem
2. $x = 4.08$
3. 47.56°
4. 102.16°
5. 30.28°
6. 149.72
7. 132.44°
8. 30.28°
9. 47.56°
10. The Hatter is telling the truth. Explanations will vary. Sample explanation: If the Dodo is lying, then the Hatter tells the truth. Therefore, March Hare tells the lies and he said that both the Dodo and the Hatter tell lies. Since March Hare is lying, Dodo could be lying and the Hatter could be telling the truth. This does not contradict the assumption. Therefore, the Hatter is telling the truth.

Answers

Lesson 3.1
Level A

1.

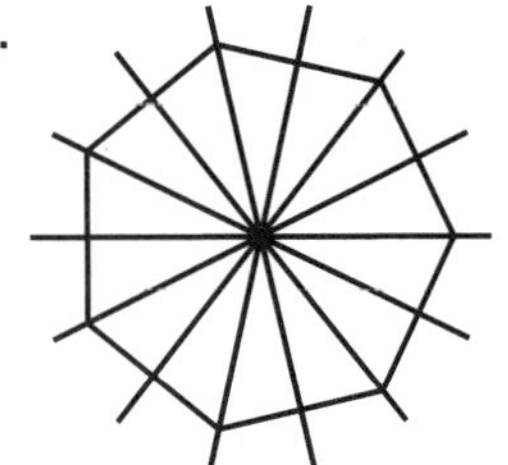

2.

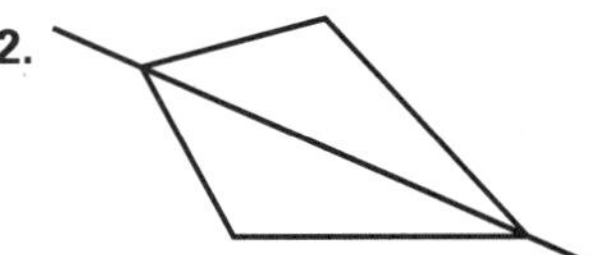

3.

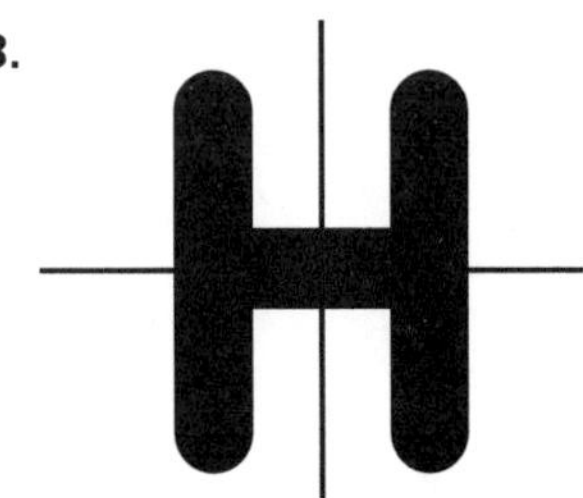

4.

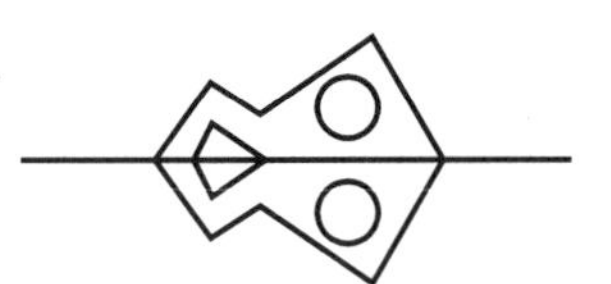

5.

6.

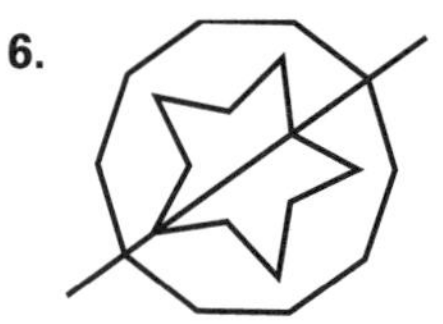

7. #5 and #6 have rotational symmetry.

8. e

9. f

10. g

11. a

12. c

13. d

14. b

Lesson 3.1
Level B

1. 2-fold rotational; reflectional symmetry; 2 axes of symmetry; measure of angle of rotation = 90°

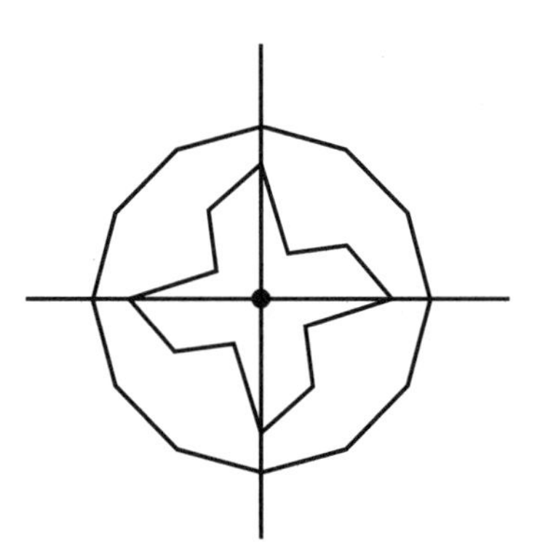

2. 5-fold rotational symmetry; reflectional symmetry; 5 axes of symmetry; measure if central angle = 72°

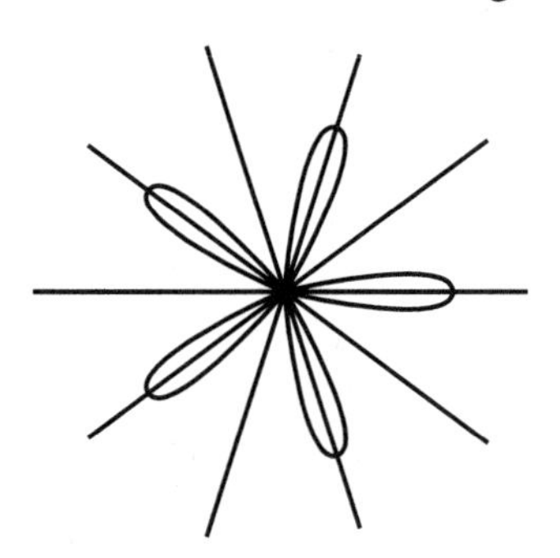

3. reflectional symmetry only; axes of reflection at $x = 2$

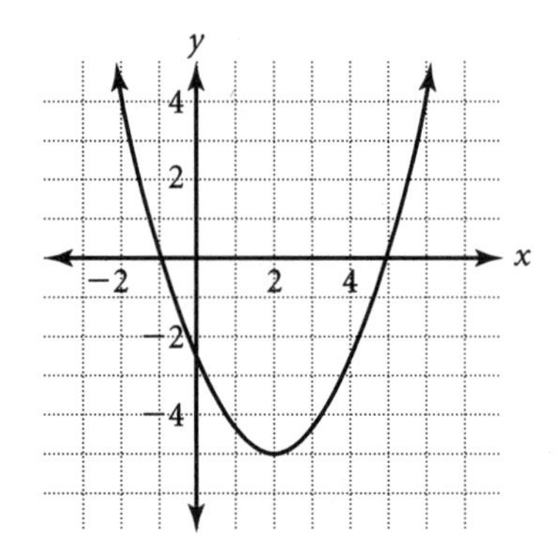

Answers

4. 

5. 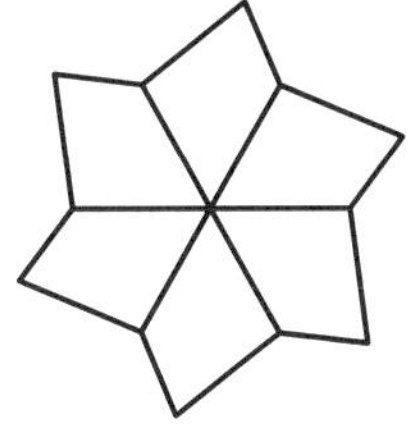

6. False, a (non-rectangle) parallelogram has 2-fold rotational symmetry only.

7. True

8. False; A rhombus is equilateral, but not equiangular.

9. False; The measure of the central angle of a regular n-gon is $\left(\frac{360}{n}\right)^\circ$.

10. True

11. False; A polygon needs at least 3 segments.

12. True

Lesson 3.1
Level C

1. The figure has two-fold rotation symmetry about the origin.

2. The figure has reflection symmetry. The axis of symmetry is the y-axis.

3. The figure has both reflection symmetry and two-fold rotation symmetry about the origin. The axis of symmetry is the y-axis.

4. 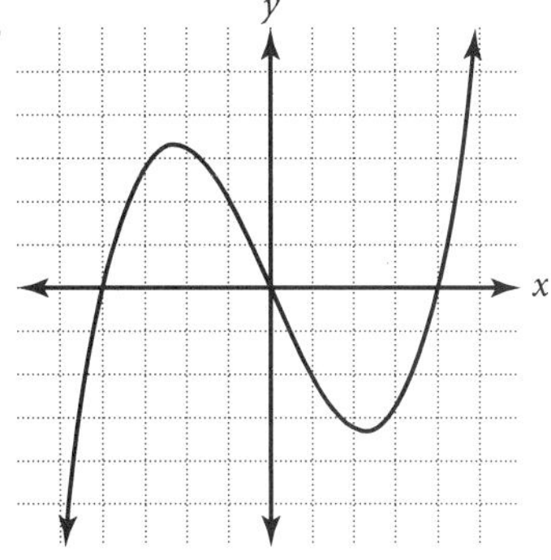

5. 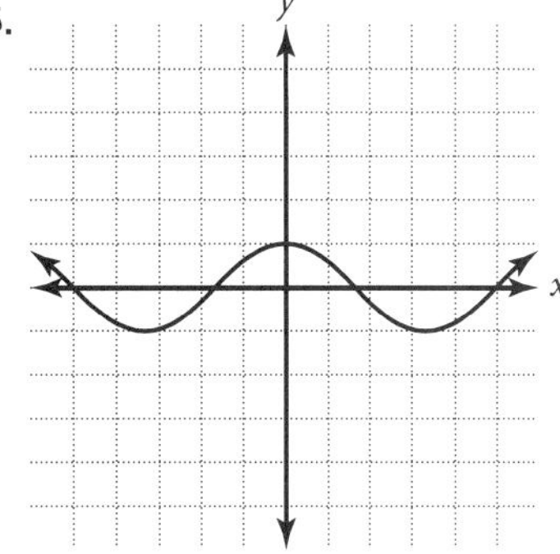

6. 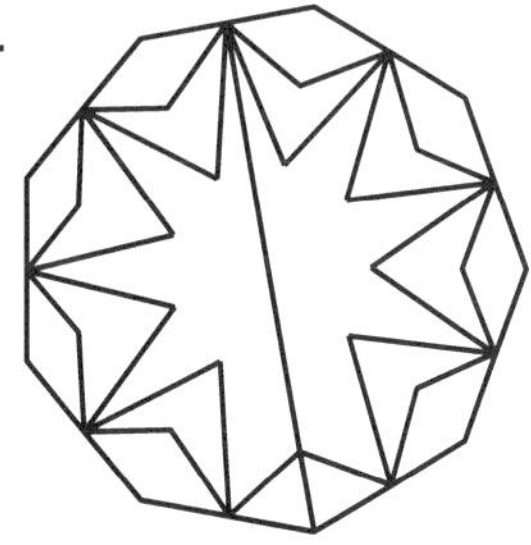

7–12.

Number of sides	Number of axes of symmetry	Measure of central angle
7. 3	3	120°
8. 4	4	90°
9. 5	5	72°
10. 7	7	$\frac{360°}{7}$
11. 8	8	45°
12. 9	9	40°

13. If n is odd, the axes of symmetry are the perpendicular bisectors of each side, through the opposite vertex. If n is even, the axes of symmetry lie on the line containing the opposite vertices.

Answers

Lesson 3.2
Level A

1. f
2. c
3. d
4. b
5. a
6. e
7. 49°
8. 23.4
9. 120°
10. 12
11. 71°
12. 11.7
13. 53.13°
14. 63.43°
15. 126.87°
16. 80.5
17. 161
18. 72
19. 70.6°
20. 9
21. 90°
22. 35.3°

Lesson 3.2
Level B

1. trapezoid
2. four, four
3. quadrilateral, congruent
4. parallelogram
5. quadrilateral
6. four, right angles
7. $x = 7$
8. 78°
9. 78°
10. 102°
11. 44°
12. 59°
13. $x = 2.4$
14. 16.8
15. 33.6
16. 16.8
17. 16.8
18. 33.6
19. False; it could be a rhombus.
20. True
21. True
22. False; some counterexamples: square, kite.

Lesson 3.2
Level C

1. rectangle
2. rhombus
3. kite
4. Quadrilateral *ACBD* is a kite because it has two pairs of congruent, adjacent sides.
5. $\overline{AB}$ is the perpendicular bisector of $\overline{DC}$.
6. The resulting quadrilateral will be a rhombus. It has four congruent sides.

7. The resulting quadrilateral will be a kite, it has two sets of adjacent, congruent sides.

8. The resulting figure is a square, it has four congruent sides and perpendicular congruent diagonals.

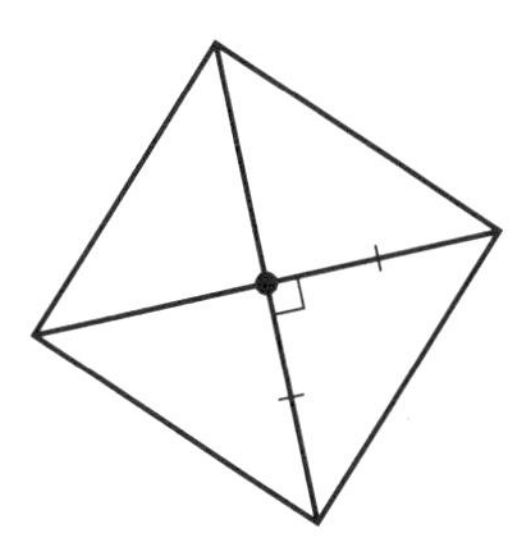

9. The resulting figure is a rectangle, it has four right angles.

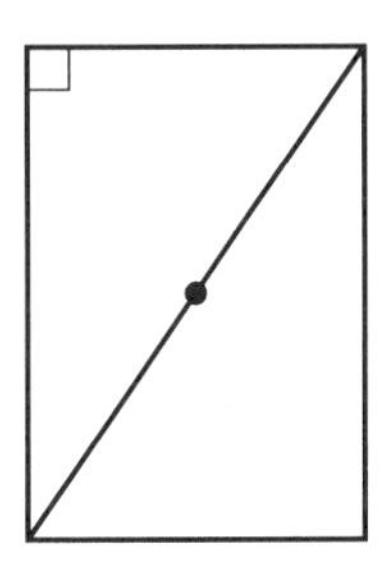

Lesson 3.3
Level A

1. e
2. a
3. b
4. c
5. d
6. 60°
7. 80°
8. 40°
9. 80°
10. 100°
11. 80°
12. 60°
13. 120°
14. 60°
15. 120°
16. 100°
17. 80°
18. Given
19. If parallel lines are cut by a transversal, then corresponding angles are congruent.
20. Vertical angles are congruent.
21. Transitive Property of Congruence

Lesson 3.3
Level B

See students drawings. Sample sketch:

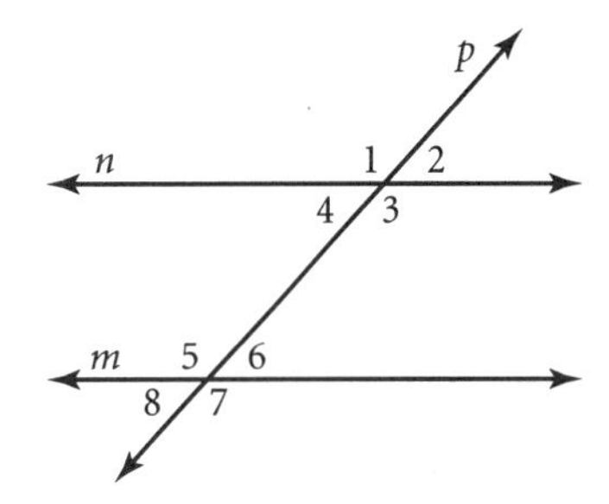

1. $\angle 3$, $\angle 5$ and $\angle 4$, $\angle 6$ are alternate interior
2. p is the transversal
3. $\angle 4$, $\angle 5$ and $\angle 3$, $\angle 6$ are same-side interior
4. $\angle 1$, $\angle 7$ and $\angle 2$, $\angle 8$ are alternate exterior
5. $\angle 1$, $\angle 5$ and $\angle 4$, $\angle 8$ are same-side interior (also $\angle 3$, $\angle 7$ and $\angle 2$, $\angle 6$)
6. 22°
7. 70°
8. 22°
9. 40°

10. 70°

11. 88°

12. Alternate Interior Angles Theorem

13. Vertical Angles Theorem

14. Linear Pair Property

15. Corresponding Angles Postulate

16. Same-Side Interior Angles Theorem

17. Alternate Exterior Angles Theorem

18. Given

19. Alternate Exterior Angles Theorem

20. Corresponding Angles Postulate

21. Transitive Property of Equality.

Lesson 3.3
Level C

1. Two nonadjacent interior angles that lie on opposite sides of a transversal.

2. A line, ray, or segment that intersects two or more coplanar lines, rays, or segments, each at a different point.

3. Interior angles that lie on the same side of a transversal.

4. Two nonadjacent exterior angles that lie on opposite sides of a transversal.

5. Two nonadjacent angles, one interior and one exterior, that lie on the same side of a transversal.

6. $x = 4$

7. 32°

8. 58°

9. Sample answer: $\angle 5$ and $\angle 4$ are supplementary because they are same-side interior angles. $m\angle 5 + m\angle 4 = 180°$ by the definition of supplementary angles. $\angle 5$ is congruent to $\angle 7$ because they are vertical angles. $\angle 4$ and $\angle 2$ are congruent because they are vertical angles. By substitution, $m\angle 2 + m\angle 7 = 180°$. So, because of the definition of supplementary angles, $\angle 2$ and $\angle 7$ are supplementary.

10. $\angle E \cong \angle A$, parallelogram *ABCD* given

11. Sample answer: $\angle A \cong \angle DCB$, opposite angles in a parallelogram are congruent.

12. $\angle E \cong \angle DCB$, Transitive Property of Congruence

Lesson 3.4
Level A

1. Given

2. Converse of Alternate Exterior Angles Theorem

3. Given

4. Same Side Interior Angles Theorem

5. Congruent Supplements Theorem

6. Converse of Corresponding Angles Postulate

7. Alternate Interior Angles Theorem

8. Corresponding Angles Postulate

9. Alternate Exterior Angles Theorem

10. Same Side Interior Angles Theorem

Answers

11. $\overline{BA} \parallel \overline{ED}$, because if two coplanar lines are perpendicular to the same line, then the two lines are parallel to each other.

12. $\overline{BA} \parallel \overline{BC}$, because if two coplanar lines are parallel to the same line, then they are parallel to each other.

Lesson 3.4
Level B

1. Linear Pair Property
2. Given
3. Congruent Supplements Theorem
4. Converse of Corresponding Angles Postulate
5. Corresponding Angles Postulate
6. Vertical Angles Theorem
7. Transitive Property of Congruence
8. Given
9. Transitive Property of Congruence
10. Converse of Corresponding Angles Postulate
11. Lines r and s are *not* parallel. Since $\angle 1 \cong \angle 2$ because they are vertical angles, $x = 4\frac{2}{3}$. When you plug that back in, $m\angle 1 = m\angle 2 = 28°$, and $m\angle 3 = 154°$. But $154° + 28° \neq 180°$.
12. $m \parallel n$, Converse of Alternate Interior Angles Theorem
13. $t \parallel q$, because if two coplanar lines are perpendicular to the same line, then they are parallel to each other.
14. $s \parallel t$, because if two coplanar lines are parallel to the same line, then they are parallel to each other.
15. $t \parallel q$, Converse of Alternate Exterior Angles Theorem

Lesson 3.4
Level C

1. $\overline{SP} \parallel \overline{RE}$, if two coplanar lines are perpendicular to the same line, then the two lines are parallel to each other.
2. $\overline{IP} \parallel \overline{RN}$, since they are alternate exterior angles. (Converse of Alternate Exterior Angles Theorem)
3. $\angle P \cong \angle R$. Reasons for conjecture will vary. Sample answer: congruent triangles
4. Since m and n are parallel, $\angle 1 \cong \angle 2$, and $x = 3.875$. If $x = 3.875$, $\angle 1 = 72°$, $\angle 2 = 72°$, and $\angle 3 = 71°$. Thus r and s are *not* parallel because $\angle 3 \not\cong \angle 2$.
5. $t \parallel v$, because vertical angles are congruent and same side interior angles are congruent. k is not parallel to l because $122° \not\cong 120°$.
6. Cannot be determined

Lesson 3.5
Level A

1. 70°
2. 55°
3. exterior angle
4. 56°
5. 56°
6. 34°
7. 12

8. 31°

9. 82°

10. 67°

11. 67°

12. 67°

13. 46°

14. One line can be drawn because of the Parallel Postulate.

15. 61°

Lesson 3.5
Level B

1. 75°

2. 17.5°

3. 27.5°

4. 72°

5. 54°

6. 54°

7. 360°

8. 51.5°

9. 51.5°

10. 38.5°

11. 51.5°

12. 51.5°

13. 38.5°

14. 124°

15. 81°

16. 155°

17. One line can be drawn because of the Parallel Postulate.

Lesson 3.5
Level C

1. 29°

2. 44°

3. 105°

4. 58°

5. 92°

6. 61°

7. 59°

8. 118°

9. 31°

10. 128°

11. He starts his journey at the North Pole.

12. The triangle-sum of a triangle on a sphere is greater than 180° (and less than 540°).

Lesson 3.6
Level A

1. 118°

2. 125°

3. 159°

4. 135°, 45°

5. 144°, 36°

6. $n = 9$

7. $n = 15$

8. $n = 12$

9. $n = 18$

10. 100°

11. 80°

12. 100°

13. The exterior angle measure is 18°; the interior angle measure is 162°.

14. The polygon has 20 sides.

Lesson 3.6
Level B

1. $x = 78°$

2. $x = 126°$

3. $x = 150°$

4. $x = 27$

5. 115°

6. 90°

7. 130°

8. 55°

9. 150°

10. 69°

11. 119°

12. 133°

13. 39°

14. 135°, 45°

15. 30 sides

16. 18 sides

Lesson 3.6
Level C

1. 6 sides (hexagon)

2. 3 sides (equilateral triangle)

3. 8 sides (octagon)

4. 5 sides (pentagon)

5. $x = 7$

6. $y = 4$

7. 152°

8. 97°

9. 105°

10. The shapes can be described as quadrilaterals (trapezoids). The angle sum is greater than 360°.

11. no

12. 32°

13. 125°

14. 89°

15. 112°

16. 23°

17. 36°

Lesson 3.7
Level A

1. 54

2. 48

3. 40

4. 25, 12.5, 6.25

5. 48, 72

6. midsegments

7. parallelogram, since $\overline{FE} \parallel \overline{BD}$ and $\overline{FD} \parallel \overline{EB}$.

8. trapezoid, since $\overline{FE} \parallel \overline{AB}$.

9. $\overline{CE} \cong \overline{EB} \cong \overline{FD}$, $\overline{FE} \cong \overline{AD} \cong \overline{DB}$, $\overline{ED} \cong \overline{CF} \cong \overline{FA}$

Answers

Lesson 3.7
Level B

1. The triangles are congruent. Explanations will vary
2. $x = 2$
3. $y = 4$
4. 10
5. 5
6. 18
7. 8
8. 23
9. 11.5
10. 12.5
11. 7.6
12. 30.4
13. rectangle; The sides of the rectangle are parallel to the diagonals of the rhombus, which are perpendicular.
14. trapezoid; $EH \parallel BD$.

Lesson 3.7
Level C

1. 18.75
2. 11.25
3. 3.75
4. 10
5. 20
6. 30
7. $x = 7$
8. 39
9. 52
10. 65
11. rectangle; Since *ABCD* is a kite, it has perpendicular diagonals. Each segment formed by *EFGH* is parallel to one of the diagonals, *EFGH* is a rectangle (four 90° angles).
12. parallelogram; Each side of *MPNT* is parallel to one of the diagonals of *QUAD*.

Lesson 3.8
Level A

1. slope = 1, midpoint: $\left(\frac{1}{2}, -\frac{1}{2}\right)$
2. slope = $\frac{1}{6}$, midpoint: $(-1, 7)$
3. slope = -2, midpoint: $(-4, -6)$
4. neither; The product of the slopes is -3.
5. parallel; Both slopes are 3.
6. perpendicular; The product of the slopes is -1.
7. The figure is a parallelogram. The slopes of both pairs of opposite sides are equal.

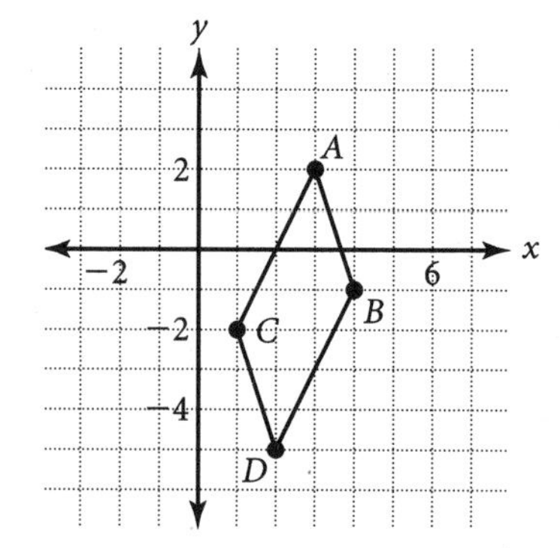

8. The figure is a kite. Diagonals are perpendicular and it has a pair of adjacent congruent sides.

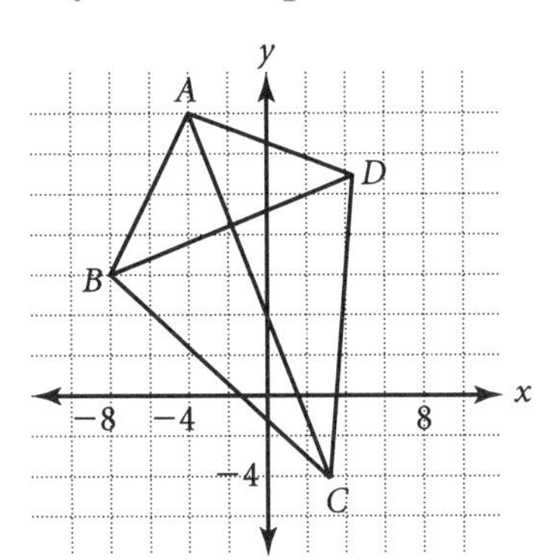

Lesson 3.8
Level B

1.

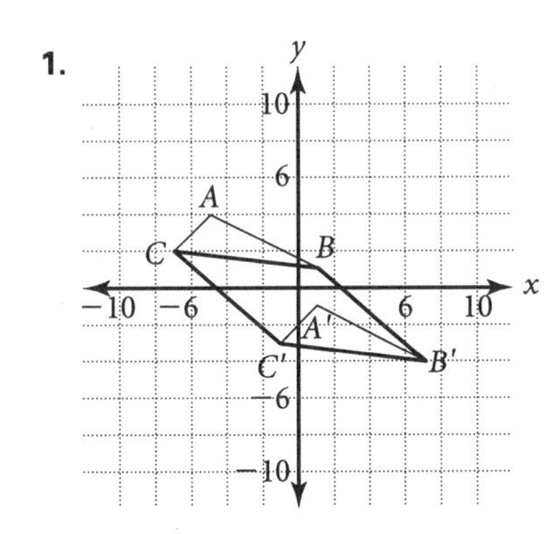

2. $(1, -1)(7, -4)$ and $(-1, -3)$
3. $CC'B'B$ is a parallelogram: the slopes of the opposite sides are equal.
4. Three parallelograms can be formed.
5. $(-1, -1)$, $(-5, -1)$, $(3, 3)$
6. $(4, 5)$
7. $-\frac{1}{2}$
8. possible solutions: $(-6, 4)$ or $(2, 10)$
9. Yes, it is a right triangle. Slopes are $\frac{2}{7}$, $-\frac{7}{2}$, and $-\frac{12}{11}$, and two sides are perpendicular.

Lesson 3.8
Level C

1. $(-3, 7)$
2. $-\frac{2}{11}$
3.

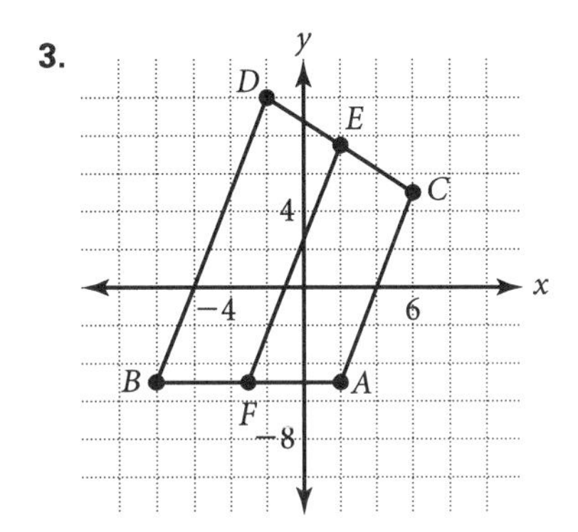

4. trapezoid; The slopes of one pair of opposite sides are equal.
5. Coordinates: $(2, 7.5)$ and $(-3, -5)$; The slope is $\frac{5}{2}$.
6. $(5, 1)$, $(5, 7)$ and $(-1, 7)$ respectively.
7. 45°
8. slope = 1
9. three; Sample explanation: There are three points that are possible that will yield two sets of opposite sides with equal slopes.
10. $(0, 5)$, $(-10, -1)$, and $(4, -9)$

Answers

Lesson 4.1
Level A

1. $\triangle ABC \cong \triangle DEC$
   $\triangle BCA \cong \triangle ECD$
   $\triangle CAB \cong \triangle CDE$
2. $AFEB \cong CDEB$
3. $\triangle ADC \cong \triangle BCD$
   $\triangle DAB \cong \triangle CBA$
   $\triangle ADE \cong \triangle BCE$
4. $\angle L$
5. $\angle M$
6. $\angle R$
7. $\angle Q$
8. $\angle U$
9. $\overline{LM}$
10. $\overline{TU}$
11. $\overline{KJ}$
12. 60°
13. 75°
14. 45°
15. 9
16. 6
17. 10
18. 90°
19. 90°

Lesson 4.1
Level B

1. $\triangle BAC \cong \triangle DBC$
   $\triangle ACB \cong \triangle BCD$
   $\triangle CBA \cong \triangle CDB$
2. $ABCDE \cong RQUTS$
3. 70°
4. 35°
5. 45°
6. 27
7. 15
8. 17
9. Yes, alternate interior angles are congruent.
10. $\angle L$, $\angle T$
11. $\overline{PQ}$, $\overline{EF}$
12. $\overline{ML}$, $\overline{CB}$
13. $\overline{DC}$, $\overline{WV}$
14. $\angle F$, $\angle Y$
15. $\angle D$, $\angle N$
16. $\overline{GF}$, $\overline{ZY}$
17. $\angle C$, $\angle V$

Lesson 4.1
Level C

1. $\triangle ABC \cong \triangle RQP$
2. $\triangle KLM \cong \triangle RPN$
3. $\triangle RNP \cong \triangle EFD$
4. $\triangle GHJ \cong$ cannot tell; insufficient information
5. $\triangle BDE \cong \triangle CAG$
   $\triangle BAG \cong \triangle CDE$
   $\triangle CFE \cong \triangle BFG$
   $\triangle BCE \cong \triangle CBG$

Answers

6. 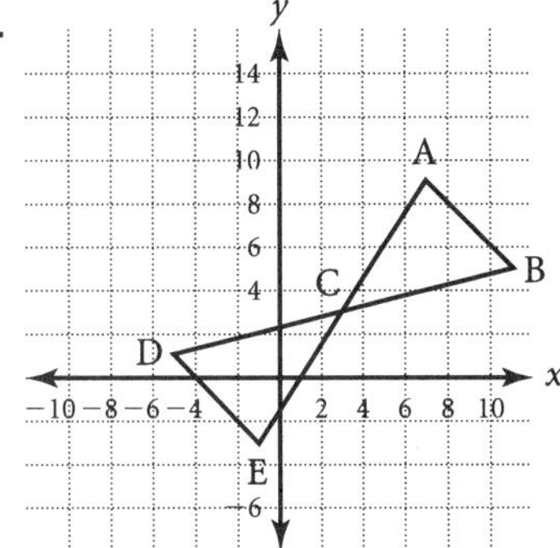

7. Yes:
   $AB = ED = 4\sqrt{2}$
   $AC = EC = \sqrt{13}$
   $BC = DC = 2\sqrt{17}$
   C is the midpoint of $\overline{DB}$
   C is the midpoint of $\overline{AE}$
   $\angle DCE \cong \angle ACB$
   $\triangle ABC \cong \triangle EDC$

8. Polygon Congruence Postulate

Lesson 4.2
Level A

1. Yes, SAS
2. Yes, ASA
3. Yes, SSS
4. No, AAA does not work.
5. $\angle ABC \cong \angle DCB$
6. $\angle ACB \cong \angle DBC$
7. $\overline{AB} \cong \overline{DC}$, $\angle ABC \cong \angle DCB$ or $\overline{AC} \cong \overline{DE}$, $\angle ACB \cong \angle DBC$
8. Yes, SAS
9. Yes, ASA
10. Yes, SSS
11. No, SSA does not work.
12. Yes, AAS
13. Yes, SAS
14. No, AAA does not work.

Lesson 4.2
Level B

1. No, AAA does not work.
2. Yes, ASA
3. Yes, ASA
4. Yes, AAS
5. $\overline{AC} \cong \overline{DF}$, $\angle A \cong \angle D$
6. $\overline{FE} \cong \overline{CB}$, $\overline{AC} \cong \overline{DF}$
7. Yes, SAS or HL
8. no
9. no
10. $x = 22°$
11. $y = 18$
12. $\angle ABD \cong \angle CBD$
13. Reflexive Property
14. $\triangle ABD \cong \triangle CBD$
15. SAS

Lesson 4.2
Level C

1. No
2. Yes, SAS
3. No, SSA does not work.
4. Yes, AAS
5. Always, SAS
6. Sometimes, yes if SAS, no if SSA
7. Always, ASA or AAS

Answers

8. Always, ASA or AAS
9. Sometimes, If a pair of corresponding legs are congruent; ASA
10. $x = 23$
11. $y = 2$
12. Reflexive property
13. Segment addition
14. Addition property of equality
15. Transitive property
16. If 2 sides of a triangle are congruent, angles opposite are congruent.
17. SAS ≅ SAS

Lesson 4.3
Level A

1. No, the sides could be different lengths.
2. Yes, AAS
3. Yes, HL
4. Yes, AAS
5. Yes, AAS
6. No, AAA
7. No
8. ASA
9. AAS
10. ASA
11. SAS
12. AAS

Lesson 4.3
Level B

1. No, AAA
2. Yes, SAS
3. No, SSA
4. Yes, HL
5. Yes, SSS
6. No, SSA
7. Yes, SAS
8. Yes, AAS
9. Yes, HL
10. No
11. Yes, AAS
12. Definition of perpendicular lines
13. $\angle RQX \cong \angle STX$
14. Right angles are congruent.
15. Given
16. Vertical angles are congruent.
17. $\Delta RXQ \cong \Delta SXT$
18. AAS ≅ AAS

Lesson 4.3
Level C

1. $\Delta ABC \cong \Delta ADE$, ASA
2. $\Delta DAB \cong \Delta DEB$, AAS
3. $\Delta DFC \cong \Delta BFE$ or $\Delta DFB \cong \Delta EFC$, SAS
4. $\Delta DBC \cong \Delta BDE$ or $\Delta DCE \cong \Delta BEC$, SSS
5. Sometimes
6. Never

Answers

7. Sometimes

8. Always

9. Sometimes

10. $\angle EAD \cong \angle CAD$

11. Definition of angle bisector

12. $\overline{FE} \cong \overline{FA}, \overline{BC} \cong \overline{BA}$

13. Definition of midpoint

14. $\overline{AF} \cong \overline{AB}$

15. $\frac{1}{2}$ of equal segments are congruent.

16. $\overline{AG} \cong \overline{AG}$

17. Reflexive property

18. $\triangle GAF \cong \triangle GAB$

19. SAS $\cong$ SAS

Lesson 4.4
Level A

1. 32°

2. 60°

3. 46°

4. 100°

5. $3\frac{1}{2}$

6. 17.3

7. 3

8. 12

9. 8.5

10. 54°

11. 125°

12. 54°

13. 98°

14. 27°

15. 126°

16. 13; CPCTC

Lesson 4.4
Level B

1. 70°

2. 29

3. 19

4. 36

5. 3, 1

6. 49°

7. Y, Z trisect $\overline{DC}$

8. congruent sides $\overline{XD} \cong \overline{XC}$

9. $\angle XYD \cong \angle XZC$ by CPCTC

Lesson 4.4
Level C

1. Sometimes

2. Always

3. Always

4. Sometimes

5. $AB = 10, BC = 8, AC = 8$

6. 9

7. 20°

8. Since $\overline{BD}$ is congruent to itself, $\triangle ADB \cong \triangle CDB$ and corresponding sides $\overline{AD}$ and $\overline{CD}$ must be congruent. If two sides of a triangle are congruent, it is an isosceles triangle.

Answers

Lesson 4.5
Level A

1. 40°
2. 40°
3. 110°
4. 8
5. 10
6. 70°
7. 22
8. 69°
9. 37°
10. 74°
11. 106°
12. 37°
13. 74°
14. 53°
15. 109°
16. 122°
17. 46
18. $\triangle AED \cong \triangle CFB$, $\triangle AEB \cong \triangle CFD$, $\triangle ADB \cong \triangle CBD$

Lesson 4.5
Level B

1. 127°
2. 7
3. 2, $8\frac{1}{2}$
4. 68
5. 122°
6. 108°
7. 72°
8. $(3x + 10)°$
9. Opposite sides of a parallelogram are parallel.
10. Corresponding angles are congruent.
11. Opposite angles are congruent.
12. Transitive property
13. Opposite sides of parallelogram are congruent.
14. Given
15. SAS ≅ SAS

Lesson 4.5
Level C

1. 68°
2. 99°
3. 81°
4. 81°
5. 18°
6. Sometimes
7. Always
8. Sometimes
9. Sometimes
10. Sometimes
11. Sometimes
12. Always
13. 19
14. 25 by 3
15. *ABCD* is a parallelogram.

16. Alternate interior angles are congruent.

17. Diagonals bisect each other.

18. $\angle AGE \cong \angle CGF$

19. $\triangle AGE \cong \triangle CGF$

20. $\overline{EG} \cong \overline{FG}$

21. CPCTC

Lesson 4.6
Level A

1. angles *BAD, ADC, DCB, CBA, AEB, BEC, DEA*, and *CED*

2. angles *DAE, EAB, ABE, CBE, ECB, ECD, CDE*, and *ADE*

3. $\overline{AB} \cong \overline{DC}$, $\overline{AD} \cong \overline{BC}$, $\overline{AC} \cong \overline{BD}$, $\overline{AE} \cong \overline{BE} \cong \overline{CE} \cong \overline{DE}$

4. angles $BAE \cong BCE$, $DAE \cong DCE$, $BEA \cong BEC$, and $CED \cong DEA$

5. $\overline{AB} \cong \overline{CB}$, $\overline{AD} \cong \overline{CD}$, $\overline{AE} \cong \overline{CE}$

6. angles *AEB, BEC, CED, DEA*

7. angles *AEB, BEC, CED, DEA*

8. $\overline{AB} \cong \overline{BC} \cong \overline{CD} \cong \overline{DA}$, $\overline{AE} \cong \overline{EC}$, $\overline{BE} \cong \overline{ED}$

9. *ABCD* is a parallelogram, both pairs of opposite sides are parallel.

10. no conclusion

11. *ABCD* is a square. Congruent diagonals prove it is a parallelogram and diagonals are perpendicular, which prove it is a square.

12. *ABCD* is a parallelogram. Both pairs of opposite sides are congruent.

Lesson 4.6
Level B

1. no conclusion

2. parallelogram; The diagonals bisect each other.

3. parallelogram; Same pair of sides are congruent and parallel.

4. parallelogram; Both pairs of opposite sides are parallel.

5. rectangle; It is a parallelogram with right angles.

6. parallelogram; Diagonals bisect each other or one pair of sides are both congruent and parallel.

7. no conclusion

8. No, it could be a rectangle.

9. Yes, it is a rectangle with congruent consecutive sides.

10. No, it could be a rhombus.

11. Given

12. Opposite sides of a rhombus are parallel.

13. $\triangle ABC \cong \triangle DCB$

14. $\angle ABC \cong \angle DCB$

15. CPCTC

16. Same side interior angles are supplementary.

17. $\angle ABC$ and $\angle DCB$ are right angles.

18. parallelogram with right angles

Answers

Lesson 4.6
Level C

1. neither
2. no conclusion
3. parallelogram; Opposite sides are congruent.
4. parallelogram; Opposite sides are congruent.
5. no conclusion
6. 25 units
7. 60 square units
8. rectangle
9. 88 units
10. 110 units
11. 27
12. $287\frac{2}{5}$
13. rhombus, parallelogram with congruent adjacent sides

Lesson 4.7
Level A

Check student's constructions.

Lesson 4.7
Level B

Check student's constructions.

Lesson 4.7
Level C

Check student's constructions.

Lesson 4.8
Level A

1.

2.

3.

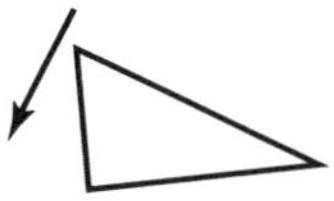

4.

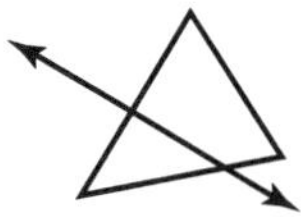

5. 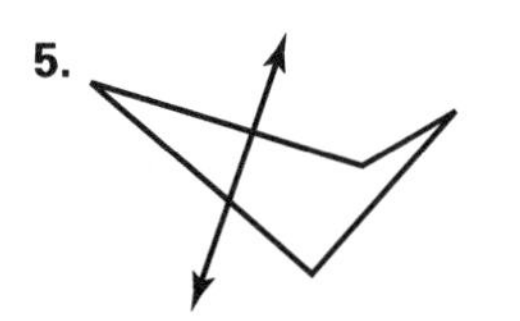

6. 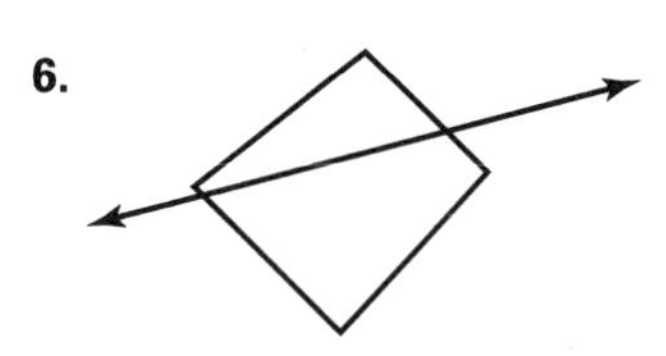

7. Yes, the sum of the two smaller sides is greater than the largest side.
8. No, the sum equals the third side.
9. Yes, the sum of two smallest sides is greater than the third side.
10. No, the sum equals the third side.
11. No, the sum of the two smallest sides is less than the third side.
12. Yes, the sum of the two smaller sides is greater than the third side.
13. No, the sum of the two smallest sides is less than the third side.

Lesson 4.8
Level B

1.

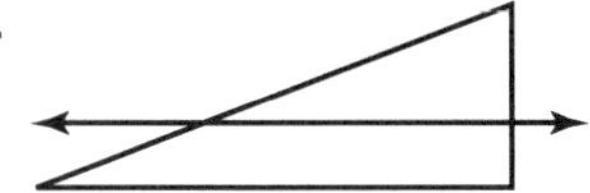

2.

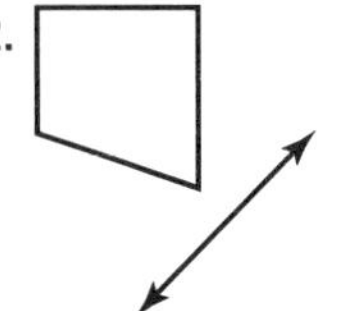

3.

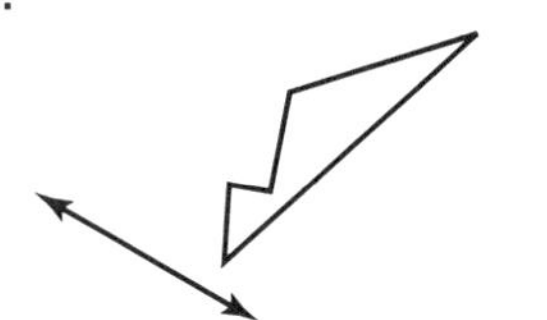

4.

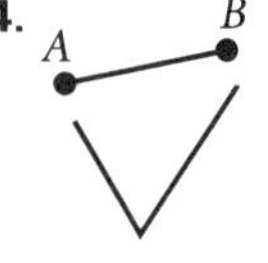

5.

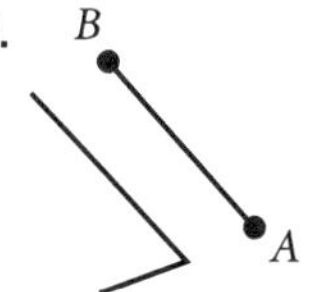

6.

7. $4 < AC < 14$

8. $285 < WY < 549$

9. $7.2 < DF < 15.6$

10. $5\frac{5}{8} < GJ < 13\frac{1}{8}$

11. $2\sqrt{3} < MN < 8\sqrt{3}$

12. $0.08 < PR < 0.14$

13. $2 < PQ < 16$

14. If the sum of the lengths of $\overline{AB}$ and $\overline{BC}$ is equal to the length of the third segment $\overline{AC}$, which is given, then the endpoints are collinear by the Betweeness Postulate and thus ABC is not a triangle.

Lesson 4.8
Level C

1.

2.

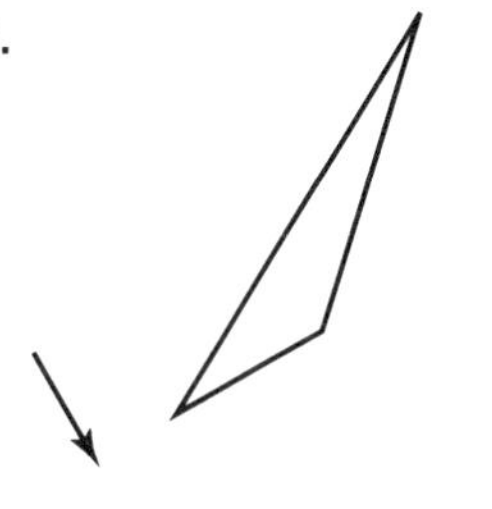

3.

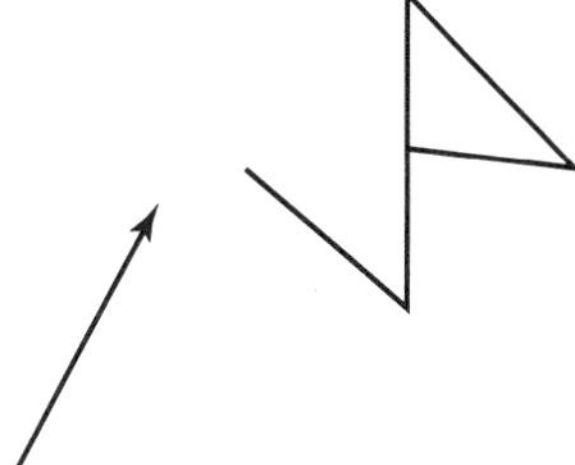

4.

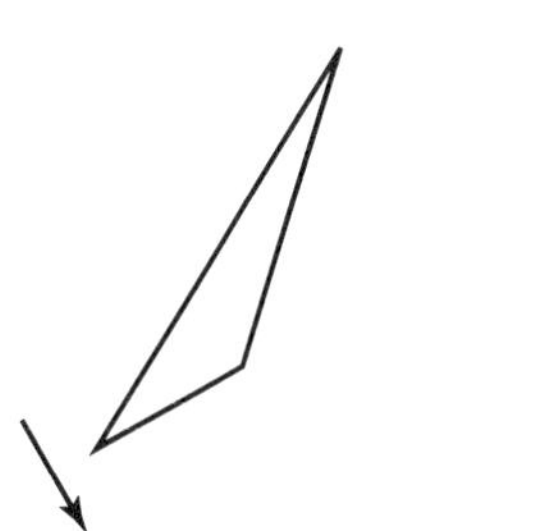

5. yes

6. $4 < x < 18$

7. $1 < x < 4$

Answers

8. $5\frac{2}{3} < x < 17$

9. $1\frac{1}{5} < x < 2$

10. $1 < AB < 10$

11. $4 < AB < 12$

12. $3 < AB < 22$

13. $0.9 < AB < 15.5$

Answers

Lesson 5.1
Level A

1. $P = 56$ units, $A = 106$ square units
2. $P = 40$ units, $A = 62$ square units
3. $P = 76$ units, $A = 124$ square units
4. $P = 40$ units, $A = 11$ square units
5. 80 inches
6. 400 square inches
7. 44 inches
8. 80 square inches
9. 1 meter, 5 meters, 5 square meters
10. 2 meters, 4 meters, 8 square meters
11. 3 meters, 3 meters, 9 square meters

Lesson 5.1
Level B

1. $P = 52$ units, $A = 71$ square units
2. $P = 24$ units, $A = 21$ square units
3. 126 units
4. 810 square units
5. 90 units
6. 198 square units
7. 315 square units
8. 120 units
9. $202\frac{1}{2}$ square units
10. $364\frac{1}{2}$ square units
11. 1 unit, 24 units, 50 units
12. 2 units, 12 units, 28 units
13. 3 units, 8 units, 22 units
14. 4 units, 6 units, 20 units

Lesson 5.1
Level C

1. 60 units
2. 160 square units
3. 87 units
4. 224 square units
5. 119 units
6. 288 square units
7. 80 square units
8. *ABKP* and *PKJN*
9. *NACH*
10. *TNLR*
11. *NLEDH*
12. *TNMGFR*
13. 1 foot, 48 feet, 98 feet
14. 2 feet, 24 feet, 52 feet
15. 3 feet, 16 feet, 38 feet
16. 4 feet, 12 feet, 32 feet
17. 6 feet, 8 feet, 28 feet

Lesson 5.2
Level A

1. $A = lw$, 84 square units
2. $A = \frac{1}{2}bh$, $7\frac{1}{2}$ square units
3. $A = bh$, 390 square units
4. $A = \frac{1}{2}h(b_1 + b_2)$; 48 square units

5. $A = \frac{1}{2}h(b_1 + b_2)$; 32 square units

6. $A = 2\left[\frac{1}{2}h(b_1 + b_2)\right]$; 441 square units

7. 96 square units

8. 21.9 square units

9. 69 square units

10. 10 units

11. 6 units

12. 23 units

13. 7.5 units

14. 216 square units

15. 216 square units

16. 432 square units

17. 648 square units

18. 432 square units

Lesson 5.2
Level B

1. $A = \frac{1}{2}bh$; 32.1 square units

2. $A = \frac{1}{2}h(b_1 + b_2)$; 38.8 square units

3. $A = bh$; 200 square units

4. 6 meters

5. $3\frac{3}{7}$ meters

6. 8 meters

7. 135 square units

8. $12\frac{1}{2}$ square units

9. 95 square units

10. 27.5 square units

11. 122.5 square units

12. 107.5 square units

13. 13 feet

Lesson 5.2
Level C

1. 36 units

2. 1350 square units

3. 486 square units

4. 13

5. 18

6. 9 units by 11 units

7. 77 square units

8. 7 units

9. 13 units

10. 3 units

11. 24 units

12. 432 square units

13. 384 square units

Lesson 5.3
Level A

1. $\frac{17}{\pi} \approx 5.4$

2. 5

3. $\frac{27}{8\pi} \approx 1.1$

4. $\sqrt{\frac{23.78}{\pi}} \approx 2.75$

Answers

5. $\sqrt{\dfrac{157}{\pi}} \approx 7.1$

6. $\sqrt{\dfrac{314}{\pi}} \approx 10$

7. r is cut by $\frac{1}{3}$

8. 43.96; 153.86

9. 75.36; 452.16

10. 37.68; 113.04

11. area is $\frac{1}{4}$ of the original area

12. 12.56; too big

13. 54.5

14. 21.5

15. 21.5

16. the same

Lesson 5.3
Level B

1. $\dfrac{576}{\pi} \approx 183.35$ square units

2. 75 feet by 75 feet by 75 feet; 2437.5 square feet

   25 feet by 87.5 feet; 2187.5 square feet
   56.25 feet by 56.25 feet; 3164.1 square feet

   $r = 35.8$ feet; 4027.5 square feet

3. 7.5 meters

4. 469 feet

5. 37.7 feet more

6. $160\pi \approx 502.65$

7. 25.73 square units

8. 18π or 56.55 square units

9. 4π or 12.57 units

Lesson 5.3
Level C

1. 10

2. 12

3. 106 miles per hour

4. 705 revolutions per minute

5. 8.22 square units

6. 32.6 feet

7. $164\pi \approx 200.96$ square units

8. 8 cookies

9. 48 square inches

Lesson 5.4
Level A

1. 13.9 units

2. 10.2 units

3. 16.4 units

4. 10.9 units

5. 10.1 units

6. acute

7. obtuse

8. right

9. obtuse

10. 6.36 units

11. 9.9 units

12. yes

13. 17.5 inches

14. 18.5 inches

Lesson 5.4
Level B

1. 7.9 units
2. 3 units
3. 0.7 units
4. 2 units
5. 1.6 units, 4.7 units
6. 3.1 units, 6.2 units
7. obtuse
8. acute
9. obtuse
10. 7.2 units
11. 41.6 units
12. 9 units, 17.9 units
13. 60 square units
14. 72.3 square units

Lesson 5.4
Level C

1. $2:\sqrt{2}$
2. 7.5 square units
3. acute
4. $\frac{3}{8}$
5. 7
6. 25
7. 84 square units
8. 24
9. $4\sqrt{2}$ or 5.7 acute
10. $12\sqrt{2}$ or 17.0
11. 16
12. 18
13. 312 square units
14. 210 square units
15. 8.5 miles
16. 5.7

Lesson 5.5
Level A

1. $6\sqrt{3}$, 12
2. $2\sqrt{3}$, $4\sqrt{3}$
3. 3, $3\sqrt{3}$
4. $7\sqrt{2}$, 14
5. 4, 4
6. 9, $6\sqrt{3}$
7. 3, 6
8. $\frac{3\sqrt{3}}{2}$, $\frac{9}{2}$
9. $4\sqrt{15}$
10. 15
11. $3\sqrt{3}$
12. $9\sqrt{3}$ square units

Lesson 5.5
Level B

1. 4, $2\sqrt{3}$, 6, 2, $4\sqrt{3}$
2. 4, $\sqrt{3}$, 3, 1, $2\sqrt{3}$
3. $\frac{\sqrt{3}}{3}$, $\frac{8\sqrt{3}}{3}$, $4\sqrt{3}$, $\frac{4\sqrt{3}}{3}$, 8

Answers

4. $12, 6, 3\sqrt{3}, 3, 6\sqrt{3}$
5. $40, 20, 10\sqrt{3}$ $30, 20\sqrt{3}$
6. $8\sqrt{3}, 4\sqrt{3}, 6, 6\sqrt{3}, 2\sqrt{3}$
7. 18
8. $9\sqrt{3}$
9. $32\sqrt{3}$
10. $15 + 3\sqrt{3} + 3\sqrt{6}$
11. $\frac{27\sqrt{3} + 27}{2}$
12. $4\sqrt{2}$
13. $4\sqrt{6}$
14. $8\sqrt{2}$

Lesson 5.5
Level C

1. $2\sqrt{55}$
2. $3\sqrt{3}$
3. 13
4. $5\sqrt{2}$
5. $10\sqrt{2}$
6. $5\sqrt{2}$
7. $5\sqrt{2}$
8. $\frac{441\sqrt{3}}{2}$
9. $3x\sqrt{2}$
10. $3x + 3x\sqrt{3}$
11. $3x$
12. $6x$
13. $\frac{9x^2\sqrt{3}}{2}$
14. $\frac{9x^2}{2}$
15. $\frac{9x^2 + 9x^2\sqrt{3}}{2}$
16. $2\sqrt{14}$
17. 6, 8, 24, 10
18. $4\sqrt{2}$

Lesson 5.6
Level A

1. 2.24 units
2. 8.54 units
3. 13.34 units
4. 9.90 units
5. 4.61 units
6. 18.86 units
7. ≈ 15

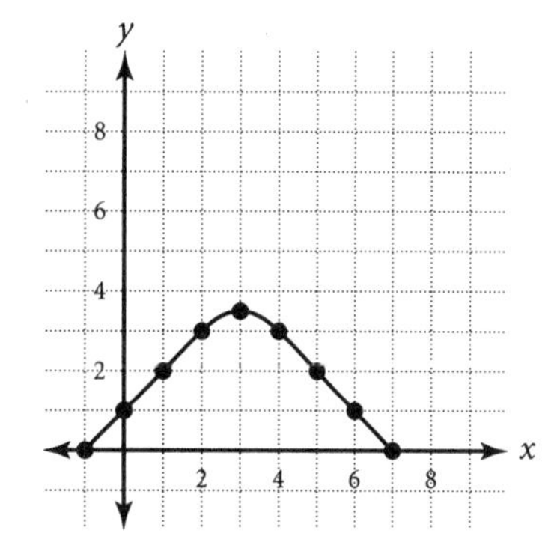

8. ≈ 56

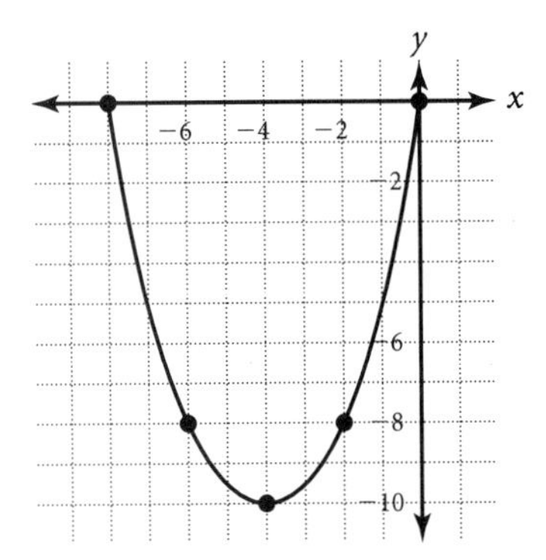

9. 7.81

10. 9.85

11. 15.03

12. obtuse

13. isosceles

Lesson 5.6
Level B

1. $2\sqrt{5}$

2. 5

3. $\sqrt{37}$

4. acute

5. $5\sqrt{2}$

6. $\sqrt{89}$

7. obtuse

8. parallelogram; The distance between corresponding points is the same, so $HF = CD$, and $HC = FD$.

9. ≈ 30 square units

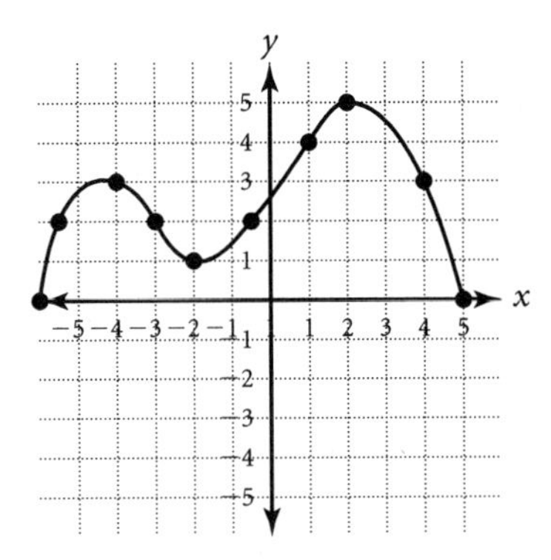

10. ≈ 14 square units

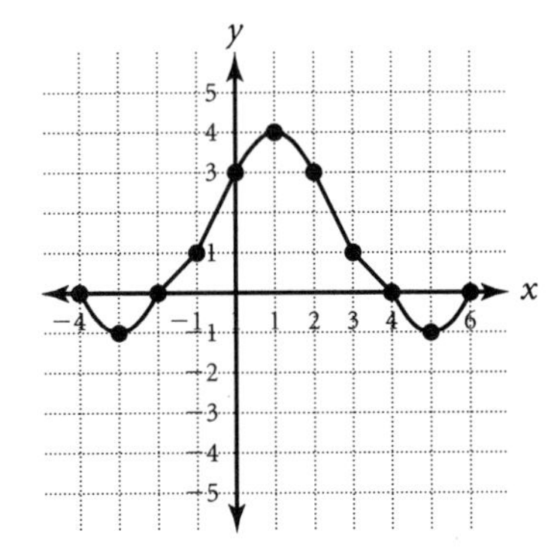

11. $\sqrt{20}$

Lesson 5.6
Level C

1. 5 or −3

2. 6 or −2

3. (10, 1)

4. (−11, −8)

5. (5, 0)

6. 13

7. 13

8. congruent, right, isosceles

9. 12

10. 4

11. *GHJ* is larger than *HMK*, but both are isosceles triangles.

12. y-values are 10, $8\frac{1}{4}$, 7, $6\frac{1}{4}$, 7, $8\frac{1}{4}$, 10

    $A \approx 58\frac{1}{2}$

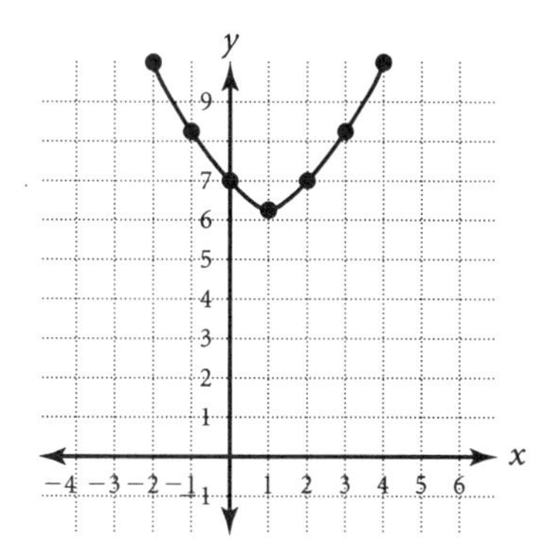

Lesson 5.7
Level A

1. $(2a, 0)$
2. $N\ (0, -2a)\ P\ (-q, 0)$
3. $D\ (a, 0), B\ (b, c)$
4. $Q\ (p, q), W\ (b, 0)$
5. congruent
6. $(1, 3)$
7. $(2, -3)$
8. reverse

Lesson 5.7
Level B

1. $(a + c, b + d)$
2. (c, d)
3. $\frac{b}{a}$
4. $\frac{b}{a}$
5. They are parallel
6. $\sqrt{c^2} + d^2$
7. $CB = 2\sqrt{c^2} + d^2$

   It is one-half of CB.
8. It is parallel to the third side and equal to one-half its length
9. $EF = HG = a$

   $EH = FG = \sqrt{b^2} + c^2$

   $EG = EG$

   $\therefore$ the triangles are congruent by SSS.
10. Using the distance formula, the diagonal JL and KM both equal $\sqrt{2a^2}$, so the diagonals are congruent.
11. The congruent diagonals intersect at the midsegment of each diagonal. Therefore, the diagonals of a square bisect each other into two congruent segments.
12. The two diagonals have slopes that are negative reciprocals of each other. Therefore, they are perpendicular to each other.
13. perpendicular; -1

Lesson 5.7
Level C

1. $D\ (-2, 1), E\ (3, 0)$; slopes equal $-\frac{1}{5}$;

   $DE = \sqrt{26}\ \ AC = 2\sqrt{26}$
2. $\sqrt{13}$
3. $\sqrt{65}$
4. $\sqrt{65}$
5. $\sqrt{13}$
6. kite

Answers

7. diagonals are perpendicular
$\frac{10}{-2} \cdot \frac{1}{5} = -1$

8.

x	3	1	−7	0
y	5	8	7	3

; (3, 5), (1, 8), (−7, 7), (0, 3)

9. $\sqrt{13}$

10. $\sqrt{65}$

11. $\sqrt{65}$

12. $\sqrt{13}$

13. kite

14. midpoint $(-a, b)$
$AM = MC = MB = MD = \sqrt{a^2 + 4b^2}$
$AB = BC = CD = DA = \sqrt{a^2 + 9b^2}$

Lesson 5.8
Level A

1. $\frac{4}{11}$

2. $\frac{10}{11}$

3. 0

4. $\frac{2}{5}$

5. 1

6. 405 square feet

7. 80 square feet

8. $\frac{16}{81}$

9. a. $\frac{65}{81}$
b. $\frac{11}{405}$
c. $\frac{1}{8}$

10. 12.5%

11. 0.125

12. $\frac{7}{16}$

13. 0.65

14. 0.375

15. 0.0025

16. 0.417

17. $\frac{9}{20}$

18. $\frac{7}{8}$

Lesson 5.8
Level B

1. $\frac{9}{13}$

2. $\frac{6}{13}$

3. 0

4. $\frac{1}{2}$

5. K and L, or L and N, or N and Q, or P and R

6. K and M, or L and P

7. K and N

Answers

8. 12.56 square inch
9. 37.68 square inch
10. 62.8 square inch
11. 200.96 square inch
12. $\frac{157}{360}$
13. $\frac{263}{720}$ or 0.37
14. $\frac{13}{720}$ or 0.18
15. $\frac{7}{80}$ or 0.88

Lesson 5.8
Level C

1. $\frac{1}{8}$
2. $\frac{3}{4}$
3. $\frac{7}{8}$
4. $\frac{1}{2}$
5. $\frac{3}{8}$
6. $\frac{1}{4}$
7. $\frac{1}{8}$
8. $\frac{7}{8}$
9. 0.07
10. 0.31
11. 0.41
12. 0.21
13. 0.42
14. 0.58

Answers

Lesson 6.1
Level A

1. 15 cubic units

2. 49 square units

3.

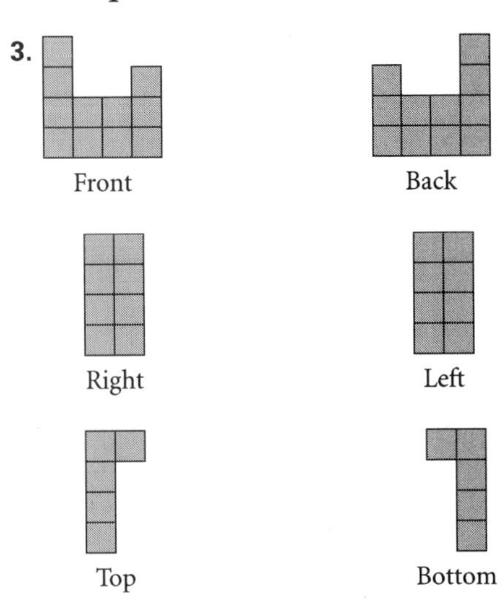

4. Check student's drawing.

5. Check student's drawing.

Lesson 6.1
Level B

1. 14 cubic units

2. 56 square units

3.

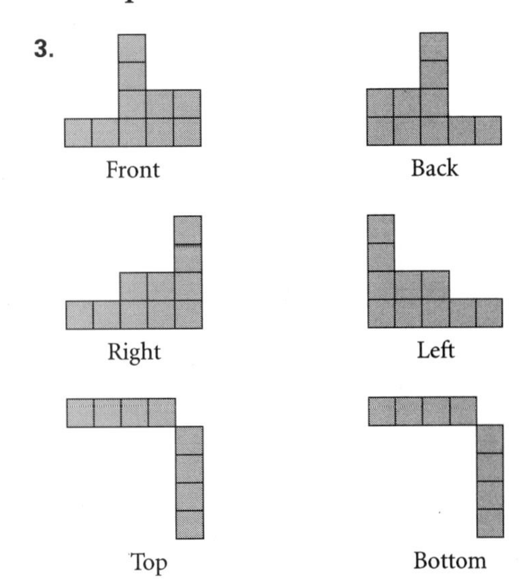

4. Check student's drawing.

5. Check student's drawing.

Lesson 6.1
Level C

1. 25 cubic units

2. 71 square

3. Check student's drawing.

4. Check student's drawing.

5. Check student's drawing.

Lesson 6.2
Level A

1. Sample answers: $\overline{AD}$, $\overline{BC}$ or $\overline{AC}$, $\overline{BD}$ or $\overline{EH}$, $\overline{CB}$ or $\overline{GJ}$, $\overline{KN}$

2. Sample answers: $\overline{AD}$, $\overline{BF}$ or $\overline{GJ}$, $\overline{DH}$

3. Sample answers: $\overline{AE}$ or $\overline{HD}$ or $\overline{FB}$

4. 14

5. Sometimes

6. Sometimes

7. Sometimes

8. *DC*; *DC*

9. definition of perpendicular lines

10. definition of right triangle

11. $\overline{DC}$

12. HL Congruence Theorem

Lesson 6.2
Level B

1. *CAFÉ* and *LMN*, or *ACG* and *EFJK*

2. $\overline{CG}$ or $\overline{EK}$ or $\overline{FJ}$

3. $\overline{CG}$ or $\overline{DH}$ or $\overline{EK}$

4. Never

5. Never

6. Sometimes

7. $\overline{BC}$; $\overline{FC}$

8. given

9. Two lines perpendicular to the same line are parallel.

10. If one pair of opposite sides of a quadrilateral are parallel and congruent, then the quadrilateral is a parallelogram.

Lesson 6.2
Level C

1. Sample answers: *JKRP* and *BFGD*, or *DCHG* and *MTU* or *ABDC* and *EJLG*

2. Sample answers: $\overline{JE}$ and $\overline{LG}$, or $\overline{KR}$ and $\overline{LS}$, or $\overline{DC}$ and $\overline{AB}$

3. Sample answers: $\overline{KF}$ and $\overline{LS}$ or $\overline{DC}$ and $\overline{BE}$; they are noncoplanar and do not intersect

4. 90°

5. $\overline{AB}$

6. *N*

7. Infinite

8. Infinite

9. Always

10. Always

Lesson 6.3
Level A

1. *ABCD*, *EFGH*

2. *DCGH*, *CBFG*, *BFEA*, *AEHD*

3. no; The bases are not regular.

4. rectangles

5. 7.1

6. 14.4

7. 17.5

8. 6.2

9. 10.5

10. 8

11. 17.6

12. 9.6

13. yes; It is proven by the Pythagorean Theorem and the substitution property of equality.

Lesson 6.3
Level B

1. 30–60–90 right triangle

2. *ACFD*, *ADEB*, *BEFC*

3. rectangles

4. $36 + 12\sqrt{3} + 12\sqrt{3} \approx 77.57$ feet

5. no

6. $\overline{EF}$, $\overline{DF}$, $\overline{DE}$, $\overline{CF}$, $\overline{BE}$

7. 12.8

8. 49.0

9. 19.0

10. 10.0

11. 11.0

12. 11.0

13. No, oblique prisms do not have right angles so it cannot be proven with the Pythagorean Theorem.

Answers

Lesson 6.3
Level C

1. *ABCDEF* and *GHJKLM*; regular hexagon
2. Sample answers: *AFMG* and *CDKJ*; *BCJH* and *FELM*
3. 48 inches
4. 16 inches
5. 7.1
6. 18.8
7. 9, 10.8
8. 2, 3.5
9. $4\sqrt{3}$, 13.9
10. $4\sqrt{3}$, 6
11. Sample answer $\sqrt{4a^2 + l^2}$; explanations will vary.

Lesson 6.4
Level A

1. Bottom–Front–Right
2. Top–Back–Right
3. Top–Front–Left
4. Top–Back–Left
5. Bottom–Front–Left
6. Bottom–Back–Right
7. Top–Front–Right
8. Bottom–Back–Left
9. 6.6; (3, 4, 6)

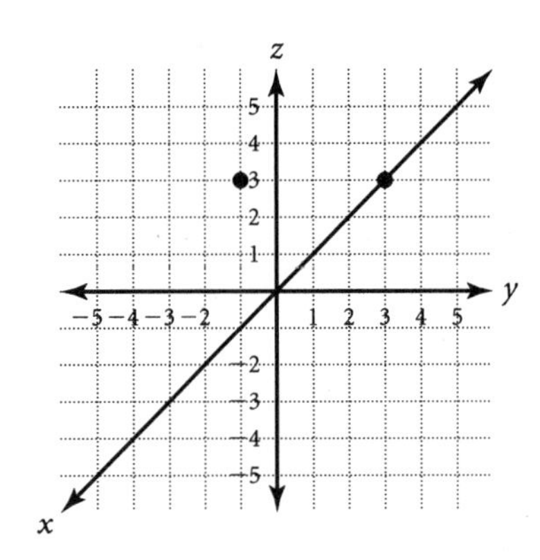

10. 5.4; $\left(7\frac{1}{2}, 1, 4\right)$

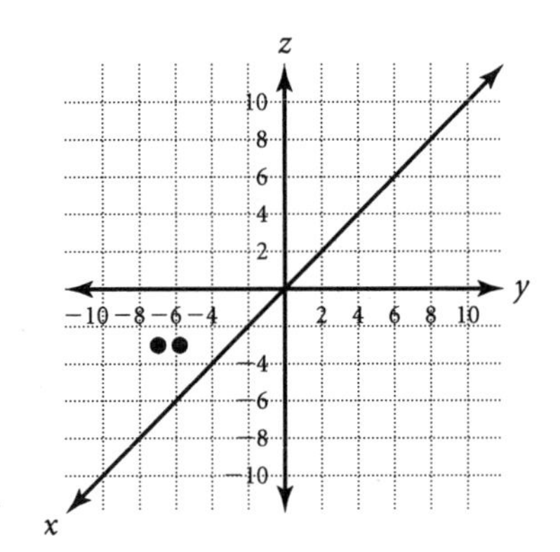

11. 10.7; $\left(\frac{1}{2} 0, -1\frac{1}{2}\right)$

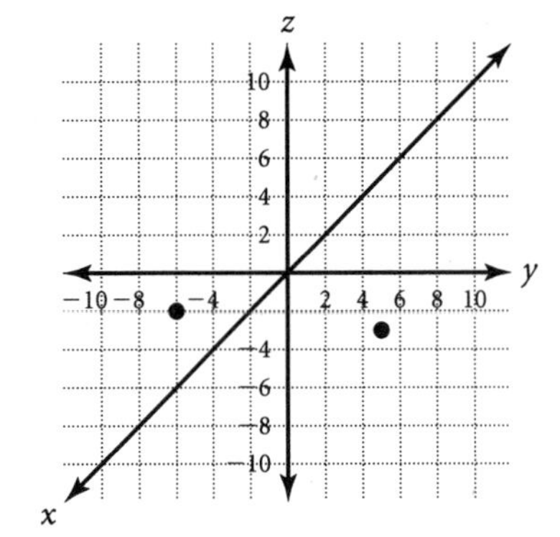

Answers

12. 9.6; $\left(-5, -4\frac{1}{2}, -5\right)$

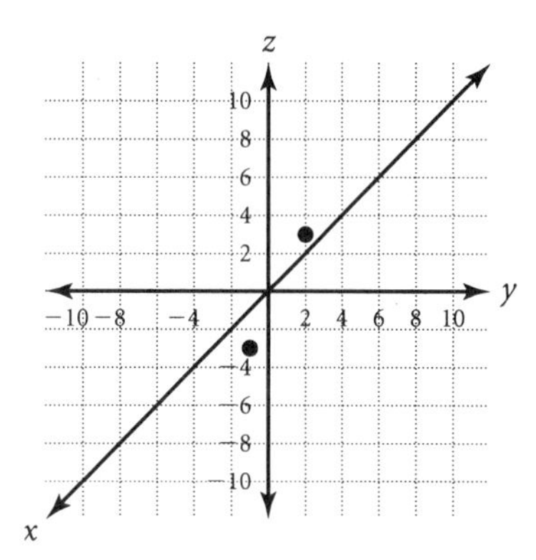

Lesson 6.4
Level B

1. Top–Back–Right
2. *yz*-plane
3. Top–Front–Left
4. Top–Back–Left
5. Bottom–Front–Left
6. negative *x*-axis
7. 8
8. *xy*-plane
9. 11.6, $\left(-\frac{1}{2}, \frac{1}{2}, 4\right)$

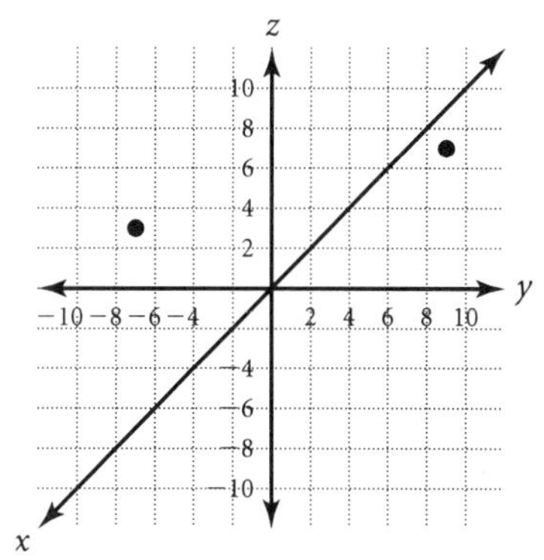

10. 13.9 $\left(4\frac{1}{2}, -2, 6\frac{1}{2}\right)$

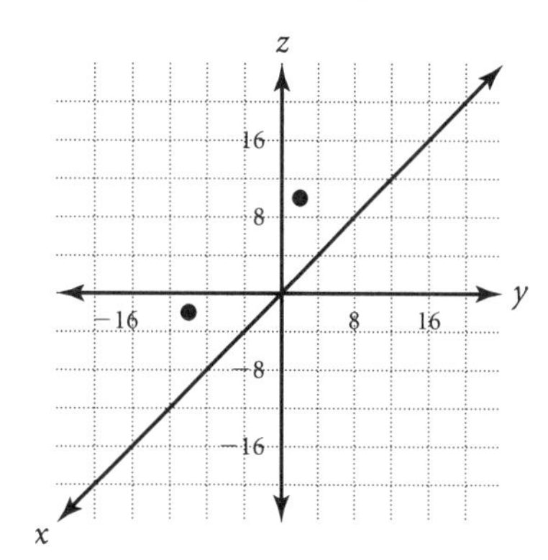

11. (18, 0, 12)
12. (18, −15, 12)
13. (0, −15, 0)
14. (0, −15, 12)
15. 18
16. 15
17. 26.3
18. 12
19. 19.2
20. 21.6
21. 216

Lesson 6.4
Level C

1. *yz*-plane
2. Top–Front–Left
3. negative *x*-axis
4. *xy*-plane
5. Top–Front–Right
6. positive *z*-axis

Answers

7. 22.3, $\left(10, -5, 16\frac{1}{2}\right)$

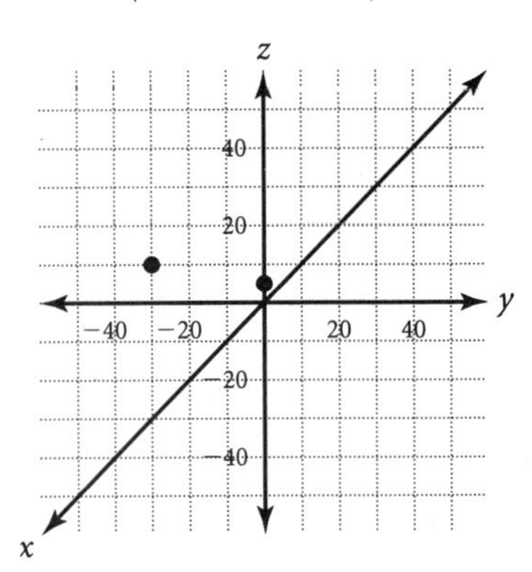

8. 13.2, $\left(\frac{1}{2}, 9, -12\right)$

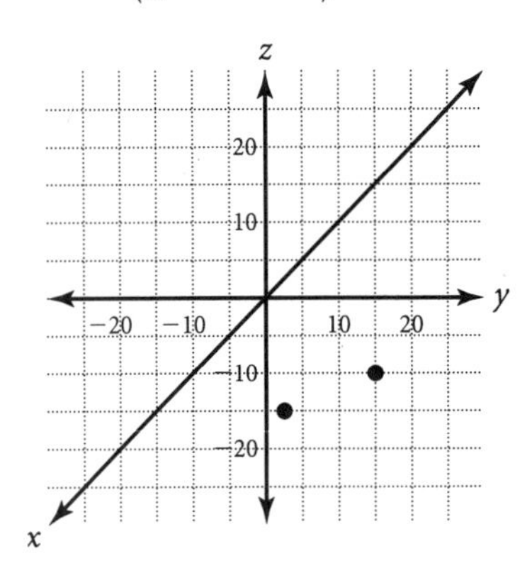

9. $(-10, 12, 1)$; 11.7
10. $(-6, 19, -4)$; 27.9
11. $(40.5, 30, -4)$; 52.2
12. $(0, 4, 0)$
13. $(0, 4, -4)$
14. $(8, 0, -4)$
15. 4
16. $\sqrt{96}$
17. $4\sqrt{5}$

Lesson 6.5
Level A

1. $\frac{9}{5}, \frac{9}{2}, -3$
2. $-6, \frac{-12}{5}, 12$
3. $\frac{8}{3}, -4, 8$
4. $-12, 3, -3$
5. The graph of the equation $3x + 5y = 9$ lies on a two-dimensional coordinate system and the graph of the equation $3x + 5y + z = 9$ lies on a three-dimensional coordinate system.
6.

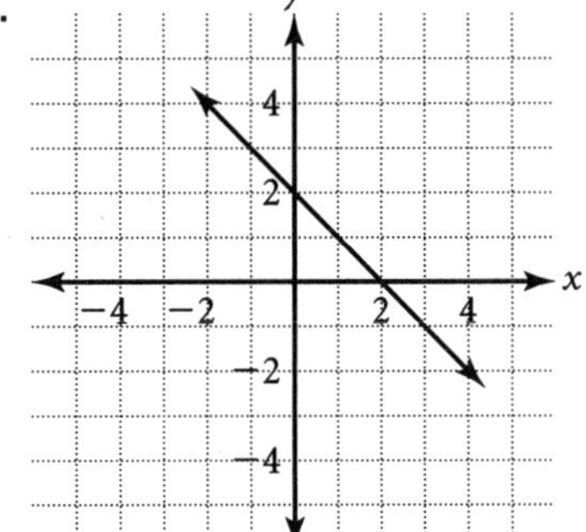

7.

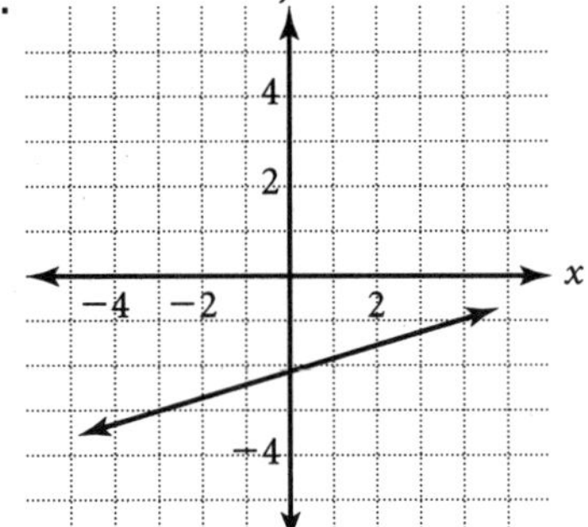

Answers

8.

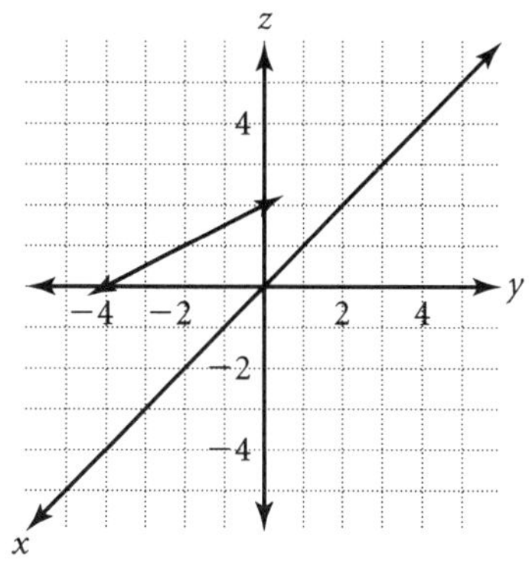

9.

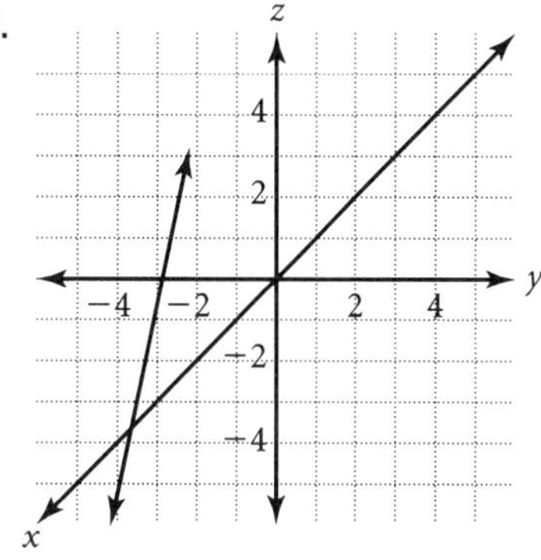

Lesson 6.5
Level B

1.

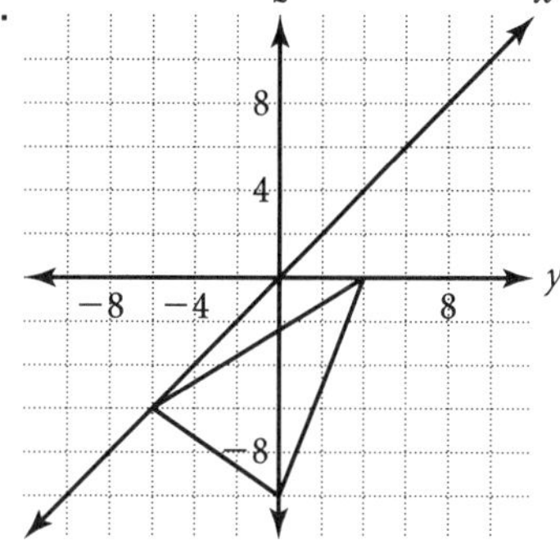

2.

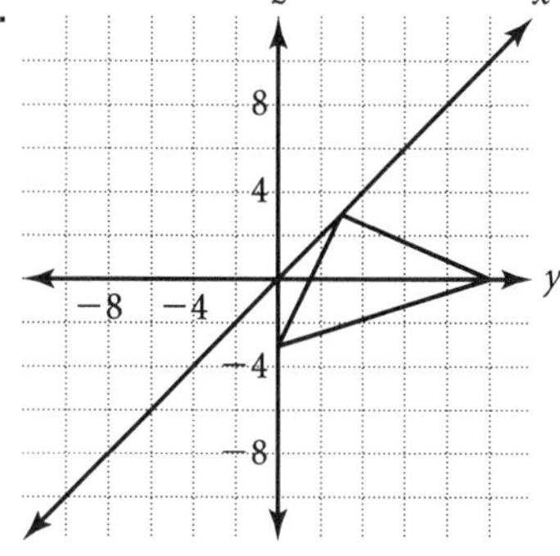

3.

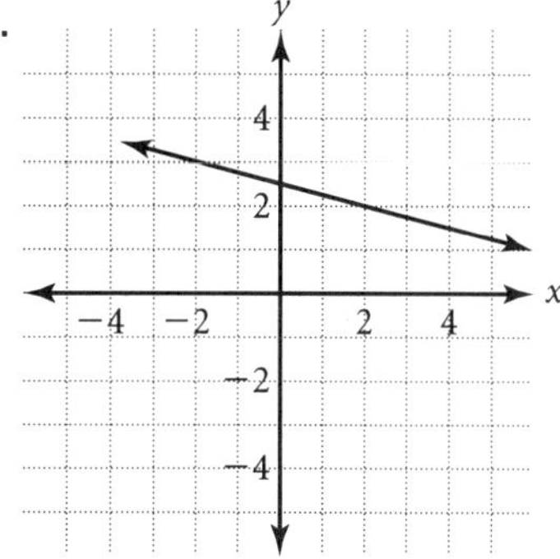

4.

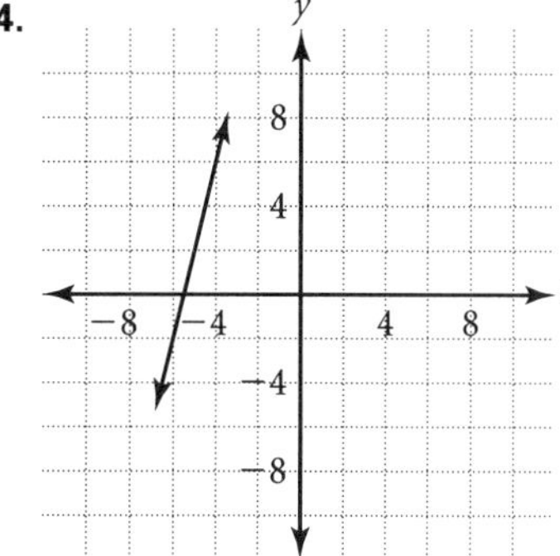

5.

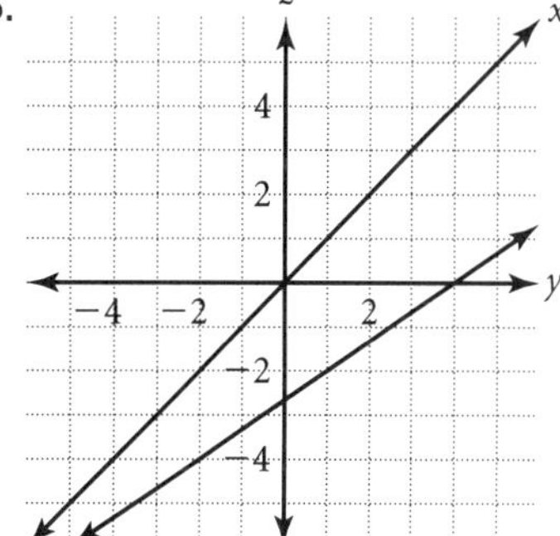

6.

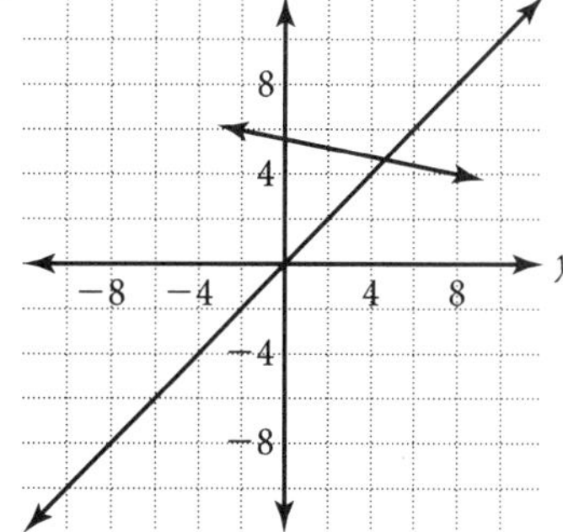

Lesson 6.5
Level C

1.
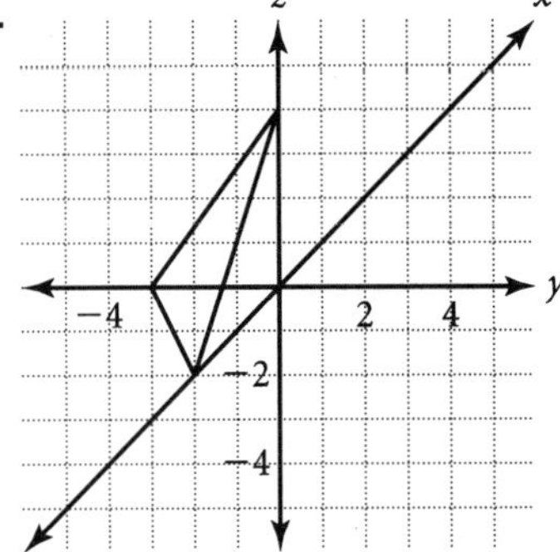

2.
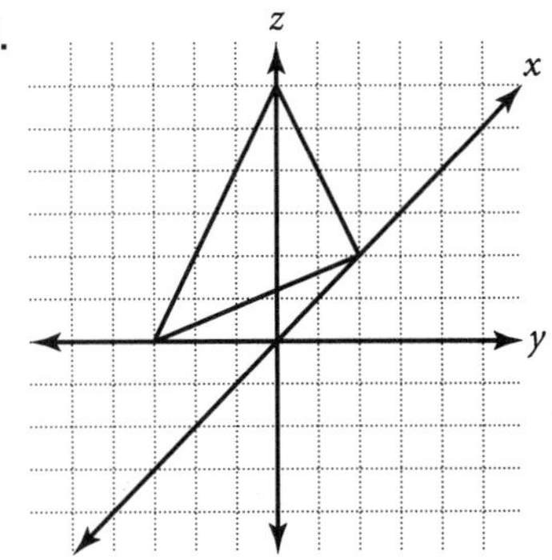

3.
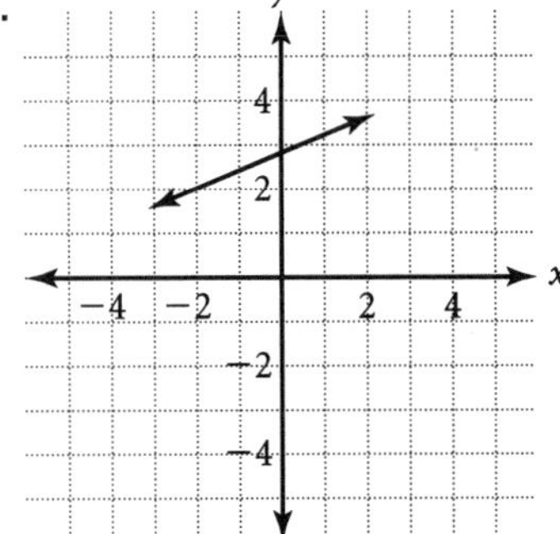

4.
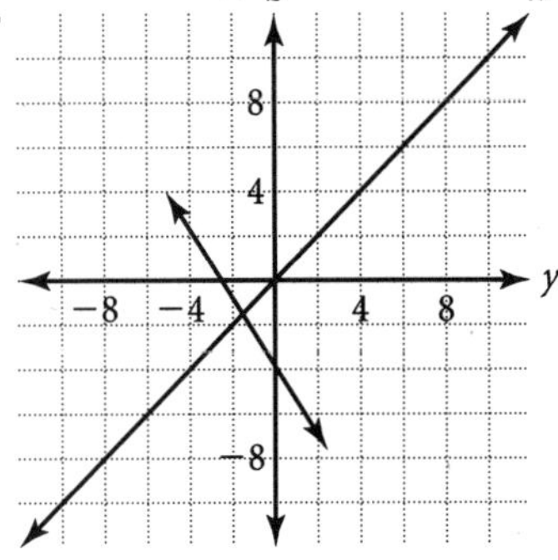

5. $x = -4t - 6$
$y = -3t + 4$
$z = 2t - 2$

6. $x = 12t - 6$
$y = 3t + 1$
$z = -5t + 5$

7. $\dfrac{x - 3}{4} = \dfrac{y - 3}{-1} = \dfrac{z - 1}{3}$

8. $\dfrac{x + 4}{-1} = \dfrac{y}{2} = \dfrac{z - 3}{-1}$

Lesson 6.6
Level A

For all exercises, see student's work.

Lesson 6.6
Level B

For all exercises, see student's work.

Lesson 6.6
Level C

For all exercises, see student's work.

Answers

Lesson 7.1
Level A

1. $S = 6$ units^2; $V = 1$ unit^3
2. $S = 58$ units^2; $V = 20$ units^3
3. $S = 46$ units^2; $V = 14$ units^3
4. $S = 94$ units^2; $V = 60$ units^3
5. $S = 24$ units^2; $V = 8$ units^3
6. $S = 48$ units^2; $V = 20$ units^3
7. $S = 90$ units^2; $V = 50$ units^3
8. $S = 150$ units^2; $V = 125$ units^3
9. $S = 54$ units^2; $V = 27$ units^3
10. $S = 66$ units^2; $V = 36$ units^3
11. $S = 66$ units^2; $V = 36$ units^3
12. $S = 80$ units^2; $V = 48$ units^3
13. $S = 80$ units^2; $V = 48$ units^3
14. $S = 96$ units^2; $V = 64$ units^3
15. 6 to 1
16. 3 to 1
17. 2 to 1
18. 3 to 2
19. 6 to 5
20. 1 to 1
21. 2 to 1
22. 1 to 2
23. 2 inches
24. box A

Lesson 7.1
Level B

1. 11 to 15 = 0.733
2. 5 to 4 = 1.25
3. 53 to 110 = 0.4818
4. 13 to 6 = 2.166
5. 3721 to 2856 = 1.303
6. 7 to 10 0.7
7. Sample answer: minimize the surface area since the volume is constant
8. Sample answer: maximize the volume since the surface area is fixed and you want to create the maximum amount of space
9. Sample answer: maximize the volume since you want the most storage capacity
10. Sample answer: minimize the surface area since you may need to save on construction materials
11. 1 to 3
12. 25 centimeters
13. 3
14. 0
15. 26 square inches
16. 184 to 158 = 1.16

Lesson 7.1
Level C

1. Sample answer: minimize the surface area since the volume is limited and since children's hands are small

Answers

2. Sample answer: maximize the volume since the surface area is fixed and you want to create the maximum amount of space
3. $1 + \frac{4}{s}; \frac{6}{s}$
4. 174 inches^2
5. 720 inches^3
6. 4 to 5 or 0.8
7. 7 to 6
8. 4 centimeters
9. 6 centimeters
10. $s = 15$
11. $s = 1.8$
12. 96 units^2
13. 512 units^3

Lesson 7.2
Level A

1. 100 inches^3
2. 72 centimeters^3
3. 48 centimeters^3
4. 225 inches^3
5. 120 inches^3
6. 120 inches^3
7. 33 centimeters^3
8. 33 centimeters^3
9. 70 centimeters^3
10. 144 units^3
11. 12 units^3
12. 16 units^3
13. 45 units^3
14. 36 units^3
15. 800 units^3
16. 240 units^3
17. 630 units^3
18. 472 units^2
19. 346 units^2
20. 104 units^2
21. 164 units^2
22. 94 units^2
23. 94 units^2
24. 1020 units^2
25. 1020 units^2
26. 250 inches^2
27. Cavalieri's Principle

Lesson 7.2
Level B

1. Divide the volume by the base area.
2. Find the area of each part of the net. Find the sum of the areas.
3. $S = 186$ meters^2; $V = 126$ meters^3
4. $S = 175.2$ inches^2; $V = 124.7$ inches^3
5. $S = 179.1$ centimeters^2;

   $V = 166.3$ centimeters^3
6. $S = 558.1$ meters^2; $V = 580.6$ meters^3
7. $S = 100$ centimeters^2; $V = 50$ centimeters^3
8. $S = 41.1$ units^2; $V = 15.8$ units^3

9. $S = 320$ inches^2; $V = 300$ inches^3

10. $S = 221.2$ feet^2; $V = 62.4$ feet^3

11. 22 ounces, by Cavalieri's Principle

12. 5.2 centimeters

Lesson 7.2
Level C

1. 430.2 units^2

2. Find the surface area of the cube and subtract the area of the two bases of the triangular prism, then add the lateral area of the triangular prism.

3. 1525.2 units^2

4. 567 units^3

5. 2177 units^3

6. 64 units^3

7. 575.12 centimeters^3

8. 30 meters^3

9. 500 meters^3

10. They must be equal.

Lesson 7.3
Level A

1. 114 inches^2

2. 114 inches^2

3. 114 inches^2

4. 114 inches^2

5. 144 inches^2

6. 600 inches^2

7. 864 inches^3

8. 24.36 meters^2

9. 32.625 meters^2

10. 105.705 meters^2

11. 54.375 meters^3

12. 64 units^3

13. 156 units^2

14. 96 units^3

15. 33 units^2

16. 480 units^2

17. 479.9 units^2

Lesson 7.3
Level B

1. 40 inches

2. 692.82 inches^2

3. 15 inches

4. about 22.91 inches

5. 1374.6 inches^2

6. 2067.42 inches^2

7. 4 meters

8. 9 meters

9. 4.5 meters

10. about 6.02 meters

11. 36 meters

12. 108.36 m^2

13. 189.36 m^2

14. 229.35 m^2

15. 25.8 feet

16. 240 units^3

17. 106.67 units^3

18. 35 units^3

Answers

Lesson 7.3
Level C

1. 210 $feet^2$
2. 432 $feet^2$
3. 822 $feet^2$
4. 1680 $feet^3$
5. 13 feet 5 inches
6. 11.4 $yard^2$
7. 3034.3 $centimeters^3$
8. 1152 $centimeters^2$
9. 7.28 inches
10. 18 feet

Lesson 7.4
Level A

1. volume
2. surface area
3. 175.9
4. 150.8
5. 12.6
6. 50.3
7. 113.1
8. 201.1
9. 25.1
10. 75.4
11. 3
12. 1
13. 7
14. 10
15. 4
16. 2
17. 6
18. 5
19. π
20. 8π
21. 27π
22. 64π
23. 36π
24. 48π
25. 5
26. 10
27. 1
28. 8
29. 6
30. 2
31. 6
32. 3
33. 1
34. 10

Lesson 7.4
Level B

1. $V = 90\pi - 48\pi$
2. 12.6
3. 4.7
4. 9.4
5. 3.1
6. 46.2
7. 104.7

8. 434.3

9. 144.8

10. 1.8

11. 7

12. 3.5

13. 10

14. 4

15. 2

16. 12

17. 15

18. 0.2π

19. 0.04π

20. 7.2π

21. 30π

22. $an^3\pi$

23. $a^2n^3\pi$

24. 3

25. 1.5

26. 20

27. 0.6

28. 4.9

29. 1.2

30. 5.5

31. 85

Lesson 7.4
Level C

1. 1437.4 inches^3

2. 7.2 centimeters^3

3. 345.58 feet^3

4. 64π feet^2, or about 201 feet^2

Lesson 7.5
Level A

1. 36π meters^2

2. 312π centimeters^2

3. 210π feet^2

4. 1500π inches^2

5. 1176π meters^2

6. 90π centimeters^2

7. 4π feet^3

8. 216π meters^3

9. 192π inches^3

10. 27π feet^3

11. 81π feet^3

12. 81π feet^3

13. Sample answer: The surface area of a right cone is the sum of the lateral area and the sum of the base area. The volume of a right cone is one-third of the base area times the height of the cone.

Answers

Lesson 7.5
Level B

1. 1304.2 $feet^2$
2. 1288.4 $centimeters^2$
3. 3246.4 $meters^2$
4. 11 meters
5. 6 feet
6. 13.2 meters
7. 2171 $meters^3$
8. 68,094 $centimeters^3$
9. 42,412 $inches^3$
10. 141.99 $inches^2$
11. $\sqrt{3} : 1$
12. 3 : 1

Lesson 7.5
Level C

1. 1322 $centimeters^3$
2. 110 $centimeters^3$
3. 1240 $centimeters^3$
4. 8 centimeters
5. 3 : 1
6. 144.9 $meters^2$
7. 6283.19 $inches^3$
8. 1986.92 $inches^2$
9. 6,283.19 $inches^3$
10. 1997.79 $inches^2$
11. Answers may vary. Sample answer: The volume stays the same no matter where the base of the cones is located, but the surface area increases as the base moves from the center of the "height."

Lesson 7.6
Level A

1. sphere
2. sphere
3. surface area
4. 113.1 $units^2$
5. 50.3 $units^2$
6. 314.2 $units^2$
7. 615.8 $units^2$
8. 50.3 $units^2$
9. 314.2 $units^2$
10. 254.5 $units^2$
11. 1,017.9 $units^2$
12. 1,809.6 $units^2$
13. 76 $units^2$
14. 176π $units^2$
15. 960 $units^2$
16. 200π $units^2$
17. $1{,}232\pi$ $units^2$
18. 4.2 $units^3$
19. 33.5 $units^3$
20. 904.8 $units^3$
21. 113,097.3 $units^3$
22. 523.6 $units^3$
23. 523.6 $units^3$
24. 5575.3 $units^3$
25. 5575.3 $units^3$
26. 65,449.8 $units^3$

Answers

Lesson 7.6
Level B

1. 64π units^2
2. 16π units^2
3. 100π units^2
4. 163.84π units^2
5. 17.64π units^2
6. 302.76π units^2
7. 1.4
8. 0.9
9. 2.8
10. 3
11. 2.6
12. 2.2
13. 11,494.04 units^3
14. 124.79 units^3
15. 65.45 units^3
16. 523,598.78 units^3
17. 1124 units^3
18. 0.08 units^3
19. 0 units^3
20. 0 units^3
21. 5,728.72 units^3
22. 94 units^3
23. 523.6 units^3
24. 14.1 units^3
25. 43.4 units^3
26. 22 units^3
27. 897.6 units^3
28. 2538.9 units^3
29. 432.2 units^3
30. 530.7 units^3
31. The volume increases by a factor of 8.
32. The surface area increases by a factor of 4.
33. 725.25 units^3
34. 293.60 units^3

Lesson 7.6
Level C

1. cone: 2250π centimeters^3
   sphere: 4500π centimeters^3
   hemisphere: 2250π centimeters^3
   cylinder: 6750π centimeters^3
   greatest volume: cylinder
2. 12
3. 1.8
4. 192π centimeters^2
5. $r = \dfrac{s\sqrt{\dfrac{6}{\pi}}}{2}$
6. 6 units
7. They are equal in length.
8. 15,308.33 centimeters^2
9. $54{,}000\pi$ centimeters^3
10. 50 centimeters long

Lesson 7.7
Level A

1. $(1, 2, -3)$
2. $(2, -5, -6)$

3. $(-1, -7, 2)$
4. $(2, 8, 5)$
5. $(2, 3, -1)$
6. $(-5, -1, -8)$
7. $(9, 9, 8)$
8. $(-2, 0, -4)$
9. $(-5, 3, 7)$
10. $(0, -4, -10)$
11. $(-2, -4, 1)$
12. $(-11, 8, -6)$
13. circle
14. sphere
15. cone
16. cylinder
17. hemisphere
18. donut
19. y-axis
20. z-axis

Lesson 7.7
Level B

1. octants: back, right, top
   coordinates: $(-4, 7, 3)$
2. octants: front, right, bottom
   coordinates: $(2, 5, -9)$
3. octants: front, right, top
   coordinates: $(8, 6, 1)$
4. octants: back, left, top
   coordinates: $(-7, -4, 2)$
5. octants: back, left, top
   coordinates: $(-9, -9, 1)$
6. octants: back, right, bottom
   coordinates: $(-4, 3, -6)$
7. 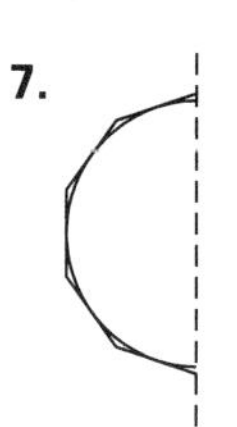
8.
9.
10.
11.
12.

Lesson 7.7
Level C

1. xy-plane, or y-axis
2. xz-plane, or z-axis
3. x-axis
4. cylinder
5. 12 units

Answers

6. 5 units
7. 720π units^3
8. 408π units^2
9. cylinder
10. 5 units
11. 300π units^3
12. 170π units^2
13. circle
14. 13 units
15. 169π units^2

Answers

Lesson 8.1
Level A

1.

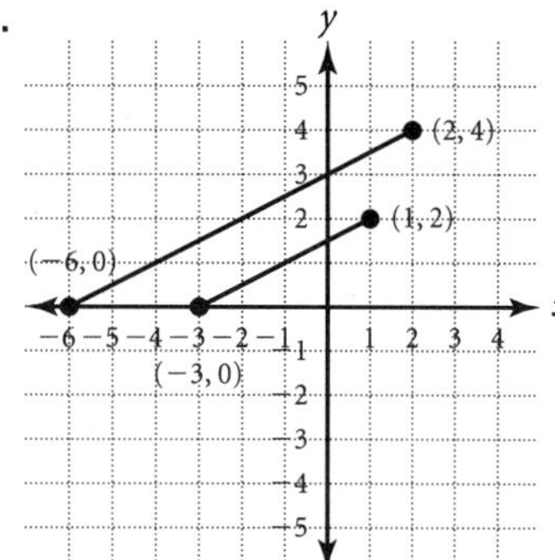

2.

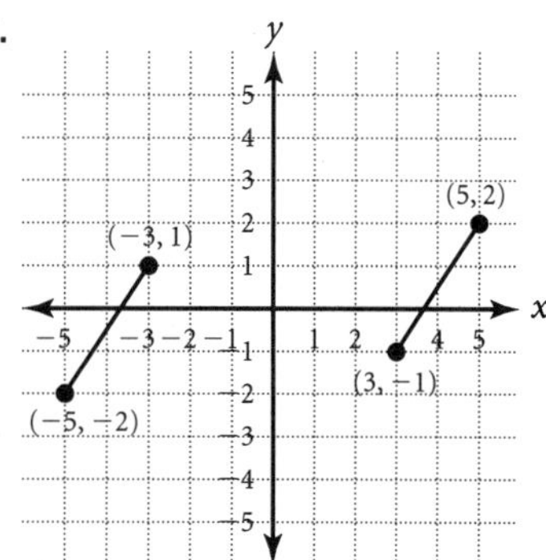

3.

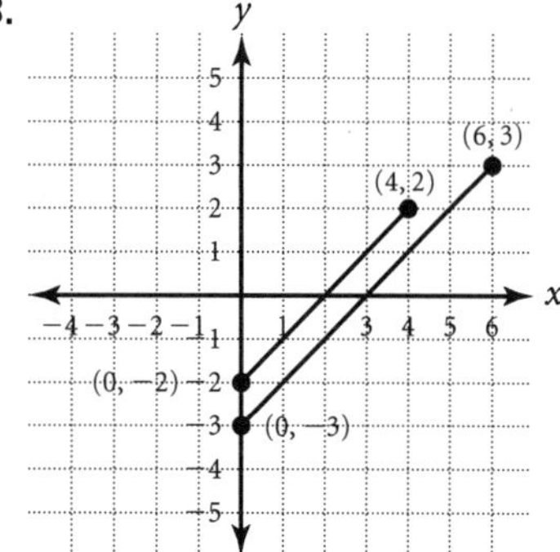

4.

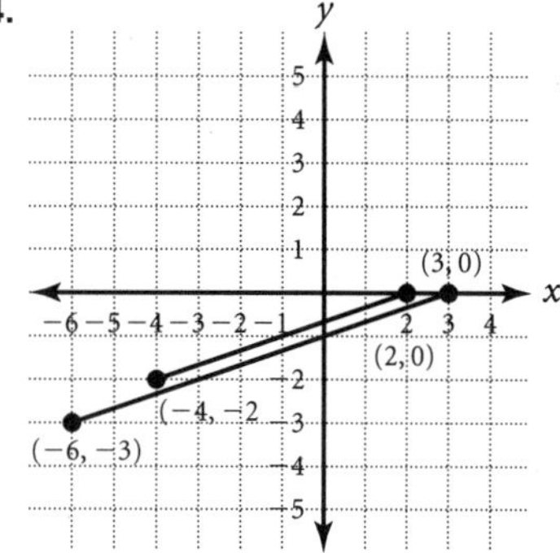

5.

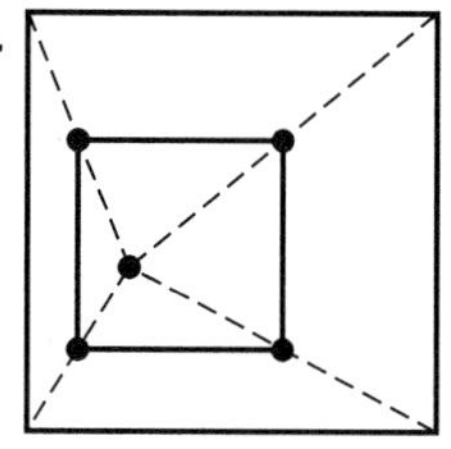

6.

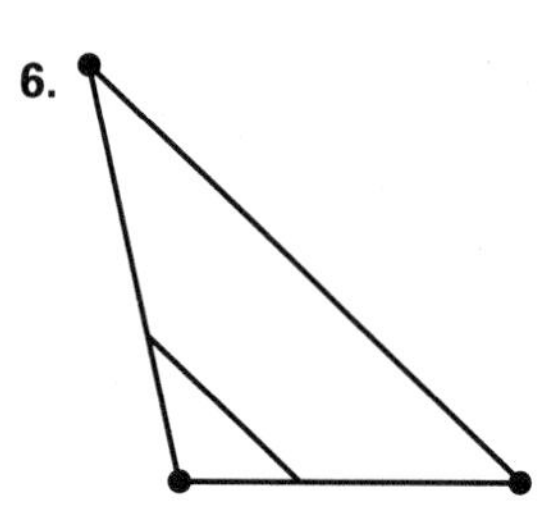

Lesson 8.1
Level B

1.

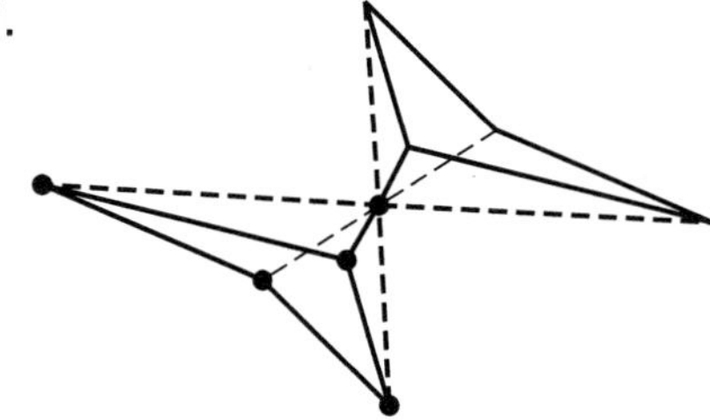

2.

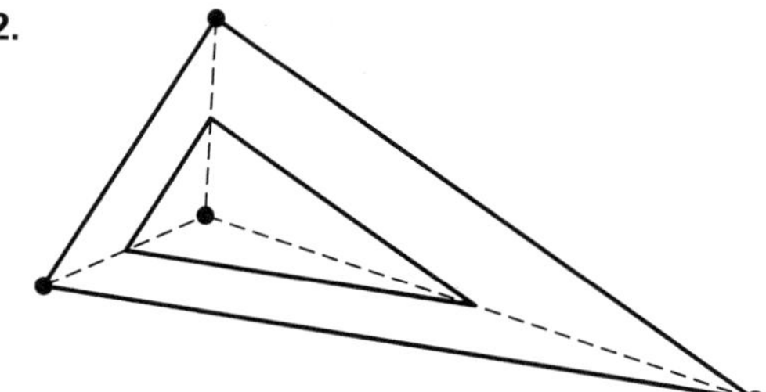

3.

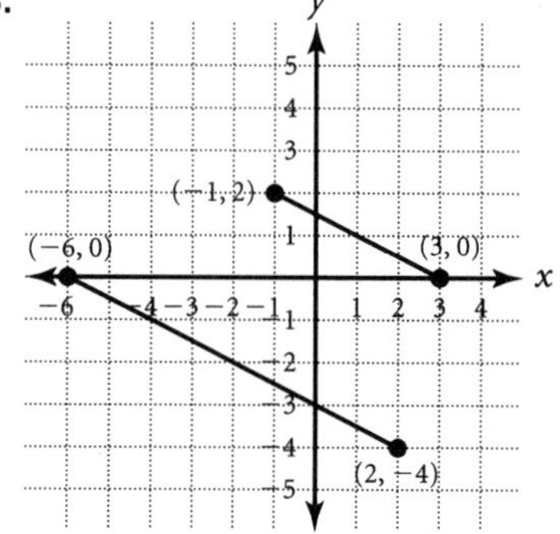

Answers

4.

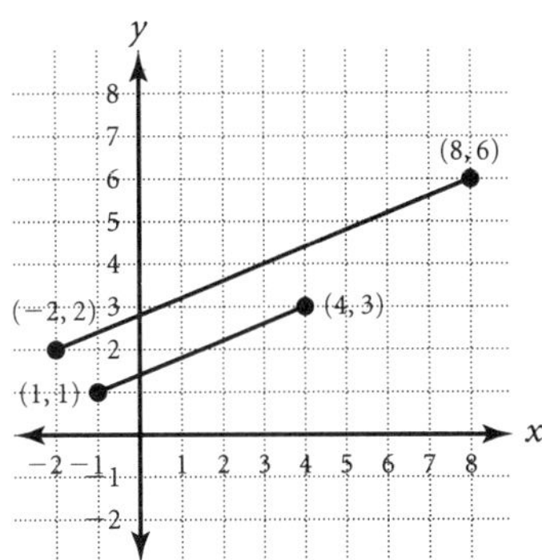

5. a. They meet at the origin.

   b. They meet at the origin.

   c. at the origin

6. $n = -\frac{2}{3}$

7. $n = 2$

Lesson 8.1
Level C

1.

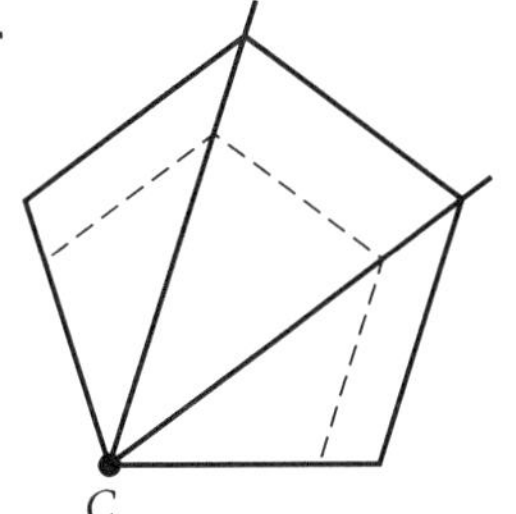

2.

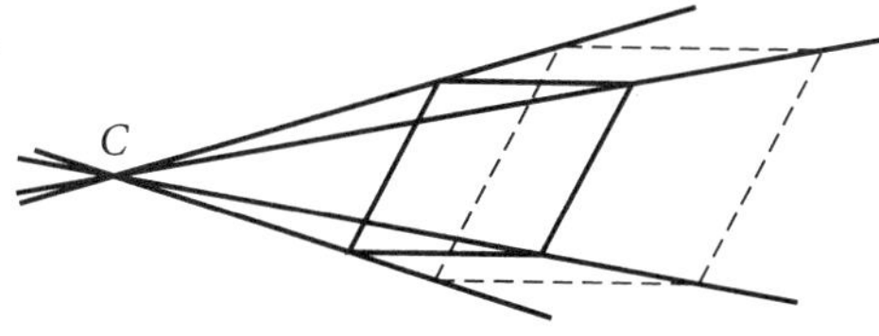

3.

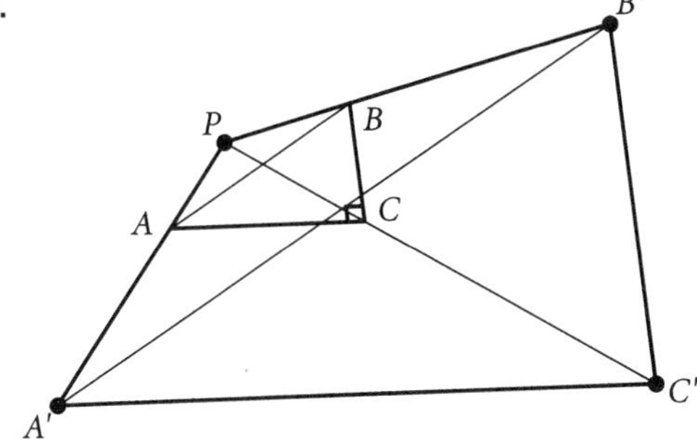

4. 3:1

5. 9:1

6. 1:1

Lesson 8.2
Level A

1. $\frac{8}{6} = \frac{8}{6} \neq \frac{6}{4}$ so triangles are not similar.

2. Corresponding angles are not congruent, so triangles are not similar.

3. Corresponding angles are congruent and corresponding sides are equal (proportional). Sample answer: $\triangle RTC \sim \triangle CER$

4. Corresponding angles are not congruent, so triangles are not similar.

5. $x = 7.5$

6. $x = 70°$

7. $y = 1.64$

8. $y = 4$

Lesson 8.2
Level B

1. Sample answer: $\frac{9}{12} = \frac{12}{16} = \frac{12}{16}$ and corresponding angles are congruent. So $\triangle PQR \sim \triangle STU$.

2. $m\angle X = 90 - 62 = 28 = m\angle B$
   $m\angle A = 90 - 28 = 62 = m\angle W$
   $m\angle C = m\angle R = 90$ and
   $\frac{7.5}{6.0} = \frac{6.5}{5.2} = \frac{3.5}{2.8} = \frac{5}{4}$ so corresponding sides are proportional.
   $\triangle ABC \sim \triangle WXR$

3. $x = \sqrt{2.5} \approx 1.58$

4. $x = \frac{10.4}{2.2} \approx 4.7$

5. $y = \frac{11}{13}$

6. $y = \pm\frac{5}{6}$

7. 1 inch by $2\frac{1}{4}$ inch

8. $\frac{2 + 5}{5} = \frac{7}{5}$ and $\frac{6 + 15}{15} = \frac{21}{15} = \frac{7}{5}$

Lesson 8.2
Level C

1. Not enough information; Do not know about angles.
2. $\frac{6}{12} = \frac{8}{16} = \frac{10}{20}$, so sides are proportional. The right angles are congruent, and the pair of alternate interior angles are congruent. Thus the remaining pair of angles must be congruent. $\Delta XRP \sim \Delta PAM$
3. $x = 4.1, y = 126°$
4. $x = 3\frac{1}{3}, y = 7\frac{1}{2}$
5. \$9.60
6. a. \$329

   b. \$38.25
7. Multiplication Property of Equality
8. Multiplicative Inverse Property

Lesson 8.3
Level A

1. Sample answer: $\Delta ABC \sim \Delta PXR$, by SAS
2. Sample answer: $\Delta DML \sim \Delta ZMY$, by AA
3. Cannot be proven similar.
4. Sample answer: $\Delta ABC \sim \Delta ADE$, by SAS
5. Sample answer: $\Delta PQR \sim \Delta OMN$, by AA
6. Sample answer: $\Delta ABC \sim \Delta EFD$, by SSS

Lesson 8.3
Level B

1. Sample answer: $\Delta PQR \sim \Delta PTS$, by AA
2. $\frac{10}{12} \neq \frac{15}{17}$; The triangles are not similar.
3. Not enough information.
4. $AB = 13$ and $EF = 24$, using the Pythagorean Theorem. Sample answer: $\Delta DEF \sim \Delta BCA$, by SAS or SSS
5. Vertical angles are congruent.
6. $\overline{AB} \parallel \overline{XC}$
7. If two lines cut by a transversal are parallel, then alternate interior angles are congruent.
8. AA Similarity Postulate
9. a. $\angle B \cong \angle Y$,

   b. $\angle C \cong \angle Z$,

   c. $\frac{BC}{YZ} = \frac{20}{24} = \frac{5}{6}$

Lesson 8.3
Level C

1. Given
2. Reflexive Property of Congruence
3. AA Similarity Postulate
4. ΔCPB (in that order only)
5. Yes, because they are both similar to ΔACB so $\Delta APC \sim \Delta CPB$ by the Transitive Property of Similarity.

6. The altitude to the hypotenuse of a right triangle splits the triangle into two right triangles so that each is similar to the original triangle and to each other.

7. $PQ = 8$, $QR = \sqrt{41}$, $RP = 5$

8. $P'(-6, -2)$, $Q'(10, -2)$, $R'(0, 6)$

9. $P'Q' = 16$, $Q'R' = \sqrt{164} = 2\sqrt{41}$, $R'P' = 10$

10. yes; because of SSS

Lesson 8.4
Level A

1. $x = 6.4$

2. $x = 24$

3. $x = 10.5$

4. $x = 7.5$

5. Sample answer: $\triangle AGB \sim \triangle AFC \sim \triangle AED$ by either AA or SAS

6. Sample answer: $\triangle PQR \sim \triangle TSR$, by AA

7. $x = \frac{27}{8} = 3\frac{3}{8}$

8. $x = \frac{2}{3}$

Lesson 8.4
Level B

1. $x = \frac{50}{11} = 4\frac{6}{11}$

2. $x = \frac{15}{4} = 3\frac{3}{4}$

3. $x = 1.5, y = 9$

4. $x = 9.6, y = 7.5$

5. $x = 18, y = 18\frac{1}{3}$

6. $x = 3.2, y = 11\frac{2}{3}$

7.

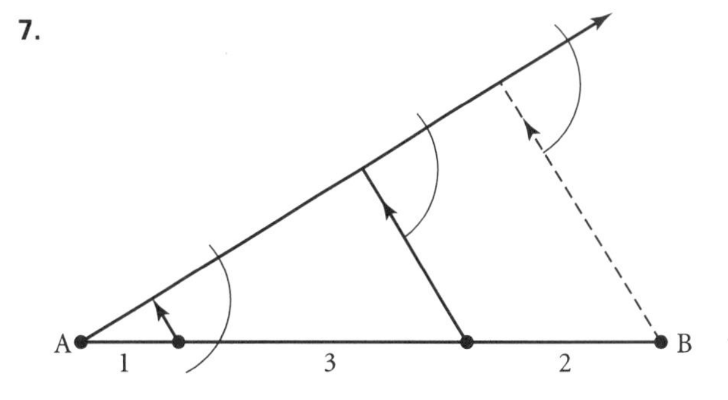

Lesson 8.4
Level C

1. $x = 9, y = 7.5$

2. $x = 5\frac{5}{8}, y = 9\frac{3}{8}$

3. $x = \sqrt{12} = 2\sqrt{3}, y = 6.3$

4. $x = 15, y = 26$

5. a. $x = 1\frac{2}{3}$ cm, $y = 3\frac{1}{3}$ cm, $z = 5\frac{1}{3}$ cm, $p = 8$ cm

   b. Corresponding sides are not proportional, so the trapezoids are not similar.

Lesson 8.5
Level A

1. $\frac{PQ}{XZ}$ or $\frac{PR}{XY}$ or $\frac{QR}{ZY}$

2. $\frac{AC}{AD}$

3. $x = 2.4$

4. $x = 8\frac{1}{3}$

5. $x = 7.5$

6. $x = 40$

7. $h = 9.9$ feet

Lesson 8.5
Level B

1. $x = 6\frac{2}{3}$

2. $x = 4$

3. $x = 9.6$

4. $x = 15$

5. $x = 20$ meters

6. $h = 13\frac{1}{3}$ feet

Lesson 8.5
Level C

1. $x = 13\frac{1}{3}$

2. $x = 12.6$

3. $x = 4.5$

4. *APC*

5. Definition of perpendicular lines

6. Reflexive Property of Equality

7. *CPB*

8. AA Similarity Postulate

9. Transitive Property of Similarity

10. Corresponding sides of similar triangles are proportional.

Lesson 8.6
Level A

1. a. $\frac{4}{5}$

   b. 8 centimeters

2. c. volume

3. b. area

4. b. area

5. a. linear dimensions

6. 13.5 pounds

7. $247.50

8. a. $13.69 = 3.7^2$

   b. $50.653 = 3.7^3$

9. $6\sqrt{3}$

10. An animal ten times the size of a normal animal and similar to it would have legs with cross sectional area (and thus ability to support weight) of 100 times normal. Its actual weight would be 1000 times normal. Its bones could not support its weight.

Lesson 8.6
Level B

1. a. $\frac{3}{4}$

   b. 8 centimeters

2. b. area

3. a. linear dimensions

4. c. volume

5. c. volume

6. $1\frac{1}{3}$ gallons

7. about 8.7 inch

8. about 0.8 yard^2

9. a. \$15.92

   b. about 4 people

Lesson 8.6
Level C

1. a. $\left(\frac{10}{8}\right)^2 \approx 1.56$

   b. $\left(\frac{10}{8}\right)^3 \approx 1.95$

2. kl, kw

3. $A = lw$

4. $A' = (kl)(kw) = k^2lw = k^2A$

5. $\frac{A'}{A} = k^2$

6. kr

7. $A = \pi r^2$

8. $A' = \pi(kr)^2 = k^2(\pi r^2) = k^2A$

9. $\frac{A'}{A} = k^2$

10. Area formulas all seem to involve the product of two linear dimensions. Therefore, when each linear dimension is multiplied by k, the product will be multiplied by k^2.

11. ks, ks, kh

12. $V = \frac{1}{3}s^2h$

13. $V' = \frac{1}{3}(ks)^2(kh) = \frac{1}{3}k^3s^2h = k^3V$

14. $\frac{V'}{V} = k^3$

15. kr, kh

16. $V = \pi r^2h$

17. $V' =$

    $\pi(kr)^2(kh) = \pi k^3r^2h = k^3(\pi r^2h) = k^3V$

18. $\frac{V'}{V} = k^3$

19. kr

20. $V = \frac{4}{3}\pi r^3$

21. $V' = \frac{4}{3}\pi(kr)^3 = \frac{4}{3}\pi k^3r^3 = k^3V$

22. $\frac{V'}{V} = k^3$

23. Volume formulas all seem to involve the product of three linear dimensions. Therefore if each of the dimensions is multiplied by the same scale factor, k, the product will be multiplied by k^3.

Answers

Lesson 9.1
Level A

1. MA, MB, MD
2. DB
3. AB
4. $\overparen{HFG}$, $\overparen{FGH}$, $\overparen{GHF}$
5. $\overparen{HF}$, $\overparen{FG}$, $\overparen{GH}$
6. 122°
7. 48°
8. 3.93
9. 15.70
10. 8.38
11. 41.89
12. 6.90
13. 41.05
14. 82°
15. 40°
16. 119°
17. 115°
18. 31°
19. 40°
20. 100°
21. 20°
22. 260°
23. 260°

Lesson 9.1
Level B

1. 154°
2. 86°
3. 240°
4. 170°
5. 206°
6. 240°
7. No; The way the arcs are labeled decides the direction of the arc. These could be two different arc measures.
8. 3.14
9. 29.53
10. $(0.79x + 2.36)$
11. $0.12x$
12. 71.62°
13. 79.20°
14. 11.27
15. 5.65
16. 8.30
17. 4.77

Lesson 9.1
Level C

1. 1.25
2. 0.12
3. 8.04
4. 4.28
5. −80.11
6. 1.08
7. 114.59°
8. 171.89°
9. 6 : 1
10. 2.24

11. 3.58

12. 27°

13. 53°

14. 127°

15. 127°

16. Since all radii of a circle are congruent, then $BP = PD$. If a diameter of a circle is perpendicular to a chord, then it bisects that chord, $BR = \frac{1}{2}(AB)$ and $DS = \frac{1}{2}(DC)$. Since $AB = CD$, then by the transitive property of equality, $BR = DS$. By the HL congruence theorem, $\triangle BRP \cong \triangle DSP$. Hence, CPCTC, $RP = PS$.

Lesson 9.2
Level A

1. 24

2. 4

3. 12

4. 8

5. MT

6. PT, PQ

7. $\sqrt{27}$

8. $\sqrt{15}$

9. both equal 12

10. 6

11. 25.55

12. 16.49

13. 20

14. 8

15. 15

16. chord

17. secant

Lesson 9.2
Level B

1. tangent

2. perpendicular

3. diameter

4. 2

5. 14.55

6. 6

7. a. 128

b. 40

c. 40

8. 4

9. 5.66

10. tangent

11. 6

12. 10.58

13. 5.48

14. 14.14

15. 10

16. 10

Lesson 9.2
Level C

1. 21.17

2. 142.45

3. 10, 18

4. 18.48

5. 46.96

6. 10.82

7. 10.18

8. 36.14

9. 21.17

10. 15.7

11.

Statements	Reasons
1. Draw PM	1. 2 points determine a line
2. $MA \perp PA$ $MB \perp PB$	2. Tangent to circle is ⊥ to radius at point of tangency
3. PAM and PBM are right angles	3. Def of ⊥ lines are right angles
4. $PAM \cong PBM$	4. All right angles are ≅.
5. $AM \cong BM$	5. Radii of same circle are ≅.
6. $PM \cong PM$	6. Reflexive
7. $\triangle PMA \cong \triangle PMB$	7. $HL \cong HL$
8. $PA \cong PB$	8. CPCTC

11. 62°

12. 118°

13. 62°

14. 31°

15. 59°

16. 128°

17. 90°

18. 52°

19. 232°

Lesson 9.3
Level A

1. a

2. $\angle AVC$

3. $\overset{\frown}{AC}$

4. 65°

5. 90°

6. 78°

7. 156°

8. 102°

9. 102°

10. 118°

Lesson 9.3
Level B

1. 51.5°

2. 77°

3. 154°

4. 103°

5. 26°

6. 38.5°

7. 39°

8. 39°

9. 102°

10. 51°

11. 90°

12. 90°

13. 10.58

14. 25.61

15. 90°

16. 90°

17. 30°

18. 120°

19. 29°

20. 134°

21. 135°

22. 38°

23. never

24. never

25. always

Lesson 9.3
Level C

1. 51°
2. 25.5°
3. 78°
4. 51°
5. 28°
6. 104°
7. 180°
8. 24°
9. 90°
10. 76°
11. 41°
12. 82°
13. 41°
14. 49°
15. No; alternate interior angles are not congruent.
16. 74°
17. 19°
18. 37°
19. 106°
20. 106°

Lesson 9.4
Level A

1. 122.5°
2. 145°
3. 147.5°
4. 70°
5. 124.5°
6. 52.5°
7. 30°
8. 60°
9. 90°
10. 47.5°
11. 42.5°
12. 47.5°
13. 220°
14. 150°
15. 210°

Lesson 9.4
Level B

1. 122°
2. 251°
3. 62°
4. 100°
5. 115°
6. 35°
7. 65°

8. 14°
9. 40°
10. 86°
11. 43°
12. 24°
13. 118°
14. an acute angle

Lesson 9.4
Level C

1. 120°
2. 80°
3. 40°
4. 200°
5. 62°
6. 20°
7. 22°
8. 70°
9. 48°
10. 90°
11. 88°
12. 110°
13. 111°
14. 69°
15. 137°
16. 90°
17. 44°
18. 46°
19. 47°
20. 44°
21. 23°
22. 44°
23. 30°
24. 37°
25. 23°
26. 104°
27. 114°
28. 81°
29. 29°
30. 70°
31. 76°
32. 66°

Lesson 9.5
Level A

1. AB, EB
2. AC, EC
3. $\angle AHM$, $\angle BHM$, $\angle JDK$, $\angle JDC$
4. CD
5. CD
6. 72.66
7. 50
8. 35
9. 4
10. 10
11. $6\sqrt{5}$

12. 5
13. 6
14. 4
15. 2.66

Lesson 9.5
Level B

1. It is equal to the product of the length of the other secant segment and its exterior segment.
2. exterior of $\odot Q$
3. 120°
4. 120°
5. 30°
6. 90°
7. 30°
8. 60°
9. 4
10. 3.42
11. 3.45
12. 1
13. 1
14. 3
15. $\angle EMA$ and $\angle FMD$
16. chord
17. 7.5
18. 37.5
19. 12 inches
20. 20

Lesson 9.5
Level C

1. 38°
2. 90°
3. 128°
4. 52°
5. 5
6. 8
7. 36°
8. 144°
9. 144°
10. 36°
11. 16
12. 15
13. 8
14. 18
15. 10
16. 8.33
17. 3.67
18. 10.25
19. 20

Lesson 9.6
Level A

1. (0, 0), 13
2. (0, 0), $6\sqrt{2}$
3. (−2, 0), 6
4. (0, 4), 1
5. (6, 2), 5
6. (−2, −7), $2\sqrt{6}$

Answers

7. $(x + 3)^2 + (y - 4)^2 = 9$
8. $(x - 2)^2 + (y - 13)^2 = 64$
9. $(x - 4)^2 + (y - 6)^2 = 64$
10. $(x + 5)^2 + (y + 4)^2 = 36$
11. $(13, 0), (-13, 0), (0, 13), (0, -13)$
12. $(-6\sqrt{2}, 0), (6\sqrt{2}, 0), (0, -6\sqrt{2}), (0, 6\sqrt{2})$
13. $(0, -5.6), (4, 0), (0, 5.6), (-8, 0)$
14. none, $(0, 5)$, $(0, 3)$
15. none, none
16. $(-9, 0)\ (9, 0)\ (0, -9)\ (0, 9)$

Lesson 9.6
Level B

1. $(0, 0), 3\sqrt{3}$
2. $(1, -2), \sqrt{3}$
3. $(-\sqrt{5}, \sqrt{2}), \sqrt{3}$
4. $(m - n), \sqrt{w}$
5. $(x + 2)^2 + \left(y - \frac{1}{2}\right)^2 = 2$
6. $(x - 2)^2 + (y - 11)^2 = 9$
7. $(x + 6)^2 + (y - 3)^2 = 36$
8. $\left(x - \frac{3}{4}\right)^2 + \left(y - 1\frac{1}{2}\right)^2 = 2.25$
9. $(x - 3)^2 + (y - 4)^2 = 25$
10. $(x + 4)^2 + (y - 4)^2 = 16$
    $(x + 4)^2 + (y + 4)^2 = 16$
    $(x - 4)^2 + (y + 4)^2 = 16$
    $(x - 4)^2 + (y - 4)^2 = 16$
11. $(x + 3)^2 + y^2 = 25$
12. $(x + 4)^2 + (y - 4)^2 = 4$
13. $(x - 4)^2 + (y + 2)^2 = 16$

Lesson 9.6
Level C

1. $(-2, 5), 3$
2. $(6, 1), 1$
3. $(-1, -2), 4$
4. $(2, -3), 4$
5. $(x - 2m)^2 + (y - 4)^2 = 49$
6. $(x + 3)^2 + y^2 = 184$
7. $(x - 5)^2 + (y - 2)^2 = 1$
8. $(x - 7)^2 + (y + 3)^2 = 9$
9. $(x - 7)^2 + (y - 3)^2 = 49$
10. $(x - 5)^2 + (y - 5)^2 = 25$
11. $(x - 1)^2 + (y - 2)^2 = 1$
12. $(x - 7)^2 + (y + 2)^2 = 41$
13. The coefficients are the same.
14. The signs on the coefficients are equal.
15. moved 1 unit right, 3 units up;
    $(-2, 7), (4, -1), (1, 8)$

Answers

Lesson 10.1
Level A

1. $\dfrac{BC}{AC} \approx \dfrac{3.3\text{cm}}{2.4\text{cm}} = 1.375$

2. $\dfrac{BC}{AC} \approx \dfrac{1.7\text{cm}}{3.8\text{cm}} = 0.447$

3. a. $54°$

   b. $\tan^{-1}\left(\dfrac{3.3}{2.4}\right) \approx 53.97°$

4. a. $24°$

   b. $\tan^{-1}\left(\dfrac{1.7}{3.8}\right) \approx 24.1°$

5. $\dfrac{2}{\sqrt{21}} \approx 0.4364$

6. $\dfrac{3}{1} = 3$

7. a. 1.48

   b. 0.21

   c. 11.43

   d. 1.73

8. a. $34°$

   b. $22°$

   c. $55°$

   d. $77°$

Lesson 10.1
Level B

1. $\dfrac{3}{2\sqrt{10}} \approx 0.4743$

2. $\dfrac{5}{2} = 2.5$

3. a. 3.27

   b. 0.32

   c. 57.29

   d. 0.19

4. a. $39°$

   b. $35°$

   c. $70°$

   d. $75°$

5. $\tan 38° = \dfrac{x}{25}, x = 19.53$

6. $\tan 20° = \dfrac{40}{x}, x = 109.90$

Lesson 10.1
Level C

1. $\tan 55° = \dfrac{x}{12}, x = 17.14$

2. $\tan 70° = \dfrac{20}{x}, x = 7.28$

3. $\tan \theta = \dfrac{5}{\sqrt{11}}, \theta \approx 56°$

4. $\tan \theta = \dfrac{5}{2}, \theta \approx 68°$

5. $x \approx 4.5, y \approx 9.5$

6. $\theta = \tan^{-1} \dfrac{8}{10} \approx 39°$

Lesson 10.2
Level A

1. a. $\dfrac{3}{5}$

   b. $\dfrac{3}{5}$

   c. $\dfrac{4}{3}$

Answers

2. a. $\angle E$

   b. $\angle E$

   c. $\angle D$

3. a. 0.9613

   b. 0.9272

   c. 1.1106

4. a. 30°

   b. 60°

   c. 72°

5. $h = 12 \sin 58° \approx 10.18$

6. $h = 5 \tan 29° \approx 2.77$

7. $x = 100 \sin 55°, x + 4 \approx 85.9$ feet

Lesson 10.2
Level B

1. a. $\frac{2}{\sqrt{5}}$

   b. $\frac{2}{\sqrt{5}}$

   c. $\frac{1}{2}$

2. a. $\angle D$

   b. $\angle E$

   c. $\angle D$

3. a. 0.8290

   b. 0.8290

   c. 0.3249

4. a. 42°

   b. 45°

   c. 67°

5. $h = 4.5 \tan 23° \approx 1.91$

6. $h = 4 \sin 65° \approx 3.625$

7. $\theta = \tan^{-1}\left(\frac{12}{5}\right) \approx 67°$

8. $\theta = \sin^{-1}\left(\frac{2}{5}\right) \approx 24°$

Lesson 10.2
Level C

1. $x = 35 \cos 27° \approx 31.2$

2. $x = \frac{40}{\sin 50°} \approx 52.2$

3. $x = 15 \tan 72° \approx 46.2$

4. $x = 24 \sin 37° \approx 14.4$

5. $\theta = \cos^{-1}\left(\frac{4}{5}\right) \approx 37$

6. $\theta = \tan^{-1}\left(\frac{6}{4}\right) \approx 56°$

7. $\theta = \tan^{-1}(1.5) \approx 56°, d \approx 1.8$ mile

8. Not an identity. Sample answer: For a counterexample, you could use $\theta = 20°$ $\tan 20° \cdot \sin 20° \approx 0.1245$, but $\cos 20° \approx 0.9397$.

9. Is an identity. Proof:

$$\tan(90 - \theta) = \frac{\sin(90 - \theta)}{\cos(90 - \theta)} = \frac{\cos\theta}{\sin\theta}$$

Answers

Lesson 10.3
Level A

1.

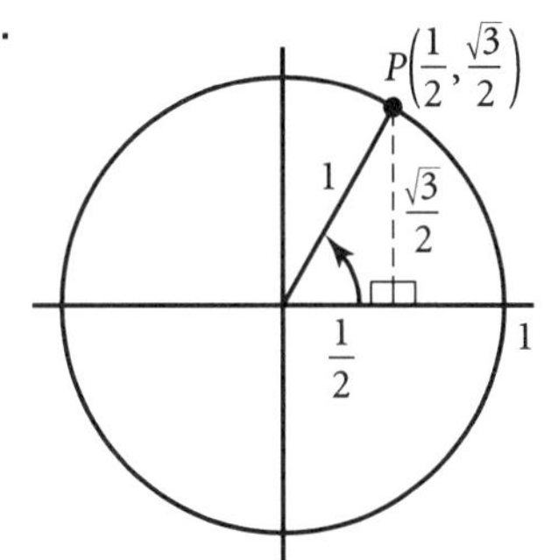

$\sin 60° = \frac{\sqrt{3}}{2}$

$\cos 60° = \frac{1}{2}$

2.

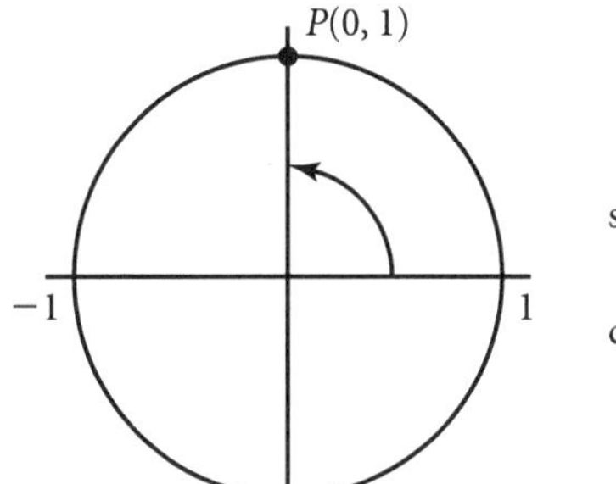

$\sin 90° = 1$

$\cos 90° = 0$

3.

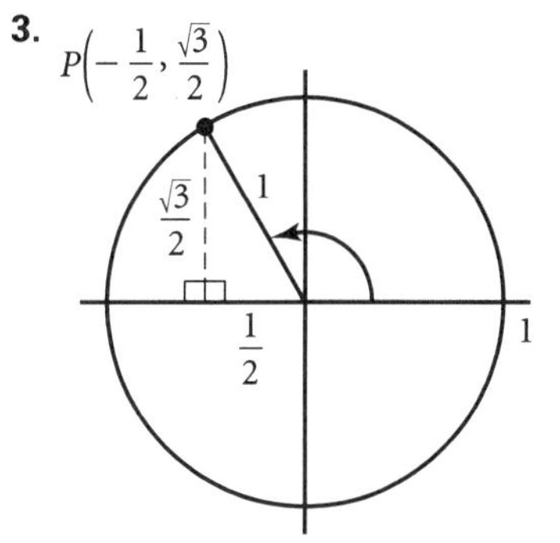

$\sin 120° = \frac{\sqrt{3}}{2}$

$\cos 120° = \frac{-1}{2}$

4.

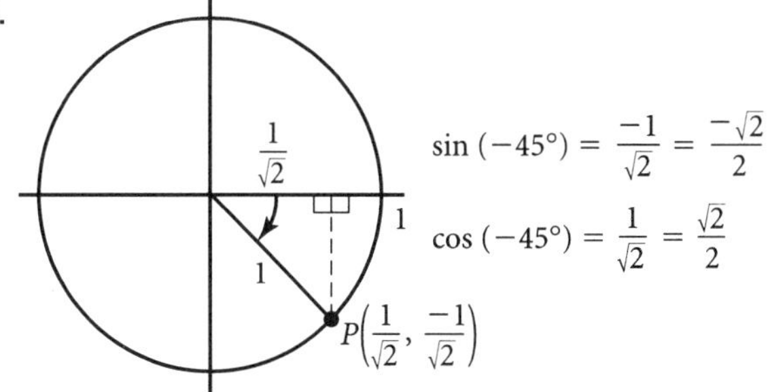

5.

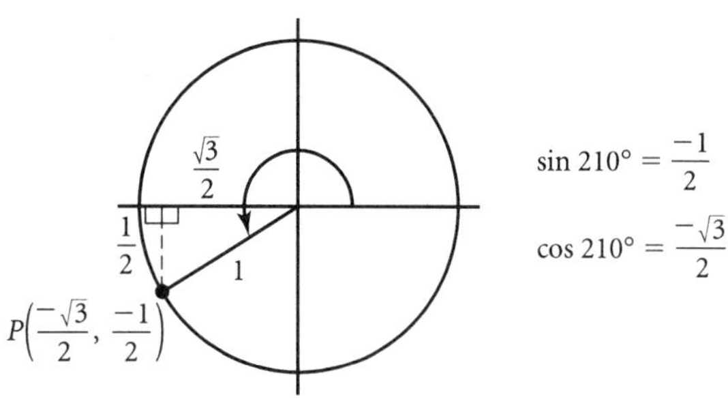

6.

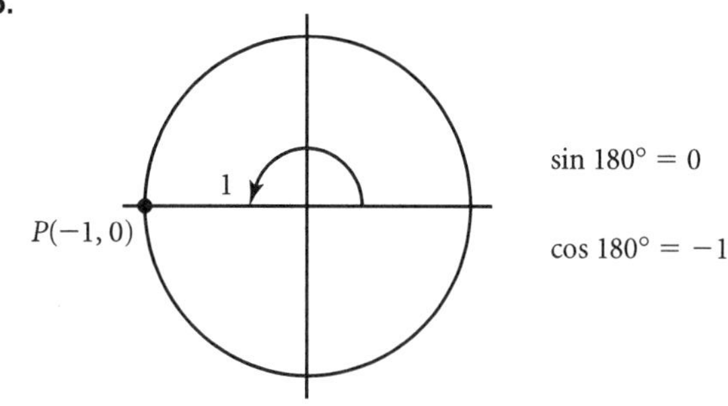

7. $\sin 60° = 0.8660$, $\cos 60° = 0.5$

8. $\sin 90° = 1$, $\cos 90° = 0$

9. $\sin 120° = 0.8660$, $\cos 120° = -0.5$

10. $\sin(-45°) = -0.7071$, $\cos(-45°) = 0.7071$

11. $\sin 210° = -0.5$, $\cos 210° = -0.8660$

12. $\sin 180° = 0$, $\cos 180° = -1$

13. a. Quadrants I or II

 b. $\theta = 49.7°$ or $130.3°$

14. a. Quadrants II or III

 b. $\theta = 104.3°$ or $255.7°$

Answers

Lesson 10.3
Level B

1.

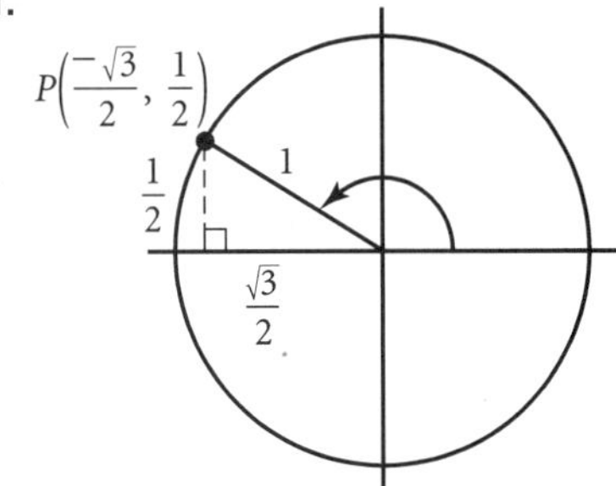

$\sin 150° = \frac{1}{2}$

$\cos 150° = \frac{-\sqrt{3}}{2}$

2.

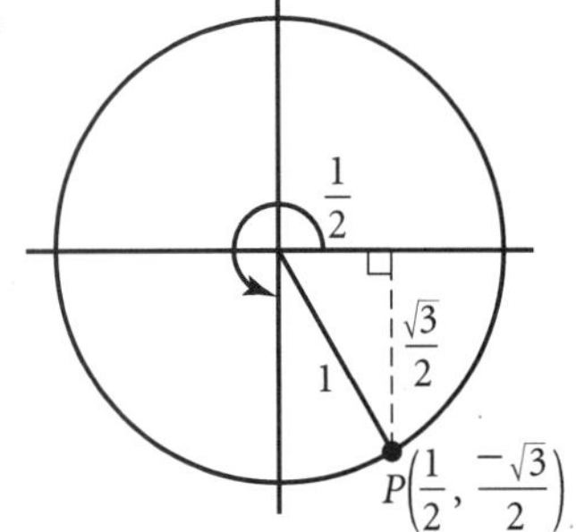

$\sin 300° = \frac{-\sqrt{3}}{2}$

$\cos 300° = \frac{1}{2}$

3.

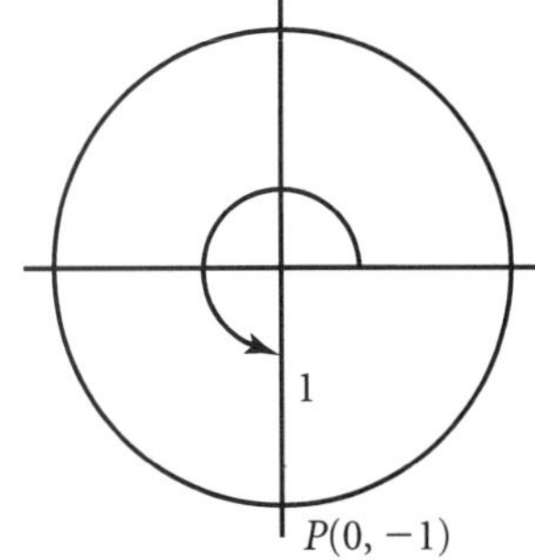

$\sin 270° = -1$

$\cos 270° = 0$

4.

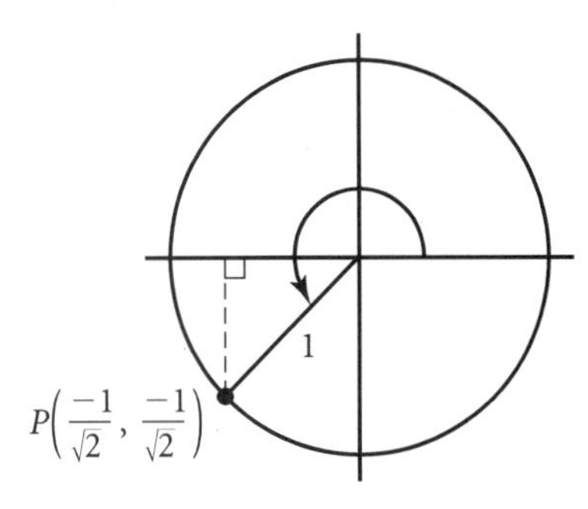

$\sin 225° = \frac{-1}{\sqrt{2}} = \frac{-\sqrt{2}}{2}$

$\cos 225° = \frac{-1}{\sqrt{2}} = \frac{-\sqrt{2}}{2}$

5. $\sin 150° = 0.5$, $\cos 150° = -0.8660$

6. $\sin 300° = -0.8660$, $\cos 300° = 0.5$

7. $\sin 270° = -1$, $\cos 270° = 0$

8. $\sin 225° = -0.7071$, $\cos 225° = -0.7071$

9. a. Quadrants I or II

b. $\theta = 12.7°$ or $167.3°$

10. a. Quadrants II or III

b. $\theta = 148.5°$ or $211.5°$

11. $y = \sin(-6t)$

Lesson 10.3
Level C

1.

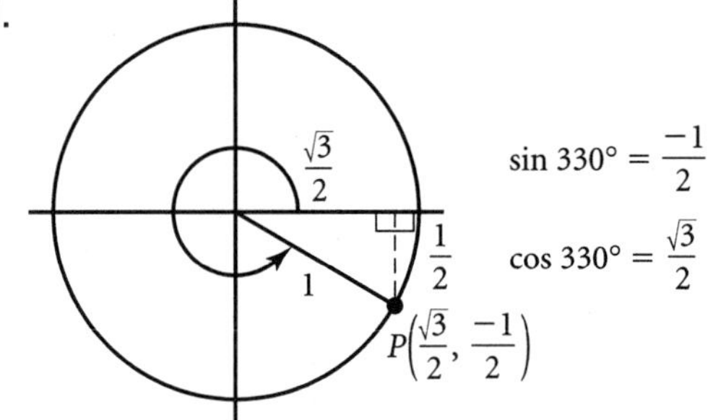

$\sin 330° = \frac{-1}{2}$

$\cos 330° = \frac{\sqrt{3}}{2}$

2.

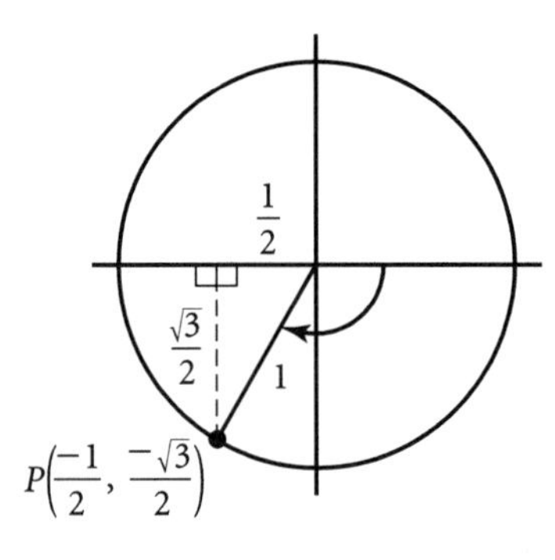

$\sin(-120°) = \frac{-\sqrt{3}}{2}$

$\cos(-120°) = \frac{-1}{2}$

Answers

3.

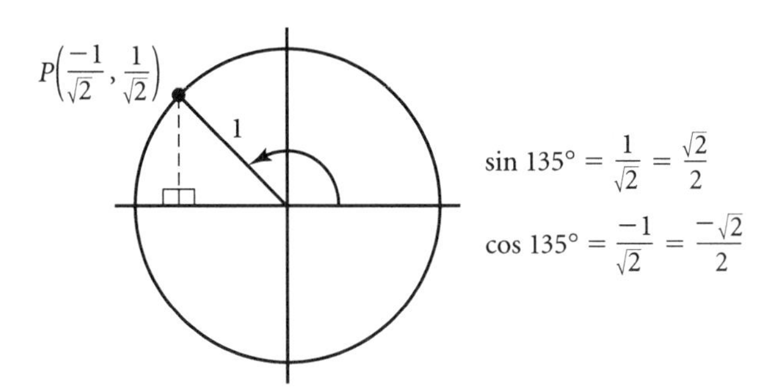

4. $P(0, 1)$

1

$\sin 90° = 1$

$\cos 90° = 0$

5. $\sin 330° = -0.5, \ \cos 330° = 0.8660$

6. $\sin(-120°) = -0.8660,$
$\cos(-120°) = -0.5$

7. $\sin 135° = 0.7071, \ \cos 135° = -0.7071$

8. $\sin 90° = 1, \ \cos 90° = 0$

9. define $\tan\theta = \frac{y}{x}$ or $\tan\theta = \frac{\sin\theta}{\cos\theta}$

10. a. Quadrants I or III

11. a. Quadrants II or IV

12. $OP = 1 = \sqrt{(\cos\theta - 0)^2 + (\sin\theta - 0)^2}$

$1^2 = 1 = (\cos\theta)^2 + (\sin\theta)^2$

Lesson 10.4
Level A

1. $c = 25.7$

2. $b = 20.6$

3. $c = 12.2$

4. $m\angle C = 60.7°$

5. $m\angle A = 27.1°$

6. $m\angle B = 46.6°$ or $133.4°$

7. $m\angle B = 68.5°$ or $111.5°$

8. One triangle is possible. The given angle, *A*, is obtuse, so there can be only one possible acute size for angle *B*.

9. No triangles are possible. It is indicated that the smaller side, *a*, is opposite the obtuse angle. This would be impossible.

10. One triangle is possible. The side opposite the given angle is greater than the side adjacent.

11. No triangles are possible. The side opposite the given angle is shorter than the altitude, so it is too short to reach the adjacent side.

12. $m\angle C = 51°, b = 7.4, a = 6.4$

13. $m\angle B = 16°, m\angle C = 139°, c = 19.6$

Lesson 10.4
Level B

1. $b = 23.2$

2. $b = 21.0$

3. $m\angle C = 44.7°$

4. $m\angle B = 61.3°$ or $118.7°$

5. $m\angle B = 59.3°$ or $120.7°$

6. No triangles are possible. It is indicated that the smaller side, *a*, is opposite the obtuse angle. This would be impossible.

7. One triangle is possible. The given angle, *A*, is obtuse, so there can be only one (acute) size for angle *B*.

8. Two triangles are possible. The side opposite given angle is longer than the altitude but shorter than the adjacent side, so it could hit the adjacent side twice.

Answers

9. No triangles are possible. The side opposite the given angle is shorter than the altitude, so it is too short to reach the adjacent side.

10. $m\angle C = 67°, b = 8.6, a = 7.0$

11. This is the ambiguous case. Two triangles are possible, as shown:

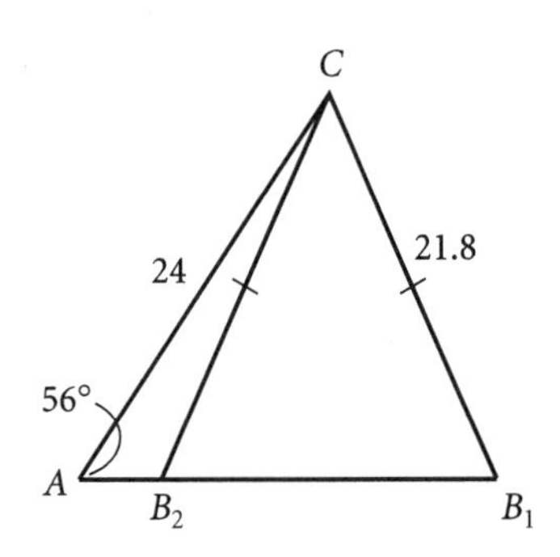

$m\angle AB_2C = 114.1°$ $m\angle B_1 = 65.9°$

$m\angle ACB_2 = 9.9°$ $m\angle ACB_1 = 58.1°$

$AB_2 = 4.5$ $AB_1 = 22.3$

12. 2378.6 feet

Lesson 10.4
Level C

1. One triangle is possible. The given angles and side fit the ASA congruence condition.

2. One triangle is possible. The given information fits in SSA, the ambiguous case, but the side opposite the given angle is longer than the adjacent side, so it can only meet that side once.

3. No triangles are possible. The shorter side is opposite the obtuse angle, which is impossible.

4. Two triangles are possible. The side opposite the given angle is longer than the altitude but shorter than the adjacent side, so it could hit the adjacent side twice.

5. $m\angle B = 65.3°$ or $114.7°$

6. $m\angle B = 33.9°$ or $146.1°$

7. $m\angle B = 83°, a = 5.1, c = 4.6$

8. This is the ambiguous case. Two triangles are possible, as shown:

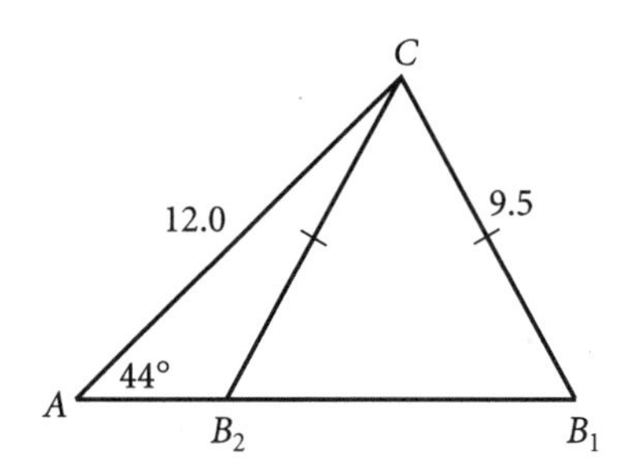

$m\angle AB_2C = 118.7°$ $m\angle B_1 = 61.3°$

$m\angle ACB_2 = 17.3°$ $m\angle ACB_1 = 74.7°$

$AB_2 = 4.1$ $AB_1 = 13.2$

9. Where $\sin C = \sin 90° = 1$, the law of sines gives $\frac{\sin A}{a} = \frac{1}{c} = \frac{\sin B}{b}$, which yields $\sin A = \frac{a}{c} = \frac{\text{leg opposite } \angle A}{\text{hypotenuse}}$ and $\sin B = \frac{b}{c} = \frac{\text{leg opposite } \angle B}{\text{hypotenuse}}$

10. 2189.5 feet

11. about 19.6 feet apart

Lesson 10.5
Level A

1. Use law of cosines because the given information is two sides and the included angle.

2. Use law of sines because the given information is two angles and a side.

3. $a = 8.9$

4. $a = 4.0$

5. m∠B = 90.2°

6. c = 9.9

7. PQ = 6.9, m∠P = 52.3°, m∠Q = 57.7°

8. m∠X = 58°, m∠Z = 27°, XY = 7.7

9. m∠M = 120°, m∠L = 27.8°, m∠H = 32.2°

10. HJ = 3.0, m∠J = 51.9°, m∠H = 93.1°

Lesson 10.5
Level B

1. Use the law of sines because the given information is two angles and a side.

2. Use the law of cosines because the given information is two sides and the included angle.

3. c = 8.1

4. c = 22.4

5. m∠C = 48.6°

6. MN = 8.1, m∠N = 20.5°, m∠M = 135.5°

7. m∠X = 60.4°, m∠Y = 27.6°, XZ = 5.0

8. m∠P = 80.2°, m∠Q = 50.3°, m∠R = 49.5°

9. m∠H = 92°, HK = 4.9, HJ = 11.4

10. 151.4 miles apart

Lesson 10.5
Level C

1. Use the law of cosines because three sides are given.

2. Use the law of sines because the given information is two angles and a side.

3. m∠C = 54.4°

4. c = 14.9

5. m∠B = 98.5°

6. ML = 3.8, m∠L = 20.1°, m∠M = 67.9°

7. m∠X = 138°, XZ = 5.0, XY = 4.6

8. m∠P = 99.1°, m∠Q = 41.0°, m∠R = 39.9°

9. m∠K = 54.2°, m∠J = 32.8°, HK = 3.5

10. AB = 5, BC = 8.1, CA = 7.6
m∠A = 77°, m∠B = 66°, m∠C = 37°

Lesson 10.6
Level A

1.
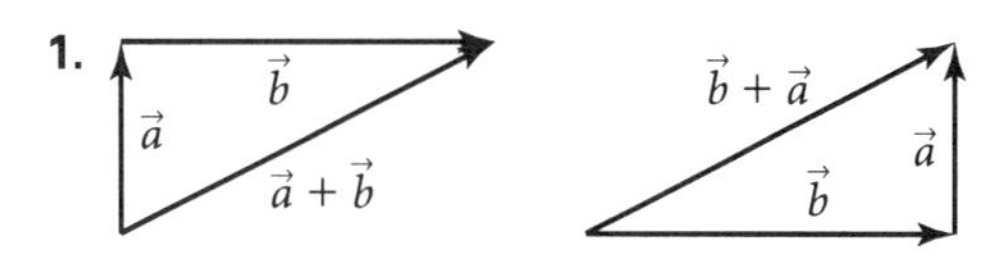

2.
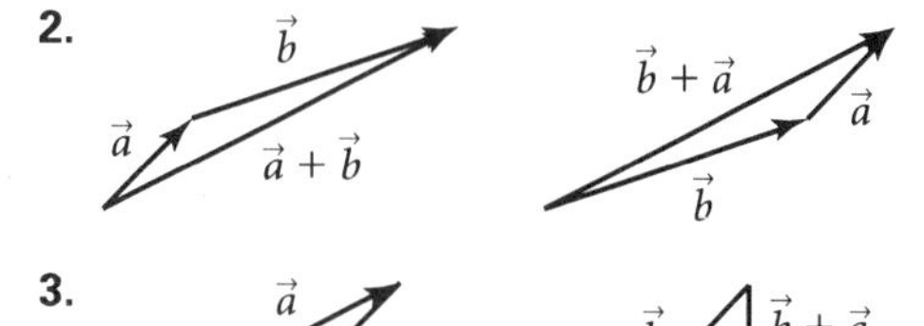

3.

4. a. It seems that $\vec{a} + \vec{b} = \vec{b} + \vec{a}$.

b. Commutative property of vector addition

5.
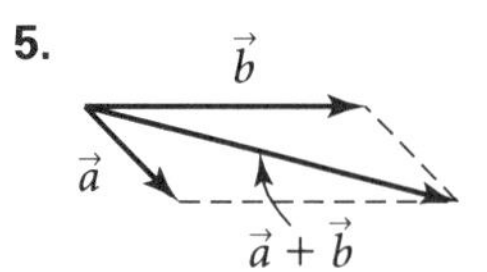

6.
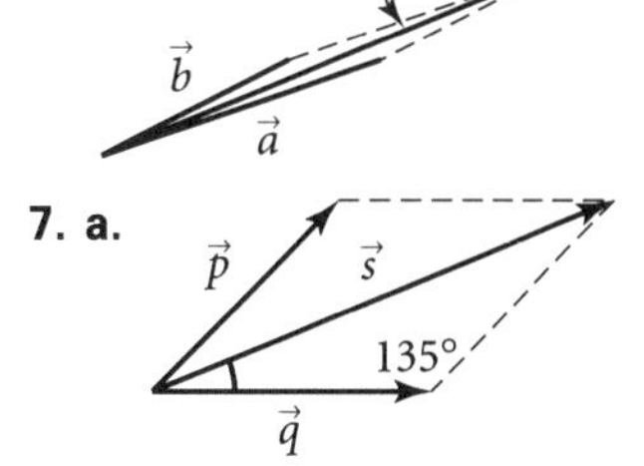

7. a.

b. $|\vec{s}| = \sqrt{29} \approx 5.385$

c. 21.8°

Answers

Lesson 10.6
Level B

1.

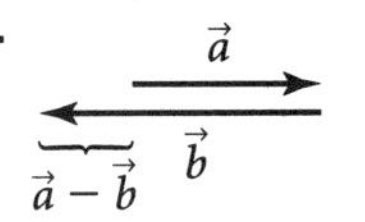

2.

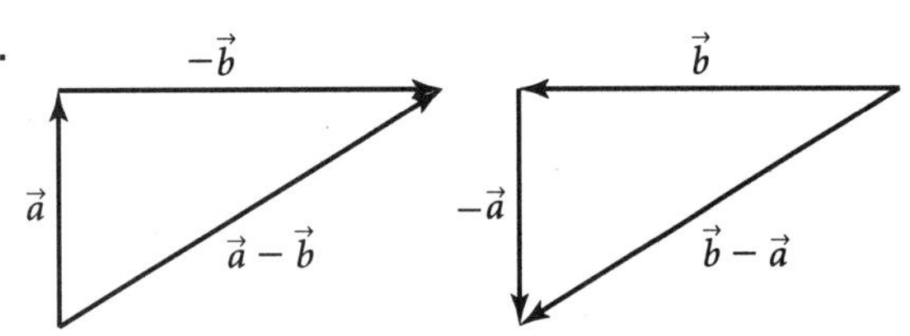

3.

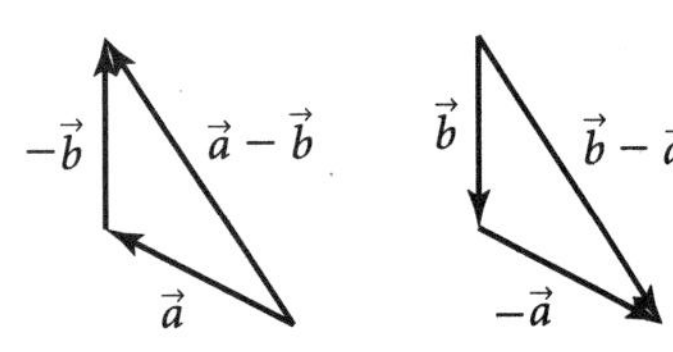

4. that they are opposite vectors

5.

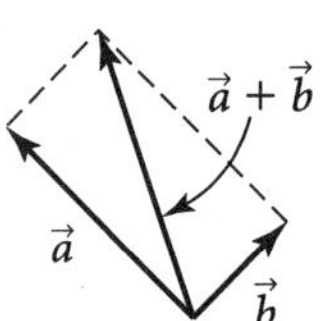

6.

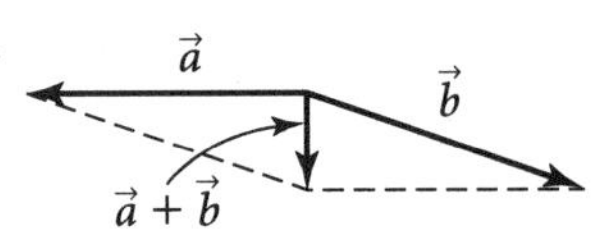

7. a.

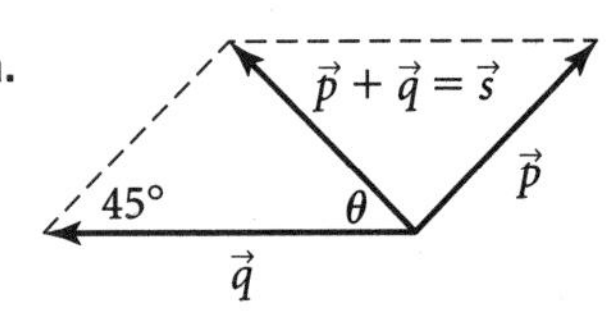

b. $|\vec{s}| = \sqrt{13} \approx 3.6$

c. 33.7°

Lesson 10.6
Level C

1.

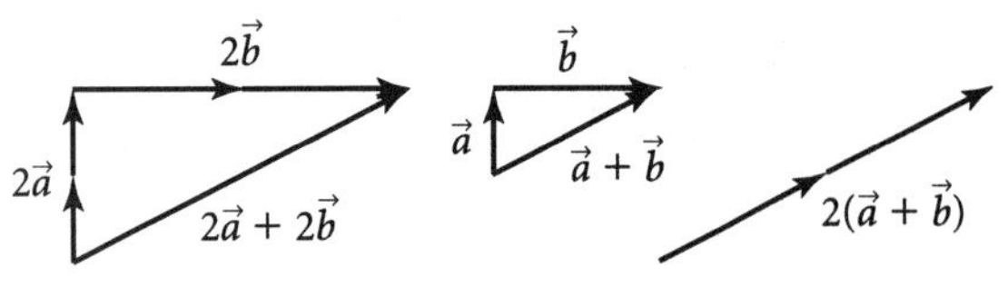

2.

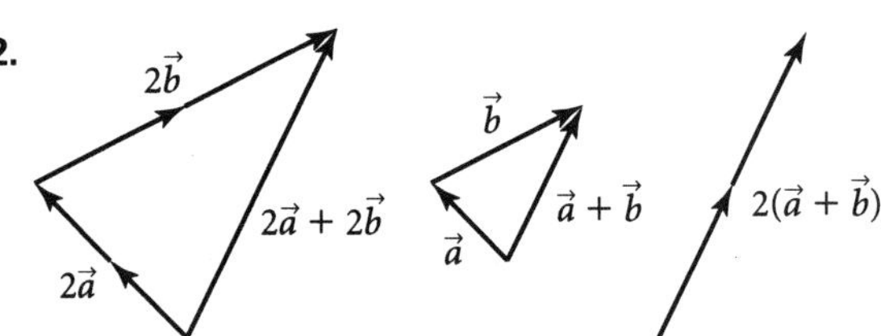

3.

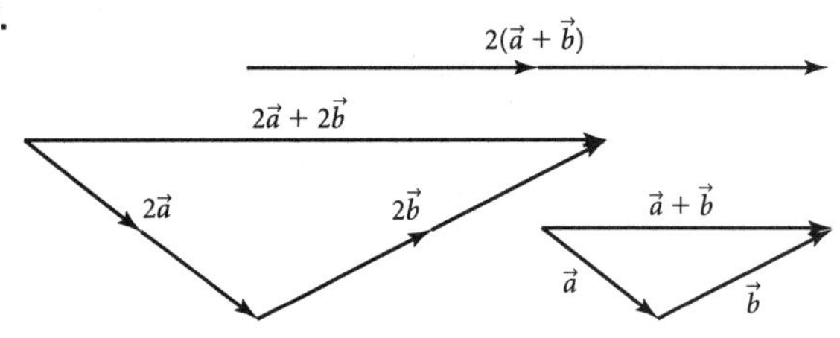

4. a. They seem to be the same vector.

b. Distributive property

5. a.

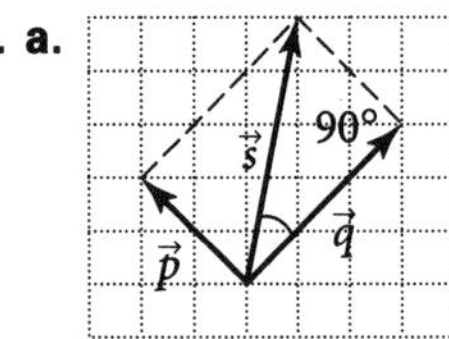

b. $|\vec{s}| = \sqrt{26} \approx 5.1$

c. 33.7°

6. a. about 3.6 mph

b. about 11.9°

Lesson 10.7
Level A

1. a. $P'(2, 2\sqrt{3})$

b. $x' = 2, y' \approx 3.5 \approx 2\sqrt{3}$

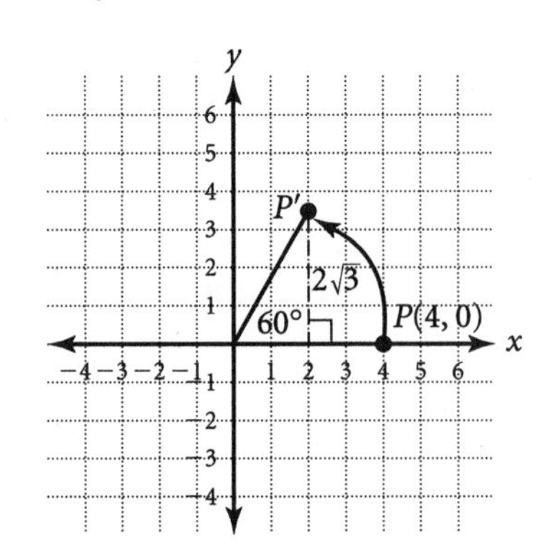

Answers

2. a. $P' = (0, 6\sqrt{2})$

b. $x' = 0, y' \approx 8.5 \approx 6\sqrt{2}$

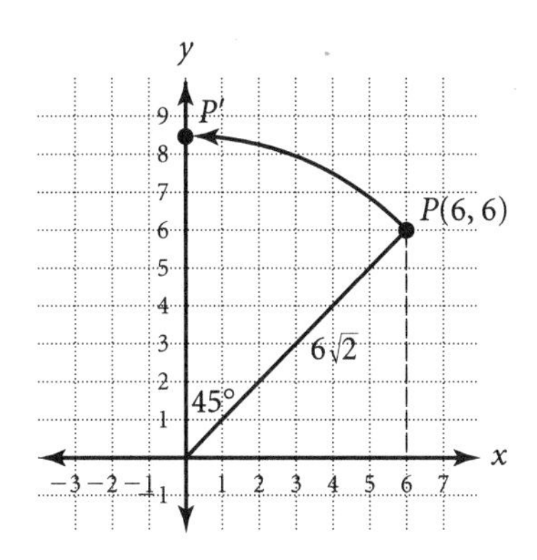

3. a. $P' = (-1, \sqrt{3})$

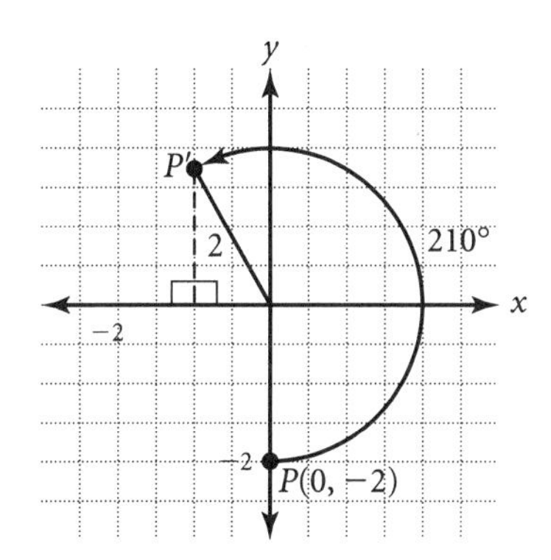

b. $x' = -1, y' \approx 1.7 \approx \sqrt{3}$

4. a. $P' = (-3, -3)$

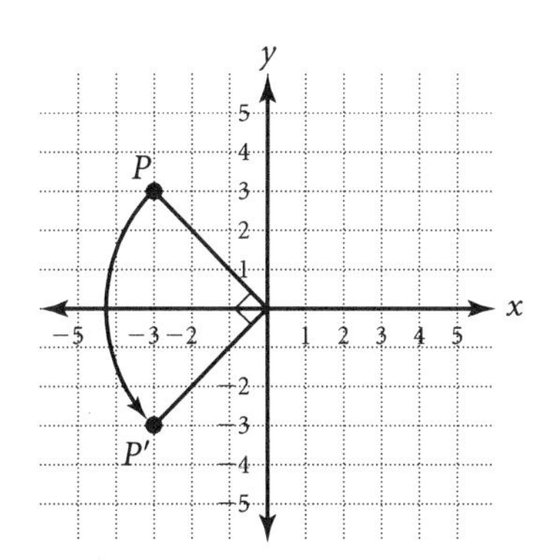

b. $x' = -3, y' = -3$

5. $x' = 3.0, y' = 4.0$

6. $x' = -2.0, y' = -3.6$

7. a. $\begin{bmatrix} 0.8660 & -0.5 \\ 0.5 & 0.8660 \end{bmatrix}$

b. $A'(4.3, 2.5), B'(-2.5, 4.3), C'(-1.9, 1.2)$

8. a. $\begin{bmatrix} -0.5736 & -0.8192 \\ 0.8192 & -0.5736 \end{bmatrix}$

b. $A'(-2.9, 4.1), B'(-4.1, -2.9),$
$C'(-1.1, -2.0)$

Lesson 10.7
Level B

1. a. $P'(\sqrt{3} - 1)$

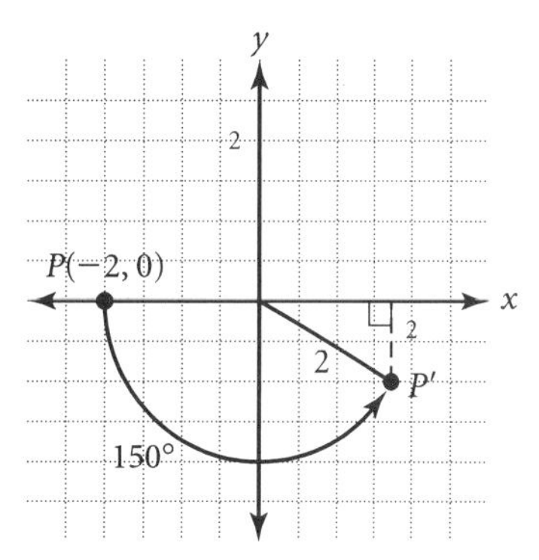

b. $x' = 1.732 \approx \sqrt{3}, y' = -1$

2. a. $P'(0, 3\sqrt{2})$

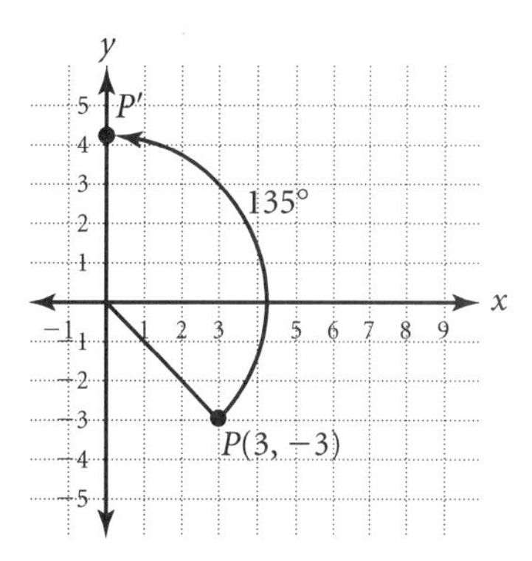

b. $x' = 0, y' = 4.243 \approx 3\sqrt{2}$

Answers

3. a. $P'\ (2\sqrt{3}, -2)$

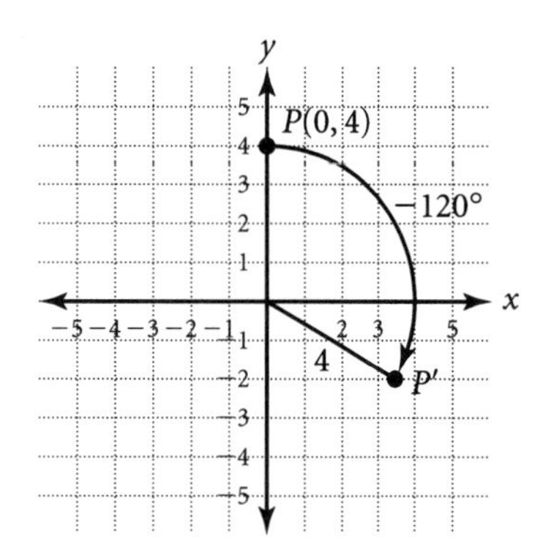

b. $x' \approx 3.464 \approx 2\sqrt{3}, y' = -2$

4. a. $P'(-5, 5)$

b. $x' = -5, y' = 5$

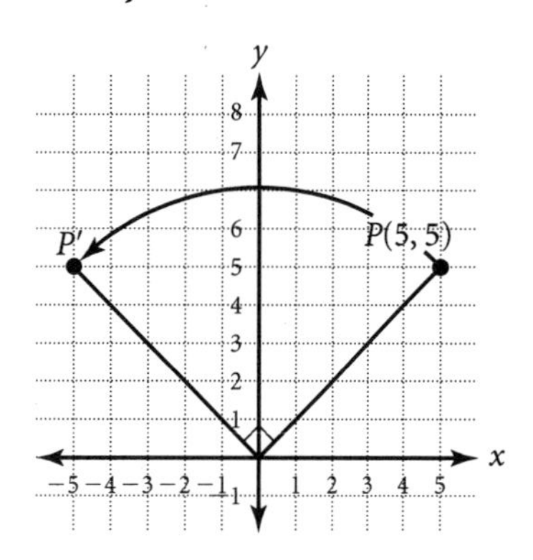

5. $x' = -4.0, y' = -3.0$

6. $x' = -4.7, y' = 2.7$

7. a. $\begin{bmatrix} -0.5 & -0.866 \\ 0.866 & -0.5 \end{bmatrix}$

b. $A'(1, -1.7), B'(-5.8, 0.1), C'(-3.7, 2.5)$

8. a. $\begin{bmatrix} 0.6428 & 0.7660 \\ -0.7660 & 0.6428 \end{bmatrix}$

b. $A'(-1.3, 1.5), B'(5.8, 0.9), C'(4.1, -1.8)$

Lesson 10.7
Level C

1. a. $P'(-7, 7)$

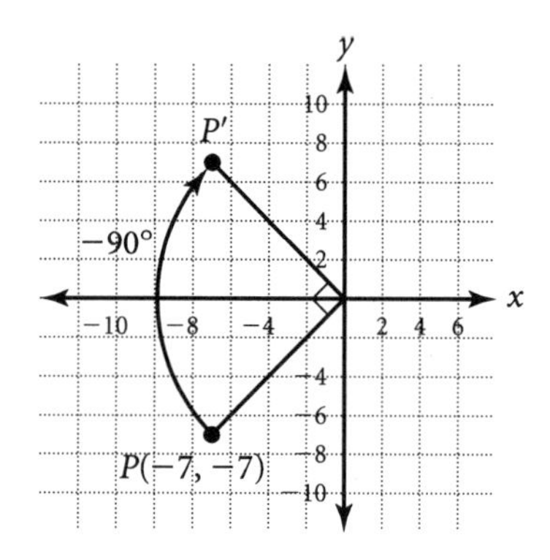

b. $x' = -7, y' = 7$

2. a. $P'(3, -3\sqrt{3})$

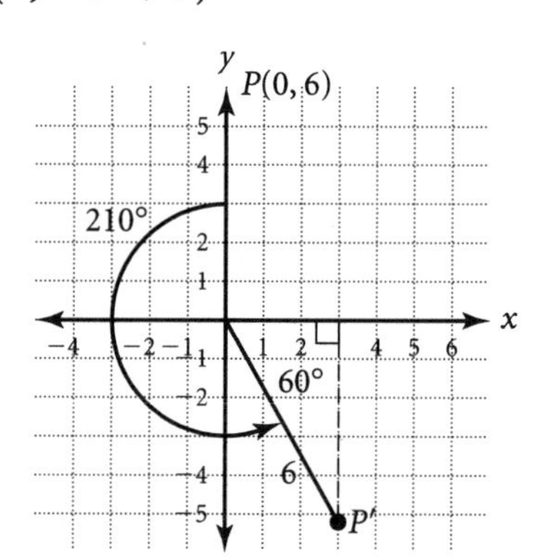

b. $x' = 3, y' \approx -5.196 \approx -3\sqrt{3}$

3. a. $P'(0, 5\sqrt{2})$

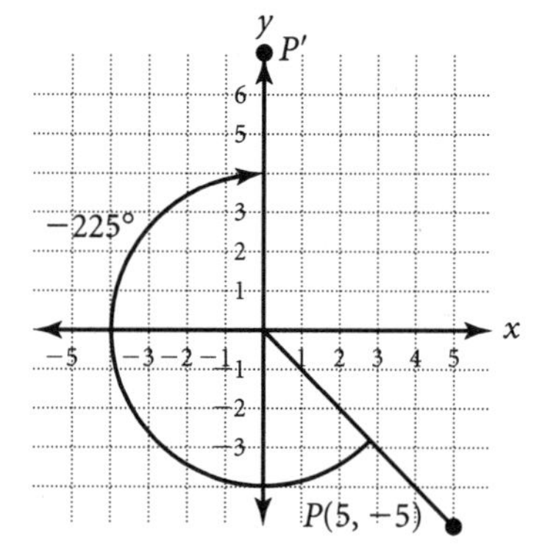

b. $x' = 0, y' \approx 7.071 \approx 5$

Answers

4. a. $P'(2, 2)$

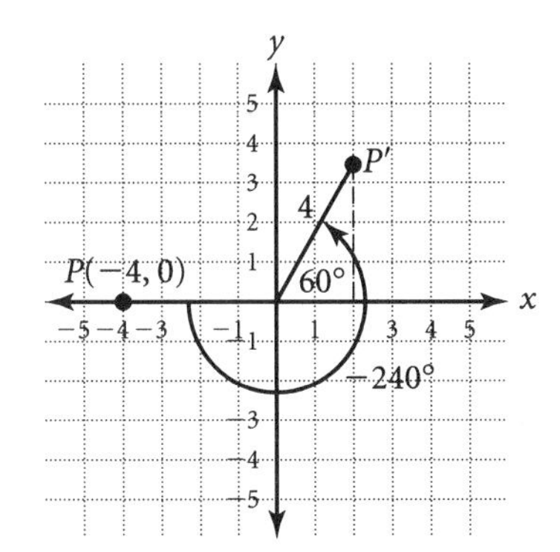

b. $x' = 2, y' \approx 3.464 \approx 2\sqrt{3}$

5. $x' = -1.4, y' = -3.3$

6. $x' = 5.0, y' = 0.8$

7. a. $(-0.3, 4.5), (-0.4, 6.7), (-2.6, 5.4)$

b. On a carefully drawn figure, measure the angle between the two positions of the pole. The angle between the initial and final positions should show a 30° counterclockwise rotation.

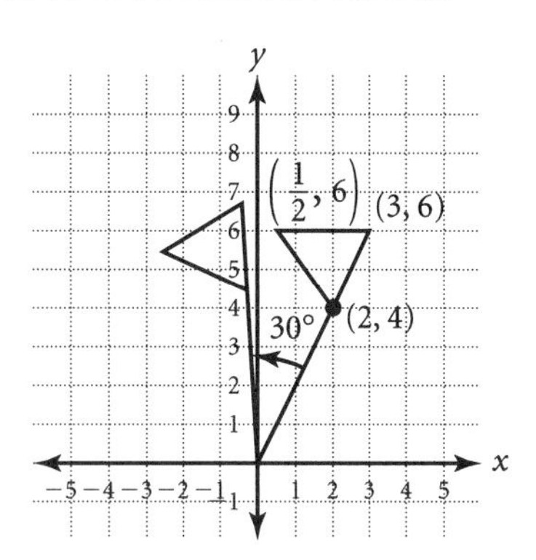

Answers

Lesson 11.1
Level A

1. 7.42
2. 11.33
3. 2.10, 3.40
4. 6.31, 10.21
5. 2.47
6. 14.56
7. 0.62, 1.62
8. 1.38, 3.62

Lesson 11.1
Level B

1. 6.18
2. 2.63, 4.25
3. 24.27
4. 17.01, 10.51
5. 3.62
6. 2
7. 1.90
8. 1.18
9. 0.53 and 0.85

Lesson 11.1
Level C

1. 12.94
2. 2.58, 4.17
3. 3.56
4. 15.42
5. $e : 1$
6. $e : 1$
7. Possible answers: *AC*, *FD*
8. $\dfrac{x(x + i)}{e}$
9. $e : 1$
10. $\dfrac{1}{2}e^2 i^2$

Lesson 11.2
Level A

1. 5
2. 36
3. 14
4. 40
5. 14
6. 12
7. Check student's drawing.
8. 5
9. infinite number
10. yes
11. 24

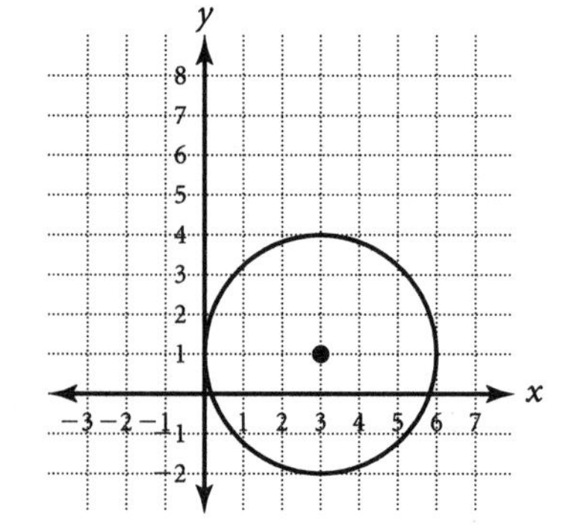

12. 64

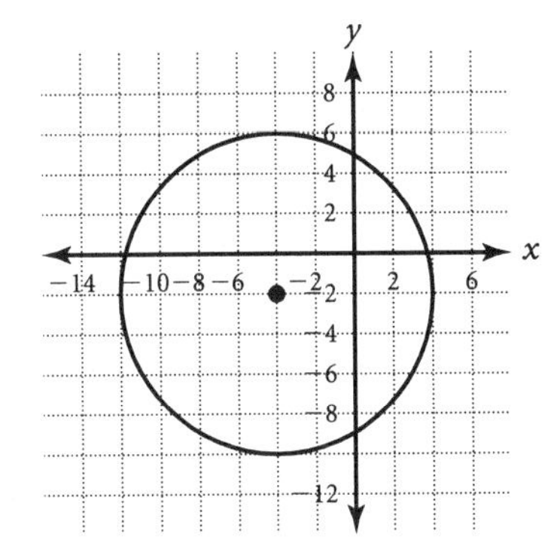

Lesson 11.2
Level B

1. 12

2. 2

3. 4

4. 4

5. 32

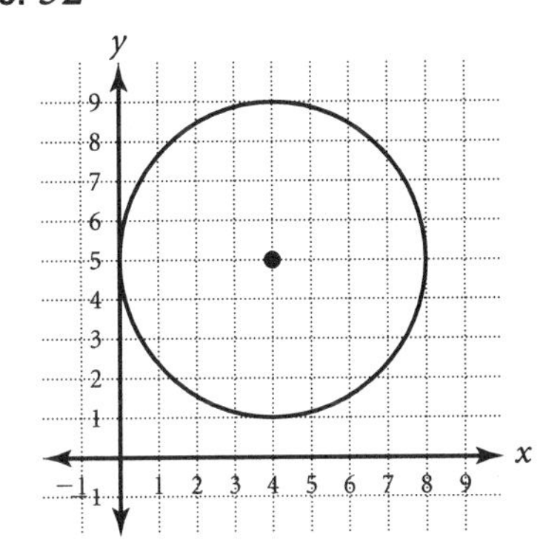

6. 48

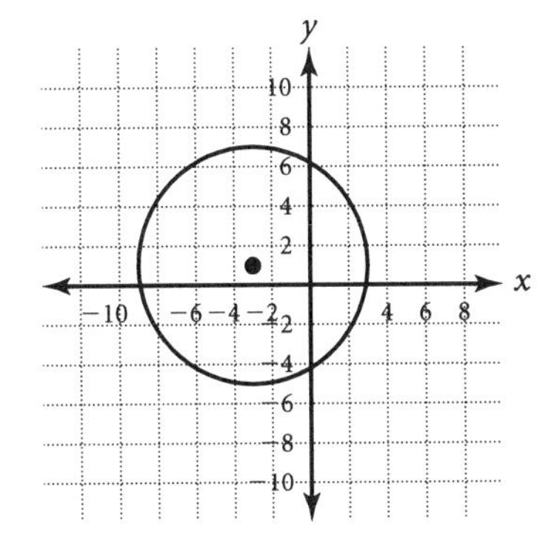

7. $-1, 1$

8. 3, 5

9. 4, 26

10. 4, 8

11. $0, -18$

12. -1

13. 24

14. 40

15. 48

16. 56

17. 24

18. 28

Lesson 11.2
Level C

1. 1

2. 1

3. 1

4. 1

5. 2

6. 3

7. 4

8. 5

9. 6

10. 10

11. 15

12. 21

13. 28

14. 36

15. 45
16. 55
17. 17, 7
18. −13, 19
19. 48
20. 56
21. 20
22. 22

Lesson 11.3
Level A

1. Euler circuit, start anywhere
2. Euler path, start at *B* or *C*
3. Euler path, start at *B* or *C*
4. Euler path, start at *M* or *C*
5. Euler path, start at *D* or *C*

Lesson 11.3
Level B

1. Euler path, start at *B* or *A*
2. Euler circuit, start anywhere
3. Euler path, start at *B* or *A*
4. Euler path, start at *H* or *F*
5. Check student's work.

Lesson 11.3
Level C

1. Euler path, start at *A* or *B*
2. Euler circuit, start anywhere
3. Check student's drawing.
4. 70 kilometers
5. 110 kilometers
6. *AE*
7. *BC*
8. 3
9. *AEC*

Lesson 11.4
Level A

1. 4
2. 4
3. C
4. B
5. none
6. If one shape can be distorted into another without cutting or intersecting with itself.
7. inside
8. outside
9. When a line is drawn to the outside of the curve, if it crosses the curve an even number of times it is outside, an odd number of times is inside the curve.

Lesson 11.4
Level B

1. 5
2. 4
3. F, G
4. C, E
5. B, E

6. outside

7. inside

8. $12 - 18 + 8 = 2$

9. $5 - 8 + 5 = 2$

Lesson 11.4
Level C

1. 7

2. 1

3. *B*, *C*

4. none

5. R

6. C, J, L, M, N, S, U, V, W, Z

7. D

8. E, T, Y

9. H, K

10. A inside, B inside, C outside

Lesson 11.5
Level A

1. arc; measurable

2. Position *D* the same distance from *C* as *A* is from *B*.

3. 2

4. 4

5. Center the figure in the circle by creating arcs exactly opposite one another.

6. no

7. great circle; measurable

8. 2

9. 270°

10. no

11. 9 centimeters

Lesson 11.5
Level B

1. $AB < CD$; The points *C* and *D* are farther apart from each other than points *A* and *B*.

2. the center

3. the edge

4. no

5. Construct tangents to points *B* and *C*.

6. 179°

7. each other

8. each other

9. 181°

10. 361°

11. parallel lines

Lesson 11.5
Level C

1. no

2. 13.91cm

3. 21.19 in.

4. 38.64 centimeters; 39.08 centimeters

5. no

6. infinitely many

7. any point except *N* or *S*

8. no

9. no

Answers

Lesson 11.6
Level A

1.

2. Each side is divided into fourths

3. Each shaded square is divided into nine congruent squares. All the squares remain shaded except for the middle square, it is white.

Lesson 11.6
Level B

1. It is the Sierpinski gasket.
2. yes
3. 6
4. Sierpinski gasket
5. 4
6. row 9; 8 units
7. row 9; 6 units
8. It would disappear.

Lesson 11.6
Level C

1. 600 centimeters^2; 1,000 centimeters^3
2. 5400 centimeters^2; 27,000 centimeters^3
3. increases
4. greater; less
5. 672 centimeters^2; 960 centimeters^3
6. two rectangular prisms; dimensions are 2 by 2 by 4 centimeters.
7. 712 centimeters^2; 928 centimeters^3
8. 768 centimeters^2; 896 centimeters^3
9. decreases
10. no

Lesson 11.7
Level A

1.

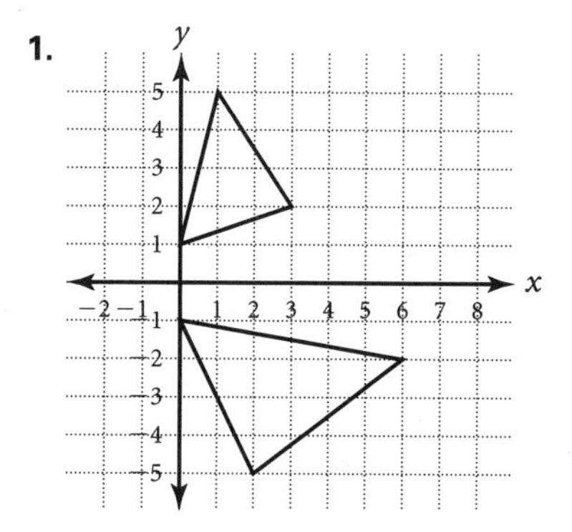

2.

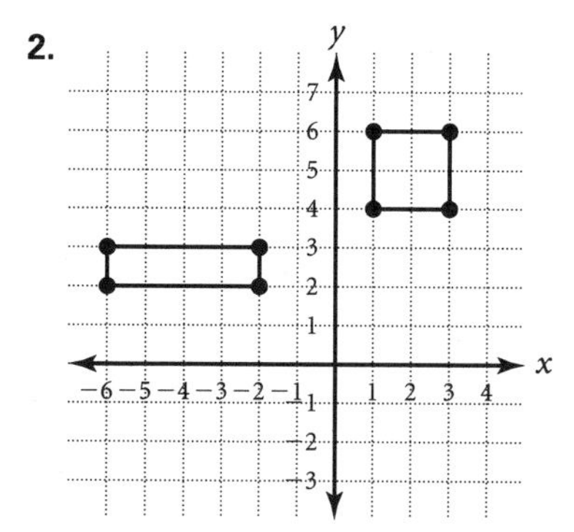

3. $\overrightarrow{PE}$ and $\overrightarrow{PD}$
4. M
5. K
6. $\overrightarrow{QA}$, $\overrightarrow{QB}$, $\overrightarrow{QC}$
7. M
8. L

Answers

Lesson 11.7
Level B

1. $A'(-1, -2)$; $B'(-1, -3)$; $C'(-2, -3)$; $D'(-2, -2)$; $2:1$
2. $A'(4, 6)$; $B'(4, 9)$; $C'(8, 9)$; $D'(8, 6)$; $1:6$
3. $A'(-2, 4)$; $B'(-2, 6)$; $C'(-4, 6)$; $D'(-4, 4)$; $1:2$
4. $A'(2a, 2b)$; $B'(2a, 3b)$; $C'(4a, 3b)$; $D'(4a, 2b)$; $1:ab$
5. $A'(-8, 6)$; $B'(3, 6)$; $C'(-36, 6)$
6. $A'(-6, 3)$; $B'(-6, -5)$; $C'\left(-6, \frac{5}{7}\right)$
7. $(-7, 2)$

Lesson 11.7
Level C

1. $x = 4$ and $x = -2$
2. $(3, -2)$ and $(3, -10)$
3. $x = 12$ and $x = -6$
4. 36 units^2
5. $A'(2, 5)$, $B'(4, 3)$, $C'(6, 1)$
6. $(-2, -3)$
7. $\left(-\frac{14}{19}, -\frac{11}{19}\right)$
8. $\left(-\frac{118}{69}, -\frac{143}{138}\right)$
9. $\left(-\frac{22}{5}, -\frac{9}{5}\right)$
10. have the same center of projection

Answers

Lesson 12.1
Level A

1. Today is the last day of John's work week; modus ponens
2. ABCD is not a square; modus tollens
3. Jose did not get three strikes in baseball; modus tollens
4. valid; modus ponens
5. not valid; affirming the consequent
6. not valid; denying the antecedent
7. valid; modus tollens
8. valid
9. not valid; x could be –2
10. not valid; y could be –2
11. valid

Lesson 12.1
Level B

1. Sally will do well; modus ponens
2. Triangle ABC is not equilateral; modus tollens
3. I'll be late for school; modus ponens; applied three times in succession
4. DEFG has four congruent sides; modus ponens
5. I did not go to the grocery superstore; modus tollens
6. valid
7. not valid; If b were 6, b would be divisible by 2 but not by 4.
8. not valid; If c were 10, c would not be divisible by 4 but would be divisible by 2.
9. valid

Lesson 12.1
Level C

1. No valid conclusion can be drawn.
2. No valid conclusion can be drawn.
3. Today is not Wednesday; modus tollens
4. not valid
5. not valid
6. valid; modus ponens
7. valid; modus tollens
8. valid
9. not valid; Triangle *DEF* could be an isosceles right triangle.
10. not valid; Triangle *GHI* could look like this:

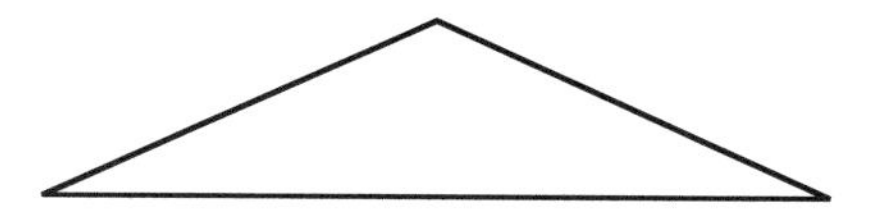

11. valid

Lesson 12.2
Level A

1. false; The conjunction of true with false is false.
2. true; The disjunction of true with false is true.
3. true; The disjunction of true with true is true.
4. All cats were once kittens and blue is a color; true
5. Six is a prime number and five divides evenly into sixteen; false
6. Elephants can fly or dogs can bite; true

7. Corn is a fruit or apples grow on vines; false.

8. $x \geq 0$

9. Dudley is a muggle or $1 + 1 \neq 3$.

10. It is not true that both $x + 3 > 4$ and $x < 0$.

11.

p	q	$\sim p$	$\sim p$ OR q
T	T	F	T
T	F	F	F
F	T	T	T
F	T	T	T

Lesson 12.2
Level B

1. true; The conjunction of true with true is true.

2. true; The disjunction of true with true is true.

3. false; The disjunction of false with false is false.

4. All paper is white and all pen ink is blue. False.

5. $2 > 3$ and $5 + 6 = 11$; false

6. $2 > 3$ or $5 + 6 = 11$; true

7. Six is an even integer or six divides into 18; true

8. $x \leq 0$

9. Dudley is a muggle or $1 + 1 = 2$.

10. It is not true that both $x + 5 < 4$ and $x > 0$.

11.

p	q	$\sim q$	p OR $\sim q$
T	T	F	T
T	F	T	T
F	T	F	F
F	F	T	T

Lesson 12.2
Level C

1. false; The conjunction of true with false is false.

2. true; The disjunction of true with false is true.

3. true; The disjunction of true with true is true.

4. Every square is a rhombus and all triangles have three sides; true

5. For all numbers x, $x^2 > 0$ and $2 \cdot 5 = 10$; false

6. Three is greater than two or $5 + 4 = 9$; true

7. Bears hibernate in winter or snakes are mammals; true

8. $x \leq 4$

9. Steve likes to play baseball or $1 + 6 \neq 5$.

10. It is not true that both $x > 4$ and $x + 3 < 5$.

11.

p	q	p OR q	$\sim(p$ OR $q)$	$(\sim p)$ AND $(\sim q)$
T	T	T	F	F
T	F	T	F	F
F	T	T	F	F
F	F	F	T	T

Lesson 12.3
Level A

1. True; It was proven true as a theorem. Converse: If two angles are congruent, then they are vertical angles. False—two angles which have the same measure need not be formed by opposite rays.

   Inverse: If two angles are not vertical angles, then they are not congruent. False—two angles that are not formed by opposite rays could still have the same measure.

   Contrapositive: If two angles are not congruent, then they are not vertical angles. True—equivalent in meaning to the original statement, a theorem.

2. True. It is the subtraction property of inequality from algebra:

   Converse: For numbers a, b, and c, if $a < b$ then $a + c < b + c$. True—the addition property of inequality from algebra.

   Inverse: For numbers a, b, and c, if $a + c \geq b + c$, then $a \geq b$. True—subtraction property of equality and inequality from algebra.

   Contrapositive: For numbers a, b, and c, if $a \geq b$, then $a + c \geq b + c$. True—addition property of equality and inequality from algebra.

3. False. When the premise is true and the consequent is false, then a conditional statement is considered to be false.

   Converse: If elephants can fly, then $1 + 1 = 2$. True—a conditional statement with a false premise is considered to be true.

   Inverse: If $1 + 1 \neq 2$, then elephants can't fly. True—a conditional statement with a false premise is considered to be true.

   Contrapositive: If elephants can't fly, then $1 + 1 \neq 2$.

   False—a conditional with a true premise and a false consequent is considered to be false.

4. If the car runs out of gas, then it will stop.

5. If an animal is a puppy, then it is cute.

6. If the weather is nice, then we'll go to the park.

Lesson 12.3
Level B

1. False. If c is a negative number, multiplication by c reverses the direction of the inequality sign. Converse: For numbers a, b, and c, if $ac > bc$ then $a > b$. False—division by a negative number would reverse the direction of the inequality sign.

   Inverse: For numbers a, b, and c, if $a \leq b$ then $ac \leq bc$. False—for example $2 \leq 5$, but multiplication by $c = -1$ gives $-2 > -5$.

   Contrapositive: For numbers a, b, and c, $ac \leq bc$, then $a \leq b$. False—again, the use of a negative number for c can provide a counterexample.

2. True—the value of the sine ratio depends solely on the angle's measure so congruent angles have equal sine values.

   Converse: If $\sin A = \sin B$, then $\angle A \cong \angle B$. False—for example, $\sin 30° = \sin 150°$, but $30 \neq 150$.

   Inverse: If $\angle A$ is not congruent to $\angle B$, then $\sin A \neq \sin B$. False—for example, if $m\angle A = 10°$ and $m\angle B = 170°$, then angles A and B are not congruent, yet $\sin 10° = \sin 170°$.

   Contrapositive: If $\sin A \neq \sin B$, then $\angle A \neq \angle B$. True—same meaning as original true statement.

3. True. A conditional with a false precedent is considered to be true. Also, notice you can get to $5 = 6$ by adding 3 to both sides of $2 = 3$, using the addition property of equality.

   Converse: If $5 = 6$, then $2 = 3$. True—a conditional with a false precedent is considered to be true.

   Inverse: If $2 \neq 3$, then $5 \neq 6$. True—a conditional with a true precedent and a true consequent is considered to be true.

   Contrapositive: If $5 \neq 6$, then $2 \neq 3$. True—a conditional with a true premise and a true consequent is considered to be true.

4. If you finish the job, then I'll pay you.
5. If an animal is a lizard, then it is a reptile.
6. If I follow this diet, then I will lose weight.

Lesson 12.3
Level C

1. True. To have the sum of their measures be 90°, each must be $< 90°$.

   Converse: If two angles are both acute, then they are complementary angles. False—for example, if each angle had measure 32° they would both be acute, but the sum of their measures would not be 90°.

   Inverse: If two angles are not complementary angles, then they are not both acute. False—for example, if the angles had measures of 10° and 20°, they would not be complementary, but they would both be acute.

   Contrapositive: If two angles are not both acute, then they are not complementary. True—if at least one were not an acute angle, it would have measured $\geq 90°$. This would not allow for any possible positive measure for the other angle in order to have a sum of measures of 90°.

2. False. The premise would be true for point-X located anywhere on a perpendicular bisector of $\overline{AB}$, but not necessarily located on $\overline{AB}$, as the midpoint would have to be.

   Converse: If X is the midpoint of $\overline{AB}$, then $AX = XB$. True, due to the definition of midpoint.

   Inverse: If $AX \neq XB$, then X is not the midpoint of $\overline{AB}$. True, due to the definition on midpoint.

   Contrapositive: If X is not the midpoint of $\overline{AB}$, then $AX \neq XB$. False—for example, when point X is the vertex of an isosceles triangle with base $\overline{AB}$, then X is not the midpoint of $\overline{AB}$ but $AX = XB$.

3. True. A conditional with a false premise is considered to be true.

   Converse: If horses have 5 legs, then a tail is a leg. True, a conditional with a false premise is considered to be true.

   Inverse: If a tail is not a leg, then horses do not have 5 legs. True, a conditional with a true premise and a true consequent is considered true.

   Contrapositive: If horses do not have 5 legs, then a tail is not a leg. True, a conditional with a true premise and a true consequent is considered true.

4. If a food is a dessert, then it is sweet.
5. If it doesn't rain, then we'll play baseball.
6. If a number is divisible by five, then it ends in either 5 or 0.

Lesson 12.4
Level A

1. $\triangle ABC$ is both an obtuse triangle and not an obtuse triangle.
2. All integers are even and not all integers are even. (Alternate answer: All integers are even and some integers are not even.)
3. $x > 3$ and $x \leq 3$
4. Two lines meet in exactly one point and two lines do not meet in exactly one point.
5. Indirect reasoning—reasoning from the negation of the consequent to the negation of the premise.
6. Not indirect reasoning. This is the form of argument called asserting the consequent, which is not valid.
7. Not indirect reasoning. This is the form of argument called denying the premise, which is not valid.
8. Distributive.
9. Subtraction Property of Equality
10. $2(5 + 6x) \neq 4(3x + 2)$ for any real number x.

Lesson 12.4
Level B

1. $x < 2$ and $x \geq 2$
2. $a = 5$ and $a \neq 5$
3. Some parties are fun and no parties are fun.
4. All people are honest and some people are not honest.
5. Not indirect reasoning. This is the form of argument called denying the premise, which is not valid.
6. Indirect reasoning—reasoning from the negation of the consequent to the negation of the premise.
7. $c^2 \leq b^2$
8. Pythagorean Theorem
9. Transitive property
10. The fact that the square of any positive number is positive.

Lesson 12.4
Level C

1. $2x + 1 = 3$ and $2x + 1 \neq 3$
2. $y^2 > 0$ and $y^2 \leq 0$
3. Some people have blue eyes and no people have blue eyes.
4. All squares are rectangles and some squares are not rectangles.

Answers

5. Indirect reasoning—reasoning from the negation of the consequent to the negation of the premise.

6. Not indirect reasoning. This is the form of argument called denying the premise, which is not valid.

7. Suppose there is another point Y and l such that $\overline{PY} \perp l$. Then PXY is a triangle that has two right angles in it. But this is impossible. It contradicts the fact that the sum of the measures of the angles of a triangle is 180°. Therefore the assumption was false and there is only one line through Point P that is perpendicular to l.

Lesson 12.5
Level A

1. Not p and q
2. Not (p or q)
3. 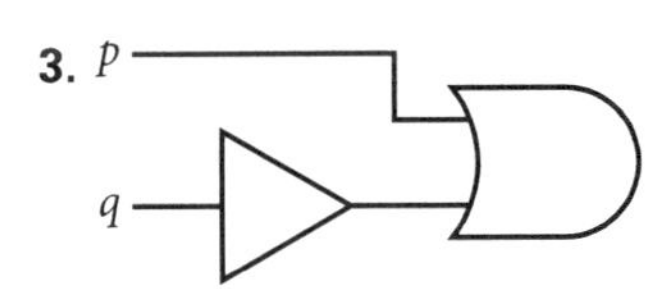

4. 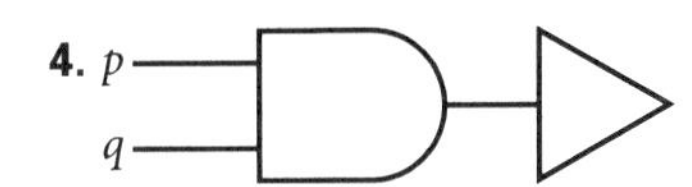

5. 0, 0
6. 0, 0
7. 1, 1
8. 1, 0
9. 1, 1
10. 1, 1
11. 1, 1
12. 1, 1
13. 1, 1
14. 1, 1
15. 0, 1
16. 0, 0

Lesson 12.5
Level B

1. p or Not q
2. Not (p and not q)
3. 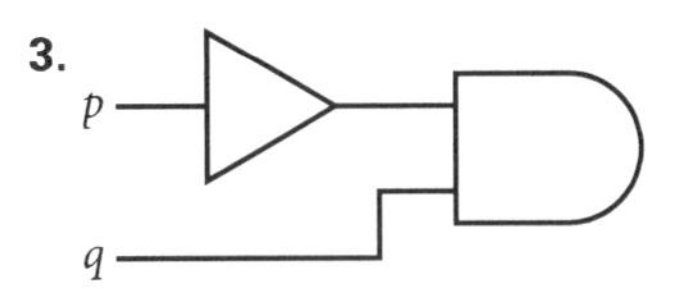

4. 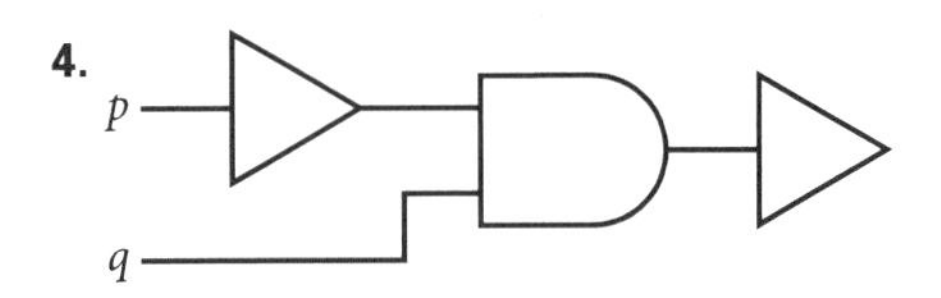

5. 0, 0, 0
6. 0, 1, 0
7. 1, 0, 0
8. 1, 1, 1
9. 1, 0, 0
10. 1, 1, 1
11. 1, 0, 0
12. 1, 1, 1
13. 1, 0, 0
14. 1, 1, 1
15. 0, 0, 0
16. 0, 1, 0

Answers

Lesson 12.5
Level C

1. 2^4, or 16

In Exercise 2–15, there may be more than one answer possible due to the logical equivalence of different expressions. One or two sample answers will be given, but other likely possibilities that students may think of should be checked.

2. NOT(p OR q). Also (NOT p) AND (NOT q)
3. (NOT p) OR q (Notice that this is equivalent to the truth table of "if p, then q", or $p \rightarrow q$)
4. NOT (NOT p OR q). Also p AND (NOTq)
5. p; If students want a reference to q as well, they could form a conjunction with a statement that is always true—for example, p AND (q OR NOT q).

For the answers to Exercises 6–15, the ten remaining truth table possibilities are listed in no particular order, with a possible logic gate for each. Again, these should be regarded as sample answers because logically equivalent statements are possibilities as well.

0 0 1 0 (NOT p) AND q
0 1 1 1 NOT(p AND q)
1 1 0 1 NOT(NOT p AND q). Also p OR NOT q
0 0 1 1 NOT p
1 0 1 0 q
0 1 0 1 NOT q
1 0 0 1 (p OR NOT q) AND (q OR NOT p)
0 1 1 0 (NOT p AND q) OR (NOT q AND p)
1 1 1 1 p OR (NOT p)
0 0 0 0 p AND (NOT p)